Robert Mayer and the
Conservation of Energy

# Robert Mayer and the Conservation of Energy

• *KENNETH L. CANEVA* •

PRINCETON UNIVERSITY PRESS

PRINCETON, NEW JERSEY

Published by Princeton University Press, 41 William Street, Princeton, New Jersey 08540
In the United Kingdom: Princeton University Press, Chichester, West Sussex


---

Library of Congress Cataloging-in-Publication Data

Caneva, Kenneth L.
Robert Mayer and the Conservation of Energy / Kenneth L. Caneva.
p. cm.
Includes bibliographical references and index.
ISBN 0-691-08758-X
1. Force and energy—History. 2. Mayer, Julius Robert von,
1814–1878. 3. Physicists—Germany—Biography. I. Title.
QC73.8.C6C36 1993 531′.6′092—dc20 92-25400

---

This book has been composed in BT Transitional 521

Princeton University Press books are printed on acid-free paper, and meet the guidelines for permanence and durability of the Committee on Production Guidelines for Book Longevity of the Council on Library Resources

Printed in the United States of America

1 3 5 7 9 10 8 6 4 2

TO THE CITY OF HEILBRONN

WHOSE SENSELESS DESTRUCTION REMINDS

US THAT WARFARE DEHUMANIZES

ALSO THE RIGHTEOUS

# • CONTENTS •

# • ACKNOWLEDGMENTS •

THE RESEARCH and writing for this book would not have been possible without the generous support of the National Science Foundation and the University of North Carolina at Greensboro. A Research Assignment from UNCG supported me during the fall semester of 1986 while I lived and did research in the Boston area; a first NSF grant (SES-8618404) funded my research there and for a month in Germany during the first half of 1987. I am especially grateful to my colleagues in the History of Science Department at Harvard who arranged for me to be a Visiting Scholar there for a year; it was a pleasure to be able to take part in some of the department's active scholarly life, as it was a privilege to work in the many Harvard libraries. A second NSF grant (SES-8722018), reinforced by a second UNCG Research Assignment, allowed me to devote the academic year 1988–89 to writing. I am also grateful to Mary Ellen Bowden and the National Foundation for the History of Chemistry for a travel grant that allowed me to work in Philadelphia libraries for a few days in March 1990.

Working for a month in German archives and libraries was a joy not only because of the sources made available to me, but also because of the unfailing competence and kindness of everyone I dealt with. I owe a special debt of gratitude to Dr. Helmut Schmolz, director of the Stadtarchiv Heilbronn, which houses the bulk of Mayer's surviving manuscripts, and to Dr. Volker Schäfer, director of the Universitätsarchiv Tübingen, who also made inquiries for me concerning the privately held Gmelin *Nachlass* and who generously continued to answer a number of research questions for me after I had returned to the United States. Hans Göbbel, director of the Kernerhaus museum and archives in Weinsberg, kindly searched their holdings for anything concerning Mayer.

The following institutions generously made available and gave their permission to quote from manuscripts in their possession: Stadtarchiv Heilbronn, Universitätsarchiv Tübingen, Staatsarchiv Ludwigsburg, Royal Institution of Great Britain (by courtesy of the Director), Deutsches Literaturarchiv/Schiller-Nationalmuseum, Bayerische Staatsbibliothek, and Staatsbibliothek Preussischer Kulturbesitz. My thanks, too, to a number of institutions that sent copies of manuscript materials I did not ultimately use: Smithsonian Institution (Special Collections Branch), Universitätsbibliothek Heidelberg, Universitätsbibliothek Tübingen, Gesellschaft Liebig-Museum, Deutsches Museum, and Germanisches Nationalmuseum. Several institutions either answered inquiries or supplied other materials: Universitätsbibliothek Augsburg, Universitätsbibliothek Giessen, Rijksuniversiteit Utrecht (Bibliotheek), Koninklijke Bibliotheek (The Hague), Columbia University Libraries, Massachusetts Institute of Technology Libraries, Center for History of Physics, and Virginia Historical Society. I am particularly grateful to the Staatsbibliothek Preussischer Kulturbesitz, Handschriftenabteilung, Zentralkartei der Autographen, for answering my inquiries concerning manuscript holdings in German libraries. A number of individuals were also kind enough to procure photocopies for me of hard-to-find items: Jodi Bilinkoff, Mark Finlay, Otto-Joachim Grüsser, Rich Kremer, John Neu, Richard Wolfe, and Bill Woodward. Finally, my thanks to Gaylor Callahan, head of UNCG's interlibrary loan department, for helping me to obtain a large number of books and articles essential to my research. To Toni Fields of UNCG I am indebted for the map and diagrams.

Three colleagues—Larry Holmes, Kathy Olesko, and Steve Turner—gave generously of their time to read and comment upon an earlier version of this work, for which I am *very* grateful. Steve's ten-page critique, in particular, was encouraging even while it made numerous general and specific suggestions for revision. An anonymous reviewer for Princeton University Press was also both encouraging and helpfully critical. I ask their indulgence if the present work does not answer all of their desiderata. I had hoped to supply both a critical bibliographical essay and a more extensive resumé of the historiography concerning the discovery of the principle of the conservation of energy, but in the end decided that I preferred not to postpone completion of the manuscript for the not inconsiderable number of months that it would have taken to research and write them. I am grateful to my skillful and conscientious editor at Princeton University Press, Alice Calaprice, for numerous stylistic improvements, for saving me from a number of embarrassing errors, and for letting me keep a few cherished peculiarities.

Last—and no one will think least—I thank Jane for love and undiminished faith during the Era of the Book, now happily behind us.

# • AUTHOR'S NOTE •

A FEW COMMENTS on conventions may be in order. In direct quotations I have always kept the original wording, spelling, punctuation, and italicizing, except that I have not italicized proper names which in the German original were as a matter of typographical convention italicized or printed in all capital letters, and I have usually silently omitted internal page or section references. Individual German words quoted in isolation from their German grammatical context have usually been reduced to a simpler form, for example nouns to the nominative case, adjectives to their uninflected form, and verbs to the infinitive or to a more compact form (as "schreitet . . . fort" to [*fortschreitet*]). Indented quotations show (by indenting or not indenting the first line) whether or not the passage began a paragraph. Omission of text from the beginning of the first sentence is signaled by a lower-cased (or bracketed upper-cased) first word; omission of text from the end of the last sentence is not shown. Shorter quotations incorporated into my text neither preserve initial capitalization nor indicate initial or final omissions, although internal omissions are always noted. If a period directly follows the last word before an ellipsis, then the previous sentence ended there; if there is a space between the last word and the first period of the ellipsis, then a portion of the previous sentence has been omitted. If an indented quote is treated as grammatically part of the sentence of my text that introduces it, its endnote applies also to any quoted passages in that introductory sentence. I have used the customary double quotation marks to denote actual quotations, single marks (a.k.a. 'inverted commas') to call attention to words I have used in some qualified sense.

Translations are my own except where otherwise noted. I thank Joachim Baer for conferring with me over some especially difficult German passages. Since I have italicized foreign-language words quoted within brackets in a portion of translated text, words that were originally already in italics (or printed in *Sperrdruck*) have in addition also been underlined. My translations tend toward the literal, though I have striven to make them readable as English as well. Thus I have rendered "todte Natur" as "dead nature," even though we might more usually say something like "the inorganic world," and the translations have more so's, thereby's, and complexities of style than is normal in English. I have chosen not to lard the quoted passages with a plethora of words in square brackets added to flesh out the English translation—for example, stylistic and's and the's, pronouns replaced by their antecedents (where the German was clear and the English would not be), verbs which the German construction clearly distributes across several phrases but which one would normally repeat in English, and the like.

In the notes and bibliography an equals sign (=) connects sources that are identical except for possible insignificant variations in spelling and punctuation; an 'approximately equals' sign (≈) means that there are some differences in wording of what is essentially still the same text. Unless otherwise specified, it can be assumed that all letters cited in the notes are either to or from Mayer. Johannes Müller is cited in the notes simply as "Müller."

The annotation "in Mayer's library" in the bibliograpahy refers to the "Verzeichnis der Handbücherei Robert Mayers" in Krusemarck 1942, 78–81; volumes I personally saw on my visit in March 1987 to the Mayer archives in the Stadtarchiv Heilbronn ("RM-Archiv") are so designated. Because of the importance of Johannes Müller's *Handbuch der Physiologie des Menschen* and the complexity of its publishing history, I have provided detailed bibliographic information on its separately published parts.

Whenever an issue (*Heft*, *cahier*, etc.) of a journal bears a date and not also a number, that date (usually a month) is put in parentheses after the page numbers of the entry. When it bears both a date and a number, the date appears in parentheses after the issue number. The parenthetic addition "(≈ May)" (for example) refers to the nominal date of the issue as determined by its numerical sequence among the (usually twelve) annual numbers of the journal in question; it should not be assumed that the issue actually *appeared* in that month.

I have used the following standard German abbreviations: Abt[h]. (*Abt[h]eilung* = part), Bd. (*Band* = volume), Hft. (*Heft* = number or issue), Jg. (*Jahrgang* ≈ the 'how manieth' year), Nr. (*Nummer* = number), and St. (*Stück* = number or issue). I have also used the handy German abbreviation *bzw.* (*beziehungsweise*), which means 'or, as the case may be.' Note that periods mark ordinal numbers (e.g., "2. Hälfte" = second half). The titles of four multivolume reference works have been abbreviated as follows:

*ADB*. *Allgemeine deutsche Biographie*. Ed. Historische Commission, Königl. Akademie der Wissenschaften. 56 vols. Leipzig: Duncker & Humblot, 1875–1912.

*DSB*. *Dictionary of Scientific Biography*. Ed. Charles C. Gillispie. 16 vols. New York: Charles Scribner's Sons, 1970–80.

*GVDS*. *Gesamtverzeichnis des deutschsprachigen Schrifttums (GV) 1700–1910*. Ed. Hilmar Schmuck and Willi Gorzny. 160 vols. plus a vol. of *Nachträge*. Munich [etc.]: K. G. Saur, 1979–87.

*NDB*. *Neue deutsche Biographie*. Ed. Historische Kommission, Bayerische Akademie der Wissenschaften. Vols. 1–14. Berlin: Duncker & Humblot, 1953–85.

# • INTRODUCTION •

ROBERT MAYER is well known as one of the several codiscoverers of the principle of the conservation of energy, the formulation of which was arguably the most important single development in physics during the nineteenth century, one which recast and unified the entire subject. Equally well known, at least in a general way, are the circumstances leading to his ostensibly sudden discovery: his voyage as ship's doctor to the Dutch East Indies and his observation there, while letting blood, of the lighter-than-expected color of the venous blood of recently arrived Europeans. Although Mayer's impact on the development of science was slight, his historiographical significance is considerable. He is a chief figure in comparative studies—such as Kuhn's pioneering "Energy Conservation as an Example of Simultaneous Discovery"—that exploit the circumstance of multiple discovery in order to isolate factors of general explanatory importance.[1] And his association with *Naturphilosophie*, now widely accepted, is regularly invoked as an example of the historical importance of this long-discredited movement, and, at least by implication, of the general importance of philosophical speculation to the history of physics.

Among these discoverers, however, Mayer occupies a peculiar position. Insofar as calculation of the mechanical equivalent of heat has, since Mayer's day, been regarded as an indispensable component of that discovery, Mayer's claim to priority of publication is solid. Yet because his work was ignored by his contemporaries until others—notably Joule and Helmholtz—advanced similar ideas, and because his command of physics otherwise looks so tenuous, it can be tempting to dismiss his accomplishment as a fluke, especially since his own route to energy conservation has remained an enigma. His sources are obscure, and commentators have repeatedly noted, with frustration and even annoyance, that his own fragmentary accounts of his discovery do not seem adequately to explain that which centrally needs to be explained, that is, how he discovered the conservation of energy. That he indeed experienced some kind of relatively sudden insight attendant upon his bloodletting experience seems certain; that this insight might have been something like the conservation of energy seems with equal certainty to be exceedingly unlikely. As this study will demonstrate, Mayer's roughly reconstructible process of discovery exhibited a complex internal structure over a period of several years, a process guided more than anything else by his penchant for analogies.[2]

Kuhn recognized that to couch his problem in terms of "simultaneous discovery" was misleading, since "only in view of what happened later can we say that all these partial statements even deal with the same aspect of nature." And Heimann insisted even more strongly that seeing Mayer simply as one of the codiscoverers of energy conservation has continued to distort our understanding of his work and intentions.[3] Yet this general appreciation has not led to a deeper understanding of Mayer's work, in large part, I suspect, because no one has known quite how to exploit it in rendering intelligible the *particulars* of Mayer's process of discovery, of whatever it was he discovered! The comparative approach itself, however valuable it may be, has perhaps facilitated the unwarranted association of Mayer's work with general factors valid perhaps for others but not especially relevant to him. For example, of Kuhn's three "trigger-factors" responsible for simultaneous discovery—availability of conversion processes, concern with engines, and *Naturphilosophie*—only the second was particularly important to Mayer, and even that one in a way quite different from its significance to others. Although Heimann correctly underscored "the conceptual individuality of Mayer's natural philosophy," he did not in fact succeed in going beyond his otherwise perceptive textual analysis to provide the context necessary for a full historical understanding. Buck's praiseworthy attempt to chart Mayer's process of concept formation went too far, it seems to me, in denying Mayer any conception of the *conservation* of energy till 1845.[4] Such a notion played a decisive role from very early on.

Perhaps the most sustained and probing historically grounded attempt to come to terms with the fine structure of Mayer's thought is Richard Kremer's. For Mayer, as well as for Liebig and Helmholtz, Kremer argued that problems relating to animal heat provided only the general context for reflections on energy conservation: Kremer, like others before him, was not able to trace a plausible route from observations on the color of venous blood to energy conservation simply via the oxygen theory of animal heat. He proposed instead that Mayer was stimulated by "sailors' chatter about storms warming sea water" and by his familiarity with chemistry, in particular the principle of conservation of matter.[5] Although Kremer was on the right general track in sensing the insufficiency of that physiological context for the explanation of Mayer's discovery, his specific proposals were weakly founded. For one, Mayer clearly reported that he *asked* the steersman about the temperature of storm-tossed waves, indicating he was already wondering about the relationship between heat and motion. For another, one could not easily have derived from contemporary chemistry any clear notion of a principle of conservation of matter. Mayer's debt to chemistry was considerable, to be sure, but it worked itself out in more complex ways.

Most commentators have tried to explain Mayer by analyzing his ideas—often ignoring those passages or papers they apparently did not know what to do with—without adequate regard to their broad historical context, to Mayer's possible sources, or to his immediate problem context. It is my goal to reconstruct as far as possible those three things and, on that basis, to render historically intelligible Mayer's formulation of what became in a wider context the principle of the conservation of energy. Many have drawn attention to close parallels between Mayer's language and Leibniz's discussions of the equality of cause and effect, the indestructibility of force, the impossibility of perpetual motion, and the principle of sufficient reason.[6] The parallels are indeed close, but in the complete absence of plausible routes by which Mayer might have confronted such considerations, it seems idle to adduce them as historically relevant to understanding him. I place no explanatory weight on general and unmediated intellectual affinities. On the other hand, I am well aware of the fragility of an historical reconstruction that necessarily depends to some extent on the chance survival of often unique references—such as Mayer's single mention of Strauss in a letter to his parents—and on the discovery or not of this or that source, and I have tried to avoid unwarranted explanatory closure around the evidence *I* have assembled.

My approach has been to try to identify sources *in the first instance* within science that Mayer might reasonably be expected to have known and that can help make intelligible the essential components of his thinking and the specific striking usages of his writings. In my view, significant theoretical reconceptualizations—conventionally but improperly included within the category of 'discoveries'—can only be understood when one has some notion of the specific issues and problems the scientist in question was *already* preoccupied with. Such 'discoveries' represent solutions to problems. Only in that way can one appreciate the attractive force, the *meaning*, of new ideas for their originator. The two questions to pose simultaneously are: What was the new idea? and, What kind of reconceptualization did it effect within the changing field of the scientist's dearest concerns? Such a process of extended 'discovery' I have termed the (progressive) crystallization of meaning, and I hope to show abundantly and convincingly how Mayer's route to something like energy conservation represented a protracted process in the course of which *what* he was discovering went hand in hand with its changing and increasingly clarified *meaning*.

My decision in the first instance to investigate sources within science—as opposed to social situation, religious convictions, more straightforwardly philosophical texts, and the like—must be understood strategically and not dogmatically. That is, I believe explanations in the history of science must begin with a reconnaissance of the technical issues, not under the assumption that such will be sufficient, but from the conviction that, methodologically, one only thus knows what needs to be explained and only thus can judge how much explanatory space remains to be filled through an appeal to extrascientific factors. In the event, I have situated Mayer's reflections within a very broad range of interconnected scientific, metaphysical, and theological issues.

In practice, my procedure was first to read all of Mayer's works, scrutinizing them for clues about his sources and wider intellectual context. I make it a general procedural rule to take a scientist's own account provisionally as valid unless and until I am forced for good reasons to question it. My experience is that such accounts, properly understood, provide some of the best evidence the historical detective has. Indeed, it amazed me when, after I had wrestled for several years with a sometimes opaque and intractable subject and had reached a point where I thought I finally clearly comprehended the problem and its solution, I discovered that my hard-won insights were little more than what Mayer had said all along! The trick, of course, was to be able to recognize what was significant, and then to have the confidence to accept it as valid. I have been struck by how many commentators dismissed or otherwise explained away the significance of the centerpiece of Mayer's own several accounts of his discovery, namely his observation of the color of venous blood while off the Java coast in 1840. No fewer than five questioned its physiological validity.[7] Heimann speculated that, when Mayer recounted his story in 1845, "his interpretation of his observation made in 1840 may well owe something to retrospective reconstruction" due to his subsequent reading of Liebig's *Animal Chemistry*. Breger was bothered that what he called the "Java anecdote" did not adequately explain Mayer's discovery of energy conservation, and suggested it continued to be repeated because of our embarrassment at not being able to explain him otherwise. Likewise bothered that "the Java story does not explain how Mayer came to formulate his version of energy conservation," Kremer adduced evidence (unfortunately for his argument, misinterpreted) that "Mayer could not have been surprised by the bright venous blood he observed in Java."[8] Kremer further speculated that Mayer might have chosen to accentuate the "Java anecdote" in order to legitimate his, a physician's, observations as having been properly *physiological*. This dismissal of Mayer's own testimony accorded with Kremer's rejection of "the claim that physiological considerations, *per se*, led Mayer to his discovery."[9] Where such commentators went wrong was in rejecting Mayer's account because they could not see how Mayer or anyone else could reasonably have inferred the conservation of energy from the reported observation.[10] In fact, he didn't—but I'm getting slightly ahead of the story.

The next stage of my investigation was to establish Mayer's context: What had he likely read or heard about? What problems might he have been preoccupied with? By following up what few specific references there were and reading broadly through the contemporary literature in physiology and physics—especially German and French textbooks—I was able to get a grounded sense of what someone in Mayer's position would likely have been confronted with, under the assumption that he was reading *anything*. It became clear that there was appreciable consensus with respect to most of the relevant issues, so it became less of a problem not always to be able to say with confidence which of the likely works or editions Mayer actually saw. The more this context unfolded, the less strange and idiosyncratic Mayer's writings appeared. All the same, several authors loomed as arguably of particular significance, namely Johannes Müller, Jacob Friedrich Fries, and Johann Heinrich Ferdinand Autenrieth, one of Mayer's medical professors at Tübingen.

Expanding Mayer's relevant context beyond the obvious domains of physiology and physics was more problematic, since judgments of relevance become more subjective the farther one gets from the clearly pertinent. For example, I devote little space to issues such as the proliferation, in the decades before 1840, of examples of the interconnections between various forces, or to the protracted debate over theories of the pile, because in my estimation they were not a significant part of Mayer's intellectual world, despite their obvious general relevance to the conservation of energy. On the other hand, I devote considerable space to topics never before considered in connection with Mayer, such as the importance of David Friedrich Strauss and theological issues, of Justinus Kerner and spiritualism, of homeopathy, and, especially, of the existence of a "common context" (to use Robert Young's apt phrase) where medicine, physiology, theology, and ontology came together around problems of the soul, the vital force, and the body.

The organization of the book reflects in some ways my research strategy, since in my view the product cannot be wholly separated from the process of production. The first chapter introduces Mayer the person by providing a focused biographical sketch of his background, education, voyage, personality, and career. Although this study is not a biography in the conventional sense, it attempts a meaningful integration of Mayer's life and work by demonstrating how the direction and quality of his work bore the clear stamp of his circumstances, his personality, and his circle of friends. The second chapter offers an analysis of his work aimed at identifying the leading ideas and peculiarities that require explanation. I pay most attention to the crucial first two years of his work, though without ignoring later developments and the insights that later writings provide with respect to the genesis of his ideas on force. From beginning to end, Mayer's work was marked by a few strongly held ideas and guiding motifs.

The third through fifth chapters attempt to establish the relevant context within which Mayer's thinking developed and in terms of which alone it can be understood. Chapter 3 describes the state of contemporary thinking on such physiological problems as the physiology of the blood, respiration, and animal heat; the relationship between organisms and the external world; the vital force and its relationship to the soul on the one hand and to the imponderables (heat, light, and electricity) on the other; and homeopathy. It briefly considers the relevance of contemporaneous technological change with respect both to the actual machines, steam engines, and railroads then increasingly visible, and to their symbolic value. One of the important determinations in chapter 3 is that almost no author of the 1830s accepted without significant reservation Lavoisier's explanation of the generation of animal heat in terms of oxidation. Hence Mayer's unqualified acceptance of that theory should be seen not as a natural starting point but as a crucial early conclusion that went hand in hand with his evolving interpretation of his observations. It also represented a crucial crystallization of meaning with respect to the primary function of respiration: from the removal of waste carbon from the blood to the production of animal heat. Chapter 4 establishes a baseline in physics and chemistry with respect to issues such as the conception of matter, force, and the imponderables (which in this context also included magnetism); the handling of central forces; and the *absence* of any explicit principle of conservation of matter. As with his 'creative' interpretation of Lavoisier's theory of respiration, Mayer did not so much draw upon an existing principle of the conservation of matter as make explicit to himself the uncreatability and indestructibility of matter *as* he forged a coherent conception of force. Chapter 5 considers discussions of the nature and scope of science and notes the transformation underway (beginning in earnest around 1840) in the study of organic systems. In it I further explore the thick web of religious and spiritualist issues woven in Mayer's immediate surroundings by the likes of Strauss and Kerner, and weigh the potential significance of the literature on natural theology.

These chapters, then, provide a basis for an historically grounded understanding of Mayer's vocabulary, conceptual framework, and problem context. What force meant to Mayer can only be appreciated if one knows what the ramifications of the term were in the full range of the literature he likely knew. One cannot understand how Mayer could have come to believe in something like the conservation of force unless one knows the locus of contemporary discussions relating to the creation and destruction of things like forces, the imponderables, and the soul. One cannot understand Mayer's creative exploitation of a host of rich analogies unless one knows the meanings and connotations of the terms involved. One cannot appreciate the circumstances surrounding the creation and elaboration of his ideas without knowing what his contemporaries were saying about, for example, *Stoffwechsel* and *Lebenskraft*: Mayer stepped onto the stage during a crucial period in the development of life-scientists' self-consciousness about the nature of science and scientific explanation. Even if, given the state of the available evidence, it may not be possible to identify with certainty all of Mayer's specific sources, it is nevertheless possible to make his work well-nigh completely *intelligible* within the contexts of historically constructed fields of meaning. A close familiarity with Mayer's broad context is thus a necessary precondition for understanding *his* words, ideas, analogies, and guiding motifs. After all, it was from that context that he derived the elements that he reshaped and rearranged in forging his own ideas. On the other hand, it would be foolish to think that Mayer can somehow be rendered solely in terms of his general context. Of crucial importance was also his preoccupation with a number of specific problems: that was what chiefly stimulated his creative drive and gave his reflections concrete form. They provided the conceptual nuclei around which his evolving thoughts continued to crystallize. If Mayer's work was decisively shaped by his historical circumstances, it was nevertheless the product of a unique creative individual seeking solutions to a particular set of problems.

In elaborating the scientific context of the 1830s and 1840s I have made a special effort to tease out the many connections among the various members of the scientific community—in letters, book reviews, and other published works. There *was* a real community of researchers who read and commented on one another's work. Indeed, despite his relative obscurity, Mayer was himself an interacting part of the community of scientists struggling to come to terms with the concepts of force and the imponderables: he derived his problems, concepts, and data from it, and he chose to address his own work to them. Nor were many of the works that are little known today—such as Autenrieth's books and various Latin dissertations—unknown to that community. Believing that the historian must pay primary attention to language and the field of meaning of particular words, I have tried to render my subjects' work in as unanachronistic terms as possible and have been generous in quoting from primary sources, since those sources are always richer in meanings than the points we often wish to distill from them. I have also paid particular attention to the dating of primary sources, since it is necessary to know who might have known what when in order to be able even to speculate intelligently about the progress of ideas—especially since Mayer was away from Europe for most of a year, and since the published date of many works does not accurately represent the timing of their appearance.

The sixth chapter situates Mayer's work within the context so established. As far as the evidence and reasoned speculation will allow, I have attempted to reconstruct the likely path of Mayer's earliest thinking and to chart its subsequent progress for the period after June 1841, as more contemporaneous evidence becomes available. My goal has been historical intelligibility: that readers be able to make sense of *all* aspects of Mayer's work in terms of a plausible, historically grounded reconstruction. In this chapter I argue that it is within the broad metaphysical-medical context identified above that the takeoff point of Mayer's thinking must be situated. Reflection on problems of animal heat and the physiology of respiration, coupled with a guiding analogy between organisms and work-performing machines, presented Mayer with the logical conclusion that there must be a constant relationship between heat and work (or "motion," as he usually put it). It is crucial to realize, however, not only that Mayer did not thereby discover energy conservation, but that he did not yet even have a concept of energy (or force—Mayer spoke consistently of *Kraft*). In concentrating their attention upon the issue of the *conservation* of energy, some historians have not fully appreciated just how important, and difficult, Mayer's creation of a tractable concept of *force* was. On the basis of evidence admittedly circumstantial, I argue that the idea of the indestructibility of force—again, not yet its conservation—derived from ontological reflections on the relationship between matter, force, and spirit. Mayer believed that his concept of force was an antidote to materialism because it vindicated the existence of something nonmaterial in the world.

At roughly the same time—and the evidence does not permit a fine-grained reconstruction—he sought clarification as to the nature of force from contemporary physics texts, an endeavor which, alas, initially created more confusion than clarity and, ironically, even encouraged him to believe for several years that force might in fact be created in both living systems and divinely ordered "organisms" such as the solar system. It is ironic that Mayer's calculation of the mechanical equivalent of heat in his 1842 paper is generally regarded as establishing his claim to the discovery of the conservation of energy, since at that time he did not in fact believe that force is conserved, only that it cannot be destroyed. Mayer's assimilation of living organisms and planetary systems is itself an example of a major theme of this study: the search for valid analogies was an essential component of the elaboration and circumscription of Mayer's evolving problem context.

The general picture, then, is one of Mayer addressing a specific set of problems within a broad context of beliefs, assumptions, and interests. Although I find no explanatory role for what one generally thinks of as social factors in this story—and it must be appreciated that I by no means reject such in principle—I would underscore the importance of Mayer's decision to pursue the physical aspects of his new and multifaceted ideas and to appeal to the physical sciences community. Given his background and position—he was a practicing physician who until then had shown no interest in strictly scientific matters—it was not an obvious choice for him to make. Considering his own interests and the fact that "metaphysical" questions were routinely discussed in the physiological literature, it is perhaps a testimony to the tenacious consequentiality of Mayer's personality that, having become convinced of the logical (and disciplinary) necessity of first grounding his theory of force in the physical sciences before seeking its application to physiology, he picked the audience he did. Having chosen to step forward as a physical scientist, he shrank his field of active concern to the mechanical theory of heat, and indeed to a single number, the mechanical equivalent of heat. This transformation is another aspect of what I call the crystallization of meaning.

The present study, on the other hand, began with a narrower focus and expanded as it progressed. The character and historical significance of *Naturphilosophie* has long been a problem of special interest to me. My first paper as a graduate student in the history of science (in Thomas Kuhn's thermodynamics seminar) looked at Kant and Schelling in order to try to understand where Mayer might have come from. Not having studied Mayer since then, I nonetheless retained strong impressions from that early encounter and had continued to think about *Naturphilosophie* when, a few years ago, I came across a casual reference to Mayer's indebtedness to *Naturphilosophie* in a general work having nothing in particular to do with Mayer, *Naturphilosophie*, or even nineteenth-century German science.[11] It struck me that this connection was well on its way to becoming one of the unproblematical, generally acknowledged facts of the discipline, yet I knew that *I*, at least, had never seen the point argued more than superficially. At that point I doubted it was true but still thought it was possible that more careful historical research might salvage something from the claim. (It has never been my task to disparage *Naturphilosophie*, and I have elsewhere argued at great length that it played a major role in shaping the character of German electrodynamics in the 1840s.)[12] In any event, I thought the topic would be a good one to investigate: I had already studied *Naturphilosophie*, German science of the period, and Mayer's works, and I thought I could exploit the study both to clarify our understanding of Mayer's route to energy conservation and to present some of the reflections I had been nurturing about the character and role of *Naturphilosophie*. My conclusion, to be argued at length in the seventh chapter, is that there is no compelling evidence, and little likelihood, that *Naturphilosophie* played a significant role in Mayer's creation of his theory of force. Nevertheless, the fact that there do exist a few terminological and conceptual resonances in Mayer's work suggestive of aspects of Schelling's philosophy of nature should remind us of the distinction between what historians can justifiably argue and what actually took place. Nor do existing weaknesses in the historiographical tradition connecting Mayer with *Naturphilosophie* necessarily mean that a reasonable case might not be built on better evidence. As far as possible philosophical roots of Mayer's ideas are concerned, a case can be made for the influence of a dilute and refracted Kantianism (via Jacob Friedrich Fries and Philipp Lorenz Geiger). But the importance of properly Kantian ideas must not be overrated, and I prefer to avoid identity labels and assignment to traditions and instead to emphasize specific ideas from specific works. It should be noted, too, that the metaphysical-medical context mentioned above cannot be assigned to any particular named philosophical movement, but rather reflects widely diffused concerns about the nature of the soul and its relation to the body, about where the vital force comes from in the generation of new beings, and the like.

Whatever the conclusion with respect to Mayer, I believe this study will contribute to an enhanced understanding of the place of *Naturphilosophie* within early nineteenth-century German science. Similarly, I believe its placement of Mayer within a broad intellectual context will shed important light on the general situation in German science at a time of major disciplinary transformation. More centrally, I hope to have made Mayer historically intelligible in a way that will contribute to a general understanding of the process of scientific discovery. Paradoxically, the more one understands Mayer, the more one sees how an appreciation of his idiosyncratic uniqueness advances in step with an appreciation of the manifold interconnections between him and his context.

# • PART I •

## The Man and His Work

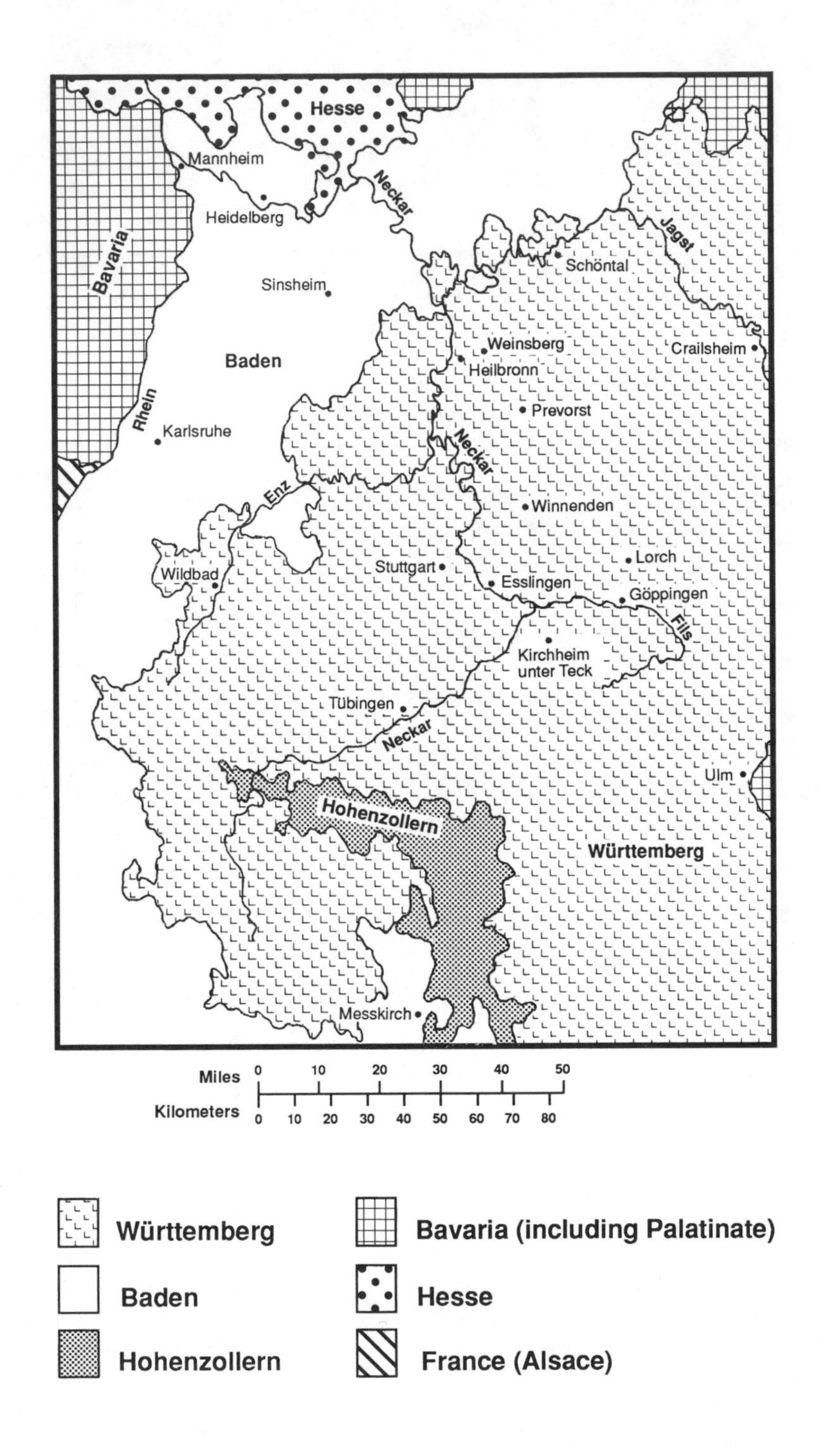
Hesse
Mannheim
Heidelberg
Neckar
Bavaria
Sinsheim
Schöntal
Jagst
Baden
Weinsberg
Heilbronn
Crailsheim
Prevorst
Rhein
Karlsruhe
Neckar
Enz
Winnenden
Wildbad
Stuttgart
Lorch
Esslingen
Göppingen
Fils
Kirchheim
unter Teck
Tübingen
Neckar
Ulm
Hohenzollern
Württemberg
Messkirch
Miles
0
10
20
30
40
50
Kilometers
0
10
20
30
40
50
60
70
80
Württemberg
Baden
Hohenzollern
Bavaria (including Palatinate)
Hesse
France (Alsace)

• CHAPTER ONE •

# Mayer the Person

## 1 Mayer's Upbringing and Education

Julius Robert Mayer was born on 25 November 1814 in the city of Heilbronn, a former *Reichsstadt* situated on the Neckar River in the southwest German kingdom of Württemberg.[1] Heilbronn, a prosperous city and one of the quickest in the state to industrialize during the nineteenth century, saw its population rise from around 7,100 in 1820 to 11,300 in 1840 and 14,300 in 1860; growth in the surrounding district kept pace, and contributed an additional population about equal to that of the central city.[2] Its inhabitants were overwhelmingly Evangelical (i.e., Lutheran), as was Mayer's family. Mayer's father owned an apothecary in which his three sons received early practical chemical experience. The eldest brother, Fritz (born in 1805), first worked as an assistant to his father, then took over the enterprise sometime around 1832 or 1833; the middle brother, Gustav (born in 1810), worked as an apothecary in the Baden towns of Messkirch and Sinsheim before emigrating to America in 1849. Gustav Rümelin, one of Robert Mayer's lifelong closest friends, reported that Mayer senior devoted his free time to scientific studies and experiments and that the Mayers' house was filled with physical and chemical apparatus and instruments, botanical and mineralogical collections, medicinal plants, and many books, especially travel accounts. He recalled that Mayer performed experiments and exhibited *curiosa* to his friends, skillfully operating an air pump and various electrical devices. Young Mayer was familiar with elementary chemical procedures, could identify the contents of most of his father's apothecary canisters and tell what they were used for, and could name plants according to the Linnaean system.[3]

In several of his autobiographical sketches Mayer told the story of his childhood fascination with machines and his abortive attempt to construct a perpetuum mobile. Although the lasting significance of the experience for Mayer has been called into question,[4] I take Mayer's account seriously and will later argue that the lesson he thereby learned played a crucial role in the development of his theory of force. Mayer recounted his early endeavors in great detail (writing of himself in the third person):

> Already as a young boy, chemical and physical experiments and the construction of watermills in his hometown were much more attractive to him than the prescribed study of Latin and Greek, which frequently drew down upon him the dissatisfaction of his teachers.
>
> Is is perhaps appropriate here to recall an in itself unimportant incident which, given the impressionable spirit of youth, made a lasting impression on our investigator. He must have been somewhat more than ten years old. It was a common diversion

> in the afternoon hours to place small waterwheels in a small brook (the Pfühlbach) which flowed into the Neckar at Heilbronn, and by means of their rotation to move other small objects as well. Now it was on this occasion that our small man hit upon the great idea of constructing a perpetuum mobile. In his mind he fastened to the axle of such a wheel an Archimedes screw, whose operation he knew from Poppe's *Physikalischer Jugendfreund,* which he had gotten from his father for Christmas. But since when such an endless screw runs over a large cogwheel what is gained in "force" is lost in speed, he accordingly restored this loss easily again by having the large cogwheel engage a small one. Now on the axle of this small cogwheel there is again a screw, which engages a large cogwheel, etc. In this way, by means of such a transference, the boy concluded, so much force must obviously be gained that arbitrarily heavy machines could be driven by a tiny waterwheel. Set straight by other, older people—namely, that by means of the transference [of motion] from a large cogwheel to a small one as much "force" is again lost as one gains in speed—he in turn quickly gave up his project, but through his error at such a young age attained the insight [that mechanical work cannot be created out of nothing].[5]

This disappointment, however, did not keep young Mayer from continuing to be fascinated by the mechanical devices that were becoming increasingly common in the area. Friend Rümelin, who had gone off to the Evangelical school in nearby Schöntal in the fall of 1828 (where Mayer would join him the following spring), recalled that Mayer wrote to him then "that he now spent his free time in the numerous and diverse mills and factories which lay next to each other along the Neckar, studying their mechanism and assisting the people in their work."[6] Mayer apparently continued to be fascinated by machines' ability to do work and by the operators' ability to control their effect by means of relatively small causes. After noting, in a late essay, that people are by nature so constituted that they like to obtain the greatest possible results by expending the least means, he went on to reminisce about his boyhood fascination with machines: "From the years of my youth I still remember very well how on free afternoons I spent many an hour in a sawmill, where by pressing on a lever and withdrawing the sluice gate the mechanism was set in motion."[7]

After attending the Evangelical school in Schöntal for two and a half years and then assisting in his father's apothecary for eight months, Mayer enrolled at the Eberhard-Karls-Universität in Tübingen in May 1832 to study medicine.[8] With the exception of a physics course during his first semester, his five-year program of study was devoted entirely to his medical training.[9] The medical faculty at the time seems to have consisted largely of relatively unprogressive and unphilosophical teachers of a practically oriented medicine, and the claim that the idealistic philosophy of Hegel "wholly determined" the intellectual climate at Tübingen while Mayer was there is without foundation.[10] The dean of the faculty and chancellor of the university, Johann Heinrich Ferdinand Autenrieth, had a solid reputation as a defender of an empirical approach to medicine allied with anatomy and physiology and thus as an opponent of the various currents of Romantic medicine that were, in the early decades of the century, gaining widespread allegiance at many other German universities.[11] Contemporary reviewers of Autenrieth's *Handbook of Empirical Human Physiology* repeatedly emphasized his devotion to empirical observation, in contrast to the speculations of those influenced by *Naturphilosophie*.[12] Of Mayer's teachers, I find only one to have shown any significant attachment to one of the then-competing major philosophical movements—Ferdinand Gottlob Gmelin, who in his *General Pathology of the Human Body* (a book Mayer owned) explicitly embraced Kant's notion of purposefulness.[13]

Wilhelm Griesinger, a close friend of Mayer's who studied medicine at Tübingen from 1834 to 1837, praised only one of his teachers as good, the anatomist (and *praeses* of Mayer's dissertation) Wilhelm Ludwig Rapp, of whom we are told by an anonymous contemporary: "In accordance with his maxim only to believe what he can convince himself of through personal observation, he must direct his efforts to such fields as offer him no obstacle in this connection, and he thus accords almost no room to speculation. He does not directly admit his generally assumed attachment to materialism; however, in his lectures he concedes the highest significance to matter and (in particular) to the blood."[14] Another member of Mayer's circle of medical-student friends, Wilhelm Roser, recalled that "the physiologist W. Rapp taught the propulsive force of the blood and the pathologist F. Gmelin the polarization of the vital force. The other pathologist, Autenrieth (junior), taught according to his own system, in which the scabious and miliary sequelae, etc., played a prominent role." Unstimulated by the instruction offered by the medical faculty, Roser and his cohort read Johannes Müller, François Magendie, and the four unauthorized volumes of Johann Lucas Schönlein's clinical lectures. But good books were hard to identify and obtain.[15] The only member of the Tübingen medical faculty to inspire members of the younger generation was Albert Friedrich Schill, *Privatdozent* from 1835 until his untimely death from typhus in 1839. Schill was a major force in introducing his enthusiastic students to the new French and English clinical medicine, not least by lending students books from his own library.[16] These students did not, however, include Mayer.[17]

In addition to Griesinger and Roser, Mayer was close to one other reform-minded and dynamic young medical student, Carl August Wunderlich. Griesinger, Roser, and Wunderlich, who were to become three of the most active voices in the reform of German medicine after 1840, had attended the Stuttgart *Gymnasium* together and actively shared an interest in natural history, botany, and chemistry.[18] Wunderlich and Griesinger, like Roser, turned to the then brand-new text of Johannes Müller to make up for the deficiencies of the classroom, in which antiquated lore was dictated from old lecture notes.[19]

The quality of Mayer's association with these young men is difficult to determine. Griesinger, Wunderlich, and Mayer were among the cofounders in 1836 of the forbidden student corps Guestphalia, having all previously been members of the corps Giovania.[20] Yet there is no evidence that Mayer shared their disenchantment with the education being offered or their enthusiasm for Schill, and neither Wunderlich's necrologue on Griesinger nor Roser's and Heubner's on Wunderlich nor any other reminiscences of his three medical-student friends mention Mayer as having shared their reformist dynamism.[21] His self-professed conception of medicine reinforces this impression:

> With regard to the principles which guided and guide me at the sickbed, I belong to those who understand medicine, the *ars medendi*, as an art and not as a science. Here one must not follow the principles of some logically thought-out system, but each individual case is to be apprehended by itself and treated according to the rules of an eclectic empiricism, whereby the [doctrine of] "ex juvantibus et nocentibus" is decisive [i.e., according to whether the treatment helps or harms]. . . . As my esteemed teacher, the ingenious [*geistvoll*] chancellor Autenrieth in Tübingen, nicely put it in his introduction to nosology, the system is like a tangent drawn to a circle, to nature; in order not to distance itself too far from the circle, the tangent must often be broken, and this inconsequentiality in the world of our thoughts is the necessary consequence of our insufficient knowledge of the objective world; otherwise the rigid system becomes a Procrustean bed.[22]

Autenrieth was the only teacher of whom Mayer ever spoke highly, calling him *geistreich* in another recounting of his comparison of an explanatory system "with a tangent drawn to the great circle of truth."[23] It is possible, too, that Mayer might have been alienated by the materialistic and reductionist approach to medicine to which his friends subscribed.

In February 1837 Mayer and a number of *Corpsbrüder* were arrested for wearing the colors of a forbidden organization. Mayer and Griesinger were expelled for a year, and Wunderlich received three weeks' incarceration.[24] During the summer and fall of 1837 Mayer traveled to Switzerland and visited clinics in Munich and Vienna, though without notable effect. Rümelin could not recall any significant details of those visits, and Diepgen's later investigations failed to turn up any traces of Mayer's presence in either city.[25] Nor are there any detectable traces of his later attendance at medical lectures and demonstrations in Munich in November and December 1838.

During his enforced absence from Tübingen, however, Mayer apparently solidified his desire to travel to the East Indies, a region of the world that had fascinated him since childhood.[26] Having passed his medical exams and completed his dissertation (on santonin, an anthelmintic) in 1838, he traveled to Amsterdam in June 1839 to take the examination to become a ship's doctor in Dutch service. Having been admitted to that service late in September, Mayer spent the next four and a half months in Paris. There he hung out with his former university friends Griesinger, Wunderlich, and Roser, and made the acquaintance of Carl Baur, a student of mathematics and mechanics who lived with Mayer and Roser in the Latin Quarter. (Baur would later become an administrator and teacher at a number of Württemburg secondary schools before obtaining a professorship in mathematics and practical geometry at the Polytechnic School in Stuttgart.)[27] Drawing apparently from information supplied to him by Baur, Weyrauch recorded that Mayer principally visited demonstrations at bedside and in the operating room, without showing any particular interest in purely scientific matters.[28]

## 2 Mayer's Voyage to the Dutch East Indies

On 23 February 1840 Mayer left Rotterdam aboard the ship *Java* to sail to the Dutch East Indies. For most of the voyage his duties as ship's doctor took up little of his time, and he settled into a routine that included periods of "scientific activity" once or twice a day.[29] Several letters written after the voyage allow a sharper specification of what those scientific activites were. In recounting the circumstances surrounding his discovery, he portrayed himself as "taking as my starting point physiological and pathological investigations" and as "occupying myself . . . almost exclusively with the study of physiology"; he said he made his discovery "through the unremitting study of a specialized area of physiology," from "having occupied myself zealously and unremittingly with the physiology of the blood."[30] This study and Mayer's reflections thereon set the stage for his observation of the lighter-than-expected color of the venous blood of Europeans recently arrived in the tropics. Mayer's first recorded account of this episode dates from 1845: "During a 100-day sea voyage there had occurred no appreciable incidence of disease among the 28-man crew; however, a few days after our arrival at the Batavian roads there spread in epidemic fashion an acute (catarrhal-inflammatory) affection of the lungs. In the copious bloodlettings I performed, *the blood let from the vein in the arm had an uncommon redness, so that from the color I could believe I had struck an artery*."[31] From the entries in Mayer's medical log it appears that those venesections took place on or shortly after 20 June 1840. Subsequent accounts confirm the importance of these observations in setting off Mayer's chain of reasoning.[32]

There is, unfortunately, little evidence as to what books Mayer had with him. It would seem safe to assume he took along Johannes Müller's then new and universally acclaimed *Handbook of Human Physiology*, of which he owned the second edition of the first volume (1835) and the first (and only) edition of the second (1837–40). The third part of the second volume, although dated 1840, came out at the end of 1839, hence Mayer could have had the complete work on board.[33] Also in his library and reasonable candidates for inclusion among the books he had with him are Autenrieth's *Handbook of Empirical Human Physiology* (1801–2) and Burdach's *Anthropology for the Educated Public* (1837), which, despite its title, covered roughly the same ground as contemporary physiology texts. He might have had one of the several French or German editions of Magendie's *Elementary Sketch of Physiology*, the third edition of which was translated into German by one of Mayer's professors and published in Tübingen in 1834–36. Among other major contemporary works with which he was surely familiar one should add the first volume of Friedrich Tiedemann's *Human Physiology* (1830) and Berzelius's *Textbook of Chemistry*, most of the second edition of which was in his library. Mayer cited the fourth edition of the first volume of Geiger's *Handbook of Pharmacy* (1833) in his 1838 dissertation and undoubtedly owned a copy that might have accompanied him to Java. On the basis of evidence to be discussed in chapter 4, one can conclude that Mayer must have had at least one physics text with him, probably Jacob Friedrich Fries's *Textbook of Physics* (1826).

Given the book's title and its discussion of problems of animal heat and respiration, it is tempting to imagine Mayer having tucked away a copy of Humphry Davy's *Consolations in Travel*—German editions of which appreared in 1833 and 1839—but no evidence supports such an inference.

One of the most eventful early moments of Mayer's journey came when he retrieved the chest of books he brought with him. As he described the event in a letter to his parents:

> On the ninth [of March] I was able to obtain my chest of books from steerage . . . . Triumphantly I held up the Bible and the hymnal which I had most longed for and which give me pleasant hours every day. Removed from the tumult of the world the heart is mightily disposed to devotion, and living [amidst] the magnificent natural world one knows nothing finer than to exalt oneself to the Creator. The little work by Dr. Strauss that my dear brother put in and the star chart from my dear father also pleased me greatly; the latter is put to use every evening, and Straussian principles find such wonderful accord in a soul attuned to true piousness.[34]

Given the interest indicated here in astronomy—confirmed by numerous entries in his diary—and his ownership and later citation of a German editon of Herschel's *Treatise on Astronomy* (1838), it is possible he had this book with him, too.[35]

But what is most striking about this passage is that the three works it mentions—the only three works we can positively identify as having accompanied Mayer on his voyage—are the Bible, a hymnal, and a "Schriftchen von Dr. Strauss," which can only have been (as Weyrauch suggested in a footnote to the above passage) David Friedrich Strauss's *Two Irenic Essays* (i.e., *Zwei friedliche Blätter*) of 1839.[36] Even if one were inclined to discount the significance of Mayer's account because it was intended for his parents' ears, its spirit is corroborated by a passage in his diary for that date: "With delight I held up the sacred scriptures which I found here; a sea journey disposes one mightily to devotion and directs one's thoughts toward heaven, to the director of the universe [*zum Leiter der Welten*]."[37] Nor is this the only evidence of Mayer's deep religiosity. But how can that be rhymed with his enthusiasm for Strauss, whose *Life of Jesus* (1835–36) had recently brought its author into disrepute as an enemy of Christianity and had cost Strauss his position as *Repetent* at the theological seminar (*Stift*) in Tübingen?

## 3 Mayer's Religiosity

Neither the precise quality of Mayer's religiosity nor its relevance to his scientific work is easy to determine, but there do seem to be essential connections between the two. Although he was fond of quoting verses from the Bible, Mayer seems never to have been attached to its literal historical veracity, nor did his faith depend on revelation. In the opinion of one who knew Mayer in later years, "the Bible was for him an ingenious [*geistreich*] book of which he knew how to make ingenious use, that was all."[38] He had, however, a well-nigh unswerving belief, requiring occasional reconfirmation at times of great personal stress, in God as the providential creator of a harmonious world and in the immortality of the soul. As he expressed it in a later work, with reference to the presumed cosmic dust which, illuminated by the sun, produces the zodiacal light, "this dust, too, forms an important link in a creation where nothing [happens] by chance, but where everything has been ordered with divine purposefulness."[39]

God's benevolence and providence were constant themes in Mayer's more personal writings. From on board ship in the Sunda straits he wrote to his parents: "The infinitely good God be thanked, who so allotted everything to me—including minor vexations, which are a necessary part of life—that I never become tired of praising his wisdom." A few days later, as the ship had entered the Batavian roads, he added: "God, thy world is beautiful!" Ten days later he wrote them that "since a firm trust in God and a cheerful spirit always accompany me, you can well imagine that it is with a calm soul that I look forward to the future, which, like the past, will see its dark and bright hours pass by quickly." And from the last letter he wrote home from the Indies: "I will entrust myself with confidence still further to God's all-wise providence, which has been so visible in the vicissitudes of my fate, and if he grants me but one request, then I will bear everything with a smile; only here am I not yet able to pray sincerely 'Thy will [be done],' for the hope is to press all my loved ones to my heart again."[40]

Not surprisingly, Mayer frequently gave expression to his religious sentiments in response to death or illness. To his longtime friend, the pastor Paul Friedrich Lang, Mayer wrote: "The firm conviction, based on scientific awareness and purified of any belief based on revelation, [both] of the continued personal existence of the soul and of a higher direction of human fortunes, was the most powerful consolation to me as I held the cold hand of my dying mother in mine." Four years later, having lost one daughter less than a week earlier, Mayer wrote to his mother-in-law about the impending death of another child, adding that "only firm belief in the divine direction of all the vicissitudes of human life can give us self-composure [*Fassung*]." And to his ailing father-in-law he wrote: "May God give his blessing that the outcome [of the bloodletting] be a really good one! Look with courage into the future, dear father, for Providence surely directs everything to our benefit."[41]

During the period of Mayer's most severe mental illness, during the early 1850s, he found himself possessed by a feeling of "fanatical pietistic sentimentality."[42] After Mayer's long convalescence following a jump from his third-floor window in 1850, Rümelin reported: "After these storms there set in a beneficial calm, he was in a religiously buoyant mood, since he saw in his experiences a higher dispensation, a penance for his passionateness and a deliverance."[43] Mayer saw his problem in moral terms, as stemming from his lack of humility, which he described as his "favorite sin" and his "Achilles' heel"; nevertheless, "my faith is firm that I have been fully granted the forgiveness of sins promised to me at the altar."[44] As he confessed in autobiographical notes written in 1865 for John Tyndall,

> It is possible that the absence of any recognition, on which I had hastily counted, contributed its part to somewhat cooling my zeal for science for a while; it is certain that at that time the interest in transcendental, religious truths began to come to the

fore with irresistible force . . . . With the passionate haste, with the exclusivity, which is a lamentable defect in my temperament, I immediately threw myself into this area as well. I have now grown older and gladly let myself again be counted among the disciples of science, but the zeal for the truths of the Christian religion is nevertheless in no way becoming cooler with me.

. . .

But what I then silently forbad myself to think, I am now ready to confess without restraint. There lived in me a longing for recognition, and however much I might endeavor to fight back such a feeling as sinful pride, it nevertheless was beyond my powers to suppress in me my scientific consciousness.[45]

Reflecting on the recognition his work had begun to receive during the late 1850s, following painful periods of mental illness and brutal treatment during the early 1850s, he wrote further to Tyndall that "I regarded it as a special dispensation of God that, in order still to experience joy at the success of my intellectual children, I was able to overcome suffering that much stronger people would easily have succumbed to."[46]

Religion thus provided Mayer with faith in the providential direction of worldly affairs, with a belief in the immortality of the soul, and with a strong moral sense of sin and redemption.[47] The presence in his library of Johann Peter Süssmilch's *The Divine Order in the Fluctuations of Humankind, Proved from Its Birth, Death, and Propagation* (1765)—a work that sought to demonstrate the workings of divine providence from birth and death statistics—suggests that Mayer may also have had an ear for the arguments of natural theology common in the 1830s. As will be discussed in chapter 5, Mayer would later oppose Darwinian evolution on essentially religious grounds.

In noting that Mayer's religious views were not constant over time, Weyrauch observed that "only the opposition to materialism and atheism remained firm with him."[48] And in registering Mayer's turn to religion in the 1850s, Rümelin added that "he had earlier already been averse to the materialistic worldview widespread among young medical students."[49] Indeed it was Mayer's opposition to materialism that provides the key to the connection between his religious views and his conception of force. As will be laid out in the next chapter, one of the things his conception of force accomplished for him was to cut the ground out from under the materialistic worldview as he understood it. Although one commentator who examined Mayer's pronouncements on religion was of the opinion that Mayer deduced the existence and indestructibility of souls from his energy principle, I will argue for a relationship in the other direction.[50]

For the most part, however, Mayer did not try explicitly to link his religious views to his science. He never, for example, defended the conservation of energy in Cartesian fashion in terms of the constancy of God's action or the like. In a startlingly friendly letter of 1867 to archmaterialist Jacob Moleschott acknowledging with gratitude his election as corresponding member of the Turin Academy of Sciences, Mayer wrote:

To you before all others is due the great and lasting service of having successfully defended the principle that scientific matters and researches must not be mixed up

> with religious dogmas or even ecclesiastical questions, and even if we perhaps . . . do not agree on all points in the area of the supernatural, then I am all the less surprised by that, since in this regard I have not even been able to settle things entirely with myself, despite the fifty-three years I now have behind me, and thus for this reason alone agreement with a third party does not belong to the realm of the possible.[51]

Mayer was ambivalent with regard to whether religion and science are in close and harmonious connection or belong to totally different realms. One of his favorite mottos, which he appears to have more or less mass produced for distribution in 1868, held that "Nature, science, and religion form an eternal union."[52] And during his crisis years he had written to Lang that "my earlier intimation, that scientific truths are to the Christian religion as streams and rivers are to the ocean, has now become for me a living awareness."[53] Yet when Friedrich Dürr, who came to know Mayer in 1874, asked him how he reconciled (*vereinige*) his religious faith with the results of natural science, Mayer replied: "I don't reconcile anything; for me the physicist stops right here and the believing Christian begins."[54]

In an address late in life, Mayer urged his audience not to confuse the brain with the mind (*der Geist*), adding: "But the mind, which no longer belongs to the realm of the sensibly perceptible, is no object of investigation for the physicist and anatomist. What is subjectively correctly thought is also objectively true. Without this eternal harmony preestablished by God between the subjective and objective worlds, all our thinking would be fruitless. . . . Let me close here. With all my heart I proclaim: a correct philosophy ought to and can be nothing but a propadeutic for the Christian religion."[55] In a work published eighteen years earlier—his last major scientific publication—he had written with regard to the principle of the equivalence between heat and motion that "the experimental test of this principle, its verification in all particulars, the proof of the complete harmony between the laws of thought and the objective world, is the most interesting but also the most comprehensive task one can find."[56]

Although this belief in the correspondence between the conclusions of reason and empirically discovered truths of the physical world agrees completely with the conviction with which Mayer sometimes thought he could estabish his new theory by appealing to what he regarded to be the laws of thought (such as "nothing can be created from nothing"), it is quite at odds with another of Mayer's mottos: "Nature in its simple truth is grander and more magnificent than any creation [*Gebild*] of human hands and than all illusions of the created mind [*erschaffener Geist*]."[57] Nor does it rhyme very well with another of his reflections:

> What is insanity?
> The reason of an individual.
> What is reason?
> The insanity of many.[58]

Although, given the fragmentary character of the evidence, it is difficult to chart changes in Mayer's religious and related philosophical views over time, it seems that his earliest and deepest conviction was in a divinely preordained correspondence between the truths of sound reason and the nature of the external world, and that his reservations about the ability of the divinely created human mind to discover truths about the likewise divinely created world belong primarily to his later years, and may have been a cautionary reaction against a rationalistic scientific juggernaut seen as threatening to religious faith.

Mayer's opposition in later years to Darwinism, which he called "the modern heresy [*Irrlehre*]," represented a complex combination of religious, philosophical, and scientific motives.[59] That the subject touched him deeply is suggested by the fact that when Gustav Hüfner, the professor of chemistry at Tübingen, visited him in 1873, his nephew asked him not to speak with him about that subject or about politics: "He could not tolerate Darwin's theory, and as for politics, he was a resolute conservative [*Ultramontaner*]."[60] Mayer's fullest surviving statement of his views on the matter came in a letter to the theologian Rudolf Schmid in 1874, in which it is clear that for him, as for many critics then as now, Darwinism stood primarily for the doctrine of the spontaneous creation of life:

> From my standpoint, what I most object to in Darwin's system is the following: innumerably many new plant and animal individuals come into existence continuously before our eyes through generation and fertilization. How this takes place, however, is for the physiologist a completely incomprehensible riddle and unfathomable mystery, where the famous saying of Haller's so rightly finds its application: "[No created spirit penetrates] into the interior of nature." Seeing how we are obliged to confess our complete ignorance in these matters so near and present, all of a sudden the good Darwin, like a second Lord God, wants to give us perfectly solid information about how organisms arose on our planet in the first place! In my view, however, this goes so ridiculously far beyond the humanly possible that I would apply the Pauline text: "Professing themselves to be wise, [they became fools]." But the Darwinians are certainly zealous champions, and the matter doubtless has so many followers in Germany only because one can make capital out of it for atheism.[61]

The relationship between Mayer's antimaterialism and the changing character of the life sciences after the late 1830s will be gone into in more detail in chapter 5.

Perhaps Strauss's early appeal to Mayer was due to the way in which he sought a Christianity not based on miracles, one which accepted the lawbound causality of natural events but was nonetheless opposed to the materialism characteristic of the post-Reformation world and which could bring inner peace to the believer—in other words, a Christianity that had room for both faith (in the existence of God and in the immanent divinity of humankind) and scientific knowledge (of an orderly world). Strauss's early opposition to materialism was an important theme of his *Two Irenic Essays*, the book Mayer had with him on board ship. Part of its argument was to counter the denial (by unnamed opponents) that an idea alone—*nur eine Idee*—could be capable of major historical effect, as if an idea were (causally) *nothing*.[62] Strauss had in mind here the central notion of his theology, the identity of the human and the divine, God's immanence in the world. Some people denied that such an abstract idea could have any effective moral force, preferring rather to base religion on faith in the historical facts of the virgin birth, miracles, the resurrection, and the like.[63] Here, as in his book on Jesus,

Strauss denied the reality of miracles and rejected the traditional attempt to tie Christian faith to them. He seconded Hume's argument against them, and insisted that events be understood as due solely to natural causes.[64] All the forces of nature are equally "natural": "The animal-magnetic force, like the galvanic and electrical forces, even the force of imagination and will, insofar as it affects the corporeal organism, are just as much natural powers [*Naturpotenzen*] as the mechanical forces of collision and pressure by means of which the surgeon operates, or the chemical forces of salts and oils by means of which the physician works."[65]

In Strauss's historical analysis, the gods mingled freely with people in the oldest times but gradually became more remote, the interaction between the two realms becoming less frequent and more oblique. Then Jesus appeared, as God but at the same time a man of this world: "But the union had been achieved in an absolute fashion at only a single point; not the truly human as such in general, but this human exclusively had the dignity of the really divine; and likewise the divine had not entered completely into worldly reality, but had preserved its otherworldliness for all other points. The unity of the divine with the human in the person of Jesus was therefore an exception to the rule, a miracle."[66] But miracles are no longer acceptable, and God must be knowable in other ways. According to Strauss, the Reformation did not reject miracles outright but relegated them to the distant past, thereby opening up the possibility of discovering the divine not only through history but also in the order of nature—with portentous consequences:

> But the territories of nature and history expanded in an ever more manifold and magnificent fashion before the eyes of awakened investigation; order and rule, rational content in general, disclosed itself increasingly in both; link connected with link ever more securely in the chain of causal connection in both. If from here one glanced over toward the other world [*das Jenseits*] that the old ecclesiastical conception offered, then it appeared all the more superfluous the more the [human] spirit was satisfied by the contemplation of natural and historical connections; its intrusion into this world [*das Diesseits*] appeared all the more inadmissible the clearer the recognition of that connection became. If on the one hand the antiquated system of transcendence exerted itself as external ecclesiastical power, and if on the other hand the new principle of immanence was still strongly realistically tinged through the study of the natural sciences, which had first called it into being; then there resulted in the first instance what we find: a naturalism hostile to the church and to Christianity which, after it (as deism) had reduced the vital and richly populated other world of ecclesiastical belief to the empty and dead conception of a highest being, finally (as materialism) destroyed the ecclesiastical two-worlds system by simply denying one of the two, namely the ideal world. But the other side to this was that conversely idealism, prepared by logical and metaphysical investigations, later sacrificed the corporeal world to the domain of consciousness. In both only the antithesis of spirit and matter was immediately destroyed; mediately, however, that of the absolute and the finite was also prejudiced insofar as God on the one side was reduced to an abstraction [*Abstractum*] of matter, on the other to the pure ego or to the moral order of the world. The truth of the old dualism is misperceived in both systems: its real overcoming cannot come

about through annihilation of one or the other of its sides, but only by so recognizing the equal legitimacy of both that their separation is nullified [*aufgehoben*].[67]

Strauss's picture, then, was of an ordered and lawful natural world subject to rigid causality. In his historical view, naturalism gave rise not only directly to materialism, but also, by way of reaction, to the idealism of the day which denied the reality of the material world. Both extremes were to be rejected in favor of a revindicated dualism that accepted both *Geist* and *Materie*. Such a worldview would sanction the kind of science Mayer pursued at the same time it vindicated the existence of a nonmaterial realm.

Many years later, however, Strauss came to defend a number of positions that Mayer vehemently rejected. In his *Old and New Faith* of 1872 Strauss agreed with arch scientific materialist Carl Vogt that the assumption of a particular *Seelensubstanz* or *Seelenwesen* is a mere hypothesis unsupported by any facts and does not explain anything. Rejecting the duality of mind and body, he now proposed that there is only a single substance (*Wesen*), "which on one of its ends is extended, on the other thinking."[68] Mayer would have opposed any such monistic elimination of soul (as of force). Strauss even appealed to the law of conservation of energy (still *Kraft* for him) to justify this elimination of the sensitive soul: "If under certain conditions motion is transformed into heat, why should there not also be conditions under which it is transformed into sensation?" He said he did not object if people found in these views a "crass materialism," though it is not clear precisely what sort of ontology he stood for. He praised Darwin for having eliminated purposefulness from nature, implicitly rejected the whole natural-theological view of nature, and expressed his opinion that modern science has found the sure path that would allow it in short order to make good the Kantian challenge of creating a caterpillar out of inorganic matter.[69]

An acquaintance called Strauss's book to Mayer's attention. After reading it, Mayer remarked to him that, if Strauss wasn't a better theologian than he was a scientist, then he wasn't worth much.[70] In a brief note apparently intended as a review of Strauss's book, Mayer wrote: "In this his newest work Dr. Strauss describes Christianity as a 'belief in miracles,' as a standpoint vanquished by Darwinism. 'The resurrection is the greatest lie in the history of the world, etc.' Time will tell whether the books written by Strauss and his ilk will be able to replace the holy scripture of humankind."[71] Mayer remained unwilling to give up his theistic Christianity, even when his beliefs made him something of an embarrassment to the more generally antireligious and materialistic scientific establishment in Germany.[72]

## 4 Mayer's Circle of Friends

For someone whose profession was medicine and whose sometime passion was science, Mayer had a striking number of close friends with strong connections to theology.[73] Lifelong friend and future clergyman Gustav Rümelin was Mayer's companion both at the Evangelical school in Schöntal and then, as a student of theology, at the University of Tübingen.[74] While at Schöntal, Mayer became fast friends with the prelate's son, Paul Friedrich Lang, who also went on to study theology at Tübingen and who later delivered Mayer's funeral sermon.[75] At Tübingen other close friends included the theology student and future poet Karl Gerok—also an intimate of Lang's—and "other of our present-day highest church worthies."[76] Friend and future novelist Hermann Kurz was a student at the Tübinger *Stift*, the renowned Evangelical theological school.[77] Friend Eduard Zeller, a future professor of theology (and later of philosophy), had also attended the *Stift* before entering the university, and reported that the most important circle of his friends at the university was a group that had graduated from Schöntal in 1832, the year after Mayer.[78] Although primarily interested in mathematics and natural science, friend Gustav Reuschle and later friend Carl Baur were both formally students of theology at Tübingen since the school as yet had no natural sciences faculty.[79]

To be sure, Mayer was also closely associated with the medical students Wilhelm Griesinger, Carl August Wunderlich, and Wilhelm Roser, yet as already noted there is no evidence he was caught up in their enthusiasm for the reform of medical science. In general, Mayer gives the impression of having been a fun-loving friend who cherished socializing and card playing but showed little interest in strictly intellectual pursuits.[80] Alongside his cameraderie, however, one of his university friends remembered "the uncommunicativeness [*Verschlossenheit*] of his nature" and recalled that "his more distant acquaintances thought he had a good head, but still regarded him as something of an oddball [*Sonderling*]."[81]

To judge from the surviving evidence, Mayer seems to have remained rather unconnected with a number of significant people in his environment who were part of his larger circle of friends and acquaintances. Strauss, for example, knew Rümelin and Reuschle at least by 1837 and 1841, respectively.[82] Strauss was also close to the celebrated Weinsberg physician and spiritualist author Justinus Kerner, and he reported that every fall (for an unspecified period of years) he walked to Weinsberg and to nearby Heilbronn.[83] Yet not then nor even after Strauss moved to Heilbronn in the fall of 1843 is there any evidence that Mayer had any personal contact with him.[84] Mayer was friends with Justinus Kerner's son Theobald: they were students of medicine together at Tübingen (Theobald was there from 1835 to 1840), and the fact that Theobald later called Mayer by his schoolboy nickname—addressing a letter "Mein lieber Geist!"—suggests they were reasonably close.[85] And a letter from Justinus addressing Mayer with the familiar "Du" implies that they, too, were close.[86] Even Mayer's physics instructor at Tübingen in 1832, Ludwig Felix Ofterdinger, reminisced to Theobald about having visited the Kerner household in Weinsberg often in the years 1827 and 1833.[87] Yet there are no traces that Mayer had any significant interaction with any of these people. It has been possible to piece together a network of interconnecting relationships all around Mayer, but he himself largely escapes the evidentiary net.

## 5 Mayer's Character

Reminiscences of Mayer's closest friends afford precious insights into some of his outstanding personality traits. According to Rümelin,

> From his disposition one would call him an *anima candida*. But everything he said and did bore the stamp of originality. His train of thought, which was entirely logical—but in which he skipped or left unspoken the connecting middle terms—was always astonishing and often dumbfounding; by the time one had found the thread, he had already landed somewhere else. And since he never lacked for wit and good humor, his conversation was always delightful; he had an inexhaustible supply of quotations and maxims from Bible and hymnal, from sayings, poets, and ancient authors, and he knew how to employ them where no one else would have thought of them.
>
> . . .
>
> He possessed the most pronounced sense of play and had the liveliest interest in every kind of game; not at all for the sake of winning, but because he eagerly went after the theory of the different games and immediately searched for law and rule. He became an excellent player of chess, whist, omber, and taroc; billiards and bowling occupied him actively, too; one could compete with him only because he liked to push all rules to the extreme and easily became a stickler for principle.
>
> . . .
>
> He was so engrossed in the changing combinations and problems of the more precise games as if there were nothing else in the world, and weeks afterward he could still reconstruct in detail the state of a chess game or a turn at omber.
>
> . . .
>
> The most prominent of his intellectual gifts were always the sense for mechanical causality and the unrelenting, penetrating examination of a thought out to its farthest ramifications.[88]

After noting that Mayer had revealed no predilection for mathematics or natural science during the time he knew him in Paris in 1839, Baur reported that "on the other hand he gave free rein to his combinatorial acumen in his passion for tricks with cards and figures, in the solution of riddles and rebuses, in striking witticisms and elegant jokes. His predilection for precise and cutting satire went hand in hand without contradiction with the warm heart which he, as a reliable and self-sacrificing friend, proffered to his friends."[89] And Karl Gerok recalled that Mayer "taught us omber and entertained us with ingenious paradoxes and witty quotations out of both secular and religious literature."[90]

From these reminiscences one obtains a consistent and convincing picture of Mayer as an original, endowed with a deep sense of humor and delight in games of all sorts, with a doggedly logical mind fascinated by both rules and paradoxes, possesed of a prodigious memory and inordinately fond of citing passages from Scripture and secular literature. Even in later years Mülberger reported he had "an almost pathological mania for quoting poetry" and biblical texts, a trait also commented on by Dürr.[91] Mayer's writings, in particular his letters, are peppered with biblical and other quotations, and he left behind numerous scraps of paper inscribed with favorite aphorisms. As will be shown in the next chapter, his exposition of his ideas on force was strongly shaped by his attachment to such Latin dicta as *causa aequat effectum, nil fit a nihilo,* and *nil fit ad nihilum.* It has not been possible to trace these to any specific source in Mayer's ambit, though the traditional status aphorisms enjoyed in the medical literature since the time of Hippocrates may have reinforced Mayer's attachment to them. In one letter he wrote that "*Cessante causa cessat effectus* is a well-known medical principle."[92]

Mayer's mental breakdown after 1850, and the suggestive signs of a disposing instability of mind long before then, are well known and for the most part of no obvious relevance to the story to be told here. The most perceptive analyst of Mayer's psychological condition—and one of the most perceptive readers of Mayer's writings *überhaupt*—Heinrich Timerding, has, however, insisted on the importance of Mayer's stubborn attachment to simplistic ideas and even apparent absurdities and inconsistencies as lying at the root of both his scientific creativity and his mental pathology.[93]

• C H A P T E R   T W O •

# Mayer's Work

IN FEBRUARY 1841 Mayer returned to Heilbronn to resume the practice of medicine, his lifelong profession. His passion, however, was now the further clarification and publication of the ideas on force he had been incubating since June of the previous year. On 16 June 1841 he sent a manuscript, "On the Quantitative and Qualitative Determination of Forces," to Johann Christian Poggendorff, editor of the *Annalen der Physik und Chemie*, Germany's leading physical sciences journal. Poggendorff never replied to Mayer's three letters to him, and the manuscript remained unpublished until retrieved from Poggendorff's *Nachlaß* by Friedrich Zöllner in 1881. Although confused and quite unpublishable, this (apparently) first extended elaboration of Mayer's ideas is our earliest surviving source of information and provides indispensable clues for the reconstruction of Mayer's thought processes. No less important is the correspondence Mayer initiated on 24 July 1841 with Carl Baur, the student of mathematics and mechanics whom he had gotten to know in Paris and who was then continuing his studies in Tübingen. As will be shown more fully in chapter 6, the direction of Mayer's deliberations was influenced profoundly not only through his interchange with Baur, but also in response to his meetings with the Tübingen and Heidelberg professors of physics and, in particular, through his reading of Justus Liebig's "The Vital Process in the Animal, and the Atmosphere," published in Liebig's *Annalen der Chemie und Pharmacie* in December 1841. On 31 March 1842 Mayer sent Liebig his "Remarks on the Forces of Inanimate Nature," which Liebig promptly published in the May issue of his *Annalen*, thereby establishing the basis for Mayer's claim to priority in announcing the mechanical equivalent of heat, and for others' inclusion of him as one of the codiscoverers of the principle of the conservation of energy.

This chapter will offer an analysis of Mayer's early work in order to identify the leading ideas and peculiarities in need of historical explanation. I will concentrate on the two papers mentioned above since they constitute the earliest expressions of Mayer's ideas in the form he wished to make public. If one understands them, there are no major outstanding problems with understanding everything that came afterward. I will also draw upon his later writings to clarify and illustrate some of his abiding concerns and occasional shifts of meaning. Since the best introduction to Mayer's thinking is via his own unreconstituted words, I will begin by quoting at length from his first two papers, and then tease out and comment upon individual significant points.

## 1 Mayer's Earliest Presentation of His Ideas

### *1.1 "On the Quantitative and Qualitative Determination of Forces"*

Mayer began what he had hoped would be his scientific debut with general considerations about causality, force(s), and an analogy between chemistry and physics as dealing with conceptually comparable objects (i.e., material substances and forces):

> It is the task of natural science to explicate the phenomena of both the inanimate and the living world in terms of their causes and effects. All phenomena or events depend upon the fact that material substances, bodies, change the relationship in which they stand to each other. According to the law of sufficient reason [*das Gesetz des logischen Grundes*] we assume that this does not take place without cause, and such a cause we call force. Following the causal connection upwards, if we come to phenomena whose causes are no longer sensibly perceived, but can only be abstracted from their effects, we then call these forces in the narrow sense abstract forces.—We can derive all phenomena from a primitive force that tends to annihilate the existing differences, to unite everything that exists into a homogeneous mass in a mathematical point.—If two bodies find themselves in a given difference, then they could remain in a state of rest after the annihilation of [that] difference if the forces that were communicated to them as a result of the leveling of the difference [*behufs der Differenzausgleichung*] could cease to exist; but if they are assumed to be indestructible, then the still-persisting forces, as causes of changes in relationship, will again reestablish the originally present difference. Thus the principle that forces once given, like material substances, are quantitatively invariable, assures us conceptually of the continued existence of differences and thereby that of the material world. Thus we assume that both the science which deals with the manner of being of material substances (chemistry) as well as that which deals with the manner of being of forces (physics) have to consider the quantity of their objects as that which is invariable and only their quality as that which is variable.[1]

Although most of this introductory paragraph is relatively straightforward, it is not immediately obvious what Mayer meant by our being able to derive all phenomena from an *Urkraft* whose tendency is to annihilate or neutralize existing differences.

Considering that forces can alter the relationship of bodies either spatially separated or in contact, Mayer assigned the latter class to chemical and electrical forces and limited his attention in the first instance to the former, in which case the change in the bodies' relationship is simply motion: "Supposing we have two isolated bodies in the universe at a given difference from each other, both will then move in a straight line toward each other; the ultimate cause of the forces, or the cause that makes itself known via the leveling of the existing difference, communicates motive force to both bodies, as whose consequence or phenomenon we see motion arise."[2] Here it looks like the aforementioned *Urkraft* might be gravity, which tends to eliminate the spatial separation or "difference" between bodies, though it is not clear how "all phenomena" might be derivable therefrom. Pursuing his goal of defining forces both quantitatively and qualitatively, Mayer measured the quantity of motion (and also the motive force—Mayer's distinctions tended not to be very precise) as *MC*, the product of mass and velocity. The next task was to determine

> *how* this quantum of motion manifests itself, or *how* this motion takes place, and this we comprehend under the name *quality* of motion. This comprehends in itself first the energy [*Energie*] of the motion, or the ratio of its intensity to its extensity. Determinative for it is *n* in the expression $\frac{M}{n}nC$, in which *n* can express any whole number and any fraction, and second the direction of the motion; insofar as it is only a question of diametrically opposed directions, these can be completely expressed simply in terms of the symbols + and –.[3]

Considering now the direction of motion, Mayer imagined bodies A and *B* (now in the absence of gravity) moving with speeds *c* and *c′*. If the bodies are equal (in mass—Mayer tended to be careless about such specifications) and have equal speeds, then "the total quantity of motive force *Q* = 2A*c*."[4] "To determine the quality of 2A*c* we suppose the simplest case, that A and *B* move toward each other in a straight line; +A*c* is then equal to –*Bc*; the symbol for the united body A and *B* is then neither + nor –, but 0, since A and *B* taken together will possess motion neither toward the one nor toward the other side."[5] Mayer's expression for the combined motion of A and *B* is accordingly "0 2A*c*," in which the zero is to be taken as a qualitative symbol (*qualitatives Zeichen*). Although confusing and idiosyncratic, the intent of Mayer's usage is clear: to recognize that even when two bodies have equal and opposite "quantities of motion" they still possess a net "motive force."

> As far as the further qualitative determination is concerned, that of the energy of the motion, this lies as indicated in the determination of *n* in 0 $\frac{2A}{n}nc$; the magnitude of *n*, however, depends on the physical constitution [*Beschaffenheit*] of the bodies involved and of their surroundings, especially on the conductibility of the substances for the motive force, i.e., on the elasticity. In the case of perfect elasticity of A and *B*, *n* becomes = 1, +A*c* is simply reversed into –A*c*, –*Bc* into +*Bc*; to the degree to which the perfection of the elasticity decreases, we see less motion arise, and with complete inelasticity we see the motion cease entirely; under such circumstances a portion of the motive force 2A*c* or the whole of it is actually withdrawn as motion from observation; this quantum, [considered] as consisting of + and –, we call neutralized. In accordance with the presupposition of the invariability of the quantity of forces, the neutralized motion [*die Neutralisierte*] is equal to the originally present motion less that which remains; with the perfect inelasticity of A and *B* the neutralized [motion] is = 2A*c*.[6]

The midpoint of the straight line representing graphically the two equal and opposite motions—that is, the point at which one can imagine the neutralization of their motions to take place—Mayer called the *Nullpunkt*. If one has only a single mass in motion, then one can imagine (in some unspecified fashion) its being brought to rest by a fixed point (*durch einen festen Punkt*, to paraphrase Mayer's usage). This fixed "zero point" is capable of neutralizing the body's entire motion.

But two motions can only fully neutralize each other if they occur along the same line. If they do not, their direction and magnitude can be represented by means of a parallelogram of forces in which the neutralization is given by the difference between the sum of the components minus the resultant—with one qualification: "It goes without saying that the formation of a neutralization presupposes the existence of real motion; accordingly, there is no neutralized component of motion [*keine Neutralisierte*] in statics."[7] That is, applied to static forces, forces in the modern sense of the word, the parallelogram of forces gives the correct result; but applied to dynamics, Mayer's inchoate science of force, the graphical resultant will be either too small or too large *except* for "when, as the case requires, one allows opposing motions to become nought [*Null*] or the requisite opposing motions to arise out of nought."[8] In this passage Mayer implied that there are cases, as yet unspecified, in which opposite motions can arise from nought. He went on: "Now since in our physical apparatus forces do withdraw themselves from our observations, but can never be produced out of nought, the cases which are suitable for experimentation will be those in which a neutralized [motion] is left out, never those which presuppose the formation of such a [motion] out of nought."[9] This qualification with respect to experiments that we are capable of performing with the usual physical apparatus again left open the possibility that there might be other physical systems in which the restriction against the origination of force out of nothing does not apply.

Mayer closed his paper with some wide-ranging and perplexing applications of his ideas to heat, light, and the origin of the sun's heat and light out of the neutralization of some fraction of the gravitationally controlled orbital motion of the circumambient bodies:

> We may be permitted now to derive from the foregoing a few conclusions for physics.—Insofar as the motion does not actually take place in opposite directions, the neutralized motion [*die Neutralisierte*] 0 2*MC* is the expression for heat. Motion, heat, and, as we intend later to develop, electricity are phenomena that can be reduced to *one* force, measure each other, and pass over into each other [*ineinander übergehen*] according to definite laws. Motion passes over into heat by being neutralized by an opposing motion or by a fixed point; the heat that arises is proportional to the motion that disappears. Heat, on the other hand, passes over into motion by expanding bodies; in accordance with its formula 0 2*MC* or +*MC*–*MC* it produces opposing but radial motions in all directions; the heated body itself remains at rest, thus the qualitative symbol 0 pertains to it. Wavelike and oscillating motions constitute a special class, the transition [*Uebergang*] from simple motion to heat; insofar as they are radial, the symbol 0 applies to them; they are distinguished from heat, however, in that with them the motion retains all along its form as motion; the quantity of these motions is likewise determined by 2*MC*; they produce different phenomena according to the difference in energy. In the formula $\frac{M}{n}nC$, as indicated above,

> $n$ expresses the energy of the motion; if $n = \infty$ (at least nearly $= \infty$; we may be permitted this expression for the sake of brevity), we thus obtain the kind of motion that makes itself known to us as light or as radiant heat. Thus light obtains the formula $0\ 2\ \frac{M}{\infty}\infty C$. Light becomes heat when the motion passes over into rest, heat becomes light when the stored-up neutralized [motion] again assumes the form of motion.
>
> If we connect a body $P$ with a fixed point $c$ by means of an artificial radius vector, and if we produce a central motion by means of the motion $MC$ communicated to $P$, then $MC$ decomposes into two motions, of which one has the direction of the periphery, the other the direction $-Pc$; the latter is continuously annihilated, neutralized, by means of the fixed point $c$, from which one can see that the [motion] $MC$ communicated to $P$ gradually becomes $0\ MC$ in $c$, and that the motion of $P$ is thus a decreasing motion. In the systems of heavenly bodies, gravitation takes the place of the artificial radius vector; instead of there being subtracted from the motion $MC$ a motion in the direction $-Pc$, a like motion is communicated to it from the direction $+Pc$, and, by means of the motive forces combined according to the laws of statics and dynamics, there is obtained not only the enduring revolution of the heavenly body $P$, but there is also neutralized in $c$ a quantity of motion determinable for each rotation. Expressed in other terms, this means that in the same measure as the peripheral parts behave as falling toward the center, the latter falls toward the periphery. Accordingly, the for us insoluble task of a continuous production of force—i.e., the differentiation of 0 into $+MC-MC$—is solved by nature in the solar systems; the fruit of which is the most magnificent [phenomenon] of the material world, the eternal source of light.[10]

Note that Mayer did not specify just what happens when heat is produced by the neutralization of motion. Note, too, that his paradigm for the transformation of heat into motion is expansion (of a material body) conceived as made up of equal and opposite (i.e., "radial") motions. With respect to his treatment of light, one sees in Mayer's formula—with $n = \infty$ expressing light's infinite "energy" and the zero to be understood qualitatively— the use of mathematical notation more as a symbolic representation of his still-developing "quantitative and qualitative determination of forces" than as a conventionally understood expression of quantitative relationships. His treatment of planetary motion and the attendant production of sunlight is not only hard to make intelligible, but its clear implication is quite at odds with the conservation of energy.

Two months after submitting this essay to Poggendorff, Mayer wrote to Baur that "my first paper will contain nothing but the most succinct summary of the theory of the invariable quantity of forces, the relationship between motion and heat, application to the doctrine of the parallelogram [of forces], and a suggestion about the nature of light and the cause of the solar heat."[11] This is not a bad summary of what his paper was about, except that he in fact said very little about the "invariable quantity of forces," limiting himself to the unelaborated and rather passing invocation of "the principle that forces once given, like material substances, are quantitatively invariable." In light of Mayer's continuing acceptance of the creation, under certain circumstances, of force out of nothing, the restriction of that principle to forces *once given* was surely not casual.

## 1.2 "Remarks on the Forces of Inanimate Nature"

If one next turns to Mayer's first published paper, one encounters not only a similar constellation of issues—note the title's restriction to inanimate nature—but also a markedly transformed presentation, with different goals and a different argumentative strategy. Mayer began:

> The purpose of the following lines is to attempt to answer the question of what we are to understand by "forces," and of how such are related to each other. Whereas with the designation matter very definite properties are attributed to an object, like those of weight and of filling space, the designation force is associated principally with the concept of the unknown, the inscrutable, the hypothetical. Considering the consequences flowing from it, an attempt to understand the concept of force as precisely as that of matter, and to designate thereby only objects of actual investigation, should not be unwelcome to friends of a clear, hypothesis-free view of nature.
>
> Forces are causes, thus the principle *causa aequat effectum* applies fully to them. If the cause $c$ has the effect $e$, then $c = e$; if $e$ is in turn the cause of another effect $f$, then $e = f$, etc. [and] $c = e = f \ldots = c$. As is clear from the nature of an equation, in a chain of causes and effects a link or a part of a link can never become nought. This first property of all causes we call their *indestructibility*.[12]

Once again at the center we find the attempt to elaborate a doctrine of force as ontologically coequal to matter. Here, however, Mayer tied his conception of force more explicitly to a general principle of causality: the indestructibility of forces was a logical necessity. If matter and force—or, more precisely, material substances and forces—were before once referred to as the "objects" studied by chemistry and physics, here the repeated reference to forces as "objects" lent that word a more substantive quality. Mayer went on:

> Since therefore $c$ passes over into $e$, $e$ into $f$, etc., we must thus regard these magnitudes as different manifestations [*Erscheinungsformen*] of one and the same object. The ability to be able to assume different forms is the second essential property of all causes. Summarizing both properties, we say that causes are (quantitatively) *indestructible* and (qualitatively) *transformable* [or *mutable* (*wandelbar*)] objects.
>
> Two classes of causes are found in nature, between which experience shows no transitions to take place. One class comprises causes possessing the property of ponderability and impenetrability—i.e., material substances [*Materien*]; the other, causes lacking the latter property—i.e., forces, also called imponderables after their characteristic negative property. Forces are thus *indestructible, transformable, imponderable objects*.[13]

Considering first material substances, Mayer reasoned that "explosive gas [*Knallgas*], H + O, and water HO are related to each other as cause and effect, thus H + O = HO."[14] He insisted at length on the force of the analogy between material substances and forces as quantitatively indestructible and qualitatively variable objects, each to be understood in terms of the cause-and-effect relationships among their various manifestations. He was now ready to specify what some of those forces were:

> A cause that brings about the raising of a weight is a force; its effect, the *raised weight*, is thus likewise *a force*; expressed more generally this means that *the spatial difference of ponderable objects is a force*; since this force brings about the fall of the body, we thus call it *fallforce*. Fallforce and fall, and more generally still fallforce and motion, are forces related to each other as cause and effect, forces that pass over into each other, two different manifestations of one and the same object.[15]

In arguing against the general practice of calling gravity a force, Mayer made an essential distinction between forces and properties:

> When one regards gravity [*Schwere*] as the cause of fall, one speaks of a force of gravity [*Schwerkraft*] and thereby confuses the concepts of force and property; precisely that which belongs essentially to every force, the *union* of indestructibility and transformability [*Wandelbarkeit*], is foreign to every property; between a property and a force, between gravity and motion, one can therefore also not set up the equation necessary for a correctly conceived causal relationship. If one calls gravity a force, one thereby imagines a cause which, without itself diminishing, produces effect, and one thereby thus entertains incorrect notions [*Vorstellungen*] about the causal connection of things.[16]

In calculating the mathematical relationship between fallforce and motion, Mayer made one of his usual mistakes with even the simplest concepts of mathematical physics, writing "$v = md = mc^2$" for the force $v$ of a body of mass $m$ falling through height $d$ to acquire velocity $c$. His conclusion was that "we find the law of the conservation of vis viva to be grounded in the general law of the indestructibility of causes."[17] However, Mayer otherwise almost never used the word *Erhaltung*, and he did not regularly invoke the analogy between his principle of force and the conservation of vis viva—in part, it would seem, in order to emphasize the originality of *his* theory of force.

The heart of Mayer's paper followed: the establishment of the equivalence, as forces understood as causes, of heat and motion. He argued that we regularly observe heat to be produced as motion disappears, and are thus forced to recognize a causal connection between the two. Nonetheless, Mayer was careful to distance himself from the mode-of-motion theory of heat—*die thermische Vibrationshypothese*, as he termed it—because it places the emphasis on "discomforting vibrations" (*unbehagliche Schwingungen*).[18] He attempted to illustrate the relationship between fallforce, motion, and heat by means of a curious analogy:

> We can make clear to ourselves the natural connection existing between fallforce, motion, and heat in the following way. We know that heat appears when the individual massy particles of a body move closer to each other; compression [or densification (*Verdichtung*)] produces heat; now, what holds for the smallest massy particles and the smallest spaces between them must well also apply to large masses and measurable spaces. The descent of a weight is a real reduction in the volume of the earth, and

> thus must certainly also stand in connection with the heat that thereby appears; this heat must be exactly proportional to the size of the weight and its (original) distance. From this consideration one is led quite easily to the above-discussed equation of fallforce, motion, and heat.
>
> Nevertheless, as little as can be deduced from the connection existing between fallforce and motion that the essence of fallforce is motion, so little does this conclusion hold for heat. We would rather conclude the opposite, that in order to be able to become heat, motion—be it a simple or a vibratory motion like light, radiant heat, etc.—must cease to be motion.[19]

In order for the equation between fallforce, motion, and heat to be valid, we must also be able to transform heat into motion and fallforce: "Just as heat arises as an effect in cases of reduction in volume and of motion coming to an end, so heat disappears as a cause with the appearance of its effects: motion, increase in volume, raising of a weight."[20] It was the relationship between the change in volume of a compressed gas and the heat thereby produced that provided Mayer with both the mechanism and the data necessary to calculate the all-important numerical equivalence between fallforce and heat: "That such an equation is really grounded in nature can be regarded as the résumé of the foregoing."[21] Without indicating his method of derivation in the published paper, Mayer asserted "that the fall of a given weight from a height of around 365 meters corresponds to the heating of an equal weight of water from 0° to 1°."[22] It is primarily upon the publication of this number that Mayer's claim to discovery of the conservation of energy has been based.

## 2 The Leading Ideas and Peculiarities of Mayer's Work

### *2.1 Force*

#### 2.1.1 THE ANALOGY WITH MATTER

As Heimann rightly emphasized, "the kernel of Mayer's thought is his concept of force and his theory of the relations between forces, and its understanding is essential to our comprehension of Mayer's fundamental intentions. It is this kernel that is obscured or distorted by the search for correspondences and connections between Mayer's ideas and those of other 'pioneers' of energy conservation."[23] As abundantly demonstrated in the passages quoted above, the most essential properties of force for Mayer were its indestructibility, its transformability or mutability (*Wandelbarkeit*), and its immateriality. Although in both his unpublished essay and in contemporaneous letters to Baur he occasionally spoke in general terms of, for example, "the principle of the invariable quantity of forces,"[24] Mayer for the most part stressed only the indestructibility, not also the uncreatability of force, since he continued for several years to accept the production of force in certain nonmechanical systems. The first decisive assertion of the uncreatability of force came in the the unpublished version of what became his second publication, *Organic Motion in Its Connection with the Exchange of Matter*

(1845). In the manuscript, which probably dates from late 1844 or early 1845, Mayer introduced the principle by means of a chemical analogy: "Sulfur and mercury stand in such a genetic relationship to cinnabar; water is transformed into steam or ice, etc.; all and sundry chemical processes consist of changes in the form of given objects, whereby something expended always stands in the closest genetic relationship to something newly formed."[25] He then followed the categorical statement "Ex nihilo nil fit"—its first appearance in his writings—with the assertion, "There is no reason to restrict this axiom to ponderable matter."[26] The analogy between material substances and forces was now complete: both were rigorously invariable in quantity.

Mayer's conception of the mutability and immateriality of force illustrates the degree to which he evolved his very concept of force by contemplating its relationship to ponderable matter and the so-called imponderables. The likelihood of such an evolution in Mayer's thinking will be gone into at length in chapter 6; here I simply call attention (with Heimann)[27] to the consistency with which Mayer expounded his conception of force in terms of an explicit analogy with matter, which is similar to force in its quantitative constancy and qualitative mutability, but unlike it in its materiality. Such was the case not only in the writings he published or intended to publish, but also in his correspondence. In his first letter to Baur he wrote:

> The chemist holds firmly to the principle that substance [*Substanz*] is indestructible, and that the composing elements and the compound formed stand in the most necessary connection; when *H* and *O* disappear (become qualitatively 0 [i.e., *Null*]) and HO appears, the chemist may not thereby assume that H and O really become 0, and that the formation of HO is something accidental or nonessential; modern chemistry depends on the strict carrying out of this principle, which clearly alone could lead to well-rounded results.
>
> We must apply exactly the same principles to forces; they, like substance, are also indestructible; they, too, combine with one another, disappear accordingly in their old form (become qualitatively 0), and appear instead in a new form, [hence] the connection between the first and the second form is every bit as essential as that between H and O and HO. . . .
>
> A motion $+MC$ neutralized by an equally great directly opposing motion $-MC$ yields, because both motions (like substance) are quantitatively invariable, $2MC$, but with the symbol (qualitative determination) 0.[28]

Drawing upon a term used in one of Mayer's letters, Heimann has called Mayer's concept of force "the 'axis' of a new science of physics."[29] That is a fair reading of Mayer's words, but they can also be read as meaning that the axis of his overall system was rather the matter-force analogy:

> I will attempt first of all to exhibit to you the axis of my system; with it stand in connection upwards speculations on the nature of matter and force, downwards the explanation of the phenomena of nature; this axis was implied "in the," as you write,

"comparison between the chemical neutralization of the elements and the mechanical annihilation of the opposing forces."

. . .

It is the same with the theory of forces (physics) as with the theory of material substances (chemistry); both must be based on the same principles. . . . My first endeavor is now to secure the axis about which rotates the theory of material substances for the theory of forces as well; from thence is dated [*daher datiert sich*] the axiom of the invariable quantity of forces.[30]

In ways to be elaborated in chapter 6, I fully agree with Heimann's assessment that "the analogy he had drawn between chemistry and physics—the former as the science of matter, the latter as the science of force—was not merely a heuristic metaphor, but expressed a fundamental feature of his conception of nature: nature constituted a duality of matter and force."[31] That Mayer's exploitation of this analogy was a creative act in its own right is underscored by the fact that one side of it—the indestructibility, uncreatability, or conservation of matter—was not a principle explicitly stated in either the chemistry or physics texts of Mayer's day; hence Mayer had to do more than simply apply to forces something he obviously 'knew' about material substances.

### 2.1.2 THE CENTRALITY OF HEAT AND MOTION

Someone who approached Mayer's work after having studied the other nominal codiscoverers of the principle of the conservation of energy and *their* context would probably be struck by how unimportant virtually all the many examples of the interconvertability of the different forms of energy were to Mayer. He drew on a few for illustrative examples here and there, but for the most part his emphasis was on heat, motion, and the spatial separation of matter (the last, his "fallforce," primarily as a source of motion), with limited and undetailed (though conceptually important) inclusion of light and sound (regarded as vibratory motions), electricity and (still less often) magnetism, and (occasionally) the "chemical difference of matter."[32] His intention of applying his conception of the quantitative and qualitative definition of force to electricity did not bear significant fruit, and he had virtually nothing to say on the topic.[33] Only in 1845 did Mayer elaborate a reasonably inclusive catalog of forces and their interconnections.[34] The voltaic battery, Oersted's and Faraday's experiments on electromagnetism and electromagnetic induction, thermoelectricity, photography—none of these seem to have had any appreciable significance for him.[35] Nor was there ever talk in Mayer's writings of any grand unity of the forces of nature, although that is what his work clearly implied.

At the center of Mayer's thinking, then, was not a broad notion of the interconnections among various forces but a conception of the equivalence of heat and motion. It was not for nothing that he chose to call his collected works *The Mechanics of Heat*. Writing to the Paris Academy in 1848, what he claimed to have discovered in 1840 was "the law of the equivalence of mechanical work and heat"—though, to be sure, in 1840 he would not have spoken of "work" but of "motion."[36] In one of his autobiographical sketches he wrote of his return to Heilbronn from Java, "filled with the idea that had become clear to him on this voyage that motion and heat are only different manifestations of one and the same force, and that consequently motion or mechanical work and heat, which had hitherto mostly been regarded as entirely disparate things, must also be able to be converted and transformed into one another."[37] Whenever Mayer specified just what it was he had discovered, it was the interconvertibility or eqivalence of heat and motion—in other words, the mechanical theory of heat.[38]

### 2.1.3 THE NATURE OF HEAT (IN PARTICULAR) AND FORCE (IN GENERAL)

Despite the fact that central to Mayer's entire thinking was the equivalence of heat and motion, he nevertheless explicitly opposed reducing heat to motion. Heat and motion were causally connected via a quantitative measure of their equivalence, but, aside from their both being forces, they were not to be assumed identical in essence. This reluctance to embrace the mode-of-motion theory of heat was apparently fueled by several reinforcing considerations. In the first place, Mayer was deeply opposed to materialism, and since classical materialism recognized the existence of only matter in motion, his doctrine of force and his refusal to embrace the mode-of-motion theory of heat can be seen as a denial of one of the central components of the materialist worldview.[39] Second, in Mayer's day forces were regularly defined as *properties* of matter, whereas he was concerned to vindicate for *his* forces an existence *independent* of matter.[40] Perhaps he reasoned that taking heat to be nothing more than the motion of the smallest parts of matter would risk making it no more than a property of matter, dependent on the latter for its existence.

Finally, Mayer consistently declined to speculate about the essence of force in general, preferring instead to rest with having identified it conceptually vis-à-vis matter, having used it to explain the connections between previously unconnected phenomena, and having determined the equivalence of heat and motion. For the rest, he professed himself to be ontologically agnostic:

> I don't know what heat, what electricity, etc., is as regards its internal essence, as little as I also know the *internal essence* of a material substance or of anything at all; but I do know that I see the connection between many phenomena much more clearly than one has before, and that I can give clear and good concepts for what a force is. If one stops speaking of gravitational *force* and chemical affinity as causes of phenomena—i.e., if one wrests the name force from such things that are not forces—one thereby comes to the study of animate nature with healthfully purified concepts; one knows, among other things, what can and must be charged to the account of the forces of inanimate nature, and the vital force, the nervous force, thereby again loses considerable ground, and the babblings of the *Naturphilosophen* stand pilloried in pitiful nakedness.

* * *

> *How* heat arises out of the disappearing motion, or as I put it, *how* motion passes over into heat—to demand clarification on those matters would be to demand too much of the human mind. How the disappearing O and H yield water, why a substance with other properties doesn't perhaps arise—no chemist will likely break his head over such matters.[41]

It seems that Mayer tended to think of heat phenomenologically, in terms of its sensible manifestations—in particular, the macroscopic expansion of heated bodies—and not in terms of the imaginable motion of unseeable particles. As in the above-quoted passage, when Mayer mentioned the combustion of oxygen and hydrogen to form water he seems to have been thinking of the transformation of macroscopic quantities of the reactants, not of an imagined molecular reaction, although he otherwise accepted chemical atomism.

In his published works Mayer liked to invoke an image of positive science to justify his refusal to identify heat with motion:

> If a transformation [*Verwandlung*] of heat into mechanical effect has here been posited [*statuirt*], then it is only to express a fact, but not in any way to explain the transformation itself. A given quantum of ice can be transformed into a corresponding amount of water; this fact stands firm and independent of fruitless questions over How and Why and of empty speculations over the ultimate cause of the states of aggregation. Genuine science is satisfied with positive knowledge and willingly leaves it to the poet and the *Naturphilosoph* to attempt the solution of eternal riddles with the help of fantasy.
>
> * * *
>
> In the exact sciences one deals with the phenomena themselves, with measurable quantities; but the original cause [*Urgrund*] of things is an entity [*Wesen*] eternally inscrutable to human understanding—the divinity [or Godhead (*Gottheit*)]—while "higher causes," "supersensible forces," and the like belong with all their consequences to the illusory middle kingdom of *Naturphilosophie* and mysticism.[42]

Typically, as here, Mayer's declamations against *Naturphilosophie* came when he wished to defend his own refusal to speculate about essences. Similarly defensive of his stance on the nature of heat was his assertion that there are natural limits to human knowledge in the direction of both the infinitely small and the infinitely large.[43]

Whatever the reasons for Mayer's disinclination to accept the mechanical theory of heat, despite his fixation on the equivalence of heat and motion, this nonacceptance had profound implications: it was only Mayer's rejection of *both* contemporary conceptions of heat, whether imponderable fluid or the motion of the ponderable particles of matter, that in a sense drove him to seek a new conceptualization of the nature of force. His aim was to demonstrate the legitimacy of a new and ontologically distinct entity, not—like Helmholtz—to validate a more traditional reductionist explanation based on matter and Newtonian-style forces.

Mayer's treatment of light was curious, though entirely in keeping with his conception of heat. As he wrote to Baur in 1841:

> As regards light, from the internal necessity of my system I have to adopt the vibrations theory. . . . *Voyez-vous*, the essence of light is motion, its particular species wave motion; light is thus a phenomenon entirely analogous to sound. Light thus does not consist in the mysterious vibrations of a mysterious fluid, and its relationship to heat is not mysterious. . . . [O]nly motion which is lost through resistance will become heat. . . . As is well known, optics is able to base all the laws of the motions of light on the vibrations theory, whereas chemistry was all the more unsatisfied with the theory since it left in the dark the relationships that had to be the most important for it—light with heat, light without heat, radiant heat—and for this reason I myself was also very much disposed against it until the point in time where my theory needs instructed me that light could not be anything at all but motion, whereby the latter questions also solved themselves for me with complete clarity.[44]

Mayer thus insisted that light was a rapid vibratory motion while he denied the existence of a substance, an imponderable fluid or some kind of luminiferous aether, the vibrations of which made such a theory conceivable, at least according to the understanding of contemporary physicists: if light is a wave phenomenon, it has to be a wave *of something*. Mayer kept the waves but rejected the something. Then, too, his notion that chemists opposed the vibrational theory of light because it cannot account for radiant heat strangely misrepresented what was for many the most compelling evidence for vibrational theories of *both* light and heat.

In his book of 1845 Mayer did not even include light in his table of forces except implicitly under "II. Motion. A. *simple*. B. *undulatory, vibratory*."[45] Among the examples illustrating the twenty-five types of "metamorphoses" possible among his five basic forces—whereby electricity and magnetism (considered as *one* species of force) and heat were labeled "imponderables"—was that "the absorption of light consists in a transformation of vibratory motion into heat."[46] Once again, however, he never even hinted at what light might be the motion *of*. Perhaps he contented himself with the same kind of ontological agnosticism that characterized his understanding of heat.

### 2.1.4 THE STATUS OF IMPONDERABLES; FORCES AS OBJECTS

The thrust of Mayer's innovation can be seen as a reinterpretation of the nature of the entities that physicists of his day commonly called "imponderables," the weightless substances (often called fluids) responsible for the phenomena of heat, light, electricity, and magnetism.[47] To that class of substances (or phenomena) Mayer applied the reinterpreted word "force," which, he urged, should not be used for the *properties* of matter (such as gravity), but which should be reserved for the class of things (including, crucially, motion and the spatial separation or "difference" of heavy bodies) that act as *causes* (of motion or other changes in the physical world). He insisted that the notion of imponderable fluids, which might appear to make sense in the case of heat or electricity, was obviously inapplicable to either motion or the raising of a weight and thus that this notion was *generally* inadmissible for the new class of objects he termed forces.[48]

That, at least, is a simplified version of Mayer's intentions. In fact, however, despite the supreme importance he (usually!) attached to his new conception of force, he continued to use the word "imponderables" for his new class of objects. In his first published paper, as we have seen, he defined forces as the class of causes which (unlike matter) lack the properties of ponderability and impenetrability, and then added as a gloss on the term: "from their characteristic negative property also called imponderables."[49] We may, he argued, reasonably apply the word imponderables to "indestructible, transformable, imponderable objects . . . whose objectivity is likewise [as with ponderable objects] verified through experience (at least as well as with ponderable)."[50] Indeed, in his letters of 1842 he spoke of imponderables much more often than of forces.[51] For example, in Mayer's nine-page letter of 5/6 December 1842 to Griesinger, in which his aim was to convince his still-skeptical friend of the justice of his new theory, the word "force" appears only once, in a direct quote from Liebig's letter accepting his paper for publication, whereas he spoke of "imponderables" no fewer than nine times. Part of Mayer's difficulty with Griesinger was in getting him to see the crucial distinction between the imponderables, now including also motion, and the properties of material substances. He concluded an extended discussion of that distinction with the following words:

> If, on the other hand, one wants a practically correct mark of distinction between imponderables and (other) properties [of material substances], I would set the task of transplanting the weight, yellow color, or shape of a gold coin to a piece of silver . . . , as in general its motion, etc., can be transplanted to another material substance.
>
> I have treated the last subject somewhat extensively here because it is of special importance to me that we come as much as possible to an understanding with respect to the concept of "imponderables."[52]

Mayer was largely unsuccessful in countering Griesinger's objections to the inclusion of motion among the imponderables, which Griesinger persisted in seeing as subtle fluids, or in getting him to appreciate the distinction between imponderables and the properties of matter.[53] He responded to Griesinger's intransigence in the following terms:

> You write further: "To be sure, you do not say that motion and imponderables are the same, but you let the latter arise out of the former." But this is just the fundamental and cardinal idea: motion is an imponderable, every bit as well as is heat. Cf. p. 234 [of my paper in Liebig's *Annalen*], where in the definition I declare forces and imponderables to be one and the same, and p. 235, line 21, where I say: "Fallforce and motion are forces," etc. My assertion is indeed precisely that fallforce, motion, heat, light, electricity, and the chemical difference of the ponderables are one and the same object in different manifestations. Thus without disputing over words, when I call one of these things an imponderable it is thereby clear that I ascribe the name to the others as well.

> Why [do] I also call this class of things "forces"? That is in deference to linguistic usage; as far as I am acquainted with them, the German, French, and English writers on science all agree with one another that causes which produce a motion are forces.[54]

It appears, then, that for a while Mayer wavered as to whether he should employ the term "force" or "imponderables" for his new class of objects. In 1842, at least, he leaned decidedly toward the latter. Perhaps it was Griesinger's objections that induced him ultimately to elaborate a theory of forces. Although in his published writings he largely settled on force, after ten years he was still occasionally referring to his subject as a "theory of the imponderables,"[55] and in 1862, after the conservation of energy had been recognized and accepted, he could still write:

> Physics has recently acquired a substantial enrichment through the discovery of the law "of the indestructibility of force." . . . This law says in essence that heat, motion (i.e., the so-called living force or "work" of the mechanicians), as well as light and electricity are different manifestations [*Erscheinungsformen*] of one and the same indestructible, measurable object, so that, for example, motion can be transformed into heat and the latter in turn into the former, whereby in every case the *quantitas vis* in play remains constant. As an imponderable, heat is accordingly at the same time also a force; as a (living) force, motion is an imponderable; or in general: forces and imponderables are synonymous concepts.[56]

Mayer's terminological attachment to imponderables may have something to do with another of his characteristic usages, namely his repeated reference to forces as objects.[57] As suggested in one of the passages quoted above, Mayer may have thought that using this term was equivalent to asserting the objectivity, the reality, of force. It is probably also significant that his use of this term stood in close connection with the way in which he forged his concept of forces/imponderables by reflecting on their similarities and differences with respect to ponderable matter:

> Aside from the ponderables, however, there are still other objects (imponderables) which are likewise subject to the above law [i.e., that no given material substance ever becomes nothing (*zu Null wird*) and none comes into existence out of nothing]; . . . such an object which is not matter (an imponderable) is motion; it does not come into existence out of nothing, insofar as it must always have its cause, but, having once come into existence, it no longer becomes nothing because no cause can be conceived to have nothing as its effect [*mit der Wirkung Null*]. Thus we know that motion is *one* manifestation of an object that is not matter; it comes into existence out of another manifestation and, insofar as it ceases to be motion, becomes another manifestation of the same imponderable object. In other words, the cause of motion, the motion itself, and its effect are nothing but different manifestations of one and the same object, just as the same can be said of ice, liquid water, and steam. But just as steam can again become water, water again ice, so, too, with motion and its causes and effects; *cause and effect designate nothing at all but different manifestations of one and the same object.* . . . A kilogram's having been raised five meters and

the motion of such a weight with the speed of ten meters a second are one and the same object.[58]

In his book of 1851 he wrote that both motion and heat arise as a result of the "expenditure of a measurable object," and that his theory regarding the transformation of motion into heat required an invariable quantitative proportionality between these two "objects." In the same vein he referred to a raised weight and a moving body as "objects."[59]

Rather than calling forces objects, Mayer later sometimes spoke of the "substantiality" of force, by which he seems to have meant little more than that they *exist*, with the assumption that whatever once exists cannot cease to exist.[60] Thus he wrote that, when a given mass has been raised to a given height, "one has thus a force which, like motion, has substantiality, or the property of existing, or which can never become nought."[61] Here, too, Mayer characteristically appealed to the favored analogy with matter; in the short biographical note he composed on himself for the Brockhaus encyclopedia, he wrote that he was the first person "who enunciated and elaborated the principle that not only matter, but also living force in its different forms—that is, motion, heat, light, and electricity—possess the substantial property of quantitative indestructibility."[62] While insisting on the substantiality of motion, Mayer at least occasionally recognized that it was misleading and inappropriate to call motion an imponderable:

> While we loudly vindicate for *motion* the right to exist, [i.e.] substantiality, we must absolutely deny any materiality to *heat* and *electricity*. For would it not be really absurd to seek the essence of motion and of the spatial separation of masses in a fluid, or to wish to posit an alternately now-material, now-immaterial existence for one and the same object?
>
> Let us proclaim the great truth: "There is no immaterial matter!"
>
> We sense indeed that we have gone into battle against the most deeply rooted hypotheses, canonized by great authorities, and that we wish to banish with the *imponderables* the last remnants of the Greek gods from science.[63]

Of course, Mayer's brave rhetoric belies the strength of his own attachment to that outmoded concept. Indeed, it is highly possible that Mayer constituted *his* conception of force—in particular its inclusion not only of heat, but also of light, electricity, and magnetism—in the first instance via a notion of the substantiality it shared with the erstwhile imponderables. After all, in Mayer's day the latter were only rarely spoken of as forces.

## 2.2 *Neutralization of Differences: The Continued Importance of Chemical Analogs*

One of the striking usages of Mayer's first (unpublished) paper of 1841 is the language of the "neutralization of differences," which he applied to the canceling-out of opposing motions, to the elimination of the spatial separation (*räumliche Differenz*) between bodies, and to the qualitative change which two reacting chemicals undergo in forming a new substance. Although the first and (especially) the second application of the phrase are somewhat idiosyncratic, the third has close parallels in chemical terminology, most obviously to the neutralization of acids and bases. Yet it was not until 1844, in his correspondence, that Mayer referred explicitly to the neutralization of acids and bases as an analog to the neutralization of opposing motions.[64] His first public reference came the next year: "Just as material substances of opposite quality, an electropositive base and an electronegative acid, neutralize each other, so do motions of opposite (positive and negative) direction together annihilate each other. The given that continues to exist in changed quality but in unchanged quantity is there the neutral salt, here heat."[65] In Mayer's earliest writings his favored example was consistently the combination of oxygen and hydrogen to form water.[66] He also occasionally referred to the "sacrificing of the chemical difference of C and O" whereby motion is produced in a steam engine.[67]

A powerful analogy for him was that between the chemical doctrine of equivalents and combining weights and his developing conception of force. Indeed in 1863, in a letter to John Tyndall, he glossed the "theory of the conservation of force" as the theory of "physical stoichiometry," and in an encyclopedia entry he composed about himself ten years later he wrote that his first task had been to determine the constant relationship between work and heat—the mechanical equivalent of heat—"and thereby to lay the foundation for a kind of physical stoichiometry."[68] The image probably first suggested itself to him as he pondered the similarities and differences between chemical and physical "neutralization." In a letter to Baur in 1842 he noted that different weights of different substances are required to neutralize the same weight of a given substance. For example, it requires eight parts of oxygen or thirty-five parts of chlorine to neutralize one part of hydrogen: "If we ponder this it will not appear strange that, for diverse motions, likewise different masses are required for neutralization, or that to annihilate, to neutralize, a given motion by means of another motion of different energy not the same quantity of the second should be required as of the first."[69] Thus the motion of a body of mass $m$ and speed $c$ can be neutralized by a body of mass $\frac{M}{x}$ and speed $xc$, yet the quantity of motion of the first body is given by $mc^2$, that of the second by $\frac{M}{x}(xc)^2 = xmc^2$. In other words, chemical and physical substances (or bodies) that neutralize each other are not necessarily equal in their quantity of mass or motion. As he put it, "I believe I have made clear by means of the analogy with the combining weights of matter that from the fact that opposite motions of $m$, speed $c$ and of $\frac{M}{x}$, speed $xc$ are capable of annihilating, of neutralizing each other, the conclusion is in no way permitted that such motions are equal in quantity."[70] Part of what Mayer was involved with here was trying to reconcile his doctrine of the neutralization of motion with a measure of the quantity of motion equal to $mv^2$ instead of the $mv$ he used initially, and which he continued to use now and then for at least another two years. Perhaps one of the reasons he never elaborated this stoichiometrical analogy in any detail in his published works was that he could never make consistent sense of the proper measure of the quantity of motion. In responding to criticisms that Baur had made on a long manuscript he had sent him in 1844, Mayer wrote: "On page 45 I said: 'For the neutralization of motions the velocity plays the role of the combining weight.' Instead of velocity you want to put *mc*. I have calculated here as follows: When an acid and a base neutralize each other, their magnitudes (weights) are proportional to their combining weights; when two opposing motions neutralize each other, their magnitudes (which I, like Leibniz, measure by $mc^2$) are proportional to their velocities."[71] The ultimately published work, his book of 1845, omitted all mention of such an analogy.

As has been abundantly clear, Mayer's thoughts on forces owed a great deal to chemical analogies. Significantly, most of the chemical topics of greatest importance to Mayer—the oxygen theory of respiration, the quantitative determination of mass and the assertion of its conservation, and the composition of water from oxygen and hydrogen—were associated with the work of Lavoisier, and it is not at all far-fetched to imagine that Mayer fancied he would do for physics what he supposed Lavoisier had done for chemistry:

> The chemist is permitted to determine his objects of investigation directly by means of the balance. However, chemistry only became a science a hundred years ago when Lavoisier succeeded in discovering the invariable quantitative relations existing between the different elements, which one calls combining weights and the knowledge of which one calls stoichiometry. As is well known, Lavoisier got there by discovering the composition of water, which consists of one part by weight of hydrogen and eight parts by weight of oxygen, or one part by volume of hydrogen and sixteen parts by volume of oxygen [*sic*]. Such invariable quantitative relations as here for water—thus, for example, here the ratio 1:8—exist, however, for all other substances both simple and compound, and it was the knowledge of these invariable numerical ratios that first raised chemistry to the rank of a science.[72]

Mayer had begun this manuscript fragment by citing his calculation, twenty-eight years before, of the mechanical equivalent of heat.

### 2.3 *Causality and the Laws of Thought*

Although the particular form of his argument changed, constant in Mayer's presentation of his ideas was his identification of forces as causes and his appeal to the law of sufficient reason, to what he termed "das Gesetz des logischen Grundes" or words to that effect, as proof of the indestructibility of force.[73] As we will see in chapter 4, the identification of forces as causes was commonplace in the physics of the day. However, his interpretation of the law of sufficient reason, a law of *thought*, as applying not to the logical process of drawing correct inferences but to the necessary existence of a *physical cause* for every event, represented a misunderstanding of logic that reflected Mayer's lack of grounding in formal philosophy and his misconstrual of the ambiguous German word *Grund*, which means both "reason" and "cause."[74] Whether or not he was confused, Mayer was convinced his conclusions were sanctioned by the laws of thought, or at least he argued that way. As he explained to Baur, his demonstration of his conclusions about force presupposed "as an axiom—and by all thousand devils—not as an hypothesis, that a force is no less indestructible than a substance [*Substanz*]. This general principle admits of direct proofs as little for 'force' as for 'substance': but why the principle is to be assumed as an axiom, and how it derives from the simplest concepts of our cognitive faculty, of that one can well give an account."[75] Sixteen months later he was still arguing the same line:

> Aside from the ponderables, however, there are still other objects (imponderables) which are likewise subject to the above law [i.e., that no given material substance ever becomes nothing and none comes into existence out of nothing]; the proof for this can be derived from the general laws of human thought, from the principle of sufficient reason [*der Satz vom logischen Grunde*]; in my paper in the May issue of the *Annalen* I derived it with, as I believe, perfect precision from the axiomatically assumed principle *causa aequat effectum*.[76]

It is important to see Mayer's appeal to the laws of thought as a *strategy*—not that he was insincere, but that his mode of argumentation represented a kind of creative negotiation between his evolving perception of the truth and what he thought would convince his audience. For example, in his earliest papers he did not invoke the logical principle that nothing is created out of nothing *despite* the fact that at the time (according to Rümelin) he was quite attached to the slogan *ex nihilo nil fit*, and despite the fact that in 1845 he did not hesitate to place that slogan at the center of his exposition.[77] By then, however, he had pretty much resolved his lingering doubts about whether or not force might not be creatable out of nothing in certain nonmechanical systems (including living organisms). He did *not* resolve those doubts by repeating to himself *ex nihilo nil fit*; it was rather the case that that 'principle' acquired its probative force only *after* Mayer had convinced himself for other reasons that there are no systems in which one is required to assume the creation of force out of nothing. Once convinced, he proceeded argumentatively *as if* his factual conclusion followed unproblematically from the stated 'principle.' In those early papers he *could* fall back on the dictum *causa aequat effectum* because he limited himself to systems where force is not created and because he never entertained the possibility that forces (or material substances) might simply pass out of existence. Even here, however, his understanding of what should be regarded as *causes* itself underwent significant evolution, and whether or not an effect is equal to its cause depends, in ways Mayer only gradually made clear to himself, on what one calls a cause. The general point is that it would be a misleading oversimplification to see Mayer's conclusions as having followed straightforwardly from the principles to which he professed attachment: logically they may, but in reality these principles acquired their high status only after Mayer saw how he could exploit them to force acceptance of conclusions otherwise arrived at.

In discussing Mayer's religiosity, I have already called attention to his belief in the necessary correspondence between the conclusions of reason and empirically discovered truths of the physical world, as well as to the reservations he occasionally expressed with respect to reason's ability to fathom reality. Similar uncertainties marked his attitude toward the grounds for accepting his theory. Having in his earliest writings given unambiguous precedence to the logical unimpugnability of his conclusions, by 1845 he was insisting that their validity rested "on the laws of thought *and* on experience," even if he still thought the uncreatability of matter was more convincing as an a priori truth than as an experimental determination.[78] Six years later he even fleetingly considered the possibility that experience might disprove an axiom of thought:

> In the paper mentioned [of 1842], the pertinent natural law was traced back to a few fundamental conceptions of the human mind. The principle that a quantity that does not arise out of nothing can also not be destroyed is so simple and clear that against its correctness there is probably as little of any foundation to object as against an axiom of geometry, and we may assume it to be true as long as the opposite has not been proven, as for example by means of an indubitably established fact.[79]

In this Mayer may have been adapting himself to a scientific community that was more comfortable with empirical than with rational grounds for accepting scientific truths.

### 2.4 *Quantitative Thinking and the Measure of the Equivalence of Heat and Motion*

"Truly I say to you, a single number has more genuine and permanent value than an expensive library full of hypotheses."[80] These oft-quoted words of Mayer to Griesinger are just the kind of rhetorical flourish historians are used to discounting, often with good reason. Yet in fact from a very early period Mayer attached great significance not only to what he termed the quantitative determination of force, but also to the calculation of the numerical equivalence of heat and motion—the "résumé," as he termed it, of his ideas. Among other things, his attitude toward quantification and empirical confirmation reveals his attachment to a particular ideal of science: he saw himself vis-à-vis his chosen public as making a contribution to physical science, not to speculative natural philosophy, despite his personal conviction that his theory had profound metaphysical implications.

When Mayer first published a value for the numerical equivalence of heat and motion in 1842 he failed to indicate his method of derivation. He made good that omission in 1845 by reproducing his original calculation: Let the heat absorbed by a volume of gas in raising its temperature one degree at constant volume be $x$, the heat absorbed in raising its temperature by that amount at constant pressure be $y$. The latter case is equivalent to the raising of a weight $P$ through a height $h$, thus $y = P \times h$. A cubic centimenter of air at 0° and an atmospheric pressure equal to 76 centimeters of mercury weighs 0.0013 gram. Under constant pressure a one-degree rise in temperature increases its volume by 1/274, which under standard conditions corresponds to the raising of a column of mercury of height 76 centimeters and cross section of one square centimeter (and thus of weight 1033 grams) by 1/274 centimeters. Since the specific heat of air at constant pressure is 0.267—Mayer did not worry about units—the heat absorbed by a cubic centimeter of air is 0.0013 × 0.267 = 0.000347. To complete the derivation he needed one more piece of information:

> According to Dulong . . . the quantity of heat that air absorbs at constant volume is to that at constant pressure as 1:1.421; calculated from this, the quantity of heat that raises [the temperature of] our cubic centimeter of air at constant *volume* by 1° is = 0.000347/1.421 = 0.000244 degree. It follows that the difference
>
> $y = 0.000347 - 0.000244 = 0.000103$ degree of heat,
>
> by means of whose expenditure the weight $P = 1033$ grams was raised by $h = 1/274$ centimeter. Through reduction of these numbers one then finds that
>
> $$1^\circ \text{ of heat} = 1 \text{ gram at a height of} \begin{cases} 367 \text{ meters} \\ 1130 \text{ Parisian feet.} \end{cases}$$ [81]

In chapter 6 I will consider more closely how Mayer was led to make the only calculation then possible of that number from published data, that is, from those relative to the thermal expansion and specific heats of gases. In 1851, after his priority dispute with Joule had thrust into new prominence the significance of the numerical determination of the mechanical equivalent of heat, he was even more insistent that numbers and numerical relationships—*Zahlenverhältnisse* and *Grössenbestimmungen*—were the very foundation of the exact sciences.[82]

## 2.5 *The Measure of Force*

### 2.5.1 QUANTITY OF MOTION VS. VIS VIVA

Mayer's earliest attempts to come to terms with force as a physical concept were frustrated by the inconsistencies he encountered in contemporary textbooks relating to the definition and measure of force. Finding "quantity of motion" measured by the product of mass and velocity, Mayer involved himself in a labyrinth of confusions trying to reconcile his inchoate notion of the conservation of force with the principle that equal and opposite quantities of motion cancel each other out. Nor were his problems made easier by the fact that he tended to ignore the distinction between mass and weight. Reading the published paper of 1842, one would be tempted to conclude that by then he had at least taken vis viva—i.e., $mv^2$—as the measure of force, but in fact as late as 1844 Baur was writing to him in exasperation "that you can't be convinced to adopt from mechanics quite simply the statement that the mechanical work is equal to the vis viva."[83] In the manuscript Mayer had sent him for comments he not only still measured force by the product of mass and velocity, but attempted to extricate himself from some of the ensuing paradoxes by proving that 2 = 0.[84] In the published version of that work, and indeed still in 1851, Mayer did not make *those* mistakes, but he persisted in ignoring (for the most part) the factor of one-half by which the vis viva must be multiplied in order to have a coherent physics of force.[85]

Despite Mayer's abiding confusion as to the measure of force, he did ultimately appreciate "that it leads to a complete absurdity (or to a perpetuum mobile) if one assumes that the quantity of motion is expressed by $mc$."[86] It appears, in fact, that Mayer's rejection of $mv$ as the measure of force was prompted by his confrontation (presumably in one of the texts he had been reading, or in the course of Baur's tutoring) with the unacceptable consequence of mechanical perpetual motion if one accepted that definition. For a short while, at least, he was inclined to attach considerable demonstrative importance to the impossibility of a perpetual motion machine. As he wrote to Griesinger in 1842:

> But you will now say, with justice: "Prove the truths of your assertions." I cite in this regard: 1. The necessary consequence of simple, undeniable principles. 2. One proof which demonstrates (for me subjectively) the absolute truth of my propositions is a negative one: it is, namely, a proposition generally assumed in science that the construction of a perpetuum mobile is a theoretical impossibility . . . , but my assertions can all be regarded as pure consequences of this principle of impossibility; deny me *one* proposition, and I will immediately build a perpetuum mobile. 3. A third proof can be exhibited to science from the lessons of experimental physics.[87]

We glimpse Mayer here in the process, which extended over a period of roughly ten years, of searching for the most effective way of establishing less the truth (which he never doubted) than the scope and acceptability of his conclusions.

### 2.5.2 THE PARALLELOGRAM OF FORCES AND CENTRAL-FORCE MOTIONS

Perhaps the most baffling aspect of Mayer's 1841 paper was his application of the doctrine of the parallelogram of forces to central-force motions, whereby he believed he could explain the origin of light in the sun. As he put it to Baur:

> According to the foregoing, the parallelogram of forces requires in dynamics (not in statics) an essential correction: the theory of the neutralized component of motion [*die Lehre von der Neutralisierten*], which has been overlooked in the composition of motion and erroneously presupposed in the decomposition of the diagonal into the (taken together larger) enclosing sides. This has important consequences for the laws of central motion; briefly put, the central body behaves falling toward the periphery like the peripheral bodies behave as falling toward the center; this is the cause of the production of light.[88]

Since Mayer's ideas here are quite wrong, it is impossible to make *complete* sense of them; that is, the parallelogram of forces cannot be applied to planetary motions so as to yield the production of light in the sun. (Mayer knew nothing of the conservation of angular momentum.) Nevertheless, one can make at least rough sense of what he thought he was doing with the (relative!) clarification provided by two key passages from later letters to Baur. The key is to recognize that Mayer was treating both (Newtonian) force and "motion" (roughly momentum, though he did not have a sure grasp of the physical magnitudes involved) as forces in the sense of his new and still thoroughly confused theory. The crucial distinction for Mayer was between the case in which a single force is decomposed into two smaller, mutually perpendicular forces (e.g., a body constrained to move in a circle by means of a string—Mayer's "dead radius vector"—fastened to a central point), and the case in which two forces supposedly combine to yield a resultant larger than that given by the conventionally understood parallelogram of forces (e.g., one body moving under gravitational attraction around another). The first case is easier to render, since Mayer was more explicit about it. As he wrote to Baur two months after he had sent his first literary endeavor to Poggendorff:

> It results from my theory of the parallelogram of forces that a body that is attached to a center by means of an artificial radius vector and which receives an impulse must move with decreasing speed—ignoring, of course, friction. *The path that the body would take without the radius vector is the tangent; this represents the diagonal in the parallelogram.* The decomposition of the communicated motion into a centrifugal and a peripheral is effected by means of the radius vector; by means of the fixed middlepoint the former [component of motion] is continuously annihilated, or rather becomes heat. According to the fundamental law of the invariable quantity of forces it is at once evident that, since a portion of the formerly present stock of motion is continuously annihilated, this stock itself must be decreasing. That is, the two sides that enclose the diagonal are, taken together, equal to the diagonal, thus not as large as the figure of a parallelogram indicates, as one has quite falsely assumed up to now. Nor, therefore, can one in the case of planetary motions take away the gravitational force [*Schwerkraft*] and replace it with cords; a cord is never a force![89]

Eleven months later, despite the apparent conceptual clarification he had achieved in his published paper of 1842, Mayer was still trying to salvage his original, hopelessly confused ideas. The only difference is that now he insisted one has to consider the *square* of the lengths of the diagonal and of the (now necessarily perpendicular) sides of the parallelogram of forces in order to be able to apply the latter to dynamics:

> In this way of representing motions by lines, the quanta of motion are proportional to the squares of these lines. One *must* proceed from this proposition if one wishes to apply the parallelogram of forces to motions. . . . In general, the parallelogram, applied to motions, *must always* be a rectangle. Compare all the cases in which motions are composed or decomposed. A decomposition of motions takes place when a quantum of motion is communicated to a mass connected to a fixed point by means of a radius vector. *As a result of the communicated quantum of motion, in a unit of time the mass would cover a certain distance in the direction of the tangent.* This motion is decomposed into two others, one of which, having a centrifugal direction, is annihilated, while the other continues on in the direction of the periphery. *The communicated motion is the diagonal, the two other motions are the enclosing sides*; each of these latter is, since the parallelogram is rectangular, smaller than the diagonal. In every moment the remaining peripheral motion again becomes the diagonal motion and again as such is decomposed, in every moment the square of the centrifugal motion is subtracted from the square of the diagonal. The root of the remaining square thus becomes smaller and smaller every moment, i.e., the motion in the periphery takes place with *decreasing* speed; what becomes of the motion annihilated in the center we [already] know.
>
> In the centrifugal machine there is only one force: the first impulse. The fixed point, the mass, and the radius vector are not forces. Therefore it can never be a question here of the composition, but only of the decomposition of a motion.[90]

If one (finally) pays due attention to a phrase from Mayer's 1841 paper, quoted above, in which he applied his reasoning to planetary systems—"instead of there being subtracted from the motion *MC* a motion in the direction –*Pc*, a like motion is communicated to it from the direction +*Pc*"—then one can interpret Mayer's ideas quite well in terms of the following two figures:

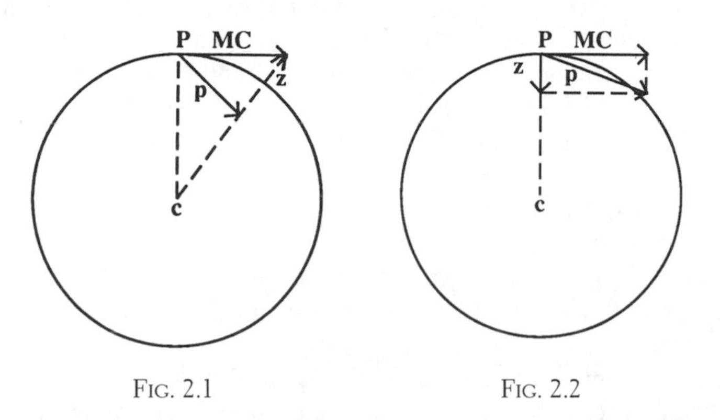


FIG. 2.1 FIG. 2.2

Figure 2.1 represents the situation of a body *P* constrained to move in a circle around a fixed middlepoint *c* by means of a "dead" (or "artificial") radius vector; in figure 2.2 the body moves about the center constrained by the force of gravity. In both figures, *MC* represents the body's initial motion, *p* and *z* the "peripheral" and "centrifugal" components of that motion (regarded by Mayer as a force). In figure 2.1 the centrifugal component *z* is understood to be continuously neutralized by the fixed middlepoint *c*, hence removing that portion of the motion from the body so that it gradually slows down. In figure 2.2 the centrifugal component is understood to represent the gravitational force exerted on *P*; combined with the body's original motion, it yields a resultant large enough to compensate for the motion otherwise neutralized (presumably after the manner of the other example), with the result that the body's revolutionary motion remains unchanged while the neutralized portion of the motion appears as light emanating from the central gravitational mass. Of course, these diagrams do not *quite* work—in particular, the centrifugal component of the motion in figure 2.1 is not applied to the actual body *P*—but I think they faithfully convey the sense of what Mayer had in mind.

### *2.6 Mayer's Restriction of His Ideas to Inanimate Nature and His Allowance for the Creation of Force Out of Nothing*

Several considerations led Mayer to announce, in the title of his 1842 paper, that he was dealing solely with the forces of "inanimate" nature. For one, he had been of the opinion from the very beginning of his reflections that he would have to secure his theory of force first in the more fundamental science of physics before seeking to apply it to physiology.[91] More importantly, the restriction to inanimate nature reflected a major continuing uncertainty in Mayer's thinking, an uncertainty revealed in his understanding of central-force motions. His treatment of them is thus important not only as an illustration of the morass of confusions he had to work himself through, but also because it was precisely those considerations which, it seems, induced him to entertain the possibility that force might be created out of nothing in certain physical systems. In a continuation the following day of the second of the letters to Baur quoted from just above, Mayer went on to explain the difference, as he perceived it, between what one might term mechanical and organic systems: "The motion of the heavenly bodies is no counterproof; it has to do here not with a dead radius vector but with the fallforce. The planetary system, stellar systems in general, are compositions ordered with divine wisdom (organisms) in which 'force' is really produced, and they thereby differ essentially and vastly [*himmelweit*] from our machines. However, my presentation will extend only up to organisms (exclusively)."[92]

Chapter 6 will include an extended exposition of a late, unpublished work of Mayer's that reveals the profound extent to which he had had to resolve extensive doubts as to the possibility that force might be created in certain physical systems before he could confidently extend his theory to include organic systems. Here I would only call attention to the fact that one of the most important general aspects of the evolution of Mayer's theory of force was his having to decide what entities his theory was a theory *of* and (concomitantly) what its *scope* was. According to Rümelin's late recollections, at a very early period Mayer wrestled with the problem of the extent to which his 'principles'—for example, that nothing can be created out of nothing—apply to organic and psychical phenomena.[93] One can freely reconstruct the kinds of questions at issue: What about generation, the mind, and the soul? What kinds of entities are involved? Is the soul a force? If it is, and if a new soul appears at the generation of each newborn individual, where does it come from? Uncertainty about these sorts of questions, combined with Mayer's belief that the solar system, as a divinely ordered organism, is itself capable of generating force out of nothing, led him to restrict his earliest claims to inanimate nature. To be sure, Mayer did not doubt that both souls and forces are imperishable once they are in existence; the question was whether forces are like chemical substances and hence to be regarded as (in the normal course of nature) uncreatable, or like souls and capable of coming into existence under certain common circumstances. Finessing the question of the creatability of force, Mayer in his early work spoke only of its indestructibility. In those few instances where he seems to imply otherwise, by speaking of its quantitative invariability, one can supply from the context his restriction of his ideas to inanimate nature.

Mayer was for the most part an individual of few thoughts long and tenaciously held, and despite the transformation his ideas underwent, issues that preoccupied him at an early period tended to reappear surprisingly unchanged years later. In a talk in 1869, "On Necessary Consequences and Inconsistencies [*Inconsequenzen*] of the Mechanics of Heat," Mayer reflected on some of the differences between the inanimate and animate realms of nature in a way strongly reminiscent of his reported conversations with Rümelin twenty-eight years earlier:

Let us now move from the domain of inanimate nature over into the living world. If necessity and the law's ever-synchronized clock rule there, so do we now enter a realm of purposefulness and beauty, a realm of progress and freedom. Number marks the boundary. In physics number is everything, in physiology it is little, in metaphysics it is nothing. Saturn, the all-devouring, has ceased to rule; time is productive in our present domain. God spoke: Let it be, and it was! The living world is not only preserved, it grows and it becomes more beautiful. Let us take the step from dead to living nature with calm deliberateness. We must guard ourselves against two mistakes. First, we must not once more immediately abandon that which has been acquired in the physical domain when we enter other fields; rather we must retain it as much as possible, even in physiology and philosophy. . . . Second, we must not, however, be all too strict [*consequent*] in the retention of physical propositions, for while we there have had to deal with laws, we now have only rules. The principle of the conservation of matter and of force is doubtless also valid in physiology. The living organism can neither create nor destroy matter or force, nor can it transform the given elements into one another; on the other hand, ternary and quaternary combinations, which as a rule cannot be produced artificially, are created in the most remarkable fashion by the plant world. Furthermore, procreation and generation truly do take place in living nature—an activity for which one in vain seeks an analog in the purely physical domain. The physically correct principle "Ex nihilo nil fit" can therefore no longer be maintained and sustained with full strictness even in physiology, much less still in the psychic [*geistig*] domain. I am here reminded of a remarkable passage in Lucian's *Demonax*. Asked whether he held the soul to be immortal, the philosopher answered: "Yes, immortal like everything else." The conservation principle, or the second principle, "Nil fit ad nihilum," is true to still greater degree in God's living creation, insofar as it is no longer limited, as in the inanimate world [*todte Natur*], by the sterile principle "Ex nihilo nil fit."[94]

### *2.7 Force as an Antidote to Materialism*

From the earliest period for which we have contemporaneously recorded evidence—the summer of 1841—it is clear that Mayer's ideas on force meant much more to him than simply a revolution in the conceptual foundations of physical science. Briefly sketching the progress of his thinking from physiology and pathology to chemistry and physics, Mayer wrote in his first letter to Baur that he had formed for himself a view of nature (*Naturanschauung*) "which completely clarified for me a vast and really infinite series of heretofore inexplicable phenomena, and which, aside from the field of science and specific medical questions, solves for me in addition the most important questions of metaphysics."[95] Ten years later he allowed himself the shortest of public allusions to this other realm: "Force and matter are indestructible objects. This law . . . is a natural foundation for physics, chemistry, physiology, and—philosophy."[96]

Unfortunately, in none of his earliest writings, either public or private, did Mayer specify just what those metaphysical considerations were, but there seems little doubt that they had to do with things like the mind, the soul, and the latter's imperishability. It is my contention that, from close to the very beginning, an essential part of the *meaning* of Mayer's developing theory of force concerned its antimaterialistic implications.[97] That is, Mayer's concept of force as an imperishable and active nonmaterial thing was part of a larger ontological picture that included similarly imperishable but inert matter on the one side and an imperishable and self-active soul on the other. As has been shown, Mayer believed for a number of years that force, unlike matter but thus in some ways like the soul, might not be wholly uncreatable. Mülberger has, I think, correctly analyzed the connection between Mayer's understanding of force and his opposition to materialism:

> As is well known, for vulgar materialism force is a mere property, a mere product of matter, of material substance. The principle of all being resides exclusively in matter. Now the law of the conservation of force, as formulated by Mayer, has proven decisively that the same properties that allow *matter* to appear in the eyes of its true believers as the highest principle—namely indestructibility and transformability—belong no less also to force. "Forces," says Mayer, "are indestructible, transformable, imponderable objects." As little as an atom of matter in the universe can become nothing, just as little can a quantum of motion, heat, electricity, magnetism, etc., become nothing. Thus force, far from being a mere property, a mere product of matter, is raised by Mayer's discovery to the status of matter. Vulgar materialism is thus decisively shattered, it no longer makes any sense.[98]

Although these observations date from late in Mayer's life—Mülberger, a doctor at the sanatorium in Kennenburg (near Esslingen), got to know him only in 1871—I suspect similar considerations were operative much earlier. Certainly Mayer was not alone in perceiving the issue in these terms. As physiologist Alfred Wilhelm Volkmann wrote three years before Mayer's voyage:

> Materialists conflate force and matter in that for them forces are nothing but properties of matter itself. The properties of matter depend on [its] form and composition, and the vital activities of organisms on [their] organic form and composition. Spiritualists proceed in the opposite fashion: either they allow matter itself to be produced out of force, or they at least regard force as in any case independent of matter, and assume an actual existence even without the filling of space. According to them the vital force is earlier than the organism and is related to it as cause is to effect. The spiritualist therefore also distinguishes the soul from the body, and regards it indeed as the cause of all vital phenomena, and as the architect of its body, as Stahl taught.[99]

Only late in life did Mayer publicly assert that his doctrines had metaphysical implications. In the preface to the first edition of his collected papers in 1867 he noted briefly, with reference to the scope of his 1851 work, *Remarks on the Mechanical Equivalent of Heat*: "The metaphysical side of the new subject has at the same time thereby been touched upon, one which is diametrically opposed to the principles and consequences of the materialistic viewpoint. We owe an extensive treatment of this theme to the excellent French physicist A. Hirn in his 'Elementary sketch of the mechanical theory of heat and of its philosophical consequences,' *Bulletin de la Société d'histoire naturelle de Colmar*, 1864."[100] Two years later, in an address before the *Naturforscherversammlung* in Innsbruck, he spelled out more explicitly what it was about Hirn he liked:

> The French physicist Adolph Hirn . . . posits—in my view in a fashion as elegant as it is true—three categories of existences [*Existenzen*]: (1) matter, (2) force, and (3) the soul or spiritual principle. If one has once arrived at the insight that there are not only material objects, but that there are also forces, forces in the narrower sense of modern science, just as indestructible as the elements of the chemist, then for the assumption and recognition of spiritual existences one has only to take *one* further logically necessary step. In the inanimate world one speaks of atoms, in the living world we find individuals. But, as we know, the living body consists not only of material parts, it consists essentially also of force. But neither matter nor force is capable of thinking, feeling, and willing. Human beings think.[101]

Hirn intended his work to constitute "the refutation of materialism and pantheism, a justification of the most absolute spiritualism."[102]

In an unpublished autobiographical fragment, Mayer cited his address of 1869 and three others as having thrown down the antimaterialistic gauntlet to an unappreciative scientific audience. His words confirm Mülberger's understanding of the connection between his conception of force and his rejection of materialism: "It goes without saying that a man who was the first to enunciate and elaborate the principle that not only matter but also vis viva in its different forms—such as motion, heat, light, and electricity—possess the property of substantial quantitative indestructibility, cannot embrace a materialism that would restrict this property to ponderable objects alone, and to that extent these addresses may not have turned out according to the taste of exclusive materialists."[103] In a letter of 1871 to the pastor Rudolf Schmid accompanying a copy of his just-published collection of addresses—which included his Innsbruck address of 1869—Mayer reaffirmed his commitment to the "antimaterialist standpoint" it defended and which he would "never disavow," whereby he invoked Jesus' words from Matthew 10:32: "Whoever then will acknowledge me before men, I will acknowledge him before my father in heaven."[104] Mayer's rejection of the materialistic turn of science that, ironically, his and others' work on the conservation of energy had only strengthened, reinvigorated the theistic strain always present in his belief system.

Although in his published works Mayer remained reluctant to elaborate on what he saw as the metaphysical and theological implications of his scientific work—and the distinction must have been a real one for him—in a late and originally long unpublished manuscript he argued at great length that empirical science justifies belief in the existence and immortality of the soul.[105] Although one cannot assume that Mayer's concerns of the late 1860s were identical to those of the early 1840s, one can reasonably conclude that a significant aspect of his early conception of force involved its ontological implications for the existence of mind and soul. That was one of the reasons Mayer was so excited by the new worldview he saw opening up in front of him.

## 2.8 The Search for Valid Analogies

To a great extent the entire course of Mayer's theorizing can be characterized as a search for valid analogies: In what ways is what like what? Absolutely central to his thinking was the rich analogy between force and matter. These two entities, which together make up the physical world, share the prime characteristics of quantitative invariabililty (or at least indestructibility) and qualitative variability; they differ with respect to their (im)ponderability/(im)materiality. The neutralization of opposing motions is like the neutralization of different chemical species, both processes exhibiting the stipulated quantitative constancy and qualitative transformability. Physics and chemistry were accordingly seen as the parallel sciences of force and matter. In a similar fashion, Mayer pondered the nature of matter, force, spirit, and soul: How are they similar, how different?

Mayer likened the fall of a body toward the earth to the compression of a gas: both processes are accompanied by the production of heat, and both can be regarded as representing a decrease in volume. He was strongly influenced by the implied analogy between living organisms and divinely ordered systems like the solar system: if, as it seemed, force could be created out of nothing in the latter, maybe the former were also capable of some sort of creation *ex nihilo*. He must have pondered, too, the extent to which animal organisms are like machines. And he was much taken by graphic metaphors, such as Autenrieth's analogy between scientific systems of thought as imperfect representations of reality, and a tangent capable of touching a circle at only one point; or the analogy between such a system and a map projection. Rümelin recounted still another favorite: "Just as little as a positive magnitude could become nought [*zu Null werden*] in the course of a calculation—this was an image and example he often liked to repeat—so little could a force disappear in the effect and vanish into nothing."[106] Other examples of the importance of analogies to Mayer's thought processes will present themselves in later chapters. The difficult issue he constantly had to confront was: Which are the *valid* analogies? Mayer seems to have believed that his theory of force was not so much illustrated by his favored analogies as proven by them.

# • PART II •

# Establishing the Relevant Context

• CHAPTER THREE •

# Physiology and Medicine

It is impossible to understand Mayer's thinking on physiological and other issues without knowing what general ideas he was exposed to from the common literature of the period. Guided by the topics identified in the last chapter as characteristic and significant in Mayer's work, in Part Two (chapters 3–5) I will seek to establish a baseline of common contemporary views in selected areas of physiology and medicine, physics and chemistry, and religion and spiritualism. I will also consider discussions of the nature and scope of science—especially the self-conscious transformation of German physiology and medicine beginning in the late 1830s—and the possible significance of railroads and the mechanization of industry to a transformed conception of the nature of organic systems. It turns out not to matter a great deal that one cannot always identify Mayer's precise sources, since the major issues were widely discussed in the most prominent (and typical) writings of the period, usually with the same conclusions. I will, however, pay particular attention to three works of special importance: Johannes Müller's *Handbook of Human Physiology* (1833–44), Johann Heinrich Ferdinand Autenrieth's *Views on the World of Nature and the Life of the Soul* (1836), and Jacob Friedrich Fries's *Textbook of Physics* (1826). Although Mayer's sources have been obscure to historians, I hope to show that he drew overwhelmingly—if, to be sure, idiosyncratically and sometimes obscurely—from the common and highly visible writings and issues of his day. Significantly, the metaphysical questions that concerned him, such as the nature of the soul and its relationship to the other agencies of nature, were then regularly treated either in the physiological literature itself or in other works written by physicians. Hence Mayer *as a physician* was confronted by a much broader range of issues than would have been the case even ten years later, after the above-mentioned transformation and disciplinary contraction of physiology and medicine.

## 1 Blood, Respiration, and Animal Heat

Of patent indispensability to life, and subject to manifold readily observable variations, the blood has, since ancient times, been of central concern to both healers and natural philosophers. In addition to the prominence regularly accorded the blood in texts, the 1830s also saw the publication of a burgeoning specialized literature on the subject.[1] It must be remembered, too, that bloodletting was then considered to be, in the words of one author, "the most important remedy in medicine, without which it is difficult, indeed almost impossible, to be a physician."[2] Hence a student of medicine such as Mayer would have had to acquire a close familiarity with the properties and physiological functions (as then supposed) of the blood. In Mayer's case, furthermore, two of his professors seem to have been especially interested in the topic: Rapp, his thesis adviser, was reported to have assigned prime importance to the blood, and Autenrieth had written his dissertation on venous blood and directed (if not written) another on fetal blood and heat.[3]

Autenrieth's *Handbook of Empirical Human Physiology* (1801–2), which Mayer owned, dealt at length with the differences between arterial and venous blood and the transformation of the former into the latter.[4] In his view it is the disappearance of vital air (*Lebensluft*) from the bright-red arterial blood—that is, its withdrawal by the organs the blood flowed through—which, under normal circumstances, transforms it into the dark-red venous blood. Autenrieth observed that the venous blood of those suffering from putrid fevers (i.e., typhus) is often bright red, as is that of healthy people during the summer. Taking, however, the blood's coagulability as an implicit measure of its oxygen content—the more oxygen, the readier the coagulation—he noted that in both those cases the brighter-than-normal venous blood nevertheless coagulates less readily, and hence contains less oxygen, than the arterial blood it superficially resembles. Thus Autenrieth's well-known[5] observation of the brighter red of venous blood in a warmer environment was *not* directly connected with any putative increase in its oxygen content compared with normal venous blood; his point was rather that *even when* that content varies in something like the normal manner, as shown by the process of coagulation, venous blood may still not take on its normal dark hue. Nor did he explicitly connect the variations in color change with the production of animal heat. Autenrieth attributed the dark color of venous blood to the production of what he called semi-acidified carbon in it, and thus the darker color of venous blood in winter to (first) the greater amount of vital (or dephlogisticated) air absorbed into the blood due to its higher concentration in colder air, and (then) to its enhanced phlogistication as the blood passes through the organs of the body.[6]

In his wide-ranging discussion of animal heat, Autenrieth assigned its origin to the absorption and subsequent decomposition of gaseous oxygen by the body—to processes whereby heat is released as oxygen leaves its gaseous state and to the lesser heat capacity of the substances thus formed.[7] Although his summary did not mention the color of the blood, he clearly saw the blood's increased oxygenation in winter and its diminished oxygenation in summer—due primarily to the greater or lesser concentration of oxygen in a volume of colder or warmer air—as the primary means by which an animal maintains a constant body temperature in a changing environment.[8] In a later work Autenrieth made only passing mention of animal heat, but clearly assigned its production to a combustion process dependent upon respiration.[9]

Treatises devoted to the blood typically discussed the effect observed on its color of such factors as ambient temperature, disease, state of the nervous system, and life-style: venous blood is lighter in the tropics, darker in polar regions and after a cold bath; high fevers and the severing of the nerves to a part of the body tend to be accompanied by lighter venous blood; and the redness of the blood is enhanced by bodily movements and a moderate life-style, while a sedentary way of life makes it darker.[10] English physician Charles Turner Thackrah's report of the brighter venous blood of a man subjected to venesection just after a warm bath was a minor *locus classicus*, though only its context reveals the point Thackrah wished to make:

> The temperature of the body is subject to such slight variation, that we can expect from this cause little change in the character of blood. In cold regions, however, the blood is stated to be darker than in temperate ones; and within the tropics, I believe, to be somewhat brighter. In this country, a horse in a straw-fold will have dark blood; and removed to a stable, he presents it considerably lighter. Such effects may be attributed less to the few degrees of variation in the temperature of the body, than to the reduced or excited state of the circulation. Cold is well known to reduce the action of the lungs and heart, heat to increase them: and to this change in the rapidity with which the blood is transmitted through the system, I should ascribe its variations of colour. On bleeding a man immediately after submersion in a warm bath, I found the blood of the basilic vein scarlet.[11]

The phenomenon, as conceived by Thackrah, has nothing to do with any change in the rate of oxidation, nor did he discuss it in terms of the economy of animal heat: "The phenomenon is not difficult of solution, if we believe the dark colour of venous blood to depend on the addition of carbonaceous matter. Blood hurried by disease through the circulation, has less time to take up this matter, and hence remains nearly of the colour it had when ejected from the left ventricle of the heart."[12] But the whole issue of the cause of the difference in color of arterial and venous blood was quite unsettled: the editor of Thackrah's posthumously published book cited two authorities as arguing "that the florid colour of the arterial blood is *not* due to oxygen, but . . . to the saline matter of the serum."[13] In reviewing the evidence on this issue, Hermann Nasse tentatively concluded that the presence of both oxygen and salts is necessary to produce the bright red arterial hue.[14] One of the very few to consider *in the same context* (albeit briefly) the dependence of the blood's color on temperature, the degree of its oxygenation, and the production of animal heat was Ferdinand Gottlob Gmelin, one of Mayer's major professors at Tübingen:

> According to the experiences of Lavoisier and Séguin, recently confirmed by Dr. Prout, an animal consumes more oxygen gas in the cold than in the heat; the arterial blood of an animal is brighter red in the cold, the venous blood darker black. There thus occurs in the cold not only a preponderance of oxygen, or the weightless agency that acts with or by means of it in the vital process, but the vital process is in general more energetic, more complete, because the difference between arterial and venous blood is greater on both sides. For just that reason, and because the principal source of animal heat is the transformation of arterial blood into venous blood, the internal source of heat is augmented in winter.[15]

Autenrieth's lectures on nosology, which Mayer attended in 1834–35, presumably dealt with other instances of pathological change in the color of the blood.

The unauthorized published text of his lectures discussed one such example in the case of bilious fevers, which are caused by exposure to a hot climate. (Batavia, the capital of the Dutch East Indies, was specifically mentioned.) Autenrieth noted the disappearance of "the sharp difference between arterial and venous blood," though without specifying whether that was because the venous blood becomes brighter or the arterial blood darker. As for therapy, he advised that "bloodletting before the outbreak of putrid fever is of some effect, to the extent that the diminished quantity of blood is more readily oxidized."[16]

Autenrieth had given a somewhat amplified account of the physiological processes thought to underlie this condition in the second volume of his *Handbook*—that is, apart from his discussion of respiration. During the first three or four decades of the nineteenth century, physiologists regularly assigned to the liver the function of removing excess carbon from the blood via the secretion of bile.[17] Autenrieth cited experiments involving the ligation of the portal vein as demonstrating that the carbon excreted in the bile comes from the venous blood. In accordance with the then common view that warmer air contains less oxygen—resulting in a decreased excretion of carbon through the lungs as carbon dioxide—he recorded that diseases involving an excess of bile predominate in very hot regions, and that in Europe during hot summers the blood often exhibits a darker yellow, sometimes even greenish serum (and an unnaturally soft crassamentum), often in association with bilious diseases.[18] Although he did not explicitly say so here, one could reasonably surmise from this line of reasoning that venous blood might be *darker* in a hotter environment, not lighter, as earlier considerations implied. On the other hand, one might conclude that such blood should not necessarily show any color change vis-à-vis the norm since the excretion of bile makes up for the decreased excretion of carbon as carbon dioxide. The whole topic was one of considerable confusion.

A much fuller and clearer understanding of the physiological and medical issues relating to the blood, its color, the ambient temperature, and the production of bile can be derived from another contemporary work in Mayer's library, Leipzig physician Moritz Hasper's *On the Nature and Treatment of Tropical Diseases*. In the section on diseases of the liver the author reasoned as follows:

> From experiments and observations performed by Crawford, Lavoisier, Séguin, Prout, Copland, Pierson in America, etc., it follows in particular that the quantity of carbonic acid formed through respiration [*Respiration*] in a given time appears diminished at a higher temperature and with a decline in the vital forces.
>
> It results from these experiments and observations that less carbon is disengaged from the blood via respiration [*das Athmen*] in a given time in hot climates than in a colder one, whereas the quantity of carbonacious material brought into the circulation of the blood is equally great, and that this substance would soon be present in excess in the blood of persons living in hot climates unless another organ substituted [*vikariirte*] for the lungs.
>
> A principal component of bile is, however, carbon and hydrogen. Therefore when the secretion of bile is increased, a greater quantity of carbon is removed from the

blood, thereby preventing the accumulation of this substance in the blood that would take place in case of diminished excretion of it through the lungs.

. . .

Since, then, as a result of the increased temperature the function of the lungs with respect to the expulsion of carbon is diminished, but the same amount of carbon is conveyed to the blood through food and mode of life, disease would thus have to result if nature did not know how to find a way to resolve this imbalance.[19]

In Hasper's view, the purpose of the oxygenation of blood in the lungs is to rid the body of the carbon—derived largely, it seems, from food—that accumulates in the blood. The body compensates for the reduced oxygenation in hot climates by increasing the secretion of carbon-rich bile. Hasper said nothing about the color of the blood in these circumstances, and simply ignored the otherwise widely discussed changes in the color of the blood of a person subjected to elevated temperatures. His later discussion of the cause of the reduced oxygenation in hot climates did contain, however, a faint clue as to what one might expect to observe:

It is a law for all of physical nature *that heat expands all bodies and cold contracts them.*

The atmosphere is accordingly in a more or less rarefied state the higher or lower the prevailing temperature in a country. As a consequence, a certain quantity of inspired air contains less oxygen the higher its degree of heat.

One therefore breathes easier and more freely in a cold atmosphere and, insofar as more oxygen enters the lungs or blood, the oxidation process takes place there more vigorously, as direct experiments performed in this regard by Séguin and Lavoisier have also shown.

. . .

Now, since a portion of the oxygen of the atmospheric air combines during respiration with the carbon of the blood, less carbon is thus disengaged from the blood in a given time in a hotter climate than in a colder one, and this principle predominates in the blood of most sick people . . . .

The force of nature [*Naturkraft*], which so often acts in an amazing manner in cases of disease, attempts . . . to correct even this imbalance of the blood arising in hot climates by stimulating an increased secretion of bile, perhaps also by means of increased evaporation through the skin.[20]

Thus it appears that an individual arriving in the tropics should show *darker* than normal venous blood. (Note that for Hasper only *part* of the inspired oxygen combines with the carbon of the blood to form carbon dioxide; nor did he ever talk about animal heat or its relationship to ambient temperature.) As for the treatment of liver disease, Hasper, too, recommended bloodletting and the application of up to thirty leeches, noting that in the East Indies often as much as three to four *pounds* of blood is withdrawn.[21]

Although his treatment was much briefer than Hasper's, Johannes Müller also remarked on the liver's function in ridding the body of excess carbonaceous matter via the secretion of bile. Comparing the liver explicitly to the lungs, he noted that Autenrieth, Tiedemann, and Gmelin had, in particular, called attention to the "reciprocal relationship between lungs and liver."[22] He noted the increased incidence of diseases of the liver and intestines in hot, humid climates, and invoked the same reasoning as Hasper with respect to the physiological function of bile secretion and the reduced oxygen content of hot air: "Tiedemann and Gmelin assert that the increased secretion of bile in tropical climates compensates for the reduced purification of the blood in the lungs, which many derive from the attenuation of the air as a result of the heat."[23] Müller said nothing further about the medical aspects of this condition, and nothing explicit about changes in the color of the blood, though the implication seems to be that such blood should be darker than normal.

On the basis of the contemporary primary literature alone it would be impossible to say just what Mayer's beliefs and expectations were as he began to let blood off the Batavian coast in June 1840, especially since it would be unwise to conclude from his later published references to Autenrieth's and Thackrah's statements about the dependency of the blood's color on ambient temperature that he was aware of them in 1840. On the basis of Mayer's surviving medical exams, however, we can be quite sure that he had well learned from Hasper and Müller what to expect. By a stroke of amazing luck for later historians, the fourth question put to Mayer on 16 August 1838 was, "What influence does continued damp and warm weather exert on a person's state of health?" His answer closely followed his presumed sources:

> Heat expands the air, it thereby becomes poorer per volume in oxygen (as well as nitrogen); at the same time the tension of the fluid is increased. As a result, therefore, of the diminished oxygen content, the respiration process is less energetic in the heat, and the air is much less decomposed during respiration, the blood becomes richer in carbon, darker, and the difference between red and black blood is less. In return the liver must enter into vicarious activity; it secretes more bile, which is a very carbon-rich substance. As a result of the increased tension of the fluids accompanying the heat, the imperceptible gaseous evaporation through the skin and the lungs is increased, whereby heat becomes latent and the body thus again becomes cooled. But when the air is nearly saturated with water vapor, then the gaseous evaporation is very restricted, nature's remedy falls away, and there arises the oppressive feeling of sultry heat; the intestines act vicariously for the evaporation through the skin and the lungs. There thus arises in damp heat an influx of humors to the liver and intestines. As a result of the deficient oxidation of the blood, people become sluggish, pale; the digestion is weakened, the intestines disposed to diarrhea. The liver is very disposed to inflammations, which are sometimes acute, sometimes extremely chronic; the former often befalls the left lobe. Structural changes in the liver are the common result, particularly suppuration. The intestines are disposed to dysentery, which often becomes very virulent and generally epidemic, and which, from the sensibility of the skin due to the heat, often arises at the slightest chill. As a result of the diminished oxidation of the blood, the vital process is much less energetic, the humoral mass disposed to sepsis, all diseases easily assume this character.[24]

Thus Mayer clearly expected to find the venous blood he let in the tropics to be *darker* than normal, hence his surprise was all the greater when he in fact found it to be bright red. In a letter to Lang a year before the start of the voyage, Mayer mentioned inflammation of the liver as one of the dangers of his contemplated trip to Java: he was mentally preparing himself for the altered physiological conditions he expected to encounter.[25] Note, too, that the evidence cited shows that Mayer regarded respiration as a means of ridding the blood of excess carbon, without any explicit consideration of the issue of animal heat.

The chief physiological function assigned to the blood was inextricably connected with respiration: the blood is the vehicle by which oxygen and nutrients are distributed throughout the body and by which waste products—some, like carbonic acid, perhaps produced in the blood itself—are conveyed to the appropriate organ for removal. Johannes Müller, for one, assigned clear physiological priority to respiration and the processes of material exchange associated with it. Repeatedly invoking an image of the body as a complex machine, Müller asserted that it was respiration that set the machine in motion.[26] He concluded from a discussion of nutrition, excretion, and general processes of material exchange

> that life in and of itself is connected with a continuous decomposition of already organized substances. Nor is this possible in any other way if it is true, as was previously proven, that the organic force in an animal only manifests itself [*sich äussert*] as long as certain vital stimuli continuously bring about material transformations in the living parts, of which the vital phenomena are only the phenomena, as fire is the phenomenon of the material transformation accompanying combustion. The impulsion [*Antrieb*] for these material transformations occurs through respiration; the blood that is continuously changed through respiration in turn brings about continuous material transformations in the organs.[27]

Respiration, in turn, stood in some close, if not entirely unproblematic, relationship with the generation of animal heat. Since its proposal in the 1780s, the dominant theory of animal heat was Lavoisier's, which attributed the heat generated by an animal to the combination of the inspired oxygen with nutrients in the body.[28] The prime chemical measure of this low-intensity combustion process was the exhaled carbon dioxide. Considerable uncertainty attended the question of whether the supposed combustion takes place in the lungs, blood, or tissues, and attempts to demonstrate experimentally the presence of oxygen and carbon dioxide in the blood gave conflicting and inconclusive results until the issue was largely settled by Gustav Magnus's work of 1837.[29] From the standpoint of the 1830s and early 1840s, however, Lavoisier's theory had two other principal weaknesses. First, experiments reported by Benjamin Brodie in 1811 and 1812 were widely used to argue that the nervous system contributes significantly to the generation of animal heat. (Brodie kept the heart of decapitated animals beating by forcing air into their lungs with a bellows, and noted that although the blood underwent the usual color change in the lungs, the animals' bodies cooled rapidly.)[30] Second, and more serious, precise experiments by Dulong and Despretz in the early 1820s failed to account for all the heat generated in terms of the combustion of oxygen (as measured by the exhaled carbon dioxide).[31] As Holmes observed, "one could either treat the difference as a small one to be explained away by secondary factors as Despretz did, or decide with Dulong that the theory had been proven inadequate."[32] Until the appearance on the scene of Dumas and Liebig in 1841 and 1842, most contemporaries concluded that there must be other appreciable sources of animal heat than simply combustion.

Thus Müller the physiologist regarded as inadequate what he termed the "chemists'" theory of animal heat.[33] He concluded from Dulong's and Despretz's experiments "that there must be other sources of animal heat than respiration, even if one subscribes to the chemical theory of respiration."[34] He explicitly rejected the auxiliary hypothesis—"one of the boldest [*gewagtest*]"—whereby some of the inspired oxygen is consumed in the production of water in the lungs, an hypothesis, he said, "which could be assumed for so long only by chemists, not by physiologists."[35] Like many of his contemporaries, Müller considered the nervous system to be a major source of animal heat.

In addition to the contributions of the respiratory and nervous systems, most German physiologists conceded some importance to general processes of material exchange such as digestion, nutrition, and secretion.[36] Although more concerned than most to trace all bodily functions ultimately to underlying physicochemical processes, the views of the Heidelberg professor of anatomy and physiology, Friedrich Tiedemann, were typical not only in their inclusiveness, but also with regard to his failure even to attempt a quantitative assessment of the relative importance of the numerous factors considered:

> As regards the cause of the generation of heat in animals, few phenomena of life have had so many diverse theories advanced about them as precisely these. . . . The physiologists who embrace the modern chemical doctrines think to find the cause of animal heat in the circumstances of respiration, which they likened to a combustion process taking place between the constituents of the venous blood and the oxygen of the inspired air, and whereby the heat liberated combines with the arterial blood and is distributed throughout the body. Other physiologists seek its source in digestion, nutrition, secretion, even the nervous system. Without here entering into the examination of those theories . . . so much is certain, that all of them afford no satisfactory explanation. Even Crawford's and Lavoisier's theory that the heat is the product of respiration—which had enjoyed the approval of most scientists—was found to be insufficient for the explanation of the heat production on the basis of Dulong's and Despretz's experiments, which sought to find out to what extent the quantity of oxygen gas consumed in respiration was sufficient for the production of all the heat that animals continuously lose, and they acknowledged still other as yet unknown sources of heat than respiration.
>
> The only thing that can be regarded as settled is that the disengagement of heat is an act of life based in the first instance upon the circumstances of the process of nutrition that determine and sustain life. The intake of food and its assimilation through digestion and respiration, the circulation of humors, nutrition and secretion, the exchange of matter accompanying life, and the persistent changes in composition of the solid and liquid parts that in animals are under the influence of the nerves, all have their share in the generation of heat, and it is erroneous to seek the source of heat

in one of these processes alone. The intensity of the disengagement of heat and the property of maintaining themselves in their own temperature relationships stand for animals as a whole in direct proportion to the greater complexity of their organization, to the sum of the manifestations of force [*Kraft-Aeusserungen*] that they exert, and to the latter's intensity.[37]

Arnold Adolph Berthold, professor of medicine at Göttingen, likewise assigned the production of animal heat to a variety of sources—the nervous system, circulation, respiration, motion, nutrition, secretion, and mental activity—without asking further after *their* causes.[38] Many, such as Volkmann, Hünefeld, and Lotze, granted prime importance to processes of respiration and combustion while invoking the need for one or more of the other suggested causes.[39] Volkmann was of the opinion that we can become warmer through bodily movement even without accelerated breathing, again declining to pursue the chain of causes and effects to some ultimate source. While Volkmann elsewhere paid appreciable attention to general questions of causality, in another particular case—whether or not the nervous system (or nervous force) is the "ultimate cause" of the heart's motion—he was inclined to let the issue be decided entirely by experiment, without invoking any energetic or other theoretical considerations.[40] Karl Asmund Rudolphi, professor of anatomy and physiology in Berlin, was one of the few explicitly to deny—on the basis of a comparison between different species—any direct role to the nervois system, preferring (in 1821) to trace the source of animal heat to changes in composition of the substance of the animal body, namely respiration, nutrition, digestion, and (in 1828) two additional processes accompanied by a heat-producing electrical process: muscular motion and nervous activity.[41]

Indicative of the changes in physiological thinking that followed Liebig's forays into the field (to be examined below), in 1842 organic chemist Franz Simon explained animal heat in terms of the respiration of oxygen, the activity of the nerves—both considered to be aspects of "the general transformation of substance [*Stoffwandel*]"—and the vigorous motion of the muscles,[42] whereas in 1843, after reviewing Liebig's epoch-making book, *Organic Chemistry in Its Application to Physiology and Pathology* (1842), he concluded that the sole source of animal heat is the combustion of carbon and hydrogen: "To consider the heat generated as the effect of the nervous system, whereby an exchange of matter [*Stoffwechsel*] is wholly excluded, means to call forth an activity out of nothing. To consider the contraction of the muscles as a source of heat is equally inadmissible, for the motion cannot take place without a certain expenditure of force, and the organs of motion obtain this force only as a result of the exchange of matter."[43] The matter-of-factness of Simon's reasoning in 1843 contrasts dramatically with the universal weakness of rigorously pursued causal and energetic considerations through the 1830s.

A minority of scientists, of whom the most important was Berzelius, took Brodie's line of reasoning a step further and sought to trace the source of animal heat to electrical activity. For Berzelius, at least, this was surely a reflection of his prior preoccupation with electrochemical theories of reaction and combination. Although it is difficult to extract a clear positive theory of animal heat from

Berzelius's influential *Textbook of Chemistry*, he was decisive in his rejection of the then most favored explanation in terms of combustion. His extensive treatment of the subject reveals just how complicated and uncertain the experimental situation was in the 1830s.[44] Given that, his failure adequately to motivate his unusually unqualified rejection of Lavoisier's theory is as noteworthy as it is puzzling. Citing the inability of Despretz's experiments to account for more than 70 to 90 percent of the heat in terms of respiration, he concluded in 1831 as follows:

> Nevertheless it is certain that both his and Dulong's experiments indicate that the development of heat in the body must have still other sources, and that, if there is such another source, the entire animal heat is certainly to be derived from it. Respiration then has no other part in it than that it, like the remaining processes in the body, prepares the conditions without which the development of heat could not take place; but an immediate development of heat through transformation of the air in the lungs appears not to take place.
>
> Accordingly, the *idea* first proposed by Brodie that the development of heat, like all vital phenomena, is determined by the nervous system appears at present most in agreement with the result of the experiments performed; and if the effects of the nervous system then depend on a method peculiar to itself of employing the influence of the opposing electricities, it thus appears to follow that the development of heat could here depend on their recombination everywhere nerves go and thus could have the same cause as with combustion, but with the difference that entering into a chemical combination is not necessary here for the liberation of the opposing electricities.[45]

In 1840 he replaced the first paragraph with the following:

> It is clear, now, from what was earlier set forth that such an oxidation process does not take place in the lungs, and that the cause of animal heat must lie in another process of heat development as yet still wholly unknown to us. These calculations are meanwhile of great value because they show that we also cannot derive the animal heat from the combination of absorbed oxygen with organic matter during the circulation of the blood, for even under the presupposition that as much heat was developed through this combination as with the . . . transformation of carbon to carbonic acid and of hydrogen to water, this would still not be sufficient. There must therefore be still another source for the heat.[46]

Berzelius's reasoning is scarcely cogent: Dulong's and Despretz's experiments demonstrate that there must be another source of animal heat than combustion; if there is, it must be the source of *all* the heat produced. In 1831 he seemed to imply that acceptance of the combustion theory of animal heat necessitated acceptance of the lungs as the site of combustion; since he (quite rightly) rejected the latter, he also (illogically) rejected the former. In 1840 he reasoned that even if the combustion took place in the circulating blood, it still would not cover all the heat produced, hence one still needs to invoke another source. He had the nerves mediate the action of electricity in producing animal heat, yet somehow he dissociated the production of electricity from any antecedent chemical change.[47] Responding to Liebig's 1841 paper, "The Vital Process in the Animal, and the Atmosphere," which regarded the combination of atmospheric oxygen with the components of the blood as the sole source of animal heat, Berzelius replied that heat is produced at many places in the body—and by processes other than chemical combinations, such as electric currents and friction; thus "it is more than a mere probability that the principal source of heat in the animal body is other than through chemical combination, although that contributes."[48] In his published review of Liebig's *Animal Chemistry* he continued to oppose Liebig's view, citing the influence of mood on the temperature of parts of the body and again appealing to Brodie's and others' results "which are irreconcilable with it and which show that the development of heat in the animal body appears to be a function governed by the eighth pair of nerves."[49] His attachment to the electrical explanation of animal heat became firmer even as he took refuge behind an agnostic what-makes-you-think-we-even-know-all-the-possible-sources-of-heat, in particular with respect to the source of the powerful electric currents generated in the nervous system of certain fish.[50] Citing experiments of Becquerel and Breschet, Berzelius had argued in 1840 that muscular contraction could be regarded as another source of animal heat.[51] In 1835 the two French scientists had explicitly connected this experimental finding with speculation as to the possible electrical origins of animal heat.[52]

Authors of physics texts of the period regularly addressed the issue of animal heat in a few sentences in the section on heat. At one end of the spectrum of opinions was Heidelberg professor of physics, Georg Wilhelm Muncke, whose wholehearted early support for Lavoisier's theory was unusual:

> The cause of animal heat does not lie in an inheritance, nor in a friction of the blood in the blood vessels, nor in a fermentation, nor in the motion of the muscles, nor even in the immediate activity of the brain, as Brodie wished to conclude from his experiments but as refuted by Emmert, even though respiration is a consequence of the activity of the nerves. It is, rather, mainly to be derived from the chemism of the vital process, whereby the consumption of oxygen gas and formation of carbonic acid through respiration is to be regarded as the chief means. . . . The vital process of animals can thus be regarded as a slow combustion process, although the oxygen gas is not decomposed exclusively in the lungs, but in the whole body.[53]

Twelve years later, however, in his long article on heat for *Gehler's Physical Dictionary*, Muncke qualified his conclusion by recognizing (unspecified) processes of assimilation as a further source of animal heat.[54]

Halle professor of physics Ludwig Friedrich Kämtz accepted the apparent limitation imposed on Lavoisier's theory by Dulong's and Despretz's experiments, and speculated that "electrical processes" also play a role, "since heat is produced by the decomposition of the nutrients and the other humors of the body"—tacitly implying his acceptance of an electrochemical understanding of chemical change.[55] At the other end of the spectrum was Andreas Baumgartner, professor of mathematics and physics in Vienna, who, while conceding something to generally recognized causes, denied that organic processes can be fully understood in physical terms.[56]

By the 1830s, very few scientists of any stature were so categorical in their denial of the applicability of the laws of chemistry and physics to the explanation of physiological processes—leaving aside more difficult problems of generation and embryological development. A prominent instance of such a denial was Humphry Davy's discussion of respiration and the problem of animal heat in his posthumously published philosophical reflections, *Consolations in Travel.* The book appeared twice in German translation during the 1830s (in 1833 and 1839), and might well have caught Mayer's attention. One of the interlocutors of the dialog, Eubathes, defended the combustion theory, which he assigned with nationalistic pride to Joseph Black:

> First, then, in all known chemical changes in which oxygen gas is absorbed and carbonic acid gas formed, heat is produced . . . . Another circumstance in favour of it is, that those animals which possess the highest temperature consume the greatest quantity of air; and, under different circumstances of action and repose, the heat is in great measure proportional to the quantity of oxygen consumed. Then, those animals which absorb the smallest quantity of air are cold blooded. Another argument in favour of Dr. Black's opinion is, the change of colour of blood from black to red; which seems to show that it loses carbon.[57]

But Davy's spokesperson, the Unknown, would have none of these standard arguments:

> I will not allow any facts or laws from the action of dead matter to apply to living structures; the blood is a living fluid, and of this we are sure, that it does not burn in respiration. . . . [T]hose animals the life of which is most active, possess most heat, which may be the result of general actions, and not a particular effect of respiration. Besides, a distinguished physiologist [Mr. Brodie] has rendered it probable, that the animal heat depends more upon the functions of the nerves than upon any result of respiration. The argument, derived from change of colour is perfectly delusive . . . .
>
> . . .
>
> [T]he moment inorganic matter enters into the composition of living organs it obeys new laws. . . . Respiration is no more a chemical process than the absorption of chyle; and the changes that take place in the lungs though they appear so simple may be very complicated; it is as little philosophical to consider them as a mere combustion of carbon, as to consider the formation of muscle from the arterial blood as crystallization. There can be no doubt that all the powers and agencies of matter are employed in the purposes of organization, but the phenomena of organization can no more be referred to chemistry than those of chemistry to mechanics.[58]

Such views would have seemed increasingly out of place to German scientists during the course of the 1830s. All the more rhetorically misleading, then, was Mayer's imputation in 1845 that the "vitalist" views of the seventy-three-year-old professor of medicine at Erlangen, Gottfried Christian Reich, were somehow typical of the conceptual confusion he had to contend with:

What bizarre ideas the so-called vitalists can even today come up with can be seen for example in Reich's *Textbook of Practical Medicine According to Chemical-Rational (?) Principles* (Berlin, 1842). Reich considers animal heat to be an inheritance given to the newborn to take with it along the way![59]

Quite the contrary; indeed, compared to the situation during the previous decade, by the early 1840s the scientific consensus had shifted dramatically in the direction of a self-conscious determination to employ only chemical and physical principles in the explanation of physiological processes.[60] One of the leaders in this movement was François Magendie, whose influential *Elementary Sketch of Physiology* was published in numerous editions and translations (including two into German in the years 1834–36). Thus it is not surprising that, among pre-1840 authors, Magendie was one of the most enthusiastic supporters of Lavoisier's chemical theory of animal heat, to the detailed exposition of which he devoted ten pages of his *Sketch*.[61] However, he, too, was moved by Despretz's experiments to regard respiration as only the principal cause of animal heat: other sources of heat in the animal economy were (probably) the friction of diverse parts of the body, the movement of the blood and the rolling of its particles against one another, and the phenomena of nutrition.[62]

It was not until 1841 that a prominent investigator of organic processes came out decisively for the unqualified sufficiency of the combustion theory of animal heat. In August of that year, in a widely published lecture given at the end of his course in organic chemistry at the École de Médecine in Paris, Jean-Baptiste Dumas asserted that the explanation of the physiological process of respiration was falling more and more under the general laws of inorganic nature.[63] To make that claim good, he had to explain away the apparent discrepancy in the results of Dulong's and Despretz's experiments:

If animals do not possess any particular power for producing organic substances [*matières*], do they at least have this special and singular power of producing heat without the expenditure of substance that has been attributed to them?

You have seen, in discussing the experiments of Dulong and Despretz, you have positively seen the contrary to be the case. These able physicists have assumed that an animal placed in a cold-water calorimeter emerges from it with exactly the same temperature it possessed on entering; a thing absolutely impossible, one knows today. It is this chilling of the animal, which they did not take into account, that in their tables expresses the excess of heat attributed by them and by all physiologists to a calorific power particular to the animal and independent of respiration.

It has been demonstrated to me that all animal heat comes from respiration; that it is measured by the carbon and hydrogen burned. It has been demonstrated to me, in a word, that the poetic assimilation of the railroad locomotive to an animal rests on more serious foundations than one perhaps thought. In both, combustion, heat, and motion, three connected and proportional phenomena.[64]

Prompted in part by Dumas's challenge to what he regarded as *his* intellectual property, Liebig soon entered the lists in defense of a modified version of Lavoisier's oxygen theory of respiration and animal heat. Liebig took steps to assure that his somewhat idiosyncratic views on the relationship between nutrition, respiration, and animal heat would reach a large audience, publishing them in progressively expanded versions first in December 1841 in two "Chemical Letters" in the widely read *Allgemeine Zeitung* and then in an important paper on "The Vital Process in the Animal, and the Atmosphere" in his *Annalen*, followed three months later by a second paper in the *Annalen* on "Nutrition and the Formation of Blood and Fat in the Animal Body." These two papers formed the first part of his epoch-making book, *Organic Chemistry in Its Application to Physiology and Pathology*—better known after the title of the second and subsequent editions as Liebig's *Animal Chemistry*—published in July 1842. After 1842 the conceptual terrain surrounding this issue had been completely relandscaped.

To be sure, Liebig had been investigating this constellation of issues before being challenged by Dumas. Kremer has argued that Liebig's *Agricultural Chemistry* of 1840 treated respiration as a kind of *Verwesung*—that is, dry decay in the presence of oxygen—both being thus regarded as slow processes of combustion at relatively low temperatures.[65] Citing data on the proportion of carbon, hydrogen, and oxygen in wood, Liebig reasoned that if atmospheric oxygen reacts directly with carbon in the wood during the decay process, then the proportion of carbon should steadily decrease, whereas it actually increases.[66] Such, he concluded, is the general nature of the slow combustion of hydrogen-rich substances: "It is evidently the hydrogen which is oxidized at the expense of the oxygen of the air, [while] the carbonic acid comes from the elements of the wood; never and under no circumstances does carbon combine at normal temperatures directly with oxygen to form carbonic acid."[67] He later explicitly referred to the *Verwesungsproceß* that takes place in the lungs: water is formed there, not carbon dioxide, though a portion of the uncombined oxygen is absorbed into the blood.[68] This work did not, however, purport to provide an adequate account of either respiration or animal heat, and Liebig addressed those topics only in passing. Strictly speaking, too, Liebig never claimed that respiration overall *was* a kind of decay process, only that such a process takes place in the lungs.

At the end of a paper on nitrogenous nutrients published in the same month as Dumas's lecture, Liebig addressed more directly the question of the sources and functions of oxygen in the animal economy. The discrepancy others had noted between the amount of oxygen inspired and the greater oxygen content of the carbon dioxide exhaled, a discrepancy that bedeviled Lavoisier's theory of animal heat, Liebig solved by arguing that oxygen must be removed from the nonnitrogenous food animals consume before it can be stored as fat.[69] The purpose of this stored fat is to protect the body against the ever-present destructive force of atmospheric oxygen through the production of carbon dioxide. He closed his paper with a brief mention of animal heat: "In a second paper I will attempt to show that the carbon of nonnitrogenous foods such as starch and sugars is employed in respiration and in the production of animal heat, that the latter stands in the closest connection with the carbon of foods."[70] Much of his work during the next year was an elaboration of the connections here briefly sketched.

Liebig's views on animal heat were only one aspect of a broad picture of the chemical changes taking place within the animal organism and between it and its environment. Nor were those views as simple or unambiguous as they might appear from his summary pronouncement that "all vital activities derive from the reciprocal action of the oxygen of the air and the constituents of the food."[71] Especially since he had just been talking about the relationship between the carbon ingested as food and that exhaled as carbon dioxide, his first explicit pronouncement with respect to the cause of animal heat seems to be a clear endorsement of Lavoisier's theory:

> The interaction between the constituents of the food and the oxygen disseminated in the body through the circulation of the blood is *the source of animal heat.*
>
> . . .
>
> This higher temperature of the animal body—or, if one will, disengagement of heat—is everywhere and under all circumstances the result of the combination of a combustible substance with oxygen.
>
> In whatever form carbon may combine with oxygen, the act of combination cannot take place without being accompanied by the development of heat; regardless of whether it occurs slowly or rapidly, the amount of heat liberated remains an invariable quantity.[72]

While it remains true that Liebig assigned the entire production of animal heat to chemical processes involving oxygen, those processes were more complex than simply the absorption of oxygen through the lungs. Hence it is not surprising that Liebig felt confident in discounting as valueless scientists' previous attempts to correlate the amount of oxygen consumed with the heat produced by an animal in a given time.[73] Those additional processes—the absorption of oxygen through the skin and the release of oxygen in the production of fat from starches and sugars—were dealt with at any length only in the second of Liebig's two related papers.[74] Before we turn to it, however, it will be useful to review what Liebig believed the function of respiration to be.

Fundamental to Liebig's understanding of the animal economy was his conception of oxygen not as the sustainer of life but as an agent against whose destructive tendencies the vital processes are in constant struggle. Thus the carbonaceous materials oxidized in the body serve not only to generate heat, but also to ward off the potentially harmful effects of oxygen on the body as a whole. Thus Liebig's otherwise strange references to "the carbon indispensable for respiration" and to "the carbon, which protects our organs from the influence of oxygen and which generates in the organism the heat indispensable to life."[75] In accordance with these views, the increased oxygen taken in by people living in the polar regions—Liebig adopted the generally accepted explanation in terms of the greater oxygen content per volume of colder and thus denser air—is compensated by the greater carbon content of their fatty foods. At the same time, the oxidation of this fat supplies the extra heat needed to keep them warm in a colder environment. This fortunate accommodation of food to climate is, in Liebig's words, the provision of "an infinite wisdom."[76]

If an animal consumes more carbonaceous matter than can be gotten rid of through oxidation, the excess is stored in the body as fat. On the basis of a comparison of the chemical composition of fats on the one hand and starch, sugars, gluten, albumin, and legumin on the other, Liebig argued that animals produce fat by removing some of the oxygen from the latter group of foodstuffs.[77]

> The production of fat depends on a lack of oxygen; only in it, in the very formation of fat, there is made available to the organism a source of oxygen, a new cause of heat production.
>
> The oxygen liberated as a result of the formation of fat leaves the body as a compound of carbon or hydrogen; whether or not this carbon or hydrogen was taken from the very substance which also conveyed the oxygen or from another compound, just as much heat must be developed through this formation of carbonic acid or water as if we had burned an equal quantity of carbon or hydrogen in the air or in oxygen gas.[78]

In support of this last claim Liebig adduced fermentation and putrefaction as examples of chemical processes in which the release of carbonic acid and water is accompanied by the production of heat, "exactly as when carbon and hydrogen combine directly with oxygen."[79] The oxygen from fat does not, it seems, pass into the animal's body as elemental oxygen. Thus, he concluded,

> The formation of fat, compared with well-known analogous phenomena of the separation of oxygen, is accordingly accompanied by a development of heat; it restores to the animal body a certain quantity of the atmospheric oxygen indispensable for the vital processes, and it does so in all those cases where the oxygen inspired through skin and lungs is not sufficient to transform into carbonic acid the carbon present and suitable thereto.
>
> . . .
>
> In the formation of fat the vital force creates for itself a means to confront the lack of [both] oxygen and the heat necessary for the vital processes.[80]

Thus when Liebig asserted (somewhat oddly) that "an oxidation process is the sole and chief source of animal heat," he had in mind a more complex process than many readers, both his and our contemporaries, seem to have understood.[81] For example, Valentin's recapitulation of Liebig's ideas made no mention of the role of fat production: "Liebig, in particular, has emphasized *the combustion process taking place in the capillaries of the body as the cause of animal heat*." Simon's review of Liebig's book presented it as rightly asserting that the sole source of animal heat is the combustion of carbon and hydrogen. And Wunderlich associated Liebig with those who *wrongly* held that oxidation is the only source of animal heat. Nor have more recent commentators noted this qualification in Liebig's acceptance of Lavoisier's theory of animal heat.[82]

In 1844 Gabriel Gustav Valentin, professor of physiology at Bern, asserted that "the principal—indeed, probably the sole—cause of human beings' proper heat is the combustion process taking place in our body."[83] Two years later he was more cautious: the combustion hypothesis seems sufficient only as long as one looks at the most general phenomena, but it runs into trouble as soon as one examines the details. He believed that the heat produced by the combustion of the elements in a compound is usually less than that of the elements burned separately—implicitly calling into question the significance of Dulong's and Despretz's experiments—and he noted an experimental anomaly with respect to the carbon dioxide produced by starving and healthy doves versus the heat they produced. He also allowed the nervous system to play some role in the production of animal heat. Nevertheless, he regarded an appeal to "still wholly unknown influences of the nervous system," as opposed to unknown phenomena of nutrition, as a "confession of ignorance," and the combustion hypothesis remained "the only certain guide capable in the future of leading to the solution of the riddle."[84]

Reservations such as Wunderlich's and Valentin's soon became increasingly less common, and by 1845 Karl Vierordt—then a physician in Karlsruhe, later professor of anatomy and physiology at Tübingen—could confidently assert that "the chemical theory of animal heat [is] completely true."[85] But whatever the remaining differences of opinion might have been, by 1845 the range of debate over animal heat had become circumscribed by a general and explicit acceptance of the need to consider all organic processes within the context of *Stoffwechsel*—an issue to which I will shortly return. It is noteworthy that the dramatically increased acceptance of Lavoisier's theory of animal heat by mid-decade was not the result of any significant new data since the time of Dulong and Despretz, but rather reflected a new way of thinking. Vierordt asserted that the idea that animal heat might derive from other than physical causes, causes belonging exclusively to animal life, no longer had any force: the new physiology had cleansed itself of all progress-impeding mysticism and romanticism.[86] He put the issue in terms of a self-conscious appeal to physics and chemistry as providing the only valid models for a scientific physiology:

> While some derived the source of animal heat from manifold causes utterly peculiar to the organism, such as the vital force or the nervous force, and even believed thereby to give an explanation of the problem at hand, others rightly transferred from physics and chemistry the facts derived through precise observation concerning the origin and distribution of heat to the circumstances of our body, without, however, having been in a position to completely overcome the extraordinary difficulties which oppose the precise solution of this complicated and difficult question on the (naturally alone true) basis of experiment. One thus finds repeated here, too, the struggle which we encounter in every individual branch of physiology, whereby nothing less is at stake than the question of whether the phenomena and processes in the organism are to be explained in terms of abstract causes existing only ideally but not in reality, or in terms of facts whose validity has been proven; in a word, whether one wishes to investigate in the spirit of the old sterile scholastic study of nature or according to the method of the exact, vital natural sciences.[87]

The issues were fundamentally ideological. Carl Vogt, too, explicitly condemned the alleged use of a vital force to explain animal heat. It was, he said, a "refuge" invoked to explain the incomprehensible, an "unknown X" lurking in the background that receded as science progressed: reliance on such a mode of explanation "no longer has any scientific value, it is only another word for ignorance [*eine Umschreibung der Unwissenheit*]." In a paper of 1845, Leibig rather overplayed his hand in creatively reinterpreting various experimental findings in order to explain away the discrepancy in Dulong's and Despretz's results. Asserting correctly that, on the basis of this discrepancy, "physiologists and many chemists have assumed an additional source of heat in the animal's organism that, alongside the inspired oxygen, maintains the animal's temperature," he announced his goal to be the demonstration that "the assumption of the existence of a cause of heat formation in the animal body inexplicable according to known physical laws contradicts the positive facts," incorrectly implying that deficiencies in Lavoisier's theory were generally being answered in a scientifically unacceptable fashion—by which his readers probably understood the (here unnamed) vital force.[88]

Although the physiological functions commonly attributed to a vital force will be the topic of a later section, it should be emphasized here that, in the decades before 1840, virtually no German scientist invoked a vital force in the explanation of *animal heat*: only the *opponents* of the vital force made that connection. As Kremer rightly insisted, Helmholtz's contention "that it was Stahl's concept of vitalism, which would have made the animal a *perpetuum mobile*, that he was assailing" rings hollow, since "in 1845, no physiological writer of note was defending Stahl. If Helmholtz had the *Lebenskraft* of Müller or Berzelius in mind, he could have chosen a better arena than animal heat to make his criticism, [since] . . . neither of these men explained animal heat in stridently vital terms."[89] The same judgment applies more generally to other writers of the preceding decades. In pretending that the vital force had been used to explain animal heat, and that thus the explanation of that heat solely in terms of chemical processes represented a defeat for the vital force—indeed *the* defeat *of* the vital force—reductionist-minded physiologists also tacitly pretended that the agency of a vital force or soul was *everywhere* disproven or (better yet) rendered scientifically illegitimate.

Despite the increasingly taken-for-granted assumption that organic processes must be explained in terms of specific chemical transformations, only gradually does one begin to encounter a principled appeal to energetic considerations. The long section on animal heat in the second edition of Valentin's *Textbook of Physiology* of 1847 reported quantities of experimental data but entertained no conservation-of-energy considerations, despite the fact that Valentin knew and cited Mayer's *Organic Motion* of 1845.[90] Indeed, in a report written the previous year he had spoken very favorably of Mayer's work, though he apparently quite missed the significance of Mayer's theory of force:

> J. R. Mayer, in a small work written with great perspicacity and profound physical knowledge, has developed a series of views concerning general physical and physiological relationships. In it he seeks to demonstrate generally how every performance [of work] is connected with a proportional conversion of agency or substance, and how this principle can also be applied to organic motions. Accordingly, the heat formation determined by the conversion of substance is expended partly on the [production of]

mechanical effect, partly on the production of [organisms'] proper heat. In the context of these considerations Mayer declares himself opposed to the existence of a special vital force, and he seeks to demonstrate that the principal seat of the combustion process of animals is in the blood itself, and that a muscle achieves more mechanical effect the more copiously and quickly it is suffused with blood.[91]

Valentin maintained that proof of the combustion theory of animal heat depends on a strict quantitative experimental demonstration of the equivalence between the production and loss of heat in the organism.[92] His insensitivity to energetic considerations is seen dramatically in his response to Mayer's criticism of his explanation—in terms of posture—of why mountain climbers, runners, jumpers, and dancers breathe harder.[93] He replied to Mayer in the context of a criticism of teleological explanations in physiology:

> Daily experience teaches, for example, that when climbing mountains we easily break out in sweat and experience throbbing of the heart and difficulty in breathing. If one says that this happens because nature intensifies the combustion process for the sake of greater mechanical action, one thereby gives not an explanation but merely a description [*Umschreibung*] appropriate for the individual case of the reckoned mode of action of the organism. If one succeeded, however, in demonstrating why the heart and the breathing muscles must just then increase their activity, then the way would be open to true understanding. All these and similar teleological conceptions do not solve the tasks which scientific investigation must set itself.[94]

Valentin had clearly missed the energetic thrust of Mayer's argument.

The physical and chemical changes in the blood in health and disease; the nature and function of respiration; the sources of animal heat: those were the most important of an interconnected nexus of issues against which one must situate Mayer's earliest physiological observations and reflections. However, there were still other issues of some contemporary visibility relating to the blood which could conceivably have played a role in stimulating Mayer's sensitivity to the central issue of causality as it related to physiological explanations. On into the 1820s and 1830s there continued to be discussion in Germany over whether blood corpuscles or the entire blood itself possesses the power of self-motion.[95] In 1828 Rudolphi protested that he could form no conception of the supposed motive force proper to the blood, and rejected the idea as an unnecessary *deus ex machina*.[96] Critics of the late 1830s typically rejected such a possibility by appealing to the unacceptability of the self-caused motion of such a fluid. Wrote Volkmann in 1837:

> Several respected physiologists have attributed to the blood a life of its own and, as a consequence, motive force. According to them the blood is supposed to be constrained by an internal principle into circulation even without the participation of the heart. The experimental grounds on which this view was based have already been discussed above, where we showed that circulation does not take place without the participation of the heart. Moreover we are unable to connect any clear concept with the view just mentioned. As will be shown below, the blood consists of serum and of

corpuscles suspended in it. To attribute so highly developed a life to such a mixture that autonomous locomotion could take place is against every analogy.[97]

Mayer's university friend and later professor of surgery at Marburg, Wilhelm Roser, found occasion the following year in his dissertation to criticize the alleged propulsive force of the blood as (in essence) inconceivable: "Despair of being able to explain the entire circulation according to physical laws led to Kielmeyer's positing of a propulsive force. However little the autonomous motion of a liquid agrees with our [general] notions, this hypothesis was nevertheless widely accepted."[98] And to illustrate people's accustomed inclination to mysticism and the neglect of the logical training of medical scientists, Schleiden quoted an anonymous passage that spoke approvingly of the autonomous motion of the blood corpuscles.[99] In contrast to these principled objections to the idea of the blood's self-motion, Johannes Müller's rejection of Kielmeyer's *Propulsionskraft* was based for the most part on his interpretation of negative experimental findings.[100] But it was still an issue that warranted discussion in the 1830s.

While not exactly advocating the self-motion of the blood, Burdach searched for another propulsive force to supply what he thought was an extra push to move venous blood from the periphery of the body to the heart. Speaking of the observed behavior of blood corpuscles outside the body, he wrote:

During the normal conditions of life they do not manifest their tendency to act adhesively and repulsively toward each other: there must therefore be something external to them which suppresses this tendency and compels them to move in a determinate, uniform direction; this impulse lies in the heart, as the central point of the blood system; but since this cannot alone be active, a second and still more essential impulse must thus proceed from the opposite point, from the periphery of the blood system or from the organic structures external to it.[101]

This passage reminds us, too, that the language of center and periphery, of centrifugal and centripetel motions, was a commonplace in discussions not only of circulation, but also of the nervous system.[102] The familiarity with such terminology from a physiological and anatomical context may have been a factor in its appeal to Mayer as he tried to come to terms (literally!) with central-force motions.

## 2 Sources of Organic Activity

The interconnected processes of respiration and the generation of animal heat are only two among the organism's manifold life-sustaining activities. Indeed, in the 1830s animal heat was typically assigned to a collection of diverse organic activities: not only respiration and oxidation, the chief agency, but also the action of the nervous system, digestion and nutrition, muscular contraction, and the motion of bodily fluids. Other organic activities—typically the more strongly teleological, such as generation, embryological development, the preservation of the body's integrity in the face of external challenges, and the direction of complex chemical transformations—were routinely attributed to the operation of some kind of vital force. Some regarded a distinct will or soul or mental force as capable of producing certain organic motions and activities, though whether this producing meant just initiating or also sustaining was a question not usually explicitly addressed. In general, physiologists of the period only gradually became sensitive to questions of what might be called energetic causality: if an organism produces some kind of material change in the world—for example, when it moves some part of itself or does work—then there must have taken place a physical change elsewhere in the system corresponding in some measure to the achieved effect. Older writers typically regarded certain organic activities, such as the production of motion, as due to causes largely ungrounded in either energetic or material transformations—a way of thinking that was to become increasingly unacceptable during the 1830s. Thus Autenrieth's highly regarded *Handbook of Empirical Human Physiology* (1801–1802) characterized the distinction between a living and a dead body in terms of the former's ability to produce motion "without this motion having been caused by visible dissociation and separation of its parts"; we thus conclude that "an originally active, motion-causing force is present within the body," a force we call the soul.[103] Autenrieth was more concerned with classifying the active forces as either internal or external, belonging either to the soul or to extrinsic stimuli, than with tracing their energetic or material conditions.

Twenty years later, in 1821, Ferdinand Gottlob Gmelin, professor of medicine and natural history at Tübingen and likewise one of Mayer's major professors-to-be, could still maintain that some organic motions are independent of material changes in the body: "A large number of animal phenomena—indeed precisely those through which life reveals itself to the outside—consist in motions which result from an internal force and [which take place] without lasting change in the material moved."[104] Following Albrecht von Haller's eighteenth-century lead, Gmelin analyzed organic motions into contractions and expansions due, respectively, to the action of two basic organic forces, irritability and what he termed the "turgor vitalis."[105] Those, however, were "forces" in the classical sense of being properties of (here organic) matter.

Such views died hard. Despite the fact that Friedrich Tiedemann was the first and most important German physiologist of the 1830s to make processes of material exchange the centerpiece of his work, he nevertheless failed to bring any energetic considerations to bear on his treatment of motion in animals, which he was thus prepared to regard as capable of generating motion via the action solely of their will, hence "every living body is to be likened to a perpetuum mobile."[106] Whether or not the animal organism is a true perpetuum mobile was indeed a relatively common theme in the German literature.

In a talk delivered before the Hufeland Medical-Surgical Society in Berlin in 1836 on the extent to which vital processes can be explained in physical terms, physician August Vetter exemplified some of the changes taking place in the basic attitude of German physiologists with respect to the *kinds* of explanations one is entitled to demand:

> If we proceed to the consideration of the chemical relations of organic beings, we find first of all that they develop if not all, then nevertheless a portion of their characteristic heat through a process of chemism by which heat is also liberated in the gasometer, etc. If, alongside the formation of carbonic acid with the help of the inspired oxygen, one ascribes the cause of animal heat with some justice to an interaction between the nerves and certain of the body's tissues (or general fluids), we nevertheless lack a physical explanation for the latter phenomenon, unless we wish to look for one in a recently proposed hypothesis of liquefied nerve-medulla. But we see nowhere in the organism that there, where according to the experiences of chemistry one expects a development of heat from an existing process, such a one does not take place.[107]

Nor, when considering the motion of the blood and the heart, did Vetter shy away from implying that such vital phenomena might be explicable on purely mechanical grounds: "We are thus dealing here with a purely physical phenomenon, and if we do not know the source of the motive force, it nevertheless does not appear more mysterious than the fall of the water or the expansivity of the steam by which wheel and piston drive a pump."[108] Using as an example the once hotly contested issue of whether an organism is capable of producing substances (such as certain compounds of phosphorus and calcium) containing elements not derived from its environment, Vetter reiterated his plea for the physical explanation of all vital processes by denying any powers of creation to the still-popular vital force:

> Now, whether there lies in the organic composition a force of chemism capable of composing out of their elements substances which we were not as yet able to decompose, can surely only be decided with certainty if it should nevertheless prove possible in at least a few cases to bring about this heretofore unattained decomposition by purely chemical means. But little would be gained if we, anxious to refute the notion of a living spontaneous generation, in turn wished to propose an hypothesis of elementary spontaneous generation for the sake of a few as yet unexplained facts, and to ascribe to the vital force a property of which we otherwise have no conception: that, namely, of making something out of nothing.[109]

The increasingly 'obvious' character of these kinds of explanatory demands was as important and fundamental a change in physiologists' thinking during the 1830s as were new techniques and experimental findings.

### *2.1 Physical and Chemical Processes: The Organism's Connection with the External World*

Foremost among German physiologists of the 1830s to emphasize the necessity of studying an organism in terms of its interrelationship—in our terms both material and energetic—with the physical world was unquestionably Friedrich Tiedemann.[110] To be sure, Tiedemann did not ignore internal and apparently nonmaterial causes of organic activity—witness his above-cited explanation of motion in terms of the action of the will—but in the event he paid overwhelmingly more attention to processes of identifiable material change and exchange. As he wrote in the introduction to his influential *Human Physiology*:

> Through the investigation of the causes lying within living bodies themselves on which depend their phenomena and manifestations of activity [*Thätigkeits-Aeusserungen*], or through the discovery of their properties and forces and the establishment of the laws according to which they act, life is nevertheless not yet known. That is to say, living bodies are not isolated creatures that have the causes of their existence only within themselves; rather they stand in reciprocal interaction with all of nature, they are component parts [*Glieder*] of all of nature. All their manifestations of activity and all their properties are also dependent upon external influences. The continued existence of organic bodies and of their manifestations of life require a certain degree of heat, the influence of the air, and the presence of water and food. . . . We must therefore subject the dependence of living bodies on external influences to closer investigation, search after the causes of this dependence, and establish the laws that prevail regarding the contingency of living bodies on these external influences.[111]

The vivid image of vital processes Tiedemann repeatedly portrayed was that of a complex of chemical transformations: "Considered from the standpoint of chemistry, living bodies appear to us as laboratories [*Werkstätte*] of chemical processes, for they undergo perpetual changes in their material substrate. They draw materials from the outside world and combine them with the mass of their liquid and solid parts."[112] It would be difficult to overemphasize the extent to which Tiedemann dwelt on the centrality of material exchange. Significantly, he also explicitly connected these material changes with the body's exertion of force: "With the continued existence of animals and with the exercising of their manifestations of force—which we call life—there is simultaneously connected a perpetual change in the material substrate of their solid parts."[113] Tiedemann discussed processes of *Stoffwechsel*—a term he more consistently used in the second half of the book—as they related to respiration, and he noted their dependence on such external factors as heat, air, water, food, and mechanical contact.[114] In this part of his work he expressed more explicitly than elsewhere the close connection between respiratory *Stoffwechsel* and the production of animal heat:

> The physiologists of the modern iatrochemical school assume that a kind of fermentative or combustive process takes place in living organs whereby the comburent principle (the oxygen) from the arterial blood conveyed to them combines with the organic compounds of the animals and transposes them into a kind of combustion. The constitution of the excreta appears indeed to indicate a process taking place in the organs through which the organic compounds of the higher or more complex kind are transformed into lower or less complex, even into inorganic compounds. . . . This course of events appears to consist in a distinctive [*eigenthümlich*] process similar to combustion. And as one of its results must also be regarded the production of animal heat, which in animals is precisely proportional to the rapidity of the exchange of

matter. It is not impossible, finally, that an agency created in the nervous system has a great share in this process.[115]

Tiedemann later gave extensive consideration to organisms' dependence on "external agencies," or "stimuli," by which he meant circumstances such as temperature, atmospheric pressure, climate, and geographical zone; such agencies act either mechanically or chemically in soliciting the organism's "motions" and "manifestations of activity" (or "manifestations of force").[116] The latter are themselves expressions of the organism's intrinsic "organic activity," and are thus entirely unlike the simple communication of motion characteristic of inorganic bodies:

> The manifestations of life and the organic motions that occur as a result of external influences are thus essentially distinct from the communicated motions of lifeless bodies in that they stand in no merely mechanical or chemical relation to the occasioning causes, the stimuli. When we speak of organic motions as occasioned by stimuli, we do not thereby assume that they are the immediate effects of the mechanical or chemical impression, but that they are always effects of the forces of the organism which were merely called into activity by the external impression. The motions occasioned in living bodies by external objects are thus not communicated but rather absolute and autonomous motions for which those extermal objects only provide the occasion. That the excitations produced in organic bodies by external objects are products of organic activity [*Wirksamkeit*], or the forces of the organisms, is clear finally from the fact that they do not occur in dead bodies.[117]

Crucially new with Tiedemann, however, was his explicit rejection of the long tradition, exemplified above by Ferdinand Gottlob Gmelin, of regarding irritability as an otherwise irreducible property of certain living tissues. He consistently sought to identify a material basis—typically some form of material exchange—for each of the organism's vital activities; of irritability (*Reizbarkeit*) he wrote:

> We must not, however, conceive of it as an independently existing force that only adheres to the organisms and is objectively distinct from the substrate of their organic matter, one which were susceptible to a change—an increase or decrease—without at the same time a change thereby taking place in that substrate. We must rather regard it as a property grounded in the distinctive constitution [*Beschaffenheit*] of the organic matter and the organization, and entirely dependent on them, a property that manifests itself in the species of living beings in just as many different ways as their organic constitution is different.[118]

Although Tiedemann did not explicity invoke considerations of causality—in particular, the requirement that the physiologist identify a physicochemical cause commensurate with every expression of vital activity or force—the tendency of his work was clearly in that direction.

Johannes Müller, whose universally acclaimed *Handbook of Human Physiology* began appearing in 1833, expressed views similar to Tiedemann's, though not as forcefully or unambiguously, despite the fact that the overall thrust of his work was toward the physicochemical investigation of vital processes. Müller did not, for example, employ the term *Stoffwechsel* in a consistent or programmatic fashion, nor did he clearly assert the underlying concept as a fundamental principle of physiological explanation. Significantly, he posed at the outset as the first question physiology has to answer: "Are the bodies that exhibit the phenomena of life different in their material composition from the inorganic bodies whose properties are investigated by physics and chemistry? And since the phenomena in both domains are so different, are the fundamental forces that produce them also different, or are the fundamental forces of organic life only modifications of the physical and chemical forces?"[119] Without anywhere giving a simple answer to this difficult question, Müller nevertheless went on to stress the *differences* between inorganic and organic compounds and the role of the organism in protecting the latter from decomposition. And as we will see in section 2.2 of this chapter, he was unwilling to give up a special "organic force" as the cause of typically vital processes, even as he believed that the working out of the effects of that cause could be understood in strictly physicochemical terms.

Here, however, we are concerned with what Müller had to say about the relationship of the organism's vital processes to externally derived physical and chemical agencies. Within *that* context, one can find any number of striking passages that place him squarely among those who took their model of physiological explanation from the experimental physical sciences. For example, on the role of external stimuli he wrote:

> In maintaining life, the external conditions necessary for life—heat, water, atmospheric air, and food—continually bring about material changes in the organic bodies in such a way that they combine with the organic bodies, while constituents of the organic bodies are in turn decomposed and expelled. These influences have been called *stimuli* or *vital stimuli*; they must nevertheless be distinguished from many other accidental stimuli that are not necessary for life, and the abiding image one should have is that these vital stimuli produce the phenomena of life through material changes, the exchange of ponderable and imponderable substances, in that they steadily maintain the mixture of humors (e.g., blood) necessary for life.[120]

In elaborating on the action of those stimuli on the body, Müller invoked powerful mechanical and chemical analogies for the nature of the ensuing bodily processes:

> These stimuli are, as it were, the external impulse for the motion [*Gang*] of the gearworks of the whole machine; however inappropriate the comparison with a mechanism may be, the organic force, which creates in the organic bodies the mechanism necessary for life, is nevertheless capable of no action without this external impulse and without continual material transformations with the help of the external so-called vital stimuli. Richerand has therefore not unjustly compared the manifestations of life to the phenomena of combustion and the flame. The phenomenon of fire lasts only as long as the combinations and separations necessary for combustion take place; oxygen combines with the burning body, heat is developed, and as long as oxygen and

combustible materials are supplied, the phenomena of fire continue. Far be it from me to make life dependent as on a combustion; I would only say that here, as there, certain continual combinations and decompositions of matter produce there the phenomena of combustion and the appearance of light, here the phenomena of the organic force; that vital stimuli are for organic bodies what the oxygen of the atmosphere and combustible material are for the phenomenon of fire, where one does not after all call oxygen the stimulus of the flame; and that, without at the same time considering the material changes thereby produced, without the continual new formation and expulsion of ponderable and imponderable substances, the name stimulus, or vital stimulus, is an empty, indeed a false concept.[121]

In contrasting plants, which are composed to a great extent of similar and independently functioning parts, to animals, which are composed of very different and mutually dependent organs, he similarly emphasized the mechanical nature of the interrelationship between the operations of brain, heart, and lungs and their joint dependence on processes of respiratory *Stoffwechsel*:

In animals, on the other hand, the reciprocal interaction between blood circulation, respiration, and nerves proves to be absolutely necessary for life. The nerves determine the respiratory motions, but the nerves do not act without blood that has respired, and blood does not flow to all the parts and thus to the nerves without the contraction of the heart, which in turn is dependent upon the bright red blood and the action of the nerves. Brain, heart, and lungs are therefore, as it were, the principal interlocking gears in the animal machine, which is set in motion by means of the exchange of matter in respiration.[122]

Despite these strong words, his subsequent discussion of the causes of animal motion was devoid of metabolic or energetic considerations.[123] And considering the force of Müller's imagery, the reader of his text might be surprised that, as we have already seen, he later rejected the sufficiency of the oxygen theory of animal heat. Here as with his rejection of the supposed propulsive force of the blood, Müller seems to have preferred to base his conclusions on empirical findings—typically contested though ostensibly decisive—rather than to invoke a universal physical or chemical principle. The other side of Müller's occasional diffidence—his reluctance to pursue a line of reasoning into its farther reaches—was an open-ended avoidance of dogmatism.[124] Müller's preference for an empirical or experimental answer to physiological questions highlights a widespread conception of science that had to be overcome, during the 1830s and early 1840s, before physiologists were comfortable appealing to general principles in order to decide issues on a more theoretical basis. Mayer, for one, showed no hesitancy in favoring a science based on univeral principles.

Burdach was another who interpreted vital functions (*das leibliche Leben*) in terms of "an uninterrupted exchange of matter [*Wechsel der Materie*]."[125] He differed from Tiedemann and Müller, however, not only in explicitly discussing relations of cause and effect—whereby he *distinguished* organic from inorganic systems in terms of the *lack of correspondence* between the kind and magnitude of the organic effect and its cause—but also in regarding the body's characteristic force of "excitability" as operating without the need for any accompanying physical changes. In that, he likened the body's response to stimuli to the mechanical excitation of the "dynamical phenomena" of electricity and magnetism in inanimate bodies, claiming that the latter process is likewise unaccompanied by any substantial physical changes, as it likewise exhibits a dissimilarity in kind between cause and effect. Excitability is the manifestation of a "force," but that meant for Burdach that it is an internal (and by implication not further analyzable) cause of motion.[126]

Writing in the same year as Burdach, Valentin held out the prospect that physiological explanations might become wholly physical, with organisms being regarded as very finely constructed and purposive machines:

> Physics—occupied for not yet a half century with the investigation of a specific number of general agencies of nature and concerned to find out their laws, mutual relations, and similarities—on the one hand dominates physiology because its norms, as fundamental rules for everything external, also embrace every object of every natural science; on the other hand, in view of its laws and relationships it can also be applied to the phenomena of life in the most manifold fashion either through the analogy of method or of fundamental principles or through actual application of its established truths. . . . Only the greatest misunderstanding of the essence of the organism could engender the insane idea that everything which occurs in living organic beings is foreign to the so-called physical laws of dead nature. . . . Without ignoring the autonomy of life, the pure mechanics (sundered, to be sure, from the true vital processes) of the circulation of the blood, of the conduction of the nerves, of digestion, of motion, etc., will show itself to be determined according to the same fundamental principles that physics teaches for its domain of phenomena, and that we apply practically in our machines and instruments. Only we shall find all of the arrangements [*Einrichtungen*] and all of the material substances employed on their behalf to be much more purposive than we are capable of imitating or constructing by art.[127]

Valentin reasserted and elaborated such views throughout the 1840s.[128] The similarity between organisms and machines which he—a student of Liebig's writings—most stressed was that both need to be supplied externally with some form of alimentation (*Speisung*) in order to continue to function:

> The expansion of a spring, the heaviness [*Schwere*] of a weight that maintains a clockwork in motion, the flow of water that drives a mill, constitute physical alimentary moments [*physikalische Speisungsmomente*], while chemical alimentary moments exist, for example, in the combustible material with which we heat a stove and in the burning hydrocarbon gas with which we illuminate; on the other hand, both physical and chemical moments are simultaneously present as nourishing agencies in steam engines. Organisms have the same need for alimentation, one which is here predominantly chemical and which for plants is supplied by the atmosphere and the assimilable materials furnished by the earth and water, for animals and human beings by the

> various foodstuffs and assimilation products. To the extent that the motion [*Gang*] of the living clockwork—if I may so put it—is thus made possible by means of this assimilation, there occurs with every manifestation of force a consumption of matter or agency proportional to it, just as this is the case with our machines and artificial apparatus.[129]

To be sure, there were also important differences: the devices we build are made of materials as stable and unchangeable as possible, whereas organic matter is changeable and subject to ready decomposition, and machines cannot repair themselves or replace their own parts, as can living organisms.[130] But if we knew how nature solved these problems of stability and transformation, we would have "a complete insight into the overall mechanism [*Getriebe*] of organisms."[131] Denying the possibility of *generatio aequivoca*—that organic beings can now be created out of inorganic matter—Valentin believed that the phenomena of life are an expression of the wonderfully contrived *organization* of organic beings deriving from an infinitely wise plan, from an original and continuously active "impulse" whose nature and whose first appearance lie beyond the reach of science.[132] Organisms' marvelous ability to build, repair, and reproduce themselves is thus a result of their organization, and does not require the application of forces beyond those recognized by physics.[133] Even organisms' capacity for self-preservation and propagation we can imagine to be solely the consequence of their organization.[134] It is only with respect to the mind and the nervous system that physicochemical explanations, in conjunction with an antecedently assumed order, are intrinsically inadequate: they imply the existence of an immortal and divinely created soul (or "spiritual principle").[135]

In addressing the question of "the fundamental force of organisms," Theodor Schwann coined the word "metabolic" in 1839 to refer to phenomena relating to the chemical transformations in and around cells.[136] In a work designed to be programmatic of a thoroughgoing physicochemical approach to the investigation of vital phenomena in both animals and plants, Schwann made only passing reference to problems of respiration and animal heat, but when he did it was as examples of physicochemical processes. Elaborating an analogy between the "metabolic force of cells" and galvanism by pointing out that the chemical changes produced by the galvanic pile are accompanied by corresponding changes in the pile itself, Schwann noted that the conditions for the occurrence of "metabolic phenomena" include an appropriate temperature and the presence of oxygen and carbonic acid, and that one result of the universal metabolic process is the development of heat.[137] Schwann coined a new term for an approach which was in the process of achieving dominance among his German contemporaries. It was the Frenchman Jean-Baptiste Dumas's lecture of 1841, however, which perhaps more than anything else first forcefully painted a picture of plants and animals as existing within a complex web of material *and* energetic exchanges: plants draw force from the sun and nutrients from their environment to create substances from which animals can in turn derive heat and force: "It is thus in absorbing without cease a true force, the light and heat emanating from the sun, that plants function, and that they produce this immense quantity of organized or organic matter, food destined for the consumption of the animal kingdom. . . . Then come the animals, consumers of matter and producers of heat and force, veritable combustion devices."[138] Dumas's lecture signaled the increasingly taken-for-granted character of a view of the organic world in terms of processes of physical and chemical transformations taking place both within each organism and among the ensemble of living things. Liebig, his contentious German rival, made such a worldview even more a matter of public debate during the early 1840s.

In his article on organic processes and the atmosphere, Liebig expounded a rich conception of chemical processes as they pertain to the two essential and characteristic functions of the animal organism: the production of heat and motion.[139] He invoked a general principle of causality—the relationship between a force and its effect—in order to link both of these phenomena to underlying processes of material exchange. His analogies were combustion, the production of motion and heat in steam engines, and the production of galvanic electricity:

> Similarly, therefore, as in the closed galvanic pile a certain something, which we call a current of electric matter, becomes perceptible to our senses through certain changes that an inorganic body, a metal, undergoes as a result of its contact with an acid, certain manifestations of motion and activity that we call life arise as a result of transformations and changes in material substances that earlier were parts of organisms.
>
> . . .
>
> Lack of a correct view of force and effect and of the connection of the phenomena of nature has led chemists to ascribe a portion of the heat generated in the animal organism to the effects of the nervous system. If one thereby excludes an exchange of matter as the condition for the effects of the nerves, then this means nothing else than to have the presence of a motion, the manifestation of an activity, come forth out of nothing. But no force, no activity can arise out of nothing.
>
> . . .
>
> One has, finally, made the observation that heat is generated by the contraction of the muscles, similarly as in a piece of rubber that one has stretched rapidly and lets contract again. One has gone so far as to ascribe a portion of the animal heat to the mechanical motions in the body, as if the motions themselves could arise without a certain expenditure of force that is consumed by these motions. Through what, however, one can here ask, is this force generated?
>
> Heat arises through burning carbon, through solution of a metal in an acid, through the combination of the two electricities, through absorption of light. In the same way heat arises when we rub two pieces of a solid against each other with a certain rapidity.
>
> We can produce a certain effect through a multitude of causes that are extremely different in their manifestations. In combustion and in the generation of electricity we have an exchange of matter, or, as with light and frictional heat, the transformation of an existing motion into a new one that acts upon our senses in a different fashion. We have a substrate, something given, which assumes the form of another substrate, in every case a force and an effect. We can produce all possible kinds of motions from the fire under a steam engine, and fire from a given quantity of motion.[140]

Liebig went on to connect his views with the conception of heat as motion, arguing the impossibility of creating even an allegedly "weightless" substance by means of mechanical action alone. His words imply the existence of a constant equivalence between heat and motion, and he came close to advancing a general and all-inclusive conception of force, one that would embrace the customary imponderables, mechanical forces, and the motion of ponderable matter. Without citing any evidence, he maintained the constancy of the heat produced by the combustion of given amounts of carbon and hydrogen under all circumstances, and forcefully reiterated his central contention that all vital functions depend on chemical processes of material exchange:

> One has only to recall that the most distinguished physicists accept the phenomena of heat solely as phenomena of motion just because the concept of the *generation* of a material substance, even a weightless one, is simply not reconcilable with its origination through mechanical causes, as through friction and motion.
>
> Everything granted, whatever part electrical and magnetic disturbances in the animal's body may play in the functions of its organs, the ultimate cause of all these activities is an exchange of matter, expressible in terms of a transition, taking place in a certain time, of the constituents of the food into compounds of oxygen . . . .
>
> Now it is simply impossible that a given quantity of carbon or hydrogen, whatever forms they may assume in the course of combustion, is capable of producing more heat than it affords when it is burned directly in oxygen gas or in the air.
>
> When we make a fire under a steam engine and use the force obtained to produce heat through friction, this heat can in no way ever be greater than the heat we needed to heat the boiler, and when we use the current in a galvanic pile to produce heat, this heat is in all circumstances not greater than what we can obtain through combustion of the zinc that dissolves in the acid.
>
> The contraction of muscles generates heat; the force needed for this manifests itself by means of the organs of motion, which receive it by means of an exchange of matter. The ultimate cause of the heat generated can of course only be this exchange of matter.
>
> An electric current arises through the solution of a metal in an acid; conducted through a wire it becomes a magnet, by means of which we are capable of producing various effects. The cause of all the phenomena so generated is magnetism, the cause of the magnetic effects we seek in the electric current, and the ultimate cause of the electric current we find in an exchange of matter, in a chemical action.
>
> There are various causes for the generation of force: a tensed spring, an air current, a falling mass of water, fire burning under a boiler, a metal that dissolves in an acid—one and the same effect can be produced by means of all these various causes. But in the animal body we recognize only *one* cause as the ultimate cause of all generation of force, and that is the reciprocal interaction exerted on one another by the constituents of the food and the oxygen of the air. The only known and ultimate cause of the vital activity in the animal as well as in the plant is a chemical process.[141]

The problem with interpreting Liebig's views is that this clearly advanced position of physicochemical reductionism was at odds with much else that he wrote not only in other works, but elsewhere in the *Animal Chemistry* itself (whose first section incorporated the article under discussion essentially verbatim). As Holmes pointed out, Liebig's claim here that chemical force is the ultimate cause of all phenomena occurring in an animal was contradicted by his invocation of a *vital* force in a later section of the book.[142] This contradiction was missed by Theodor Bischoff, then extraordinary professor of comparative and pathological anatomy at Heidelberg and one of Liebig's most faithful supporters, who read the work as asserting that a chemical process is the source of all organic motions. In approving the direction he believed Liebig was indicating for physiology—that of assimilating it methodologically to the other sciences—Bischoff gave Liebig's inchoate ideas a more chemically reductionist and causally more respectable interpretation than the text warranted.[143] Another reviewer who looked past the inconsistencies of Liebig's book to second what he took to be its insistence that all vital activities must necessarily depend on underlying processes of material exchange was the physiological chemist, Franz Simon, whose judgment reflects the self-evident quality of the emerging consensus: "To regard the heat generated as the effect of the nervous system, whereby an exchange of matter is thought to be wholly excluded, means to call forth an activity out of nothing. To regard the contraction of the muscles as a source of heat is equally inadmissible, for motion cannot occur without a certain expenditure of force, and the organs of motion obtain this force only through the exchange of matter." Chemist Heinrich Rose's review, however, made no mention at all of energetic considerations, and Otto Kohlrausch's book-length critique explicitly held open the possibility that certain organic activities might occur without any attendant chemical changes: "The action of gravity [*Schwerkraft*], the magnet, etc., as yet indicate no accompanying chemical processes. Thus nothing stands in the way of the assumption that in the living organism, too, activities *may* exist that are neither caused nor accompanied by chemical action; that these activities might *possibly* produce heat is likewise not to be denied when we see heat liberated by causes that give no indication of accompanying chemical action.—Whether or not these possibilities are true cannot of course be decided a priori."[144] The significance of Liebig's causal analysis of the general conditions for the possibility of organic functions was not at all obvious to him. In light of this lack of universal appreciation for the energetic implications of *Stoffwechsel*, it is easy to see why Mayer might have been prompted to publish an elaborated version of his theory of force in 1845 as *Organic Motion in Its Connection with the Exchange of Matter*. The disputed issues could now be decided as a matter of principle on the basis of his theory of force.

### 2.2 *Vital Forces and the Soul: The Organism's Internal Sources of Activity*

During the first four decades of the nineteenth century, German life scientists and organic chemists routinely discussed the role played by specifically *vital* forces—that is, those not reducible to the physical and chemical forces of inorganic nature—in the functioning of the animal organism. In addition, the physiologists among them regularly addressed the problem of the relationship of the vital force to the mind or soul—understood as the source of voluntary motions and the seat of consciousness—and the latter's relationship to physical processes.[145] Until the end of the 1830s one encounters relatively little explicit criticism of vital forces. Quite the contrary, a vital force was typically invoked to explain processes like embryological development, growth and regeneration, and the body's ability to synthesize complex organic compounds and to resist destructive external forces. As already mentioned, animal heat was *not* one of the phenomena commonly accounted for in this way. In general, although it is not quite true to say that the vital force was employed *only* to explain phenomena of control and direction and not also those *we* would say involve doing work, this nevertheless seems to have been the overall tendency, at least until Liebig's work of the early 1840s. Of course the nervous system was widely held to be able to produce organic motions, but the nervous system itself was typically placed under the control of the mind or soul, not the more 'vegetative' *Lebenskraft*,[146] and physiologists only slowly became sensitive to all-embracing energetic considerations.

The context of debate over these issues was to change radically in the course of the 1840s: on the one hand, the growing consensus that all vital processes should be explained in terms of material exchange and the normal laws of chemistry and physics cut all explanatory ground out from under both vital forces and the mind or soul; on the other hand, physiologists came to regard themselves as no longer responsible for the subjective experience of will and consciousness or for the possible role of mind or soul in organic functions, nor were *physiologists* (in the modern sense of the term) the ones who were most responsible for dealing with those problems, such as generation and development, that seemed most to require the assumption of a creative goal-directed agency. In addition, from the late 1830s on there was increasing methodological criticism of the vital force as a meaningless term that explains nothing but only covers one's ignorance. As we shall see, Lotze's famous *Lebenskraft* article of 1843 was a landmark in the evolution of this debate. Increasingly out of step with his contemporaries, Liebig became the most outspoken, if self-contradictory, advocate for an active vital force during the 1840s. Indeed, Mayer's *Organic Motion in Its Connection with the Exchange of Matter* of 1845 was in large part a response to Liebig's treatment of the vital force. Historians' assessment of the scientific status of the vital force in Germany during the 1830s and 1840s has, moreover, often been distorted by its being viewed from the perspective of the conservation of energy: the subsequent conviction that "force" is universally conserved triggered a general reaction against *all* special vital forces and obscured both their prior acceptability (to some) and the motives for their criticism (by others).[147] Let us first look more closely at the ways in which specific investigators dealt with these issues during the preceding decades.

Mayer's professor, Johann Heinrich Ferdinand Autenrieth, devoted a special section of his highly regarded *Handbook of Empirical Human Physiology* to a discussion of the "Relationship of the Soul to the Life of the Body."[148] His principal aim was to distinguish the soul, "the originally active force" in the body, from the vital force, "the force by means of which the body or its parts, provoked by external stimulus, produces vital motions."[149] He based his distinction on several well-known classes of phenomena. For example, some parts removed from the body continue to react to stimuli, whereas the soul is not affected by the loss of limbs; and cases of unconsciousness, where the person shows no signs of internal voluntary activity but where the normal bodily functions continue to be performed, show that the activity of the soul is not necessary for the performance of life-sustaining vital functions. Similarly, after an organism's death—that is, after the escape (*Entweichen*) of its soul—some of its parts often continue to manifest vital activity—that is, the presence of vital force.[150] Autenrieth likened the soul to an internal stimulus, capable of provoking the body into action but not the source of the resultant activity, which he assigned rather to the vital force with which the living parts are endowed. The soul is thus the source of voluntary motions, while involuntary motion was regarded as a "merely mechanical arrangement set into action by the vital force."[151] The soul's independence from the body and from the body's vital forces is demonstrated by both the irregularity of its actions and our internal sense of freedom of action.[152] Considering the body's various vital activities and the principal chemical transformations that take place in their course, Autenrieth asked: "Which is the first force that gives rise to these changes from the rest state of the animal body?" The answer was neither heat nor any of the other imponderable substances, nor any specific chemical reactions, nor food or drink, nor (of course) the independent soul. The necessary condition for the body's activity is the presence of vital force, passed on from generation to generation since the original creation of the species and capable of self-multiplication.[153]

Autenrieth subscribed to essentially these same views for his entire adult life and gave them ample exposition in his *Views on the World of Nature and the Life of the Soul*, published posthumously in 1836.[154] In that work, which Carus reported had "attracted widespread attention,"[155] Autenrieth reemphasized the independence of the directive and voluntary soul from the physical substance of the body, and now more explicitly interpreted vital force as the intermediary between the two, the means by which the soul acts on the body and through which alone the freely acting soul can make itself known.[156] Autenrieth insisted that the soul is connected to the material world only via the organic vital force; hence, even if (as may well be) spirits can exist independently of bodies, they could never make their presence felt *immediately* since they would lack any way of setting any part of the material world in motion. Thus he rejected as lies or deception all reports of spiritualist phenomena (*Geistererscheinungen*).[157] To be sure, the many involuntary vital processes that sustain life do not require the participation of the soul but are the results of the vital force acting in accordance with its proper and necessary laws.

It is thus the vital force—now identified as irritability and associated with both instinct and the formative force (*Bildungstrieb*)—which is in one way or another responsible for producing organic motions. Autenrieth did not ignore the organism's need to exchange matter with the external world—nutrients coming in and excreta going out—in order to keep functioning, but he did deny that this exchange is the source of the vital force expended in perfoming vital functions. Indeed, the vital force of an organ, once exhausted, is capable of regenerating itself without external assistance merely through rest.[158] As will be discussed in section 3.1 of this chapter, Autenrieth conceived of a "calling forth" (*Hervorrufen*) of the vital force from an invisible realm beyond the physical world.

In support of his contention that the action of the vital force is not due to any antecedent chemical changes, Autenrieth cited experimental results published by Berlin physicist Paul Erman in 1812. According to Autenrieth, Erman's experiments show that even very rapid motions can take place in an organ without any corresponding change in its ponderable matter. From this Autenrieth concluded that even in cases where such a change does take place, it is not the material change itself which generates the motion. Rather, "that which generates the motion simultaneously also occasions the increase or decrease in the ponderable matter in the individual organ of the body," and hence "we must therefore ascribe the vital motions in the first instance to a purely autonomous but involuntary force, which nevertheless is no longer a mere attribute of ponderable animal matter or a consequence of its mechanism, and which we therefore in general can call organic vital force, as opposed to our free volitive force."[159] In fact, Autenrieth seriously misunderstood Erman's results as having failed to detect any *chemical* changes in electrically stimulated muscles, when in fact Erman attempted no such determination but rather observed only the temporary decrease in the contracting muscle's *volume*.

Questions concerning the source from which new vital force comes into the world also arose in connection with generation. Autenrieth denied that a child's vital force comes to it at the expense of its parents'—a supposition easily refuted by (among other things) cases of numerous offspring—and concluded: "But if the child's share of the vital force common to all living things is really a new one, one which entered into the world of corporeal phenomena only at the child's formation, then we must also conclude that every increase in the vital force during the life of the individual comes to it in general from the same invisible source."[160] As further evidence "for a continuous stream of creation from beyond the visible world," he cited plants growing where no seeds had been present, Joseph Priestley's famous "green matter" (the alga, *Conferva fontinalis*), spontaneous generation, and the equal number of births of girls and boys.[161] Likewise demonstrating the independent presence everywhere of the same "universal dynamical source of organic life" was the existence of series of closely corresponding animal types in different climates and distant regions of the globe.[162]

Although the vital force possesses internal activity, it nevertheless acts "according to laws of necessity, albeit those proper to itself," i.e., not the laws of dead matter.[163] And although vital force can express itself only in the presence of ponderable matter, Autenrieth was at pains to deny the idea "that life is merely a *property* of a body which constructs itself in a determinate manner."[164] As evidence of the independence of life from its material substrate, he cited the reappearance in the geological record of beings similar to those wiped out by an intervening catastrophe and the revivification both of frozen fish and frogs and of dried-out rotifers like Lamarck's *Furcularia rediviva*.[165]

In addition to the lawbound vital force, however, human beings possess an immaterial *geistig* principle which is, like the vital force, the product of no merely mechanical arrangement of material parts, but which, unlike the vital force, is bound by no necessity. That principle, which manifests itself in our freedom of will, must necessarily be independent of both the body and the vital force: "For if freedom were to arise out of corporeal necessity, then it would arise out of something, absolute nonfreedom, which with regard to freedom would thus be a true nothing. But origination [*Entstehung*] out of nothing presupposes a creative force that precedes everything determinant, something that thus contradicts the concept of necessity."[166] Once again the question arises of where this principle—the soul or ego—comes from at the generation of a new being. It is not the product of the body, and neither parent is conscious of a division or diminution of *its* ego. Autenrieth concluded that it must have existed before it was "invested with a body."[167] It comes into this world from the same preterphysical realm whence come the vital force and the imponderables. Although his consideration of the soul and its relationship to the vital force within a physiological context was common in his day, Autenrieth's ideas concerning the complete self-sufficiency of the vital force and the causal independence of organic motions from any underlying physicochemical transformations were a striking anomaly by the mid-1830s. Indeed, Carus not unjustly characterized Autenrieth's views as belonging to the philosophy of the eighteenth century.[168]

We have already encountered Friedrich Tiedemann as one of the strongest advocates of a physiology based on the centrality of processes of internal and external material exchange for the understanding of organic processes. He was not, however, a physicochemical reductionist, and he believed that, by identifying and circumscribing the purely chemical and physical processes taking place in an organism, one could identify other processes peculiar to living things. As he reasoned in the introduction to his *Human Physiology* of 1830:

> The important question, whether the phenomena observable in living bodies are grounded in forces of a distinctive nature, in organic forces, or whether they are not rather to be regarded merely as effects of forces that belong to all bodies, even the lifeless, we can only answer on the basis of a future comparison between the phenomena and properties of living bodies and those of inorganic bodies. But since we hereby encounter manifestations of activity which are distinctive of living bodies and which we cannot regard as effects of general physical forces, this then justifies us in the assumption of vital forces [*vitale Kräfte*] or vital properties of their own. . . . For all essentially different phenomena we must assume particular causes and forces as long as we are unable to eliminate the distinction between the phenomena and reduce them to one another. The endeavor of physiologists must ultimately be aimed principally at discovering the conditions and laws according to which the forces manifest their activity and at investigating the relationship of the forces to each other, or their reciprocal interaction, and their mutual dependency.[169]

What the essence of force or matter might be is beyond the scope of natural science to discover—as is the nature of the close relationship of the soul to the body.[170]

In the event, Tiedemann identified the need for such special organic forces in the three principal traditional areas. One was with respect to the greater complexity of organic chemical compounds as compared with inorganic:

> There must accordingly be active in living bodies yet another force with its own nature besides the chemical affinities that determine the configuration of lifeless bodies, one by means of which is produced the specific diversity of organic forms having identical composition. Or, to express the same thing more precisely in other words, the configuration of organic bodies is not the effect merely of elective chemical attraction, as in lifeless bodies; it is the effect of a force with its own—or, if you will, higher—nature, which produces the configuration.[171]

A second area related to the organism's power of resisting destructive external influences. For Tiedemann, putrefaction and fermentation are made possible by the removal of the metamorphosing substances from the protective influence of a vital force. But unlike Liebig, who later saw them as purely chemical processes, Tiedemann placed them in a category of intermediate complexity, that of *organisch-chemische Processe*: "The chemical processes occurring after the extinguishing of the forces distinctive of living bodies—forces which during their existence opposed the chemical affinities of external objects working toward their destruction, [i.e.] putrefaction and fermentation, processes by means of which their composition and form are changed—are processes of their own kind [*Vorgänge eigener Art*] such as do not occur with the decomposition of inorganic bodies. They are organic-chemical processes."[172] A third area emerged in the course of his consideration of processes of generation, development, and growth. Not surprisingly, he found in them the need to assume the operation of a special goal-directed *Bildungskraft*:

> The principal result of the investigations undertaken is that, to judge from its effects, the formative force inherent in all living bodies—a force which acts in an internal purposeful fashion, which propagates itself from generation to generation, and which manifests itself in the generated beings under certain external conditions and circumstances—must be regarded as an autonomous force different from all other known forces. As far as experience indicates, it manifests itself only in organic substances, and in bodies composed of them, which are different from inorganic bodies in their composition, configuration, and construction as well as in their properties. This force thus appears to be a force inherent in organic substances and grounded in their distinctive material constitution, without our being able to pronounce further over its ultimate ground and the manner of its action, as is also the case with other forces.[173]

Nevertheless, despite Tiedemann's acceptance of special organic forces, he was careful not to invoke a (single) vital force, a *Lebenskraft*, as the cause of those peculiar organic processes. Although he did not discuss the issues explicity in these terms, in fact his vital forces were responsible for the direction of only a few typical and traditionally so-distinguished organic processes, while he consistently looked to explain the working out of those processes and the general course of all the other organic processes associated with the maintenance of life in terms of the exchange of matter. His attitude toward the commonly assumed vital force came out in his consideration "of the causes and forces that produce the motions of living bodies":

The motions of living bodies can be regarded as neither the effects of gravitation nor as produced by an externally communicated mechanical impulse. Just as little have they been explained up to now in a satisfactory manner in terms of the attraction and repulsion manifesting themselves in the interplay of chemical affinities. Distinguished scientists and physicians therefore regard them as autonomous motions produced by forces of a special kind grounded in their distinctive constitution. But if we ask, what are these forces, under what conditions and according to what laws do they act, we then encounter a subject over which the most diverse opinions and theories have been advanced, and the most violent controversies carried out.[174]

In his long historical sketch of the controversy, Tiedemann criticized Stahl's attempt to identify the soul as the source and cause of organic motions. Tiedemann's understanding of the physiological function of the sensing and imagining soul, whose reality he never questioned, was similar to Autenrieth's:

Against this theory [of Stahl's], which would have the animal body as such, its organization, and its material substrate to be entirely inactive and without force, one can raise serious objections. Stahl erroneously takes the soul, as one of the causes that occasion motions, to be identical with the forces that produce motions. Although it cannot be denied that animals decide autonomously upon certain motions by means of the sensitive and imaginative principle, it nevertheless does not follow from this that the soul also brings those motions to completion, and that the impulse to all motions is given by the soul. We may only regard the latter as a cause by means of which motive forces are set into action without, however, taking it to be the very force executing the motions. That the two are essentially different follows from the fact that the muscles, the heart, and all parts provided with muscles, [such as] the stomach and intestines, when removed from the body still continue their motions for a period of time if they are stimulated. If the motive force were here one with the soul, then one would have to assume a divisibility of the latter with the separation of the body, which contradicts the fundamental conception of the unity and indivisibility of the soul.[175]

Thus Tiedemann explicity rejected the conception of the soul or vital force as a motive force capable by itself of producing organic motions while granting it the ability to *direct* the expression of the motive force possessed (presumably in chemical form) by the muscles. Tiedemann further criticized others' reliance on a vital force to explain organic motions:

On the basis of their investigations of the properties of plants and animals and the comparison of them with the properties of lifeless bodies, several scientists gained the conviction that the motions and all the other manifestations of force of organic bodies are different in their causes and mode of action from those of inorganic bodies. They regarded as inadmissible every attempt to want to explain them in terms of the general physical forces. Nor were they satisfied by the forces proposed by the above-named physiologists, by the soul, Glisson's irritability, or Haller's irritability and nerve-force. They therefore believed they had to resort to the assumption of one fundamental force productive of all the manifestations of life, a force that manifested its activity in both plants and animals, which they called vital force or vital principle.

This one fundamental force inherent in organic bodies was supposed to produce all the phenomena of life by manifesting its activity in various ways and in various directions.

. . .

Zealous defenders of such a vital principle were Barthez, Fryer, Blumenbach, Hufeland, Springel, Brandis, Rosse, and others. By positing the vital force they believed they had satisfied the requirement of reason to bring unity into the multiplicity of the vital phenomena. Looked at closely, this simple fundamental force is only an occult quality of whose existence and mode of action—as well as of the reason why it manifests itself now as a formative, now a motive, now a sensible principle—those physiologists have given no account.[176]

Thus it was not the assumption of special organic forces per se that Tiedemann objected to, but the hypostasizing of a single independent energetic agency, the vital force, as a causal explanation of complex organic processes. Such pseudo-causal explanations have no place in science.

Although less critical and elaborated than Tiedemann's, Volkmann's handling of vital forces emphasized in an analogous fashion the need to assume some kind of internal self-active directive agency in order to be able to account not only for the body's power of resisting decay, but also for the gross lack of proportion between external physical stimuli and the body's complex responses.[177] In practice, however, a vital force as such did not feature prominently in his work. Volkmann also granted the existence of a self-conscious, sensitive, and volitive soul without making it the agency of appreciable physiological processes.[178] That raised the question, however, of where the souls of newly engendered individuals come from. His answer—that they derive from the parents' souls—is of less interest than the terms in which he considered the matter:

Either the divinity itself creates the soul through immediate intervention—a way of looking at things foreign to natural science—or the parents create it. If the latter is the case, then it would be against all analogies to think of a creation out of nothing, because as far as we can see, creation appears as a metamorphosis, as a transformation of something already present into something new. That the soul of the fetus arises originally as the result of an effusion of the parental soul into the germ is clearly attested by the psychic correspondence between parents and children.[179]

Thus the question of the origin of the soul, in a physiological context, led to the denial of creation *ex nihilo*—even if, to be sure, Volkmann ignored the quantitative problem raised by the implied *Vertheilung* of the soul.

Two works of physiology in Mayer's library shared Tiedemann's judgment that, even if the phenomena of organic life seem to justify the assumption of special organic forces, we must nonetheless not think that by bestowing a name on a class of phenomena we have thereby *explained* them. Accordingly, Karl Asmund Rudolphi, professor of anatomy and physiology at the University of Berlin since its founding in 1810, approved guardedly of the assumption of a vital force as the convenient designation for the unknown cause of life, but rejected it in no uncertain terms as a putative explanation or as something (*ein Etwas*) superadded to the organism and thus animating it; as he added, "To assume *several vital forces* instead of one does not improve the matter."[180] Ferdinand Gottlob Gmelin, professor of medicine and natural history at Tübingen, expressed a cognate if somewhat more antiquated opinion, recalling an older faculties-based physiology, in the same year (1821): "Since we do not yet know the immediate cause [*Grund*] of these different kinds of excitation . . . we therefore attribute forces to them, with which we by no means aim at an explanation, but merely wish to have an expression for something unknown, and [we thus] call them sensibility, irritability (vital turgor), formative force, or vegetative force." In a footnote to this passage Gmelin added that "we recognize that they all may be comprehended by one ultimate principle, which we call vital force."[181]

Similar views were advanced nearly twenty years later by the reform-minded Tübingen *Privatdozent*, Albert Friedrich Schill, who associated the vital force conceptually with other basic forces of nature:

> We notice everywhere in nature two intimately connected things, matter and activity; there is no activity without matter, nor matter without activity. . . .
>
> We do not know the internal cause [*Grund*] of activity, [just as] the principle of life is unknown to us. We call this X 'vital force' and thus understand thereby the unknown internal basis of vital activity.
>
> Physiology does this with the same right by which physics has called the unknown reason [*Grund*] why bodies are heavy 'gravity' [*Schwerkraft*] and chemistry has called the unknown cause of chemical combinations 'chemical force' or 'affinity.'
>
> The vital force is connected with every part of organized matter; in none does it have its exclusive seat. The nervous force, sensibility, is only a part; it is not identical to the vital force, for an acephalus is otherwise often well developed.
>
> The vital force cannot exist without organized matter, in the intimate linking of the two lies the basis of life. Each kind of activity of the vital force demands in turn a different organization, thence the diversity in the construction of the organs. Both always correspond to each other. Same activity, same organization.[182]

If, on the one hand, Schill's equation between activity and structure seems to evince the kind of reductionist tendencies that were soon to be prominent in German physiology, on the other hand his defense of the vital force in terms of its analogy with gravity and chemical affinity looked back to a mode of thinking that François Magendie had roundly criticized earlier in the decade.

Magendie had set himself a difficult task in his frequently reprinted and widely translated *Elementary Sketch of Physiology*: namely, to press as far as possible the application of physics and chemistry to the explanation of physiological processes while at the same time recognizing the existence of at least some distinctly vital phenomena. As he wrote in the preface, "The belief, so harmful and so absurd, that physical laws have no influence on living bodies does not have the same force any more; perceptive minds are just beginning to see that there could be different classes of phenomena in the living animal, and that simply physical acts do not exclude purely vital actions."[183] Nevertheless, Magendie utterly rejected the time-honored practice of covering one's ignorance with an empty phrase:

Since the earliest times one has seen that a large proportion of the phenomena particular to living bodies do not follow the same course, are not subject to the same laws, as the phenomena proper to inorganic bodies.

One has assigned a particular cause to the phenomena of living bodies. This cause has received different designations. Hippocrates called it φύσις (nature), Aristotle, *motive and generative principle*; Kaw Boërhaave, *impetum faciens*; van Helmont, *archea*; Stahl, *soul*; others, *vis insita, vis vitae,* vital principle, *vital force,* etc.

What do all these expressions mean? One can take two very different courses of action with regard to them: to reify them, to make of them beings to which belongs the power of producing the vital phenomena, is the first; but in following this one, would we not resemble those savages who, after having coarsely sculpted a rock, make a god out of it? The second course of action consists in recognizing that these words designate the unknown cause or causes—perhaps forever incomprehensible—of the actions of life; in that case one must agree that science hardly profited when they were invented.

Of all the illusions into which some modern physiologists have fallen, one of the most deplorable is to have believed, by coining a word *vital principle* or *vital force,* to have done something analogous to the discovery of universal gravity [*pesanteur*].

Just as attraction presides over the changes of state of inert bodies, so (they say) does the vital force direct the modifications of organized bodies; but they fall into a grave error, because the vital force cannot be compared with attraction; the laws of the latter are known, those of the vital force are unknown. At the present time physiology is at precisely the point where the physical sciences were before Newton: it awaits an intellect of the first order to come and discover the laws of the vital force in the same way Newton made known the laws of attraction.[184]

Perhaps Mayer fancied himself to be that Newton of physiology, someone whose ideas were founded on *valid* analogies.

Yet despite Magendie's apparent dismissal of the concept of vital force, he immediately introduced a vital action (*action vitale*) to which, alongside nutrition, he claimed it is possible to reduce all organic phenomena. The vital action is responsible for the production of bile and urine in the liver and kidneys and for the contraction of muscles; it was conceived of as a kind of hidden motion of (presumably) the material parts of the affected organ. Except that Magendie implicitly advanced a materialistic ontology, his vital action performed about the same functions his German colleagues assigned to the *Lebenskraft*. In fact, however, such terminological skirmishing was of little importance to the physiological work Magendie proposed to do:

All the phenomena of life can therefore in the last analysis be connected with nutrition and the vital action; but the hidden motions that constitute these two phenomena do not fall under our senses, and it is not to them that our attention should be directed; we should limit ourselves to studying their results—i.e., the physical properties of the organs, the sensible effects of the vital actions—and to investigating how they both contribute to the general life.

That, in effect, is the object of physiology.[185]

By the late 1830s, Magendie was speaking out more unambiguously and more confidently for a nonvitalist physiology. In his *Lectures on the Blood*, delivered from December 1837 to April 1838, he told his students the first day that "among these prejudices [which hinder the progress of science] there is one in particular against which I have protested since I have been engaged in teaching"—namely, "to believe that there is nothing in common between what takes place in living bodies and what happens in inert matter."[186] In his second lecture he pointed out that "it is of the greatest importance to properly determine the nature of the phenomena we wish to study," and asserted that, with regard to physiology, "our experiments have demonstrated that such and such actions that one attributed to the life force [*faisait dépendre de la vitalité*] were purely physical. These clearly demonstrated facts have scandalized many an avowed partisan of the marvelous and have provoked many an outcry; but one is beginning to get used to them."[187]

Given Magendie's stature and the wide availability of his works, it is quite likely that Mayer encountered his views urging at first the desirability, later the sufficiency of applying physicochemical principles to the explanation of the phenomena of organic life. Those views might have been reinforced by Mayer's professor of physiology and pathology, Carl Ludwig Elsässer, who published an annotated translation of Magendie's *Sketch* in Tübingen in 1834 and 1836. Unlike Autenrieth, Elsässer would have nothing to do with some kind of unitary immaterial agency diffused throughout the body and responsible for its vital phenomena. He seconded Magendie's programmatic call for the farthest possible extension of explanations based on the general forces of nature, conceding only to the mind the necessity of special immaterial forces. In his remarks to the above-quoted passage from Magendie on the "Causes of the Vital Phenomena," Elsässer wrote:

> The expression *vital force* may only be used as a name by which one designates the totality of the still unknown causes of the vital phenomena. In this sense the word is harmless, but of course is also neither here nor there [*gleichgültig*]. But if one presupposes with its use that it is only *one* force that produces the various vital phenomena, then it is objectionable, because one thereby proceeds from a highly improbable hypothesis, namely from the assumption of a particular, simple entity [*Wesen*], foreign to matter, which is distributed throughout the entire body and which brings about the various vital phenomena by acting on the intrinsically dead matter.
>
> . . .
>
> The assumption of a single vital force and of a vital principle foreign to matter is accordingly unproven, unnecessary, and even harmful, in that it declares as idle from the start all endeavors to search for the effects [or actions (*Wirkungen*)] of the general forces of nature in the sphere of life as well. . . . We have to ascribe only the psychic activities to forces that do not belong to matter, although their existence and extension is restricted to a certain size and constitution of the brain.[188]

As I will later argue, such ontological reflections on the nature and relationship of mind, force, and matter, whereby forces were typically understood to be properties of matter, probably played a central role in Mayer's earliest reflections about physiologically produced heat and work. Unlike nearly everyone else's, Mayer's forces—not imponderables, but forces—were entities *sui generis* independent of ponderable matter.

Having thus sampled German physiologists' views on the role and status of the vital force during the 1830s, we still have to examine the writings of Johannes Müller, the most prominent and influential among them. The several parts and editions of his two-volume *Handbook of Human Physiology* appeared between 1833 and 1844, spanning the crucial years of this continuing debate. Yet they provide a poor indication of the character of that debate, and Müller's own views are not always easy to extract from his often allusive and inconclusive expositions. In part, I think, this situation reflects the fact that Müller's quasi-Platonic mode of thought was coming into increasing disfavor, and he chose not to mount too aggressive a counterattack. Then, too, Müller was sensitive to the kinds of questions that belonged to natural science, and he wished to address himself to empirically grounded natural scientists. So little did he forcefully engage the issues that the first through fourth editions of the important first part of his text were essentially unchanged in all relevant particulars. Even his terminology suggests both his diffidence and his reserve vis-à-vis contemporary thinking. He only rarely used the common term "vital force," preferring for the most part to speak of the "organic" or "organizing" force, but also using as apparent synonyms a plethora of alternative words: activity, rational creative force, soul, *primum movens*, unconsciously working purposive activity, *vis essentialis*, idea, motive idea, and (a late favorite) vital principle.[189]

Müller first engaged the subject in connection with his denial of spontaneous generation (*generatio aequivoca*): organic matter never arises by itself, and only plants possess the capability of transforming inorganic matter into organic. His appended comments contained the germ of much that he would later expand on:

> Now, how organic beings have come into existence in the first place—in what way a force absolutely necessary for the formation and preservation of organic matter, but which, on the other hand, also manifests itself only in organic substances [*Materien*], has come to matter [*zur Materie gekommen ist*]—lies outside all experience and knowledge. Nor can the knot be cut by maintaining that the organic force resides in matter from eternity, as if organic force and organic matter were only different ways of regarding the same object; for organic phenomena are in fact proper to only a certain combination of elements, and even organic matter capable of life breaks down into inorganic compounds as soon as the cause of the organic phenomena, the vital force, ceases. In any event, the solution to that problem would not at all be the task of empirical physiology but of philosophy. Since conviction in philosophy and in the natural sciences has a wholly different basis, we are constrained here first of all not to leave the field of reflective experience [*eine denkende Erfahrung*]. We must therefore content ourselves with knowing that the forces that make organic bodies alive are distinctive, and then with investigating their properties more closely.[190]

We thus encounter the organic force as responsible for the formation and preservation of organisms and organic matter—typical attributes, to be sure.[191] In the event, however, when it came to the understanding of any particular vital function, Müller never contented himself with invoking the powers of this organic force, but rather sought either a phenomenological description or physicochemical explanation of it, reflecting his conviction that the satisfactory elucidation of a concept like vital force lies outside natural science.

We glimpse, too, Müller's conception of the vital force as an entity independent of matter, somehow joined to it to produce living organisms. He and Autenrieth, whom he frequently cited in a variety of contexts, were among the few to assert the possible independence of force from matter. For Müller the term "force" sometimes included the imponderables, and he exploited at least one important analogy between vital and physical forces, though at other times he was concerned to stress their differences, not their similarities.[192] The increasingly popular notion that the erstwhile vital force was nothing more than the expression of a particular arrangement of the smallest parts of matter was alien to Müller's worldview. As we shall see, the upshot of his reflections on the relationship between life and nonlife was not to take the laws of inorganic nature as adequate to explain vital phenomena, but rather to infuse all of nature with a kind of latent or potential life.

Immediately after the passage just quoted, Müller began a new section entitled "On the Organism and on Life," which revealed the Kantian origins of some of his conceptions of life. Organic bodies are distinguished from inorganic not only by the manner of their composition out of the elements, but also by the purposefulness of their activity in accordance with the laws of a rational plan.[193] For Müller the purposive behavior of the "organizing force" is what keeps it from being reducible to the mere arrangement of material parts, though some of his imagery, removed from its context, makes him look like an unalloyed mechanist:

> Some have believed that life or the activity of organic bodies is only the result of the harmony, of (as it were) the interlocking of the gears of the machine, and that death is determined by a disturbance of this harmony. This harmony, this interlocking, obviously takes place; for the respiration in the lungs is the cause of the activity of the heart, and the motion of the heart brings to the brain at every moment the blood that has been changed through respiration, whereby the brain animates all the other organs and again determines the respiratory motions. The external impulse for this mechanism [*Getriebe*] is, however, the atmospheric air in respiration. Any injury to one of these mainsprings in the mechanism [*Mechanismus*] of the organic body, every appreciable injury to the lungs, the heart, or the brain can become the cause of death, for which reason they have been called the *atria mortis*. But this harmony of the parts that are necessary for the whole does not exist without the influence of a force which is also at work throughout the whole and which does not depend on the individual parts, and this force exists earlier than the harmonious parts of the whole are present; they are created by the force of the germ only during the development of the embryo. With a purposefully constructed mechanism—e.g., a clock—the purposeful whole can exhibit an activity that proceeds from the working together of the individual parts, one that is set in motion by one cause; but organic beings exist not merely as a result of an accidental combination of their elements, they also generate the organs neces-

sary for the whole by means of their forces out of organic matter. This rational creative force manifests itself in every animal according to strict law, as the nature of every animal requires; it is already present in the germ, even before the later parts of the whole are separately present, and it is what really generates the parts that belong to the concept of the whole. The germ is the whole *in potentia*; during the development of the germ the integrating parts of the whole come into existence *in actu*.[194]

Although Müller did not explicitly say so, his conception of the organism in fact allowed the physiologist to analyze the life-sustaining operation of its organs in purely physicochemical terms. Only the *origin* of those organs and of the structure of the whole is beyond the physiologist's competency to deal with, though even there one can still provide a scientific *description* of developmental processes insofar as they follow fixed laws. Thus Müller's firmly held belief in a vital force did not necessarily entail a vitalistic explanation of most of the kinds of vital processes *we* consider to be properly physiological, that is, not including embryological development.[195]

With respect to the latter, Müller rejected the preformationist theory of *emboîtement* in favor of epigenesis, though without using either term. In reviewing the evidence for the gradual development of the embryo from an originally undifferentiated state, he associated his view of the organizing force with Stahl's *anima*—with one qualification:

If [Georg] Ernst Stahl had known these facts he would have been even more confirmed in his celebrated view that the rational soul is itself the *primum movens* of the organization, that it is itself the ultimate and only cause [*Grund*] of organic activity, that the soul purposefully constructs and maintains its body according to the laws of its operation [*Wirksamkeit*], and that the curing of diseases takes place through its organic activity. . . . Only Stahl went too far when he placed the conscious manifestations of the soul on an equal rank with the organizational force, which manifests itself purposefully but in accordance with blind necessity.[196]

That *Organisationskraft*, an "unconsciously working purposive activity," generates and animates the parts of the whole according to eternal laws and does not reside in any single organ.[197] Consciousness, on the other hand, which does not create any organic products, is itself a late creation of embryological development and is dependent on the integrity of the brain. Although plants thus lack consciousness, their material form results from the activity of the organizing force, which follows the species' eternal archetype (*Urbild*).

Müller's organizing force was thus (at least here) less the immediate cause of ongoing physiological functions than it was the original ideal creator of the organism, a form-determining idea distinct from the formed matter. Although he declined to press his reflections too far, it is clear where he saw his intellectual affinities to lie:

Organic being, organism, is the de facto [*factisch*] unity of organic creative force and organic matter. Whether the two have ever been separate, whether the creative archetypes, the eternal ideas of Plato, as he indicated in the *Timaeus*, have at some time

> come to matter [*zur Materie gelangt sind*] and from then on rejuvenate themselves in every animal and plant, is not an object of knowledge, but of unprovable myths and traditions that clearly enough indicate to us the limits of our mere consciousness [*unser blosses Bewusstseyn*].[198]

This "de facto unity of the organizing force and the organized matter" (as he also put it) would be easy to comprehend, he said, if the organizing force and all the vital phenomena were simply the result of a particular combination of the organism's material elements, of its particular form and composition (*Form und Mischung*).[199] But the insufficiency of form to determine the course of organic development is shown by the similarity in the initial embryonic form of widely divergent species, and the insufficiency of composition is shown by the material identity of an organism immediately before and after death. Hence one must assume the presence of something else in order to explain what distinguishes life from nonlife.[200] Here again Müller pleaded ignorance but defended his ignorance by appealing to the analogous ignorance of physical scientists: "Whether one is to think of this principle as imponderable matter or as force is just as uncertain as the same question concerning several important phenomena in physics, and physiology does not here lag behind the other sciences, for the properties of this principle are nearly as well known in the actions of the nerves as those of light, heat, and electricity in physics."[201] Nothing better reveals the protean nature of conceptions of the vital force than this transition, within two pages, from its being regarded as a quasi-Platonic idea to its being spoken of as if it were some kind of aetherial substance. Following now the latter path, Müller argued for the ability of this principle to act in and across space; following Autenrieth, he cited "the capability of animal parts whereby vital force is now withdrawn from, now communicated to them, and whereby the vital force often accumulates rapidly in an organ."[202] Yet despite this belief in a kind of mobility (*Beweglichkeit*) of the vital force, he had no hesitancy in dismissing animal magnetism as deception and superstition, evidence only of most doctors' incapacity to understand the nature of empirical scientific testing.[203]

Insofar as Müller's vital force was responsible not only for the organism's original creation but also for its subsequent growth and reproduction, he found himself obliged, like Autenrieth before him, to consider the problem of the creation and destruction of vital force and its relationship to the other forces of nature, the chemical and physical forces of the inorganic realm:

> Now, the organic force is multiplied during the growth and propagation of organic bodies, for from one being there come into existence many others, and from these again many others, while on the other hand the organic force of dying organic bodies appears to perish. But since the organic force does not merely pass (so to speak) from one individual to another, since (rather) a plant, after it has annually generated the germs of very many new producers of the same species, being still capable of the same production, can remain a producer, then the source of the increase in the organic force also appears to lie in the organization of new material substances [*Materien*]; and granting this, one would also have to ascribe to plants the ability—insofar as they

> form new organic substances out of inorganic elements [*Stoffe*] under the influence of light and heat—to increase the organic force from unknown causes in the external world, while animals would in turn also be able to generate organic force out of food under the influence of the vital stimuli, and to individualize it during propagation. Whether in the execution of the vital processes [*bei der Ausübung des Lebens*], besides the continuous decomposition of materials, organic force is also lost, and how it is lost, is entirely unknown. So much appears certain, however, that upon the death of organic bodies the organic force is again dissociated into its general natural causes, from which it appears to be regenerated by the plant. If one did not wish to grant the increase in organic force in the already present organic bodies from unknown sources in the external world, then one would have to assume that the apparently infinite multiplication of organic force during growth and propagation is merely an unfolding [*Evolution*] of germs enclosed one inside the other, or one would have to assume the inconceivable, that the division of the organic force that takes place during propagation does not weaken its intensity. But the fact would always remain that, upon the death of organic bodies, organic force continuously becomes inactive or dissociated into its general physical causes.[204]

Elsewhere Müller was more certain that the phenomena of weariness and exhaustion, at least, demonstrate "that the organic force is, as it were, consumed in the execution of the [vital] functions."[205]

Thus the organism requires a source from which to multiply and replenish its organic force, a source which apparently must be connected with the nutrients plants and animals take in from the external world. It would be quite wrong, however, to interpret Müller's cryptic reference to "unknown sources in the external world" as an allusion to the physicochemical forces of inorganic nature, as if the organic force were somehow a kind of higher transformation of them. Müller did not here spell out what his views were, and a later analogy could also be read as implying a causal connection between organic and inorganic forces, though the key (as we shall see presently) was in the implied latency of light and heat: "The capability of being determined [*bestimmt*] by external influences to manifestations of force is not proper to organic (in particular, animal) bodies alone. For example, many inorganic bodies develop light under certain conditions—e.g., through impact—or they develop heat. According to physicists, it is probable that the light or heat was previously bound in the bodies and became free as a result of the external influence."[206] It was not until the third and last installment of the second volume of his *Handbook* that Müller clarified his position: the intended force of his analogy was that the vital force, and perhaps even the conscious soul, must be considered to be latent in all matter, requiring only special circumstances—that is, organized matter—to become manifest. Thus the vital force was like the other forces of nature only with respect to the formal aspects of its behavior; it was not a modification of them.

Such clarification came only after an extended consideration of psychic life, or "the life of the soul."[207] Writing now in 1839, Müller felt called upon to clarify the sense in which he had spoken of the organism in mechanical terms six years earlier:

> At the beginning of this work the organism was compared to a system of parts that have been connected for the attainment of a certain purpose and whose operation depends on the undisturbed harmony of the members composing it. This comparison revealed a still greater difference than similarity. The organism is like a mechanical device [*mechanisches Kunstwerk*] in its systematic composition for the attainment of a certain purpose; but the organism generates in the germ the mechanism of the organs themselves, and [thus] propagates it. The action [*Wirken*] of organic bodies does not depend merely on the harmony of the organs; rather, the harmony itself is a result [*Wirkung*] of the organic bodies themselves, and every part of this whole has its basis [*Grund*] not in itself but in the cause of the whole. A mechanical device is created according to an idea in the mind of the craftsman, the purpose of his activity [*Wirkung*]. An idea also underlies every organism, and all organs are purposefully organized in accordance with this idea, but this idea is *outside* the machine, on the contrary *in* the organism, and here it acts [or creates (*schafft*)] with necessity and without intention. For the purposefully active [*wirksam*] cause of the organic bodies has no choice, and the realization of a single plan is its necessity; indeed for this active cause to act [*wirken*] purposefully and to act with necessity are one and the same.[208]

The "idea" associated with each separate species was, Müller asserted, unchangeable, and he renewed his denial of spontaneous generation.

The new task at hand was to distinguish what he now preferred to call the "vital principle" (usually *Lebensprincip*, sometimes *Princip des Lebens*) from the sensitive and imaginative soul of animals.[209] Both are present in extended matter and can (in a sense) be divided with the division of organic matter, but neither can be localized in the body nor is either composed of parts. In order to manifest itself, the vital principle requires only the participation of chemical forces, whereas the soul requires the presence of already organized matter and the organization of the brain. Both, finally, can be latent, and it was this property that now enabled Müller more explicitly to suggest an answer to one of the chief questions he had posed in the first installment of his book:

> And so there now finally arises the question: How is it possible that through the growth of an organic being a multiple of its organizing force is formed, and how is a capability of the psychic principle for division [*eine Theilungsfähigkeit des psychischen Princips*] to be understood in connection with this? Does it lie in the nature of the vital principle and the soul as a power [*Potenz*] that they cannot be diminished in force through apportionment [*Vertheilung*] to more matter and through division, or do more of those principles come into existence through the appropriation of more matter in a growing organism as well, such that these principles are already present, latent, in the food, but only manifest themselves in the matter they are in in organic beings?
>
> The last assumption necessarily entails a second one as well, that the principle of life and the soul are latent in all matter, for if animals can live merely from plants, so can plants increase the organic matter from inorganic substances, and without such a new formation of organic matter it would ultimately be entirely decomposed on account of the putrefying and burning of so many substances that do not enter into organic beings as food.

The empirical investigation of the relationship of the vital principle and the soul to organization and to matter does not go beyond this alternative. From here on the investigation distances itself from the domain of empirical physiology and passes over into that of hypothetical speculation and philosophy.[210]

Although Müller refused to ally himself with either of the two "cosmological systems" he was about to examine, he certainly introduced the first in a way that made it appear more problematical and less consonant with what we otherwise know about physical phenomena. In particular, he seemed more disposed to accept the possibility that vital force and soul exist in a latent form associated (somehow) with matter than that they are capable of indefinite division with undiminished strength. For my purposes, it is less important that Müller's own deepest beliefs are difficult to fathom than that one understand the kinds of issues raised concerning forces and their creation or destruction as background to Mayer's creation of a new concept of force.

The first cosmological system Müller considered was one he called the "hypothesis of motive ideas [*bewegende Ideen*] informing organic bodies as the cause of organization and of psychic life [*Seelenleben*]."[211] Although Müller did not explicitly call attention to the distinction, its chief strength lay in its ability to account for the form-determining, goal-directed, 'ideal' aspects of the vital (read: *organizing*) force. The second system, the "pantheistic view of the worldsoul and its relationship to matter," seems less well suited to an explanation of species' constancy of type than of the energetic aspects of the vital (read: *organic*) force, of organisms' material and energetic relationships with the external world.[212] Müller's terminological indecisiveness and his reluctance to declare explicitly for one or the other of the two theoretically possible philosophical alternatives thus both reflect the fact that his vital force was being asked to perform at least two quite different functions. An emerging distinction between the developmental and the energetic aspects of the vital force was still not fully self-conscious and explicit. According to Müller's first cosmological system,

Ideas of the divine mind [*göttlicher Geist*] have, to be sure, been realized in the whole ordered universe, but only in organic beings are such divine ideas at work which perpetually generate their kind and create for themselves out of matter the mechanism for the actions of organic bodies. The motive idea of an organic body is therefore an emanation of the divinity, an emanation that has lived in it and in its products since the Creation. This idea is the only thing that persists in organic bodies, for matter abandons it and new matter is ceaselessly subjected to this motive idea. Matter itself is without its inherent soul and life. Not even the power [*Potenz*] for these actions belongs intrinsically to matter. On the contrary, all phenomena of life and the soul that appear in the matter processed by organic bodies depend solely on the idea that governs the organization.[213]

Central to this viewpoint is the belief "that the soul is foreign to the physical body, is not a force thereof, is no kind of force at all of matter, and that the soul is only united with the body in the organic being."[214] Such a view strongly recalls

Müller's own Platonic notion of the "de facto unity of the organizing force and the organized matter." However, this first cosmological system cannot adequately deal with the problem of the multiplication of force—a problem with clear implications for conservation-of-force type considerations:

> But since life and the soul or the motive idea are not latent properties of all matter, the increase and division of organic beings—and with this hypothesis the same division of souls—cannot therefore be derived from the assimilation of matter through nutrition, and the multiplication of personally animated and ensouled beings must rather be explained on the basis of a property of the vital principle and the soul as a result of which—opposed and foreign to all behavior of [material] bodies—their force is not diminished and weakened through division to infinity. A property which is difficult for the understanding to conceive of.[215]

No sooner had Müller defined in general terms the essential features of the alternative "pantheistic view" than he called attention to its success in addressing just that problem:

> The theory opposed to the foregoing is that the principle of life inheres in all matter and so little has been superadded to matter that it is nothing but a force of matter itself, but one that manifests itself only for such and such a form under precise conditions and given a precise composition of matter and a precise structure. When it enters the organic body, matter encounters the conditions under which the principle of life latent in it must manifest itself in the precise form of that organic body. In this way the increase in organic force to a multiple through growth and its capability for division are then easily conceivable. But everything living that passes away loses merely the condition for the manifestation of life in the precise form [it had], and the ensouled matter capable of life returns again to the womb [*Schoos*] of nature.[216]

In order to account for the constancy of organic form, Müller appealed to Giordano Bruno's pantheistic worldsoul as everywhere active and form-determining: all things partake of the ubiquitous divinity, or Godhead (*Gottheit*), whose activity further allows organic beings to manifest the organic force that lies undetected in "the *so-called* dead matter."[217] Müller went on to treat the soul analogously to the vital force: both are latent in matter, simultaneously independent of it for their existence and dependent on it for their expression.

Müller's extensive discussions of the vital force, hedged and noncommittal as they usually were, gave no indication that many of his contemporaries were becoming increasingly critical of the whole notion. The anti-vital-force views expressed by Elsässer in a relatively obscure place in 1834 and given more expansive treatment by Vetter in an apparently little-noted talk published in 1837 were, two years thereafter, made a major topic of an important section of a highly regarded book by a major figure: the long philosophical excursus on the "Theory of Cells" in Theodor Schwann's *Microscopical Investigations of the Agreement in the Structure and Growth of Animals and Plants* of 1839.[218] (Schwann had been Müller's assistant in Berlin from 1834 to 1839.) To be sure, their motives were somewhat different. Elsässer was concerned with the nature of scientific explanation and the programmatic desirability of pursuing explanations in terms of the usual physico-chemical laws; he also rejected the assumption of forces alleged to exist independently of matter—excepting, of course, psychic activities (*geistige Thätigkeiten*). Vetter, on the other hand, was principally concerned with issues of physical causality and creation out of nothing. What Schwann most objected to was the notion that the vital force acts purposefully in the manner of a rational agent:

> The different views on the fundamental forces of the organism can be reduced to two, essentially different from each other. The first view is that every organism is underlain by a force [*daß jedem Organismus eine Kraft zu Grunde liegt*] that forms the organism in accordance with an idea present to it and joins together the molecules in such a way as is necessary for the attainment of certain purposes defined by this idea. Such a force would be essentially different from all of the forces of inorganic nature because only blind action [*ein blindes Wirken*] takes place in them. In inorganic nature a certain influence is necessarily followed by a certain qualitatively and quantitatively determinate change without consideration of a purpose. According to this view, however, the fundamental force of the organism, or the soul in Stahl's sense, would stand much closer to the immaterial principle endowed with consciousness that we must assume in human beings, in that it acts in accordance with a determinate individual purpose.
>
> The other view is that the fundamental forces of organisms are essentially in agreement with the forces of inorganic nature, in that they act blindly, entirely according to laws of necessity without regard to purpose, and in that they are forces that are assumed given with the existence of matter in the same way as the physical forces. . . . Purposefulness—even a high degree of individual purposefulness of every organism—can by no means be denied. Only according to this view, the basis [*Grund*] of this purposefulness does not lie in the fact that every organism is produced by an individual force acting in accordance with a purpose; it lies rather in the creation of matter with its blind forces by a rational being, wherein also lies the basis of the purposefulness in inorganic nature. . . .
>
> One can call the first view on the fundamental forces of organisms the *teleological*, the second the *physical* viewpoint.[219]

In Schwann's opinion, the life sciences should follow the example of physics, which has long since banned all teleological explanations such as the supposed *horror vacui*.

In opposing the "teleological viewpoint," which he associated with the assumption of the existence of a goal-directed force (or idea) as cause of the organism's vital activities, Schwann was quick to seize upon the problem of generation, of where the new force comes from:

> If one assumes that every organism is underlain by a force that forms it in accordance with an idea present to it [*nach einer ihr vorschwebenden Idee*], then a portion of this force can indeed also be contained in the egg during generation; only one must then ascribe to this portion of the original force, at the separation of the egg from the maternal body, the capability of being able to produce an organism similar to the one

> produced by the force of which it is only a portion, i.e., one must assume that this force is divisible to infinity and that each portion can nevertheless produce the same effects as the whole force.[220]

Not availing himself of Autenrieth's preterphysical realm or Müller's notion of latency, Schwann pressed home the simultaneously quantitative and causal objections to the notion of the apportionment (*Vertheilung*) of vital force which Volkmann had simply ignored. The "physical viewpoint," on the other hand, which Schwann advocated, faces no such problems: there are no forces peculiar to living things. Once again the consideration, in a broadly physiological context, of problems associated with belief in an efficient formative force led to conclusions strongly suggestive of some kind of intuitive conservation principle.

Schwann did not categorically reject the possibility that the explanation of organic phenomena might require the assumption of laws different from those of inorganic nature, insisting only, like Autenrieth before him, that any such laws must likewise operate with "blind necessity."[221] If we grant purposefulness to the original creation of things, then we can allow all subsequent processes to take place according to those necessary laws. However, teleological explanations are only to be allowed where the impossibility of a physical explanation has been demonstrated, presumably such as with respect to the original creation of both the inorganic and organic worlds.

One young scientist who took Schwann's program to heart was Mayer's close friend, the psychologist Wilhelm Griesinger, who in December 1842 urged Mayer to apply his ideas to physiology and thereby to contribute to the ongoing reformation of the life sciences toward which Griesinger and his cohort were working:

> Only I confess to you, wherever I see a possibility of wresting what goes on in the organism from the mysterious mysticism of the vitalists, etc., and of finding for it something analogous or identical in the rest of matter to which organized matter were also subjected, I regard it as progress. The development and achievement of a purely physical view of vital processes I regard as the task of the physiology of our day. You probably know what brilliant contributions Schwann, for example, has made to such.[222]

For his part, Mayer had serious misgivings about certain contemporaries' materialistic and reductionist aproach to life, one which would later seize upon the conservation of energy in defense of its position.

In retrospective remarks made toward the end of his life, Schwann recalled his earlier opposition to the notion of a vital force as a principle distinct from matter, regarded as the "architect" of the organic edifice, and with which one explains the properties of muscles, nerves, and glands in terms of their contractility, irritability, and secretory function.[223] With respect to only one phenomenon, formerly passed over in silence, did Schwann qualify his belief that all aspects of organic life can be accounted for strictly in terms of the laws of matter: "Only freedom establishes a limit where explanations in terms of forces of this kind must necessarily stop. It obliges us to admit for human beings alone a principle that is substantially distinguished from all the forces of atoms by this essential characteristic, by freedom, which is incompatible with the properties of matter."[224] But of course Schwann never regarded himself as responsible for accounting for the operation of the uniquely human free will. The reach of deterministic laws defined the scope of his science.

Not surprisingly, the role life scientists assigned to a vital force depended to a considerable extent on the specific range of phenomena with which they were most centrally concerned. Nevertheless, the fact that a scientist's choice of subject did not entail any recourse to a vital force does not mean that he rejected the reality of such an agency. Rather, what was involved for some was more a methodological restriction to certain types of problems than a principled rejection of the complete illegitimacy of vital forces. The effect of this de facto exclusion, however, was to contribute to the growing sense, in the years around 1840, that the vital force had no role to play in the scientific explanation of physiological phenomena. Jakob Henle's *General Anatomy*, published between 1839 and 1841, is a good illustration of this. Most of the 1,048-page book consists of detailed anatomical descriptions and precise physicochemical analyses, as one would expect from one of the leaders of the self-conscious reform of the life sciences in Germany. However, in a four-page section headed "Organism"—perhaps intended as a reply to Schwann—Henle revealed that he was very far from being a physicochemical reductionist. He asked what holds the organism together and determines the typical development of its individual parts—questions quite unaddressed in the body of his work. His somewhat diffident answer was to enter a plea for the necessity of positing the operation of a vital force, by which he recognized that he was going against the tendency of many of his contemporaries:

> It is only with respect to recent attempts and hopes to reduce development and the vital phenomena of the organism to physical laws that I would like to indicate in a few words the difference between the force active in the organism and the forces of dead nature. It differs already in the combinations of elements that arise under its influence, in its capability to produce multiple copies of itself [*sich zu vervielfältigen*] or to extend itself over an ever greater mass of matter without loss of intensity, but especially in its persistence alongside the exchange of matter. . . . The force active in the organism is therefore not merely the sum or the product of the forces of its individual components, for it outlasts these components.[225]

In addition to the role of this force in producing complex organic compounds, Henle assigned it the further traditional tasks of generating offspring true to the species type (even by mutilated parents) and of regenerating lost limbs—again, topics he did not otherwise address in this book. His belief that this force is capable of indefinite self-multiplication recalls Autenrieth's similar views, and reminds us that 'conservation of force' was by no means obvious to everyone, especially when one was dealing with living beings. Henle's special organic force, an *Idee der Gattung*, was in the long tradition of Platonic ideas and Kantian purposefulness, the tradition of Blumenbach's *Bildungstrieb*:

> That which forms and maintains the organism (one has called it vital force, organizing force, formative force, etc.) is therefore not a force in the physicists' sense which is assumed to be given necessarily and absolutely with the existence of matter and which is bound to matter; it does not perish with the individuals; but it manifests itself in so originally and constantly different a manner in the individual species or at least genera of animate beings that one cannot regard their specific conditions as having proceeded from the conflict between a simple and general organizing principle and the manifold agencies of lifeless creation. I therefore believe that this principle active in the organism is best designated as the *idea of the race*, and would like thereby to express that which characterizes this principle: on the one hand its spontaneity, its independence from matter; on the other hand its concrete nature. The idea of the race is, as it were, the preformed form into which grows the germ that develops into the organism.
>
> One cannot do without teleological explanations in physiology, for the processes of nutrition and regeneration can only be understood in terms of the goal they pursue.
>
> . . .
>
> The idea of the race strives toward a goal, but it strives toward it with necessity. With respect to lifeless nature, the organism is autonomous, it develops with spontaneity, but in and of itself the development is a necessary one, given from the start in the germ. . . .
>
> Physiology must distinguish and seek to discover how far the vital phenomena and reactions are determined by their original organization and the striving toward an originally set goal, how far by the influence of the external world on living substance. This goal is difficult to attain, but physiology will acquire a more dignified stature if it remains conscious of it.[226]

In rejecting just this conception of the vital force as a purposefully directing formative force, Schwann had attacked its legitimacy in its ostensibly securest domain, that of development.

As just presented, the status of vital force(s) in the life sciences of the 1830s was complex, and one of the most crucial issues—whether the vital force is responsible solely for the direction of certain complex organic operations or whether it can legitimately be invoked to explain processes that we would say involve the doing of work—was not explicitly so addressed, despite the fact that there was widespread gut sensitivity to the issue. Moreover, there was appreciable consensus over the illegitimacy of treating the vital force as if it provided a meaningful causal explanation of otherwise inexplicable phenomena. Rather, the clear tendency within physiology was to seek as far as possible to account for all vital phenomena in terms of underlying chemical and physical processes, as symbolized by the growing conceptual prominence of *Stoffwechsel*. Processes such as embryological development and the operation of the mind, which could not be plausibly so treated, increasingly fell outside the competency of physiology. Perhaps indicative of the increasingly marginalized status of the vital force in the late 1830s was its acceptance by philosopher Friedrich Fischer in his elaborate defense of spiritualist phenomena (*Somnambulismus*, which to him and his contemporaries meant not just sleepwalking but also animal magnetism, clairvoyance, possession, and the like). Fischer recognized others' reluctance to use the admittedly hypothetical concept as he urged his interpretation of somnambulistic phenomena as some kind of transformation of vital force into soul. For the rest, his conception of the functions performed by the vital force was entirely conventional: it formed, maintained, and animated the organism.[227]

Physicists sometimes felt obliged to address the issue of the existence of a force other than the well-known physical and chemical forces, at least in elementary texts and inclusive handbooks. Muncke, for example, followed the opinion of many in believing that a vital force is needed to explain the body's production of complex chemical compounds, and that we otherwise know essentially nothing about its essence or mode of existence, especially since we cannot imagine a force existing independently of a material substrate.[228] In 1831 he devoted a separate article to the vital force in *Gehler's Physical Dictionary*, the standard multivolume compendium of the day. Choosing a middle position between physiologists who denied any special life forces and those who attributed some measure of life to all bodies, Muncke believed he was with the majority in accepting the existence of special vital forces.[229] Most of the phenomena he assigned to that force were the usual ones—nutrition, growth, generation, and the creation of complex chemical compounds—but Muncke believed that other phenomena, such as the capillary rise of sap in plants, the ascent of the blood against gravity to the brain, and animals' maintenance of a constant temperature, also depend on other than physicochemical laws, indeed suspend the normal laws of nature.[230] Such a belief was not, however, very common. Physicist Baumgartner, for example, limited the role of the vital force in the usual fashion to the production of complex organic substances such as blood, flesh, nerves.[231]

The leading chemists, too, accepted a vital force as necessary to explain the body's ability to synthesize complex substances, though the terrain dominated by that force shrank as chemists succeeded in synthesizing ever more and increasingly complex organic compounds. The vital force, even when not disallowed, seemed more to mark the temporary limit of our ever-expanding knowledge than to designate an essential and peculiar agency of living things.

The progression of Berzelius's thinking on the matter is instructive. Allowing, to be sure, that the ultimate material constituents of animals' bodies are precisely those of inanimate nature, Berzelius in 1810 believed that even inorganic substances undergo transformations within the body that are fundamentally unlike those we are able to observe directly. The body is, he imagined, an "instrument" for carrying out chemical transformations in the interest of its own survival:

> But, with all the knowledge we possess of the forms of the body, considered as an instrument, and of the mixture and mutual bearings of the rudiments to one another, yet the cause of most of the phenomena within the Animal Body lies so deeply hidden from our view, that it certainly will never be found. We call this hidden cause *vital*

> *power*; and like many others, who before us have in vain directed their deluded attention to this point, we make use of a *word* to which we can affix no idea. This *power to live* belongs not to the constituent parts of our bodies, nor does it belong to them as an instrument, neither is it a simple power; but the result of the mutual operation of the instruments and rudiments on one another .... When our elementary books inform us, that the vital power in one place produces from the blood the fibres of the muscle; in another, a bone; in a third, the medulla of the brain; and in another again, certain humours, which are destined to be carried off; we know after this explanation as little as we knew before.
>
> This unknown cause of the phenomena of life is primarily lodged in a certain part of the Animal Body, viz. in the nervous system, the very operation of which it constitutes. The brain and the nerves determine altogether the chemical processes which occur within the body.[232]

Thus the vital force was, from the standpoint of organic nature, an indispensable agency, but from the standpoint of science it was an empty concept—though its inconceivability reflects our general epistemological impotence with respect to ultimate causal explanations: "And the chain of our experience must *always* end in something inconceivable; unfortunately, this *inconceivable something* acts the principal part in Animal Chemistry, and enters so into every process—even the most minute, that the highest knowledge which we can attain, is the knowledge of the nature of the productions, whilst we for ever are excluded from the possibility of explaining how they are produced."[233] Nor can we explain how blood is transformed into saliva, milk, or urine, or how our thoughts might depend on chemical transformations in the brain. Nor did Berzelius here think we would ever succeed in penetrating the secrets of life: "But is it not probable, that human understanding, which is capable of so much cultivation, which has calculated the laws of motion for distant worlds, and explored in so many instances the beauty and wonders of surrounding nature, and even attained a degree of perfection, the summit of which is concentrated in GOD, may one day explore itself and its nature? I am convinced it will not."[234]

Seven years later Berzelius repeated his conviction that the constituent parts of organic bodies follow different laws than inorganic substances. Now, however, the discovery of the nature of that difference seemed to him somewhat less decidedly beyond human capacities. In essentially unchanged portions of his influential *Textbook of Chemistry* in editions spanning ten years, Berzelius wrote that "to discover the cause of this difference between the behavior of the elements in dead nature and in living bodies would be the key to the theory of organic chemistry. It is nevertheless hidden in such a way that we are without any hope of finding it out, at least for the present. Notwithstanding that, we must strive to come closer to this knowledge; for one day we shall either succeed in wholly attaining it, or come to a halt at a definite limit, beyond which the human powers of investigation cannot be further extended."[235] But having made that small concession, Berzelius seemed immediately to take it back, appealing to the necessary creative role of a transcendent being in bringing life to our planet. When it dies, a living individual returns undestroyed to the inorganic realm the basic material components of its body, but its life is irrevocably destroyed.

> The essence of the living body is consequently not grounded in its inorganic elements, but in something else which disposes the inorganic elements common to all living bodies to the production of a certain result specific and proper to every particular species.
>
> This something, which we call vital force, lies wholly outside the inorganic elements and is not one of their primitive properties such as weight, impenetrability, electrical polarity, etc.; but what it is, how it comes into being and comes to an end, we do not comprehend. It may thus be predicted that if the globe existed with its inorganic constituents without living nature but under otherwise identical conditions, it would continue forever to be without living beings. A force incomprehensible to us and foreign to dead nature at one time brought this *something* into the inorganic mass, and not in such a way as if it were the work of chance, but with a wonderful diversity, and disposed [*berechnet*] with the highest wisdom to definite purposes and to an unending succession of perishable individuals arising out of each other, and where in a constant exchange the destroyed organization of one contributes to the maintenance of the other.[236]

The practical evidence for the operation of the vital force was, as Berzelius said, the fact that chemists are incapable of producing more than binary combinations of chemical elements.[237] Yet that reason, repeated in 1837 in words unchanged from 1827, had by the later date been rendered untenable by the progress chemists had made in the intervening years in the synthesis of more complex organic compounds. Two years later, in a reflective essay, "On Some Questions of the Day in Organic Chemistry," Berzelius reinterpreted the vital force as nothing more than an expression for the complex circumstances within which organic chemical processes take place: there are no special organic forces per se. The goal was still to discover the laws that determine the combination of the elements in organic compounds:

> But which are these laws? Does the vital force in organic nature place the inorganic elements under the influence of forces other than those that effect their combination in inorganic nature? This last question has been variously answered. One has assumed that the vital phenomena depend on a particular force foreign to inorganic nature, which one called *vital force*. Physical and chemical phenomena that are so different from those of inorganic nature that they indeed appear to justify such an explanation certainly do take place in living nature; but when we investigate the details of the effects [or actions (*Wirkungen*)] of this force, we recognize therein the interplay of the customary natural forces placed under the influence of a multitude of disparate conditions that produce the diversity of the effects. We are certainly still far from comprehending what determines all of the vital phenomena, but it is obvious from the portion of them that we have succeeded in laying bare that if by vital force something else is understood than the particular circumstances acting together in different ways under which the customary natural forces are set in operation in organic nature, if one thereby understands a particular, special natural force, [then] this natural force is an

hypothesis whose existence is as yet unproven and whose assumption is one of the many cases in which we, following Alexander's example, cut the knot instead of untying it.[238]

As far as Berzelius the *chemist* was now concerned, the notion of a vital force is as empty as it is unnecessary.[239]

Responding to essentially the same evidence as Berzelius, Mayer's chemistry professor, Christian Gottlob Gmelin, could not bring himself wholly to renounce the need for a vital force, but his position looked like a halfhearted rearguard defense of territory ultimately and inexorably to be lost. His 1835 *Introduction to Chemistry*, which Mayer owned, treated the vital force as a kind of supramaterial agency responsible for the construction of the organism. That vital force is, however, capable of regulating the reactions of only a small number of chemical elements, which form its "material substrate" as iron does for the "magnetic force."[240] It directs the formation of characteristically organic compounds and thus also of the organic body as a whole, which it animates and whose autonomous development it regulates. But the vital force ultimately separates itself from the organic body, which is then given over to the destructive forces of inorganic nature—that is, it rots and decays. All this was standard fare. Gmelin was forced, however, to recognize that the distinction between inorganic and organic compounds, as binary on the one hand and ternary and quaternary on the other, could no longer be maintained. Some binary compounds could (it seemed) only be created through the agency of the vital force, and some ternary and quaternary compounds formerly regarded as necessarily organic, such as urea, cyanuric acid, acetic acid, and alcohol, could now be artificially synthesized. But although that traditional distinction would have to be abandoned, Gmelin did not thereby surrender his belief in the vital force, which he thought still acts to facilitate certain chemical combinations in the organism and under whose influence a large number of combinations are effected "which up to now have not been able to be produced without its cooperation": "The diversity of the compounds formed from the few elements—carbon, hydrogen, oxygen, and nitrogen—which are the material substrate of the vital force is so extraordinarily large that one can confidently assert that it will never prove possible to create *all* of the organic compounds artificially, i.e., to replace the vital force by other forces such as light, heat, electricity, the nascent state, etc."[241] In the end, Gmelin's vital force was primarily only an agency of chemical change, different only in degree from inorganic chemical affinities and increasingly dispensable as chemists perfected their science. Heidelberg pharmacist and chemist Philipp Lorenz Geiger invoked a vital force in a wholly perfunctory fashion to account (as usual) for the production of organic bodies and complex organic compounds; its existence was neither a theoretical issue nor a matter of particular practical significance to his chemistry.[242]

The treatment of the vital force by chemist (and extraordinary professor of medicine at Leipzig) Carl Gotthelf Lehmann provides a curious indicator of its fast-changing status among German scientists during the early 1840s. In effect, like Berzelius in 1810, Lehmann accepted its reality by attributing to it the control of processes beyond our capacity to explain in terms of physicochemical laws, while at the same time denying it any legitimacy as a scientific explanation. As Lehmann wrote in the 1841 preface to his *Textbook of Physiological Chemistry*, his aim was "to reduce the often multiply interconnected phenomena to the simplest laws and thereby to separate that which is explicable in terms of physical and chemical laws from that which is inexplicable, i.e., from that which must remain given over to the vital force."[243] In the same vein, he later wrote that one is usually content to assume that the processes of growth and nutrition "rest on that inexplicable something which, as a general *vital force*, . . . most recent physiologists allowed to play an important though still unintelligible role."[244] For Lehmann, the vital force was only one example of the kind of pseudo-explanatory entity we are often compelled to introduce when we find ourselves at the limit of the scientifically explicable, though he declined to take the next step and categorically reject the vital force as both meaningless in principle and unnecessary in fact as science steadily claims more and more territory from the previously inexplicable:

> Even in the natural sciences [as in the science of organic nature] when we lack facts, calculation deserts us, and positive proofs quit us, we must have recourse to so-called dynamic explanations: but just because this always betrays a lack of definite knowledge, just because the dynamis [or dynamic potentiality (*Dynamis*)] is something inexplicable, we should never grant too much scope to such modes of explanation, which betray themselves as more or less pure webs of fantasy. We must therefore even in animate nature seek to push back as far as possible this dynamis, the vital force, by means of which one thought one could explain everything, and which explains nothing; we must make an effort to win ever more territory away from it, the inexplicable, and to pursue it up to the farthest boundaries, which our intellect [*Geist*] is no longer capable of crossing. We must, finally, vindicate for the all-powerful laws of physics and chemistry the rights they do not surrender even to organized bodies; in a word, we must draw over into the realm of the known and the explicable that which was previously given up as inexplicable with the words vital force, and we must confine this dynamis to only the farthest limits of human investigations.[245]

An anonymous and largely sympathetic reviewer of Lehmann's book—possibly Lotze[246]—was quick to seize on the irrationality of Lehmann's handling of the vital force. As the reviewer remarked, Lehmann had said that the dynamis, the vital force, must be pushed back as far as possible, to the outer limits of human understanding; at the same time, he was willing to assign to the vital force the explanation of problems not now soluble. What one requires instead is a demonstration that the vital force is *in principle* inadmissible in scientific explanations:

> Nevertheless, remarks such as on p. 99 and elsewhere, where the rights of the dynamis alongside the chemical laws are expressly recognized, indicate that the author does indeed wish to concede to this vital force a real, albeit very limited, sphere of activity. Nevertheless, the excellent scientist thereby gets into an odd position with respect to this dynamis. Namely, if a real vital force is anywhere assumed as an active cause of the phenomena, it then cannot be in the interest of science, given this recognition,

nevertheless to reduce the phenomena to other principles merely because that force is by its nature obscure and difficult to comprehend. If it really is the vital force that governs organic events in even only a few points, then it must by rights also be maintained, and a reduction of everything organic to mechanical principles can then no longer have the value of truth, but only that of a treatment for certain purposes by means of the substitution of imaginary but lawfully connected grounds. The proposal to assume a vital, physically lawless force only in those cases where all physical principles do not suffice for a complete explanation can therefore be regarded only as a temporary giving up and breaking off of the investigation. But one must not regard this vital force as only another explanatory principle, though one having the same legitimacy as the mechanical principles. It is therefore not only a question of pushing back, through development of the mechanical views, those groundless and—as the author himself calls them—always more or less fantastic presuppositions, of winning territory away from it, but also of showing once and for all that these assumptions are in principle inadmissible, and that the as yet unsolved riddles must be regarded simply as unexplained facts, not as facts explicable only by means of those assumptions.[247]

For the reviewer, who sought a science based on valid *principles*, the same laws of physics and chemistry govern all the phenomena of nature; organisms seem to require additional explanatory principles only because of the special circumstances under which those general laws operate, that is, the finely contrived structures of the organic machine. He insisted that "[the] difference between inorganic and organic is not an original difference in the active forces, but only in the external mechanical conditions under which the latter are active."[248] The origin of the all-important initial disposition of matter—whether of organisms or of the solar system—lies outside the scope of scientific explanation.

Lehmann's book also received warm praise from fellow physiological chemist Franz Simon, who apparently missed the ambivalence in Lehmann's handling of vital force. While conceding that organic phenomena are governed by other than purely physical laws, Simon insisted that one need not therefore relegate those phenomena to a miraculous vital force.[249] For him, the vital force is neither scientifically respectable nor physiologically necessary. Another ambivalent chemist was Berlin professor Heinrich Rose, who in 1843 shared at least in part the growing consensus that the vital force is an illegitimate scientific concept that one invokes to explain the otherwise inexplicable, but who also believed there is a limit to our ability to understand the processes of life, a limit on the other side of which lies an active vital force. As he wrote in his review of Liebig's *Animal Chemistry*:

As is well known, we call the hidden cause of the activity of living organic bodies *vital force*, and one was earlier content to regard all of the chemical processes in the living body as a result thereof. . . .

One has recently attempted to study more closely the most important processes in the living organic body from the standpoint of chemistry in order to explain their results in a way other than as deriving precisely from the vital force. Every attempt of this kind must be received thankfully; for we almost never attach a correct, clear conception [*Vorstellung*] to the words *vital force*. It is convenient to ascribe to it every process that is difficult to explain. But insofar as the phenomena whose cause it is supposed to be are represented as being so entirely anomalous with respect to all other known phenomena, the concept one attaches to it is a mystical, confused one. The more, therefore, one limits the territory of the processes that one must ascribe to the vital force, the more science wins; arbitrariness disappears when known laws can be applied.

To be sure, the application of known laws to chemical processes in the living body will remain limited, and the human intellect cannot for the present penetrate beyond a certain limit.[250]

Rose went on to quote Berzelius's thirty-year-old expression of diffident acceptance of the necessity of assuming the activity of a vital force, and of his belief in the ultimate scientific inexplicability of life.[251]

Methodological opposition to a vital force was becoming increasingly widespread during the early 1840s. Among those rejecting it as an empty verbal cover for our ignorance was Matthias Schleiden, whose *Principles of Scientific Botany* (1842–43) contained a long and important "Methodological Introduction" in which he criticized the vital force in strongly reductionist terms: "Something is as little explained by means of a distinctive vital force as the attraction between iron and magnet is explained by means of the name magnetism. We must therefore firmly insist that in the organic natural sciences, and thus also in botany, absolutely nothing has yet been explained and the entire field is still open to investigation as long as we have not succeeded in reducing the phenomena to physical and chemical laws."[252] Similar categorical denial of scientific legitimacy to the concept of vital force was expressed in 1844 by Otto Kohlrausch in his book-length critique of Liebig's *Animal Chemistry*: "At the moment the chemists' attack is principally directed against the vital force as it stands in the physiological sense. This is a weak fortress that will soon capitulate. Physiology will soon recognize that a vital force described as the sole cause of the most diverse effects is an absurdity [*Unding*] that may not continue to exist in its science under this or a similar definition."[253] From 1844 on Valentin also rejected as outmoded any appeal to a vital force: such a concept is only a verbal expression to cover our ignorance, and falsely separates organic from inorganic phenomena.[254] Vital phenomena are to be explained not in terms of peculiar vital forces, but in terms of the common physicochemical forces working within the constraints of a divinely ordained and self-perpetuating organization. We have already encountered Carl Vogt's rejection of the vital force (in conjunction with the explanation of animal heat) as no longer having any scientific value. For Vogt the vital force represented an outmoded and discredited era of physiological thinking when life was special and mysterious, whereas the new physiology seeks to explain vital phenomena in terms of analogous inorganic phenomena: only in that way will it make progress.[255] Nowadays the notion of a vital force is "ridiculous."[256] By the mid-1840s the discrediting of the vital force in Germany was almost complete, certainly among those who identified themselves with the growing physicalist movement in physiology.

One of the most prominent figures to consider the role and legitimacy of the vital force during the late 1830s and early 1840s was Justus Liebig. He is, however, a difficult figure to come to terms with since many of his expressed views are mutually contradictory. Although I will not attempt to identify the 'real' Liebig behind his variable words and opportunistic positions, some coherence can be found if one makes allowance for the different polemical purposes his writings spoke to and for the development of his ideas over time. Basically, Liebig never questioned the need to assume the agency of a vital force in protecting the organism against the multifarious influences of the environment that work to destroy the integrity of its complex chemical constituents. The extent to which a vital force is needed for the synthesis of those complex compounds was less clear. In part, at least, Liebig shared the growing conviction of contemporary chemists that there is no defensible distinction in principle between inorganic and organic compounds, that chemists could look forward to increasing success in artificially synthesizing substances produced by living things. Some of Liebig's pronouncements echoed the ambivalence of people like Tiedemann, the early Berzelius, and Lehmann, who spoke simultaneously both for the de facto necessity of the vital force as a real agency of change and for its emptiness as a scientific explanation. Liebig's statements over time also reflect the changing context of debate within the scientific community, in that Liebig became concerned with dissociating himself from the strident materialism that had been gaining ever more support since around 1840.[257] It is very possible, too, that it was at least in part Mayer's reading of Liebig's works of 1842 that prompted him to rethink his own tacit acceptance of the possibility of the creation of force in organic systems.

In 1840 Liebig began explicitly and extensively to address the *Lebenskraft* issue. In his highly polemical essay, "On the State of Chemistry in Prussia," Liebig seemed to imply that the recent progress of chemistry left no room for the operation of an unintelligible special vital force. Urea, allantoin, formic acid, and a number of other substances are, he said, products of vital processes. He asked rhetorically: "Is it not strange that the occult vital force that produced them—this conceptually wholly incomprehensible activity—can be replaced by quite ordinary chemical forces? That absolutely identical effects can be produced with them?"[258] The issues broached here in passing were expounded at great length in his *Organic Chemistry in Its Application to Agriculture and Physiology*—the so-called *Agricultural Chemistry*—much of which had appeared earlier that year (1840) in French translation as the long introduction to his *Treatise on Organic Chemistry*. If on the one hand he identified the very subject matter of organic chemistry as "the substances that are produced in the organs [of the body] by the action of the vital force [*force vitale*], and the decompositions they undergo under the influence of other substances,"[259] on the other hand he scorned those physiologists who reject chemistry and who instead "attribute to the vital force what they cannot understand, what they cannot explain."[260] After taking one of his usual grandly unspecific swipes at "die deutsche Naturphilosophie" as responsible for the allegedly sorry state of German science during the preceding decades, Liebig ridiculed the recourse some took to a vital force as the ultimate explanation for obscure organic phenomena. The position he attacked resembles that of Berzelius in 1810:

> As soon as physiologists encounter the mysterious vital force in a phenomenon, they renounce their senses and capabilities; the eye, the understanding, judgment and reflection—everything is crippled as soon as one declares a phenomenon to be incomprehensible.
>
> Before this ultimate cause there occur a multitude of last causes. From the ring where the chain begins up to us there is still a multitude of unknown links. Are these links supposed to remain imperceptible to the human intellect, which has investigated the laws of motion of the heavenly bodies [even though] only a single organ instructs it of their existence, to the intellect that has so many other aids at its disposal on our terrestrial globe?[261]

The intent of Liebig's criticism of vital forces here seems to have been twofold. First, he wished to claim for chemists the competency to deal with certain phenomena of life which, he alleged, physiologists had in their ignorant incompetence assigned to an intrinsically unknowable agency. Second, even though he himself (elsewhere) invoked a vital force as the cause of certain essential vital phenomena, he believed that physiologists had acted hastily in removing too many of these phenomena from the purview of chemical analysis. If we cannot know the first cause at the head of the causal chain, then at least we can be assured that most phenomena here on earth are due to secondary causes quite within our (i.e., chemists') capabilities to discover. The organism, Liebig said, does not create matter, but only transforms the nutrients it takes in; whatever we might call the cause of such transformation—vital force, elevated temperature, light, galvanism, or whatever—we can be sure that the process itself is purely chemical and hence accessible to our understanding.[262]

Not surprisingly, another of Liebig's apparent contemptuous dismissals of the concept of vital force came in the context of his criticism of those such as the followers of Samuel Hahnemann's homeopathic medicine, who do not understand the true character of natural science: "It annoys them and their ilk that the truth is so simple, although they cannot succeed in turning it to practical account despite great effort, therefore they present us with the most impossible views and create for themselves in the words vital force a wonderful thing with which they explain all the phenomena they do not understand. With a completely incomprehensible indefinite something one explains everything that is not comprehensible!"[263] If Liebig's words here echoed the methodological criticisms of Magendie, Elsässer, Schwann, the later Berzelius, Simon, and Schleiden, he never seemed to appreciate that they told equally well against his own vague notions of an active vital force. Liebig's vital force, like everybody else's, was a cover for phenomena he could not otherwise explain.

Aside from such strident posturing, however, Liebig made regular and matter-of-fact use of the concept of vital force. In fact, during the 1840s he was its most influential defender among first-rank German physical scientists—a defense all the more significant when one considers that by 1839 contemporaries such as

Berzelius, Magendie, and Schwann had, in various but mutually reinforcing ways, already urged the exclusion of the vital force from science, and that many others would join their ranks in the course of the ensuing decade.

Vance Hall maintained that Liebig's *Agricultural Chemistry* assigned the vital force two different functions: alongside its regulation of ordinary chemical forces "in such a manner that characteristically vital phenomena, like reproduction, could occur," it was also "responsible for the forms which seemed to characterize living systems."[264] Indeed, for the most part Liebig treated the vital force in this work primarily as the cause that regulates organic chemical transformations (*Metamorphosen*).[265] Sometimes he spoke of it as if it were an abstract controlling agency: "The vital force, in its distinctive manifestations, makes use of particular instruments, for every function a particular organ."[266] He also identified it as "the capability inherent in every individual organ of regenerating itself at every moment."[267] As Liebig wrote,

> The concept of *life* includes along with reproduction still another concept, namely that of activity *by means of a determinate form*, of creation and generation *in a* determinate form. One will be in a position to produce the constituents of a muscle fiber, skin, hair, etc., by means of chemical forces; but no hair, no muscle fiber, no cell can be formed by them. The production of organs, the collaboration of an apparatus of organs, their capability of regenerating out of the foods presented not only their own constituents, but *themselves* [with] the form, constitution, and with all their properties—this is the character of organic life, this form of reproduction is independent of the chemical forces.
>
> The chemical forces are subject to the imperceptible cause by means of which this form is determined; it itself, this cause, we only have knowledge of its existence by means of the distinctive phenomena it produces; we investigate its laws like those of the other causes that bring about motion and changes.
>
> The chemical forces are the servants of this cause just as they are servants of electricity, heat, a mechanical motion, impact, and friction; they undergo through these latter a change in direction, an increase or decrease in their intensity, a complete neutralization [*Aufhebung*], or a total reversal of their operation.
>
> It is this influence and no other that the vital force exerts on the chemical forces; but everywhere combination and separation take place, chemical affinity and cohesion are active.
>
> We know the vital force only through the distinctive form of its tools, through organs that are its carriers.[268]

These two conceptions of the function of the (singular) vital force correspond to two classes of phenomena which Tiedemann believed were under the direction of (plural) vital forces: the first is essentially identical to Tiedemann's reliance on such forces to explain the relatively greater complexity of organic compounds as compared to inorganic; the second is close to his understanding of the regulation of processes of generation, development, and growth. Not surprisingly, the anatomist and physiologist Tiedemann was more concerned with such goal-directed 'macro' processes, the chemist Liebig at most and only in passing with the general form of organs, which were of interest to him primarily as sites of production of organic compounds.

Both Tiedemann and Liebig, however, assigned still a third function to the vital force(s): the preservation of still-living vegetable and animal matter from decomposition, from the processes of fermentation, putrefaction, and decay it ineluctably undergoes once it is outside the action of those vital forces.[269] Liebig's conception recalls his image of respiration as the body's struggle against the destructive force of atmospheric oxygen: "It is the vital force that opposes to the incessant action of the atmosphere, humidity, and temperature on the organism a resistance that is up to a certain point invincible; it is the incessant leveling [*Ausgleichung*], it is the constant renewal of these activities which maintains motion, which maintains life."[270] Once removed from the protection of the vital force, organic compounds are capable of preserving their form and constitution "only as a result of the power of inertia"; they are subject to immanent disintegration upon the action of the least external influence.[271] In accordance with his antagonistic conception of the vital force, Liebig claimed that the state of health of the body and the effect of any given substance on it are determined in large part by the relative balance between the *Lebenskraft* and the *chemische Kraft* of external agencies.[272] In a later work, in which he criticized Henle's and others' conception of stimuli as agencies necessary for the eliciting of particular organic functions, Liebig defended a conception of the organism as always active, not merely reacting to external stimuli.[273] Such a view was of a piece with his conception of an active vital force constantly effecting chemical transformations and resisting forces of decay. As he said in a major antimaterialistic address of 1856, "On Inorganic Nature and Organic Life": "The inorganic forces forever create only that which is inorganic; by means of a higher force at work in the living body, one whose servants are the inorganic forces, there comes into being the organic, distinctively formed matter different from the crystal and endowed with vital properties."[274]

In previous sections of this chapter we encountered Liebig's 1841 paper on "The Vital Process in the Animal, and the Atmosphere" as a forceful advocate of a somewhat idiosyncratic oxygen theory of animal heat and of the more general position that all vital functions depend on chemical processes of material exchange. Nevertheless, it, too, posited the existence of a vital force in no uncertain terms. The article began:

> In the animal egg, in the seed of a plant we recognize a remarkable activity, a cause of increase in mass, of the replenishment of spent material, a force in the state of rest. The static moment of this force is destroyed [*aufgehoben*] by external conditions, fertilization, and the presence of humidity and air; the force that passes over into motion manifests itself in a series of formations [*Formbildungen*] that, even if occasionally enclosed by straight lines, are nevertheless far removed from geometrical shapes such as we observe with crystallizing minerals. This force is called *vital force*.[275]

Here the vital force was responsible for the growing organism's increase in mass (a newly identified function) and its developmental differentiation. Here Liebig was concerned to emphasize its *difference* from all other (inorganic) forces. He called the vital force an independently existing force (*eine für sich bestehende Kraft*), and insisted "that all phenomena in the organism of plants and the animal must be ascribed to an entirely distinctive cause wholly different in its manifestations from all other causes that determine changes of state or motions."[276]

This article, later incorporated almost unchanged into his *Animal Chemistry*, also contained one of Liebig's few discussions of mind (*Geist*) and its essential difference from the "vegetative" vital force. The mind can influence only the rate of the vital processes, which for the most part—and in plants and nonhuman animals perforce entirely—take place independently of the mind. Although much less elaborated, Liebig's views recall those of Autenrieth and Tiedemann and the traditional distinctions between sensibility and irritability, animus and anima:

> The phenomena of higher psychic life [*geistiges Leben*] cannot, in the present state of science, be reduced to their immediate, much less to their ultimate, causes; we do not know anything more than that they are present. We ascribe them to an immaterial activity, more precisely—insofar as its manifestations find themselves bound to matter—to a force that is wholly different from, and has nothing in common with, the vital force.
>
> As cannot be denied, this distinctive force exerts a certain influence on the vegetative vital activity, similar to the way in which this happens with other immaterial powers, with light, electricity, heat, and magnetism, but this influence is not of a determinative nature, rather it manifests itself only as an acceleration, disturbance, or retardation of the vegetative vital processes; in an entirely similar fashion the vegetative vital activity exerts certain effects backwards on conscious psychic life.[277]

But "science has a definite limit, which it must not overstep," and that limit does not include mind or consciousness.[278] Occupation with such subjects only harms science, which can get along quite well by ignoring them: "The continuously reinvigorated endeavor to wish to discover the relationship of the psyche to animal life [*die Beziehungen der Psyche zu dem animalischen Leben*] has long since retarded the progress of physiology; it was a continuous withdrawal from the domain of science into the kingdom of fantastic forms."[279] The mind has nothing to do with the development of either embryos or seeds. But if Liebig was prepared to exclude consciousness and mind from his science, if not yet from his world, he was not disposed to eliminate the vital force from either.

So far the consideration of Liebig's *Animal Chemistry* has concentrated on the first of its three major sections, the one entitled "The Chemical Process of Respiration and Nutrition," which consisted of minimally revised reprintings of two earlier journal publications. In the book's difficult third section, "The Phenomena of Motion in the Animal Organism," Liebig presented a radically reworked interpretation of the nature and function of the vital force. Where before he had been concerned to emphasize the differences between the vital force and all other forces, here his agenda was to assimilate as far as possible the former to the latter. Where previously the vital force had served to direct the synthesis of complex organic compounds, to determine the characteristic forms of organisms, and to preserve them from destructive external influences, here the emphasis was on its role in increasing the organism's body mass and in affording resistance against external agencies. Although Liebig did not now dwell on the still-accepted role of the vital force in synthesizing organic compounds, his overriding concern with establishing its affinities with chemical forces went hand in hand with his total exclusion of its more developmental and form-determining aspects. Whereas the first section of the book argued overwhelmingly—if, to be sure, with a few contradictions—that the two essential and characteristic functions of the animal organism, the production of heat and motion, depend on underlying processes of material exchange, the third section argued primarily—if, to be sure, with even more contradictions!—that the vital force itself is the ultimate cause of the organism's production of mechanical effects. (Heat he still assigned to the chemical process of oxidation.) As he wrote: "We are acquainted with only one source of motive force in the organism of the animal, and this source is the same cause that determines the increase in mass of animated body parts and that gives them the ability to resist external actions; it is the *vital force*."[280]

Earlier his analogies had been with the steam engine and the voltaic pile; now he likened the vital force to chemistry's peculiar catalytic force, which seems capable of producing effects indefinitely without any diminution of the causative agent. The slogan that no force or activity can arise out of nothing belonged to the first section, not the third. Still another innovation was Liebig's weak attempt to elaborate a pseudotechnical set of concepts (*Kraftmoment* and *Bewegungsmoment*), which, he hoped, would further cement the connections between vital and physicochemical forces. Given the content of his book and the timing of its publication, it is not impossible that Liebig was at least in part prompted to undertake a more general analysis of force in response to Mayer's work, which he had received and accepted for his *Annalen* in April 1842, and about which he might also have known through his correspondence with Philipp Gustav Jolly, the professor of physics at Heidelberg with whom Mayer had discussed his work around November 1841.[281]

Examined more closely, this third section of Liebig's *Animal Chemistry* is itself not a unified piece. The first and longer of its two major subsections, designated only by the Roman numerals I and II, contains more or less the views sketched in the preceding paragraph; the second is closer to Liebig's earlier writings with respect to the importance of *Stoffwechsel*.[282] Although Liebig expressly claimed that his new views were "necessary conclusions" of those developed in the book's first section,[283] I have not been able to resolve the apparently glaring contradictions between them. It should be noted that Liebig faced a real dilemma: how, now, to demonstrate the close affinities between the vital and other forces while at the same time preserving the former's privileged irreducible (and higher) status. Exacerbating Liebig's problems, and ours in understanding him, was the fact that, despite much talk about causality, in practice he regularly obscured the question of whether the organism's chemical activities are the cause, the effect, or the concomitant of its vital activities.[284]

Liebig declared at the outset that his conclusions with respect to "the nature and essence of the distinctive cause . . . which must be regarded as the ultimate basis [*letzter Grund*] of the phenomena that characterize animal and plant life" would lose their significance if it could be proven "that the cause of vital activity has, in its manifestations, nothing in common with other known causes that bring about motion or changes in the form and constitution of matter."[285] Although a *distinctive* force, the vital force's effects must still be in accord with the same general laws of resistance and motion that govern the solar system and bring about material change. The principal physiological function of the vital force is (here) to counteract the forces of attraction between the molecules of the nutrient matter and redirect their chemical forces toward the production of organic substances.

The unifying conceptualization of force that Liebig advanced was to see forces as agencies which both cause and resist motion. With respect to the force of gravity, for example, one must distinguish the pressure (*Druck*) exerted by a stone resting on a table from the motion it acquires in free fall.[286] Liebig called the intensity of the gravitational force both pressure and motive force; its effect in causing a body to fall was proportional to what he termed the *Kraftmoment*, the product of the "pressure" and the time of fall. The poorly defined *Bewegungsgröße* considered also the mass of the falling body, and was equal to the product of either the mass and the *Kraftmoment* or the mass and velocity of the body. The *Bewegungsmoment*, finally, measured the effect of the force in terms not of elapsed time but of distance moved, and was effectively what we would call work. Liebig concluded:

> *Moments of force* or *moments of motion* are accordingly expressions or measures in mechanics for effects of force that are related to a velocity acquired in a given time or to a given space; in this sense they can be applied to the effects of all other causes of change in motion, form and constitution . . . .
>
> Accordingly, every force manifests itself in matter as resistance to external causes of changes in place, form, and constitution; it manifests itself as motion-generating force when no resistances oppose it or in the overcoming of resistances.
>
> . . .
>
> We observe both manifestations of activity in the force that gives animated body parts their distinctive properties.[287]

Insofar as the vital force resembles other forces—at least in terms of its susceptibility to analysis in analogous terms—it must have a certain *Bewegungsmoment*, in which case "we know that this moment of motion of the vital force is available [*verwendbar*] in an animated body part to impart motion to resting substances (to bring about decomposition, to neutralize resistances), and if in its manifestations the vital force behaves like other forces, then this moment of motion must be able to be communicated or propagated by substances that in and of themselves do not neutralize its free manifestation by means of an opposing activity."[288] The analogy he had in mind was between the propagation of galvanic electricity through the connecting wire, which he believed offered no resistance to the current, and the propagation of motive force via the nerves from one part of the body to another.[289] Liebig made contradictory statements as to whether or not motion can be annihilated without effect—and under "effects" he included such nonenergetic factors as the dent made by a falling body when it is stopped by a piece of wood. His example of the coming to rest of two colliding inelastic bodies and his image of forces opposing each other in a state of dynamic equilibrium, and thus canceling out each other's effect, leave the strong impression that there are common conditions under which forces *can* be destroyed.[290]

One of the most important similarities between vital and chemical forces was that they both act only upon contact or at immeasurably small distances.[291] In choosing to liken vital forces to chemical ones, however, Liebig singled out a special class of the latter, which included those that Berzelius had identified six years earlier as "catalytic" forces, although Liebig did not use that term here: "Transformations—or, if you will, phenomena of motion—can . . . be brought about by the free and available chemical force active in another chemical combination, and indeed without its manifestation being exhausted or neutralized by resistances."[292] Liebig's examples were the use of sulfuric acid in the production of grape sugar from cane sugar and of glucose from starch and water. He reiterated the outstanding distinguishing characteristic of these unique forces: "Entirely different from the manifestation of the so-called mechanical forces, we have recognized in the chemical forces causes of phenomena of motion, of changes in form and constitution, without perceptible exhaustion of the force by which they were produced; but the reason [*Grund*] for the continuing manifestation of activity always remains the same, it is the lack of an opposing activity (a resistance) which is capable of neutralizing or balancing it out."[293] As close as Liebig sometimes came to an apparent intuitive grasp of the conservation of energy, on many other occasions he showed himself to be quite oblivious to the implications of any such principle. Here, of course, his goal was to assimilate the vital force to chemical forces, among which the experimentally demonstrable (if conceptually murky) catalytic force exhibited the closest affinities.

In a passage continuing the one just quoted, Liebig went on to offer an interpretation of the vital force in terms of the arrangement of the elementary particles of the organ:

> Just as the manifestations of the chemical forces (the moment of force of a chemical compound) appear to be dependent on the particular configuration in which its elementary particles are in contact with each other, in a similar fashion experience indicates that the vital phenomena are inseparable from matter, that the manifestations of vital force in an animated body part are determined by a certain form of the carrier and by a certain configuration of its elementary particles; if we destroy [*aufheben*] the form or composition of the organ, then all of the vital phenomena disappear.
>
> Nothing prevents us from regarding the vital force as a special property that belongs to certain material substances and becomes perceptible when their elementary particles have come together in a certain form.
>
> . . .
>
> So conceived, the vital force unites in its manifestations all of the distinctive features of the chemical forces and of the no less wonderful cause that we regard as the ultimate basis of the electrical phenomena.

The vital force does not manifest itself at infinite distances like the force of gravity or the magnetic force; rather, like the chemical forces, it is active only on immediate contact; it becomes perceptible through a complex of material parts.[294]

This is probably as close as Liebig ever came to offering what might appear to be a reductionist interpretation of the vital force, although his interpretive enterprise might also be understood as an exercise in legitimation-via-analogy—and, as I will show presently, he soon went on to treat the vital force more like an independent cause *sui generis*. Reading this passage carefully, one notes that Liebig only asserted that the vital force is a manifestation or expression (*Aeußerung*) of matter having an appropriate form or configuration (*Form, Ordnung, Ordnungsweise*), and as such his conception of the vital force might reasonably be compared with Johannes Müller's notion of a special force latent in all matter but only manifesting itself in appropriately configured organic matter.

According to Liebig's crude theory, the various parts of the body, in particular the muscles, possess a certain quantity of vital force, a portion of which goes to preserve the equilibrium of their substance against forces seeking their destruction—in particular, oxygen. When some of that vital force is expended in the performance of mechanical effects, that part of the body is no longer capable of counteracting the destructive chemical force of the oxygen circulating throughout the body in the blood. The result is that a portion of the once-living tissue combines with oxygen to form lifeless compounds, a process Liebig termed *Stoffwechsel*.[295] Thus this exchange of matter was the result of the production of animal motion, not its cause. Liebig made abundantly clear what the underlying cause was in a passage Mayer quoted and criticized in his aptly titled *Organic Motion in Its Connection with the Exchange of Matter* (1845):

No other conclusion can be drawn from this very definite connection between the exchange of matter in the animal body and the [vital] force consumed by mechanical motions than that the vital force active or available in certain animated body parts is the cause of the mechanical effects of the animal body.

The motive force doubtless derives from animated body parts. They possessed a moment of force or motion that they lost to just the degree that others have received a moment of force or motion; they lose their capability for increase in mass, their ability to resist external causes of disturbances. It is clear that the ultimate cause, the vital force, from which they obtained these properties, has served to produce mechanical force; it has been consumed as motion.[296]

Significantly, Liebig did not explicitly equate this motion-producing vital force with the chemical forces contained in the animal's *belebte Körpertheile*. Except for a brief key phrase, the following conclusion might wrongly be read as asserting a causal relationship between oxygen consumed and work done in a sense opposite to the one Liebig intended:

Exchange of matter, manifestation of mechanical force, and uptake of oxygen stand in so close a connection to each other in the animal body that one can put the quantity of motion and the amount of transformed animate matter into one and the same ratio with respect to a certain amount of oxygen taken up and consumed by the animal in a given time. For a determinate measure of motion, *for a proportion of vital force consumed as mechanical force*, an equivalent of chemical forces becomes manifest, i.e., an equivalent of oxygen becomes a constituent part of the organ that lost the vital force, and an equal proportion of the matter of this organ leaves the body part in the form of a compound of oxygen.[297]

Liebig's principles led him to the curious conclusion that, when a large quantity of vital force is consumed in the production of mechanical motion, the body compensates by slowing down the motion of the heart.[298]

To the obvious question, How is the continuously expended vital force replenished?, Liebig gave only the sketchiest of unsatisfactory replies. Somewhat at odds with his more usual claim that the organism's growth is a *result* of a directive and form-creating vital force was his passing suggestion that the vital force derives from the assimilated nutrients, making it either a kind of higher-level chemical force or a latent force *sui generis* à la Müller: "When we consider that the capability for increase in mass has almost no limit in the plant, and that a hundred willow branches taken from one tree become a hundred trees, then one can scarcely doubt that, with the uniting of the elements of the food to a constituent part of the plant, to the already present moment of force, a new moment of force is added to the newly formed part of the plant, that with the increase in mass the quantity of vital force increases."[299] Liebig confessed his complete ignorance as to the manner in which the vital force produces mechanical effects, and which will no more be able to be discovered through experiment than the connection between the chemical action in a galvanic pile and the phenomena of motion it produces.[300] Abandoning, at least for rhetorical purposes, his earlier expressed attachment to the mode-of-motion theory of heat, he here invoked the imponderable-fluid theory as another analogy-of-ignorance in defense of his protean notion of vital force:

We do not know how an intrinsically invisible, imponderable something, heat, imparts to certain natural substances the capability of exerting the most enormous pressure on their surroundings, or how at all this something is produced when we burn wood or coal.

It is the same with the vital force and the phenomena presented by animate bodies. Their cause is not chemical force, not electricity, not magnetism; it is a force that possesses the most general properties of all causes of motion and of change in the form and constitution of matter, and [yet it is] a distinctive force because it has manifestations that none of the other forces possess.[301]

As far as this passage goes, the vital force may for the most part be analogous to the usual physicochemical forces, but it is not identical to them. It remains a distinctive force, an *eigenthümliche Kraft*.

One's frustration at discovering what Liebig 'really' believed is only heightened by his confused and contradictory discussion (in the next unnamed subsection of the third part of his *Animal Chemistry*) of the relationship between the vital force, ambient temperature, food consumption, respiration, the exchange of matter, organic motions, growth, sleep, and age-dependent differences in the body's economy of force and substance. Sometimes it seems clear that the cause of the production of force is the exchange of matter.[302] But if on one page he attributed the production of animal heat to processes of oxidation, on another he allowed that "heat can be produced in the animal body without any exchange of matter."[303] In the previous subsection of his book Liebig had argued that the expenditure of vital force on mechanical effects exposes the animal's tissues to the destructive effects of the ever-present oxygen, leading to an increased exchange of matter; here he argued that cooling a part of the body reduces the strength of its vital force with a similar effect, and that the increased exchange of matter serves to raise the temperature of the affected part. Mayer singled out the following summary paragraphs for citation and criticism in his 1845 book:

> However closely connected the conditions for the production of heat and for the production of force for mechanical effects may present themselves to observation, the development of heat can by itself in no way be regarded as the cause of the mechanical effects.
>
> All experience proves that there is in the organism only one source of mechanical force, and this source is the transition of animated body parts into lifeless compounds.
>
> . . .
>
> The cause of the consumption [of matter in the body] is the *chemical action of the oxygen*; its manifestation is dependent upon a withdrawal of heat, as well as on the employment [or expenditure (*Verwendung*)] of vital force for *mechanical effects*.
>
> *The act of consumption is called exchange of matter, and it occurs as a result of the assimilation of oxygen into the substance of animate body parts; this assimilation of oxygen only takes place when the resistance that the vital force of animate body parts opposes to the chemical action of the oxygen is smaller than this chemical action itself, and this weaker resistance is determined by the withdrawal of heat or by the employment for mechanical motions of the [vital] force active in the body parts.*[304]

The effect on Mayer of his critical reading of Liebig's confused physiology of the vital force will be addressed in chapter 6. Mayer's university friend and later professor of medicine, Carl August Wunderlich, ridiculed Liebig's conception of vital force as a "completely abstract and meaningless word," an abstraction supposedly capable of opposing the influence of a ponderable substance, oxygen.[305] Liebig had not, in fact, clearly addressed the question of the ontology of the vital force—whether it is a property of matter, like other forces; or a particular arrangement of the smallest parts of matter; or an imponderable substance *sui generis*; or an abstract but causally active idea. That Liebig himself was dissatisfied with his handling of the vital force is suggested by the fact that—perhaps in silent response to Mayer's telling criticisms—he dropped the long third section of his *Animal Chemistry* from the third edition of 1846.[306]

If Liebig can be taken as the most influential spokesperson in Germany for a vital force during the early 1840s, then surely Hermann Lotze can stand for a growing and outspoken antivitalist strain among German scientists during that period. Indeed, with the publication in 1842 of Lotze's *General Pathology and Therapy as Mechanical Sciences* and his famous *Lebenskraft* article at the head of

Rudolph Wagner's *Dictionary of Physiology* in 1843, an assertive materialism thrust itself onto the contemporary biological stage, one whose voice demanded to be heard. After around 1842 the vital force was a contentious and symbol-laden topic in ways that it had not been only a few years earlier. Though materialist in their orientation, Lotze's writings continued the tradition of the multifaceted "common context" identified earlier. That is, they dealt with the interconnected problems of life and the vital force and their relationship to nonlife and the nonvital forces of physics and chemistry; with the soul (or mind or will) and its relationship to the body and the vital force; with the relationship—whether substantial or analogical—between the imponderables and all the rest; with spiritualist phenomena; with general issues of causality; and with God. This was, I contend, also Mayer's broad problem context, and if a self-professed materialist and antivitalist like Lotze could still allow for the effective creation and destruction of force out of nothing, then Mayer's achievement looks all the more impressive, all the less simply an expression of contemporary tendencies.

At the center of Lotze's worldview was his contention that exactly the same laws govern both organic and inorganic phenomena: there are no forces special to living things. Life is "strictly speaking a summation of nonliving processes"; it belongs not to the parts of the organism but to "the totality of the processes which the *entire* body produces."[307] Lotze demanded nothing less than "the universal domination [*Herrschaft*] of the mechanical viewpoint."[308] Alongside such ontological assumptions, Lotze maintained the methodological principle that science can only treat of the laws of nature as they apply to already given arrangements of things: science cannot account for the initial conditions (to use our term) we find ourselves confronted with, whether the object of investigation is the solar system or a living organism. Yet it is in part on the arrangement of the points of application (*durch die Anordnung der Angriffspunkte*) that any actual course of physical events depends. Lotze did not deny all difference between organisms and inorganic nature, but he placed their difference in the complex purposive organization of organisms, which he characterized as natural purposes (*Naturzwecke*) and whose original creation he attributed to God.[309] As he forcefully put it in his *Lebenskraft* article:

> What happens in the living body is not distinguished from what happens in the inanimate physical [world] by a difference in principle in the nature and mode of operation of the efficient forces [*vollziehende Kräfte*], but by the arrangement of the points of application available to them, on which depends, here as everywhere in the world, the configuration of the ultimate outcome. The coming together of the individual masses into that determinate arrangement out of which proceed, as mechanical results, all of the phenomena of life is nothing accidental, but is only preserved and propagated by descent from the race. The coming into being [*Entstehung*] of an organism (at least of a higher one) is therefore never an object of observation in such a way that it might be possible to demonstrate how, in the course of the changes in the earth's surface, chemical elements might have come together by chance in certain proportional quantities and under certain external circumstances in such a way that a series of vital motions, an individual, might have been formed from them as an

automatic result. For that reason physiology has, in this continuity in the development of the germ out of earlier organisms and of later organisms out of the germ, its ultimate fact, one which it must take into consideration, and it can, as the first cause [*Grund*] of this continuously unfolding series of developments resulting from the process of reproduction, only assume a creation lying beyond the realm of natural science, but not a chance coming into being, even one that takes place according to mechanical principles. Thus here, as in every natural science, it is only a question of deriving the laws in accordance with which every natural phenomenon can be deduced *from its presuppositions* and in turn determines new ones. . . . Moreover, from the principles [*Grundlagen*] adduced here there follows by itself the refutation of the objections directed against mechanical theories from a religious standpoint, as if life would lose something of its dignity and sanctity if it were conceived to be the result of a mechanism. One forgets that this mechanism did not come into being through its own virtue, but that the wisdom of God created it, and has charged it, as the most reliable servant who never abandons himself to his own pleasure, with the realization of the ideas of nature [*Naturideen*].[310]

The principal object of physiological investigation should be the processes of material exchange by which the organism maintains itself. Lotze saw in the exchange of matter "the centerpiece of the organic mechanism, around which all the other processes of the animal economy can be connected."[311]

Having thus declared his intention to regard vital functions as the result of the lawbound working out of general physical laws, it is clear that Lotze would have nothing to do with the notion of a vital force as cause of a special class of phenomena. Noting, in his *General Pathology*, that the word "force" was used in two senses—in physics as the constantly acting cause of simple motion and in engineering as the capability of a machine to perform a certain amount of work—he argued that by "vital force" one must not understand a supposed active cause of organic motions, but rather only a measure of the mechanical effects the organism produces.[312] Mayer's friend and confidant, Wilhelm Griesinger, took exception to Lotze's restriction of the term. As he wrote to Mayer:

You have occupied yourself with the concept of *force*. In his—good—*General Pathology* Lotze has . . . also called attention to the different senses this word can have in organic nature; [but] however much I am an opponent of the views he combats and in this regard entirely agree with him, nevertheless his view of it will not satisfy me either. He believes that one can speak of force (e.g., vital force) in organisms only in the sense of mechanics, namely to designate the *magnitude of the work performed* that arises from an entire system of masses, but that one can never so understand a cause [i.e., as a force]. But I do not know where he then wants to go with the causes of the first development of the germ.[313]

Once again it is the developmental functions of the vital force that seem most resistant to reduction to mechanical principles, even to someone like Griesinger who otherwise saw himself as advocating a physicalist approach to the life sciences. As we have just seen, Lotze answered Griesinger's question in his more elaborated *Lebenskraft* article, where generation was interpreted as the passing on from generation to generation of a particular material disposition. In that work, too, Lotze subjected the concept of force to closer philosophical analysis, and concluded that force in general is a fiction, an abstraction we wrongly project into nature: what we have knowlege of is only the lawlike relations of things.[314] As for the vital force, he further dismissed the misguided practice of regarding it as the unknown cause of phenomena—indeed, as the unitary cause of a multiplicity of diverse phenomena—and criticized the related concepts of "formative force" and "self-preservation force," of the alleged "forces" of sensibility, irritability, and reproduction, as providing at best only a classification of phenomena, not their explanation.[315]

Still other criticisms touched upon others' false ontology of force. Accepting the use of the term in physics to describe the inherent properties of things, Lotze objected to attempts to regard force not only as a cause but also as an independently existing entity:

> But the definition which regards force as a cause at once brings with it the error either that force is identified with some substance whose sole property consists in possessing this force, or that forces are regarded as distinctive existing entities that presuppose nothing further, but exist every bit as well by themselves as things. Both errors have been championed by two famous men, to mention whose celebrated names suffices to indicate how necessary a thorough investigation of this conceptual definition [*Gedankenbestimmung*] is. The first error was defended by Treviranus, the second by Autenrieth.[316]

Treviranus, said Lotze, taught that a universal and inherently formless vital matter (*Lebensstoff*) requires the influence of external forces to determine its manifestation as this or that specific creature. But if those external forces are so powerful, argued Lotze, then the supposed vital matter is effectively reduced to the status of just another material substance.[317]

Citing the first essay in Autenrieth's *Views on the World of Nature and the Life of the Soul*, Lotze objected to the author's identification of the vital force as an independent entity analogous to the imponderables, the weightless fluids assumed to underlie the phenomena of heat, light, electricity, and magnetism. For him, the imponderables are either inactive weightless fluids or properties of matter, and since forces cannot exist independently of matter, the very concept of an Autenriethian vital "force" is empty. Lotze also criticized the pro-vital-force views expressed in Henle's *General Anatomy*. As we have seen, Henle had distinguished the vital force from the forces of inanimate nature in terms of the former's supposed unique ability to propagate itself without loss of intensity. Lotze discounted this alleged difference by citing the behavior of a physical agent, magnetism: "Didn't that ingenious observer consider that exactly the same thing takes place with a magnet, whose force can be propagated without weakening to many iron rods so that they exhibit the same polar form of action?"[318] Having thus accepted the possibility of the multiplication of force, Lotze went on to acknowledge the possibility of its destruction, too. Hadn't Henle further considered that, especially for a few lower classes of animals, the act of propagation appears to entail a fatal exhaustion of its forces? Johannes Müller has also discussed these ideas; nevertheless it appears to me that one thereby embraces the phenomena altogether too grossly in an abstraction. Forces are in general not communicated from substance to substance in nature, but only velocities and (in general) changes, or individual diffusible fluids. What takes place with the transmission of magnetism we do not very precisely know; but vital phenomena offer here, as I believe, no real difficulty at all.[319]

The act of generation, Lotze said, involves not the transmission of a vital force but the imposition of a particular arrangement of parts in the germ, which then grows by drawing into itself the ambient forces of inorganic nature.

Lotze criticized still another of the differences Henle had identified between vital and inorganic forces, namely that the vital force, in surviving the material changes that take place in the living organism, cannot be regarded as merely the sum or the product of the forces of the individual parts. Strictly speaking, Henle may have been right, but the antimechanist conclusion he drew was unwarranted:

From it [i.e., experience] we know in the first place nothing at all about whether an organic *force* has been conserved in the exchange of the [organism's] constituent parts, but only that the *form* of the body and the *sum of its vital phenomena* do not change *markedly* during this exchange. . . . That the vital force should be the sum or the product of the forces of the particles—surely no one demands that; rather it is only supposed to be some function or other of them. When we recall that even the diagonal of the parallelogram of forces only translates into the sum of the two component forces [*Seitenkräfte*] in the single case when both component forces have one and the same direction, . . . then we naturally only require that the vital force vary in some, perhaps very complicated functional form proportional to the individual forces. Someone who wished to suppose here that it grew in accordance with the sum of the molecules, and thus the weight of the body, would not be proceeding more accurately than someone who wished to determine the resultant of two component forces merely from their velocities and without considering the angle they form.[320]

Thus in choosing analogies to refute Henle's belief that the organic force follows laws different from those of the normal physical forces—analogies with the process of magnetization and with the combination of forces in the parallelogram of forces of mechanics—Lotze demonstrated how far even a dedicated materialist reductionist could be from an intuitive notion of the conservation of energy, especially when the concept of force still embraced Newtonian forces, velocities, and a variety of crudely conceptualized natural powers. His example reminds us, too, of the extent to which everyone's thinking in these matters revolved around the exploitation of favored analogies. Mayer's dogged and ultimately successful pursuit of a coherent understanding of force seems all the more impressive by comparison.

Among the topics dealt with in Lotze's *General Pathology* was the relationship between the soul and the body, between ideas (either an individual's or 'Platonic') and matter. Ideas by themselves can never function as active forces: "The idea has no force of limbs to move masses; if this is to happen there must exist a system of masses which is already ordered in accordance with its laws and which, therefore, [even] acting purely mechanically, nevertheless produces the result that, as predetermined and necessarily existing, corresponds to the idea of life. It follows from this that ideal relationships never explain the modes of realization of some phenomenon."[321] The one huge exception to this principle, Lotze later noted, was the original creation of the world and all its physical arrangements and organic forms in accordance with the divine idea; however, such an ideal influence is not a possible object of human experience, and can thus be excluded from consideration by science, which has to do only with the completely mechanical processes that govern the working out of the system thus established. The problem is that it certainly *appears* in the organic realm—with us, at least—that ideas do indeed effect changes in the material world. How is that to be reconciled with "our mechanical way of regarding things"?[322] Lotze's response to this dilemma was to change the venue for its solution from the natural sciences to metaphysics: he opted for occasionalism.[323]

Apparent influences in two directions had to be considered. After laying out his occasionalist interpretation of the relationship between physical motions and our ideas (*Vorstellungen*), he broached the question of the connection between our ideas and the motions they appear to cause:

> The other incomparably more difficult point is the commencement of corporeal motion under the influence of the mind. The mechanical theory demands in all of its other applications that for the initial excitation of a motion there be given either a communication of velocity on the part of another moved mass or the elimination of the obstacles that oppose the free action of the given forces. The latter can itself not occur without prior physical motions. The well-known principle that the quantity of motion in the universe must always be the same stands opposed to every free commencement of a corporeal motion by means of ideal influences, which would indeed continuously change that sum of motion.[324]

Lotze interpreted this line of reasoning as leading to an unacceptable denial of freedom of the will. His solution represented nothing less than a denial of one of his fundamental principles, the identity of the laws governing both the organic and inorganic realms: "There is . . . no other assumption necessary than that of a physical force in the substance of the central organ whose intensity is variable, and in such a way that its variations correspond to the ideal events in the mind in accordance with certain general arrangements."[325] He assigned the question of the acceptability of such a *variable Kraft* to metaphysics—which, he found, has no objection to it. So much (again) for the conservation of energy! He was well aware that this maneuver was tantamount to an "exclusion of the concept of causality," and he took refuge behind the ostensible need for a special force to account for a special set of circumstances: "We are forced to assume a variability of force only by the influence of the mind," by the "influence of the ideal on the material."[326]

Lotze did not use the term variable force in his *Lebenskraft* article. He instead located the distinctiveness of the organism, considered as a machine (*Mechanis-*

*mus*), in the fact that it possesses a "principle of immanent disturbances which follow no mathematical law of their strength or recurrence at all," a principle that endows the organism with the capability for "an absolutely new initiation of mechanical motion."[327] In that article, too, he did not give such explicit or unqualified acceptance of occasionalism, but instead toyed rather inconclusively with the notion that the soul might be able to affect the body insofar as both are "substances."[328] Accordingly, he was also willing to entertain at least the possibility that one person's soul might be able to affect another's directly, or even to affect inanimate objects, as in alleged cases of what was then called, quite generally, "animal magnetism." Though skeptical about the reality of such phenomena, he refused to rule out their possibility on a priori grounds. Lotze had expressed a similar judgment in his earlier work, though he was more dismissive there of the possibility of certain spiritualist phenomena such as the displacement of one of the senses, the substitution (*Vicariren*) of one organ for another, or the movement of an external physical object.[329] Lotze's views on these difficult matters were clearly in flux.

(The last section of this chapter contains a summary of this section's findings with respect to the vital force and its significance as part of the background to Mayer's formulation of a concept of an autonomous, uncreatable, and indestructible force.)

## 3 Leading Analogies

As we saw in chapter 2, the search for valid analogies played a key role in the development of Mayer's ideas. Indeed, the very concept of force and its characteristics were both arrived at and presented in terms of the central analogy between uncreatable, indestructible, imponderable *force* and uncreatable, indestructible, ponderable *matter*. This section will first examine the rich analogies elaborated by contemporary German physiologists—the chief authors in Mayer's presumed intellectual background—with respect to the relationship between the imponderables, vital force, and the soul. In Autenrieth's hands, this analogy was extended to include parallel (if not always evenhanded) treatment both of ponderable matter and of gravitational attraction and motion at opposite ends of a rough spectrum of entities that embraced the imponderables, vital force, and the soul. As was the case with authors' handling of the vital force alone, such discussions inevitably involved reflection on problems of real or apparent creation and destruction.

Similarly important to the course of Mayer's thinking were the image of the organism as a machine and the analogy between the solar system and living organisms. Not that Mayer necessarily believed that the organism *is* a machine—he was, after all, fervently opposed to the reductionist materialism of mid and late nineteenth-century Germany—but in order to analyze its material and energetic relationships he thought quite naturally of the organism *as* a machine. That metaphor, too, was widespread in the scientific literature with which Mayer was likely familiar, and it was a metaphor freely employed by people who were very far from regarding organisms to *be* machines.

One of the most perplexing features of Mayer's work was his application of the parallelogram of forces to central force motions, leading him to the conclusion that force is generated in the sun. The sources of those notions of force and motion in the contemporary physics literature will be explored in chapter 4. Here I will present evidence that Mayer's conclusion was facilitated by the widespread practice of regarding the solar system as a kind of divinely ordered "organism." Although there is no evidence that Mayer, early on, paid any special attention to the concept of a motion-producing vital force per se, he nevertheless did not hesitate to entertain the possibility that all "organisms" might contain within themselves the ability to create (though not to destroy) force, once he had convinced himself that force is created in the solar system as a result of the gravitational interaction between planets and sun. The ready availability of the metaphor of solar-system-as-organism enabled him temporarily to bracket those systems together as acceptable exceptions to the laws he believed held for the purely mechanical systems of inanimate nature. Mayer's thinking was shaped at every turn by the images and usages he encountered all around him.

### *3.1 The Relationship between the Imponderables, Vital Force, and the Soul*

So intimately did German physiologists of the 1830s relate the concepts of vital force and soul to that of the imponderables that it was impossible not to anticipate those connections in the preceding section. The issue was of such contemporary significance, however, and was so significant an aspect of the conceptual background to Mayer's creation of a new concept of force, that it warrants full and separate treatment. It must constantly be borne in mind that if Mayer was to discover the conservation of energy—at least in the form in which *he* conceptualized it—he first had to 'discover' an entity about whose conservation he could ask pointed questions. It is my contention that the identification of such a new entity, Mayer's "force," took place within the context of the issues to be examined in this section. I believe that a crystallization of meaning with respect to force could have taken place only if Mayer had already been actively aware of a rich context of issues and associations within which a relatively small change of perspective could have brought about a major reconceptualization of the nature of force and its ontological relationship to the imponderables. Although the physiological "common context" dealt centrally with the vital force and the soul, it is well to recall that Mayer, in his earliest papers, specifically limited the scope of his ideas to the inorganic realm; the possibility, indeed the positive belief, that force can be created in organic systems was an idea Mayer continued to hold until at least the second half of 1842.

Although these issues were widely explored in the physiological literature, one author above all others made them the subject of an amazingly ramified and extended discussion: Johann Heinrich Ferdinand Autenrieth, Mayer's most famous and personally most esteemed professor at Tübingen. His posthumously published book of 1836, *Views on the World of Nature and the Life of the Soul*, was devoted overwhelmingly to the investigation of the phenomenological and ontological similarities and differences among a series of entities ranging from the soul to matter and motion via the vital force and the imponderables. There is, to be sure, no direct evidence that Mayer read this work. Fortunately, given the prominence of the issues in works we can be sure Mayer knew, such as Autenrieth's and Müller's physiology texts, or in one he stood a good chance of knowing—his teacher Elsässer's 1834 edition of Magendie—the thesis to be argued here does not absolutely require him to have done so. Nevertheless, the assumption that Mayer read Autenrieth's *Views* renders his problem context and initial insights more vividly intelligible than any other known to me.

Carl August Wunderlich, Mayer's university friend and later a professor of medicine at Leipzig, wrote a *History of Medicine* in which he identified several schools of physiological thinking current in Germany during the decades around 1800. One *Anschauungsweise*, which he associated with Reil, Humboldt, and others, sought to understand life in terms of the general laws of nature applied to the particular combinations and arrangements of the familiar substances of the physical world found in organic (read: organized) beings. He attributed a second way of regarding life to Autenrieth:

> A second point of view regards the principle of vital events [*Lebensvorgänge*] as something *distinctive, though comparable and analogous to the other imponderable substances*. It is not electricity, not magnetism, but a specific principle that only has analogies with them. This was already a not unimportant step forward. It stems from Autenrieth and was expressed in many places in his *Handbook of Empirical Human Physiology* (1801). An extensive consideration of the relationship between the vital principle and the imponderables is found in the dissertation of Autenrieth's (Matthes *respondens*), *On the Difference between the Nature of the Organic Force and the Imponderable Fluids*. Autenrieth demonstrates at length there that the vital principle can be confused with no other imponderable, but is completely specific.[330]

Let us look first, then, at the two works Wunderlich mentioned.

Autenrieth devoted a separate section of his *Handbook* to the "Similarity between the Vital Force and the Imponderable Substances [*Stoffe*]."[331] He began by noting that since a person's body neither gains nor loses weight upon death, the vital force must consist either of an imponderable substance or in the arrangement of the material parts of the body. The most telling analogy he invoked was between the indefinite multiplication of vital force in the process of reproduction and the ability of imponderables to undergo a similar indefinite increase, as evidenced by a number of common phenomena:

> Just as a magnet can make another piece of iron magnetic, both together a third, and so on without end, without this propagation of magnetism being a mere apportionment [*Vertheilung*] of a determinate quantity of the magnetic matter of the first; for the last piece of iron made magnetic (and likewise every other piece in the whole collection), if it is rubbed by the whole collection of all the pieces previously made into magnets, can become incomparably stronger than the first magnet, which set all

this in motion; just as here there is a true generation [*Erzeugung*]; just as furthermore a small quantity of electricity can cause the unlimited production of a larger quantity by means of the voltaic condensor; just as, finally, a burning kernel of gunpowder can set fire all at once to a whole heap of them without this flame being merely apportionments of the original flamelet:

In the same way the vital force of one organic body produces others ad infinitum without this progeny consisting merely of an apportionment of the first force.[332]

Against this common capability of the vital force and the imponderables to regenerate themselves as they are passed from one body to another, Autenrieth contrasted the behavior of "mechanical" systems, in which the propagation of force is always accompanied by its division:

On the other hand, in the mechanical world every propagation of force is merely apportionment. A moved ball can, to be sure, simultaneously set in motion two other equal balls; but each of the latter will then proceed with only half the force of the first.

Just this apparently greater size of the effect than the cause—insofar as the number and bulk of the organic progeny is greater than the number and mass of the organic ancestor—by which besides the vital force the imponderable substances are in general distinguished from the mechanical effect of impact and counterimpact in the case of ponderable substances, is also evident with the vital force in the manner in which external stimuli act on an animate part.[333]

For example, stimulating only a few fibers of a muscle will cause its entire mass to contract, and an insignificant prick of a thorn can cause a horse to leap several feet. Autenrieth called explicit attention to the "lack of proportion between effect and cause" in animals' response to stimuli.[334]

These and other phenomena, such as the return of the vital force to once-frozen body parts, Autenrieth believed justified the conclusion that the capacity to produce vital motions must lie not in a particular arrangement of parts or solely in a particular chemical combination of the conventional imponderable substances in an organ. Rather, as the first cause of these motions we must assume the existence of a distinctive imponderable substance that follows the same laws as the other imponderables, just as one assumes that it is the presence of a distinctive galvanic fluid or magnetic substance that decomposes water or moves iron filings.[335] Among the analogies Autenrieth elaborated between the vital force and the imponderables was the instantaneous propagation of electrical, magnetic, and nervous actions.[336] He interpreted the appearance of "Priestley's green matter" and infusoria in distilled water as evidence for "the probability that vital force in general can also come into existence where previously none was evident, just as galvanic fluid newly comes into existence, at least for the observer, in the case of heterogeneous metals and evaporating liquids, and as do electricity and heat by means of friction and magnetic fluid in iron by means of an impact in a certain direction."[337] Where those substances come from, beyond the ken of the observer, Autenrieth did not there say; his purpose was simply to make his root analogy more acceptable.

The issues broached in 1801 were further explored ten years later in the medical dissertation of one of Autenrieth's students, *On the Difference between the Nature of the Organic Force and the Imponderable Fluids.* While repeating some of the similarities Autenrieth had insisted upon, this short work was more concerned with their differences. The imponderables, for example, can pass through empty space and, once in existence, can continue to exist independently of ponderable matter, whereas the organic force (*vis organica*) is never found without its associated ponderable matter.[338] And whereas the protean organic force is constantly changing within a given organism, and is subject to mutation in the production of fertile plant hybrids, "the imponderable fluids declare the immutability of their kind": "There is no common *tertium quid* between galvanism and electricity, nor between the latter and magnetism; galvanism always remains galvanism, light light, etc. etc. No imponderable passes over into another by means of a hybrid. The development of one imponderable fluid from another never takes place by means of an intermediate species or an insensible transition."[339] The organic force, furthermore, contains within itself a "first cause" of its "primitive activity," whereas the imponderable fluids, like ponderable matter, only change as a result of an externally applied force.[340]

Matthes's dissertation differed somewhat from Autenrieth's *Handbook* in its treatment of the capacity of the organic force and the imponderable fluids to undergo indefinite increase. Accepting parents' ability to produce offspring without suffering a "distribution" of their organic force or a decrease in its intensity—thus increasing the amount of organic force in the world—Matthes argued that heat, light, and electricity show a decrease in intensity when communicated to other bodies or into a larger space.[341] Only in another context did he invoke Autenrieth's examples of the indefinite multiplication of the imponderables, namely with respect to the ability of the organic force to set limits to its own increase—as shown, for example, by the fact that a hare and an elephant who eat the same food nevertheless assume vastly different sizes. Imponderables, on the other hand, are incapable of setting any limit to their own increase: "Fire . . . would burn to eternity if new wood were constantly supplied; nor would a constantly moved electrical machine lack for the electric fluid, nor would magnetism cease its expansion if you continuously rubbed new steel rods with the old magnet; rather, it would increase without bounds."[342] Noting that the peculiar properties of both ponderable matter and the imponderables are an expression of their own nature and not something one class derives from the other, Matthes asked rhetorically at the end: "Why shouldn't the cause of the organic force . . . be taken to be distinct from the nature of ponderable and imponderable [substances]?"[343] In a similar fashion, he further suggested that the soul, conscious of its freedom, is an entity *sui generis* distinct from the organic force.

Although Autenrieth stressed similarities and Matthes differences, they were in substantial agreement that the vital or organic force is *like* the imponderables in certain regards but nevertheless *essentially* distinct from them. The issue of the indefinite (self-)multiplication of both vital force and the imponderables received similar prominence and was illustrated by essentially the same examples in both works. Autenrieth returned to these issues with a vengence in the work published posthumously in 1836. Before examining it, however, let us look at the writings of two other German physiologists, Müller and Elsässer, who addressed the issues from a similar standpoint during the intervening years.

In Wunderlich's account, "the idea of a certain parallelism between the vital principle and the imponderables became quite popular in German physiology," especially with Karl Friedrich Burdach and Johannes Müller.[344] Wunderlich believed that this *Parallelisiren* had benefited physiology by encouraging the application of precise methods of investigation developed in physics. Be that as it may, his inclusion of Burdach in the company of Autenrieth and Müller requires some qualification. Having adopted and retained certain features of the *Naturphilosophie* that had engaged many German life scientists more substantially during the first decade or two of the century, Burdach defended views that found little resonance in the late 1830s. Unlike most writers of that decade, who regarded the imponderables as weightless fluids, for Burdach vital activities resemble electricity, magnetism, and heat insofar as they are all "dynamical activities," that is, manifestations of a force, an "internal ground," which is neither spatially extended nor endowed with material substantiality.[345] Burdach's conceptions of both the vital force and the imponderables were quite outside the mainstream when he published them in his *Anthropology* in 1837 and in the last volume of his *Physiology as an Experiential Science* in 1840—when he was in his sixties—nor did he press their analogy with respect to issues of causality or quantitative increase. His way of thinking about the world had few affinities with the problem context I have identified as providing the background necessary to understand Mayer's work.

Müller, on the other hand, exploited certain analogies between the vital force, the soul, and the imponderables in ways, reminiscent of Autenrieth, that raised the kinds of ontological questions I have put forward as central to the contextual understanding of Mayer's achievement. Addressing the problem of apparent creation, Müller called repeated attention to the latency of the imponderables—the concept of latent heat being the original historical exemplar—as an analog for both the calling forth of an energetic vital response to a slight external stimulus and for the replenishment of the organic force from ingested nutrients. With respect to the first likeness he wrote, in a passage already cited: "The capability of being determined by external influences to manifestations of force is not proper to organic (in particular, animal) bodies alone. For example, many inorganic bodies develop light under certain conditions—e.g., through impact—or they develop heat. According to physicists, it is probable that the light or heat was previously bound in the bodies and became free as a result of the external influence."[346] The second likeness provided one of the strongest reasons for Müller's favorable estimation of the "pantheistic view" he sketched in the second volume of his *Human Physiology*:

> But as soon as so-called dead matter comes into interaction with the existing organism, and is transformed by it into the same structure and subjected to the vital princi-

ple of the organism, the capability latent in it for life in a determinate form also becomes manifest, and the form of its activity is contained by the already existing organization within its limits. In this way, by means of the appropriation of matter by an organic being, the organic force is increased along with the appropriated and organized matter, and through the increase in force a division of it is again possible. As phenomena analogous to the manifestation of the general principle of vitality [*Lebensfähigkeit*] latent in matter one might then cite the physical phenomena in which an existing force—latent, however, as far as appearance is concerned—such as electricity or light makes its appearance under determinate conditions of the interaction of bodies.[347]

As already seen, the assumption of the latency of organic force in all matter solved the vexing problem of the multiplication of organic force exhibited in growth and generation.

Thus the organic force is *like* the forces of electricity, light, and heat insofar as all are capable of existing in both latent and manifest forms. But the organic force is not thereby simply a transformation of the physical forces of nature. The analogy was designed to render plausible a particular conception of the organic force, not to assimilate it to the physical forces of nature. Müller hastened to make that clear early on in his compendium. After discussing the phenomena which he believed support the assumption of an independently existing and, as it were, communicable organic force, and after invoking the analogy with the known imponderables of physics, he cautioned: "As certain, with all these facts, as the existence of an often quickly acting and spatially extending [organic] force or imponderable substance is, so little is one justified in taking it to be identical to the known imponderable substances or general forces of nature such as heat, light, and electricity, a comparison that is rather confounded by every closer investigation."[348] Müller reinvoked this guiding analogy and its pivotal notion of latency when he considered the interaction between the soul and the organism. His introductory paragraph is worth quoting in full:

The relationship between the soul and the organism can in general be compared with the relationship between any physical general force and the matter in which it manifests itself—e.g., light and the bodies in which it makes its appearance. What is puzzling about the connection remains the same in both cases. Light makes its appearance in bodies partly through merely mechanical change in their matter—e.g., pressure and impact—partly through a chemical change in them. Light is in turn also capable of bringing about material changes in bodies. Likewise, electricity makes its appearance through material change in bodies and here in turn produces material changes in bodies. Psychic actions [*geistige Wirkungen*] occur in organic bodies as long as their matter changes, and the psychic actions here in turn change the matter. For the germ contains, along with its indwelling vital force, at the same time the latent force for the psychic actions of the subsequent animal being; before a determinant structure of the brain has been created, the organic activity of the germ also remains without ideas [*Vorstellungen*]. Along with the structure is given the activity of the already existing force, which in its ultimate ground [*Grund*] is therefore not depen-

> dent on the structure of the brain, but with respect to its manifestation is dependent on that structure. Up to that point the relationship between the psychic forces and organization is not more puzzling than the relationship between every other natural force and the material state of bodies, or rather both are equally puzzling. The relationship between the psychic forces and matter differs from the relationship between other physical forces and matter only because the psychic forces are present only in organic bodies—in particular, those of animals—and because they propagate themselves only to their same products, [whereas] the general physical forces, which one also calls imponderable substances, have a much more general action and distribution in nature. Since, however, organic bodies are also rooted in inorganic nature and live from it (in that animals nourish themselves from animals and plants, plants however nourish themselves in part from inorganic substances and thereby grow and multiply), it therefore remains uncertain whether the disposition [*Anlage*] to psychic actions itself is not also present and becomes manifest in a determinate way by means of the existing structures.[349]

Müller has here amplified his analogy with the latency of force: not only is force, whether organic or physical, released when ponderable matter is appropriately disturbed, but that now-manifest force can in turn act back on matter. He has also exploited a common function of such suggestive and tentative analogies by appealing to the puzzling and unexplained aspects of the nevertheless generally accepted conception of the physical forces of nature: my vital force is no more problematic than the forces *you* accept. Although Müller closed the paragraph with his usual diffidence toward expressing metaphysical views in a scientific work, it is clear what kind of ontological ideas he thought would answer some of the most difficult questions about life. Note, too, that although the soul, like the vital force, requires the presence of an organ of appropriate structure in order to manifest itself, its existence is not dependent on that structure, but rather the other way around: the structure is the manifestation of the operative idea or force.

We have already encountered Carl Ludwig Elsässer, Mayer's professor of physiology and pathology at Tübingen, as an early and severe critic of the concept of vital force. A central part of that critique involved a brief discussion of the status and legitimacy of a range of entities considered to underlie the phenomena of both organic and inorganic nature. Is one dealing with imponderable substances, with forces (either organic or physical), with some immaterial principle, or with the properties of ponderable matter? In opting for the last-named, Elsässer not only denied a special vital force, but, in accepting a world in which only ponderable matter and its properties are real, called into question the existence both of imponderables and of any independently existing forces:

> The whole controversy over the causes of life turns in general on the question: are the distinctive phenomena in living bodies results of the forces of their matter, thus of their chemical and mechanical composition, such that their differences would be explained simply in terms of the differences of the latter (as is the case in inorganic nature), or do they derive from a principle which is to be sought outside the properties of matter? One saw in this principle either something wholly immaterial, spiritual

[*Geistiges*], or one assumed a substrate for the general vital force related to the imponderables. . . . Against the assumption of an imponderable matter as substrate for the general vital force is the variety of vital processes that this one substance would have to produce, besides the circumstance that physicists still leave it questionable even for the customary imponderables whether they really exist or are not rather mere properties or forces of the ponderable substances in which their effects [or actions (*Wirkungen*)] are manifest.—One would be able to decide in favor of the acceptance of such hypotheses only if it were proven positively impossible that the different activities of living bodies can be the result of their chemical and mechanical composition. In inorganic nature we always see matter and force change simultaneously: if two bodies have different composition and form, they also have different forces; if a body exhibits a distinctive force, then we also find a distinctive state of its matter or can at least infer such (if the imponderables are not real sustances, then their effects must likewise be derived from a change in composition or at least in the mutual position of the atoms). In accordance with these experiences we everywhere regard the forces of inorganic nature not as entities in their own right [*eigene Wesen*], but as mere properties of matter. Now should this relationship between matter and the activity-determining forces in living nature be something else? Don't the living bodies endowed with distinctive forces also have a distinctive composition and form of their matter?[350]

One of Mayer's prime tasks would be to vindicate for force the status of an immaterial substance existing independently of matter: it is not just a property of something else. Although the legitimacy of the vital force per se was not Mayer's central concern, the ontological issues raised in the context of its treatment in contemporary physiological works—typically involving comparison with the imponderables—set the conceptual stage on which he would shape his innovative doctrine of force.

The high-water mark of attempts to elucidate the nature of the vital force by exploiting the analogy between it and the imponderables was, in the Germany of the 1830s, Autenrieth's *Views on the World of Nature and the Life of the Soul.* In many ways the entire 552-page book was an extended—indeed, a tediously repetitious—elaboration of that analogy, which Autenrieth exploited to justify the assumption of the existence of both a vital force and a soul as entities independent of matter and capable of indefinite increase. If you're willing to accept thus-and-so for the imponderables, even when you can't explain it, then you shouldn't balk at accepting an analogous vital force. But whereas Müller explained the apparent creation of the vital force and the soul in terms of their latency in ponderable matter, Autenrieth used the same phenomena to justify belief in a realm of existence beyond the physical world. Significantly, both solutions were premised on the implicit absurdity of granting the creation of something out of nothing.

So central to Autenrieth's conception of the vital force was the analogy with the imponderables that he scarcely ever spoke of the former without invoking a comparison with one or more of the latter. After introducing the topic of his first essay as the nature of life and its vital forces, he proceeded to identify a number of things common to all organic beings: their forms remain within distinct boundaries; they constantly exchange substances with the external world; they respond in a complex fashion to external stimuli; they grow and reproduce. With that, Autenrieth had cataloged pretty much the standard set of phenomena assigned to the vital force. He went on:

> This common something that expresses itself in all living bodies can be considered not only by itself, separate from that which otherwise distinguishes every species, but one can also ask whether it is not also underlain by a common and real force possessing an autonomy of its own, apart from the existence of the organic body in which it acts, or whether life, without an antonomous basis [*Grund*], is merely a property of visible body mass assembled in a determinate fashion into a plant or an animal.
>
> Such a question may well be raised because even the continually agitated force of heat in heated bodies is not visible to us separately; but we perceive here that its action communicates itself to even the most varied bodies with always the same characteristics, and that [on the other hand] its phenomena in the same corporeal substance are now present, now absent, can now strongly express themselves, now be present to a diminished degree . . . . We therefore consider ourselves justified in ascribing the phenomena of heat to the presence of a particular autonomous force, which is to be sure capable of combining with all space-filling bodies, but which is nevertheless still distinct from them all.[351]

The ability of heat and the "magnetic fluid" to pass through space devoid of ponderable matter demonstrates that heat and magnetism are independently existing entities, not mere properties of matter; and he believed (without apparent logic) that certain common phenomena of magnetization demonstrate the preexistence of the magnetism before it becomes manifest in a particular body.[352] In driving home the force of his analogy, Autenrieth invoked the equivalent inexplicability of what happens to electricity, light, and life when each disappears from the visible world:

> Since we already recognize activities in the rest of nature whose cause cannot be merely a property of the bodies in which they are active, we may likewise investigate, without undertaking anything contradictory or frivolous, whether the basis of life, even if it is perceptible to us only in the living body, does not nevertheless lie in an autonomous force that—like the magnetic fluid, when it comes to the iron makes it into a magnet, and when it again leaves it leaves it behind as the mere unmagnetized iron it was earlier—in the same way combining with organic matter animates it, withdrawing from it leaves it again dead, in this regard, then, again the same as every other ponderable corporeal substance. It is no objection that heat and light only become perceptible to us in a state of uninterrupted motion, since we likewise also recognize life only through its manifestations of activity. Just as we do not perceive in the case of electricity where it again withdraws to when it again suddenly disappears for all our sense through the reunification of its opposing forms, the positive and negative electricities, and just as we do not know where light goes when it is extinguished, in the same way we are also not capable of perceiving where life escapes to when the animal or plant dies, and only its visible body remains behind for a time in its exterior form.[353]

So, too, does the sudden appearance of a disease-producing contagion demonstrate the existence of an "extrasensual world," the same "indifferent source" from which its opposite, the vital force, also comes.[354] When a candle burning in a mirror-lined room is extinguished, the light previously present does not become latent in the walls, nor do the positive and negative electricities that recombine in a wire become latent in the wire. In both cases the once-present imponderables have literally withdrawn from the phenomenal world, from space itself.[355] The fulminates of silver and gold (*Knallsilber* and *Knallgold*) explode with great force upon being struck, yet they can be chemically decomposed without exploding or releasing a comparable quantity of electricity, heat, or light. The latter's nonlatency in the original compound had profound implications:

> Those imponderables are therefore not latent in the small quantity of fulminate, otherwise they would more or less also have to become free when the latter is dissolved. Now, since one also cannot assume that they could have been suddenly gathered together in such an immense quantity out of the surrounding air at the moment of the explosion in order thereby to create the latter—something that could be raised to an incontrovertible certainty through observation of an electrometer and thermometer in the neighborhood of the explosion—one must thus assume that these imponderable fluids come into the spatial world [*Raumwelt*] only at the moment of the explosion. In the same way it can also be shown that the withdrawal of electricity from the spatial world when its two opposing forms mutually combine with each other is not a redistributing of the previously accumulated electricity in the neighborhood of the surrounding air—a distribution that could only occur gradually, not suddenly be achieved. In the same way even physics can prove that there really is, in the case of the imponderables, a coming to the spatial world and again a withdrawal from it, and that therefore—unless one wishes to assume a perpetual passing over of one part of creation into nothing and its reappearance out of nothing—a real world of determinate diverse forces must occur beyond or outside all space, forces that, given the appropriate constitution of the corporeal world when it (as it were) unlocks the world beyond [*das Jenseits*], are capable of slowly or suddenly coming into it, but likewise also of again withdrawing from it.[356]

Thus the assumption of what might be called a preterphysical realm is the only way to avoid having to grant creation *de nihilo* and destruction *ad nihilum*, a point Autenrieth made again and again. That realm, of whose existence physics assures us, is also the abode of vital force and soul.[357] Autenrieth likened the appearance of the ego (*unser geistiges Ich*) in connection with the generation of an appropriate organ, the brain, to the appearance of magnetism in a piece of iron struck while being held along the magnetic meridian.[358] Both the soul and the magnetic force exist before their manifestation in the material world. Autenrieth embellished the analogy by likening the fertilization of toads' eggs by extremely dilute sperm to the ability of a small spark to set off a large mass of gunpowder: "The mass of fire which this produces was not in the small fiery spark which ignited the powder, but without this little spark it would never have appeared by itself. The animation of organic germs likewise appears to be an opening up of the spatial world for the supervening of force from the nonspatial Beyond, an opening up already hinted at by phenomena of even inanimate nature."[359]

Autenrieth cited a variety of phenomena which he thought demonstrate "that the vital force is something *sui generis* [*etwas Eigenes*], which can have by itself a kind of continued existence even without being joined to an organic body."[360] Such phenomena, which reinforced the analogy between the vital force and the imponderables, included things like spontaneous generation, the regeneration of snails' heads, and the revivification of frozen animals. For Autenrieth, the last-named phenomenon was an instance of the vital force's leaving and then returning to the affected organism. Here, too, belonged the common ability of the vital force and the imponderables to undergo indefinite multiplication as long as the appropriate substances or motions were provided, though the vital force distinguishes itself by its ability to create new self-contained and independent entities (i.e., individual organisms) as a result of its quantitative increase.[361] Autenrieth pointed out that although the imponderables thus share with the vital force a trait which distinguishes both of them from the quantitative invariability of ponderable matter, it is only the vital force that possesses the capability of *self*-multiplication: the activity of an external agency is required to bring about a multiplication of the imponderable fluids.[362] So, too, does the analogy break down with respect to the ability of the vital force to manifest itself in a diversity of ways and in ever higher stages of development, whereas electricity, the magnetic fluid, heat, and light remain just what they have always been.[363]

Both the vital force and the imponderables—with the exception of heat—manifest intrinsic polarities (*Polaritäten*) or antitheses (*Gegensätze*), whose constant ratio in the imponderables indicates that they represent a "division of a common neutral activity or indifference [*Indifferenz*] of the particular fluid."[364] On the other hand, only with respect to the vital force are those polarities always active, never in complete equilibrium.[365] Organisms thus possess a "not further explicable basis" of autonomous internal activity:[366]

> In the indifference of that which, when divided, expresses itself as two kinds of electricity there appears no self-determining activity . . . . Even when divided electricity has been called forth, its remaining divided does not depend on itself but is entirely dependent on the bodies with which it entered into combination. With the organic vital force, on the other hand, not only are the antitheses indeed produced in the same body . . . , but they also separate themselves by themselves more and more from each other in very different ways especially determined for each different species . . . . Thus here, too, the organic vital force, although its appearance is likewise still dependent on the influence of the corporeal external world, nevertheless already reveals vis-à-vis the imponderables an active self-determination, and thereby in general a certain independence of its motion, motion whose continuance and change does not therefore also depend solely on the external cause that called it to appear in the spatial world.[367]

Such a hidden source of self-movement, Autenrieth suggested vaguely, might also explain the continued motion of the planets: "In the same way only a continuing source of all motion, not an impulse imparted once in remote antiquity, most likely . . . explains the continuation of the motion of all of the stars, the sun, and our earth, since space does not appear to be any kind of absolute vacuum."[368]

Thus, to use terminology which Autenrieth here did not, both microcosmic and macrocosmic organisms harbor within themselves an internal source of motion, and are thereby distinguished from the merely reactive world of ponderable *and* imponderable matter. Here, as so often elsewhere, the validity of a guiding analogy was assessed and circumscribed by exploring the ramifications of its outer reaches, though occasionally Autenrieth seems to have lost control of his images, as when he cited the ability of *motionless* ponderable matter to entice (*entlocken*) not motion but the force of attraction from the same immaterial preterphysical source from which the imponderables are enticed as a result of certain physical *activities* (as electricity through the rubbing of amber or light and heat through the collision of steel and stone).[369] In another context, Autenrieth described the sun and stars as sources of light they draw from a universal invisible source.[370] Autenrieth's image of the solar system, the very paradigm of gravitational attraction, as containing somehow within itself a source of motion or force, has close affinities with Mayer's attempt to relate the creation of heat and light in the sun to the gravitationally constrained motion of the planets around it. Both men sought to justify the plausibility of their ideas by appealing to an analogy between organisms as creators of force and the solar system as an organism. Autenrieth's notion of the origination of opposing activities out of a primitive state of "indifference" also recalls Mayer's use of a similar conception.

Autenrieth's richly elaborated analogy between the imponderables, the vital force, and the soul was part of an overall strategy designed not only to render the latter two entities scientifically legitimate by assimilating them conceptually to certain essential characteristics of generally accepted substances, but also, more particularly, to render plausible the idea of independently existing substances that do not occupy space and the idea of what I have called a preterphysical realm of existence beyond the visible spatial world. Light, the force of gravitational attraction, and the vital force are all such really existing yet immaterial and nonspatial entities:

> But already a closer consideration of light . . . does not allow us to apply the concept of filling space [*Raumerfüllung*] . . . to everything that we must undeniably regard as something real, as an independently existing entity. If the properties of light—which, despite its reality, still only has the capability of filling space in one of the latter's three dimensions—give our thinking ability a point of support in order to rise gradually to the concept of autonomous and nevertheless nonspatial entities, and to wean us from imagining everything that exists only in essentially corporeal form; so can the prosecution of this consideration also serve to acquire for us some conception not only of how the heavenly bodies scattered in infinite space could, according to their different position in space, exert a different influence on a vital force that, although not itself existing in space, is nevertheless universal and autonomous, but also of the fact that these different external influences are even capable of exciting in such a nonspatial source of force internal changes united into a particular universal system. Insofar as

> we imagine the source of the light a body calls forth or attracts as itself extraspatial, nothing can prevent us from nevertheless simultaneously assuming that this source of light, despite its nonspatiality, can still retain all of the particular characteristics that have nothing to do with space, thus for example the capability of being able to be induced to exhibit now red, now green, yellow, or blue light. Not every property of an autonomous entity follows simply from the way in which it fills space . . . . Such properties of light . . . [are] entirely independent of its filling space . . . . Instead of making use (as here) of the fiction of light, we can also use as our conception any other imponderable fluid, accordingly even the universal vital force itself, as long as it is still only a question here of the formation of intelligible modes of conceptualization.[371]

Although we cannot form a mental image of such a nonspatial entity, nothing prevents us from defining it in terms of concepts drawn from experience.[372] Despite its nonmaterial and nonspatial character, Autenrieth frequently referred to it as a substance (*Substanz*).[373] One is reminded here of Mayer's efforts to vindicate the "substantiality" of force.

Autenrieth devoted a great deal of attention to the existence of a graduated series of entities running from ponderable matter at one end through the imponderables, the vital force, and the soul at the other.[374] In the sometimes inconsistent and always tedious actuality of Autenrieth's exposition, he also elaborated a slightly different series consisting of ponderable matter, the imponderables, gravitational attraction, and motion. In neither case did he always treat every member of the series with respect to the particular trait under consideration, and my account represents an attempt to bring somewhat more order to these issues than the text itself possesses. Seen from this perspective, Autenrieth's book was first and foremost an ontological tract: what kinds of entities exist in the world, and how are they related? We have already looked at the fundamental properties Autenrieth attributed to some of those entities. Of further interest to us are the terms in which he characterized and compared the entities comprising those series. For example, one measure of difference was in their degree of activity (*Thätigkeit*):

> This self-division of the vital force, as far as it arises out of its own activity, indicates most decisively its higher standpoint in the series of entities. Ponderable substances [*Stoffe*] have no activity of their own; the imponderable fluids exist, to be sure, only in an uninterrupted activity, but this is not yet a free self-activity insofar as its arousal depends entirely on the motions of the corporeal world that obeys the law of gravity. But even the self-activity that . . . first expresses itself in the organic vital force is not yet a voluntary self-activity deriving from the freedom of the will; rather, it likewise still originates out of the law of a necessity.[375]

Only the soul possesses the capacity for *free*—that is, freely willed—self-activity.

Another feature which distinguishes the different forces—that is, excluding ponderable matter from the comparative series—is their degree of autonomy (*Selbstständigkeit*).[376] The motion of a body is inseparable from the body in motion and extends no farther than the body; it thus possesses no independence.

Regarded as a source of motion, universal gravitation also requires the presence of a body in order to act, but its action extends beyond the limits of the body; it thus possesses a degree of independence. The imponderables, although likewise manifest only in the presence of corporeal matter, enjoy an *existence* completely independent of matter, and unlike gravitation they can be communicated from one body to another. The vital force, finally, has the same independence from its material substrate as the imponderables.[377] Autenrieth's inclusion of motion and the imponderables within the same class of "forces" was decidedly unusual for his time, and even he did not otherwise consistently so employ the terminology of force. As far as I can tell, his more usual characterization of the imponderables as "weightless fluids" did not imply the inclusion of motion.

Closely related to the quality of independence was the somewhat oddly termed personality (*Persönlichkeit*), which characterizes the degree of self-contained individuality manifested by the several classes of entities. As Autenrieth saw it, arbitrary aggregations of ponderable matter have no particular individual identity, although crystals show the first stage of separation into distinct "personalities."[378] Like uncrystallized ponderable matter, heat exists only in arbitrarily aggregated amounts, but magnets manifest a distinct "personality." That is, each magnet possesses a central *Indifferenzpunkt* which in a sense joins the opposing magnetic polarities streaming out from it in closed vortices; although the range of action of the magnet extends indefinitely far out, the ensemble of *Indifferenzpunkt* and magnetic vortex forms a kind of unity.[379] Even more does the vital force manifest itself in separate individuals, though one "organic personality" may, under certain circumstances, be divided—as when a plant is propagated by cuttings—just as two previously separate trees may unite to form a "common tree individual."[380] If such phenomena demonstrate the spatial localizability of the vital force, other phenomena reveal its nonspatial quality. As attesting the latter, Autenrieth cited his favorite example of the equal number of male and female births, the supposed increased fertility of women after many people have died in a plague, and the system of relationships that connects similar but geographically separated species through space and simpler to more complex species through time: all of these phenomena, he maintained, can be explained only in terms of some kind of spatially nonlocalized, universally diffusable law-bound vital force. At the end of the series are free-willed spiritual beings whose personalities are inseparable from their essence. These souls or egos—Autenrieth used both words—possess the further end-of-the-series trait of complete nonspatiality, just as at the beginning of the series a complete absence of personality characterized extended ponderable matter, which exists only in sharply circumscribed regions of space.[381] That the soul is not extended in space is shown by the fact that our consciousness is undiminished when even a large portion of the brain is injured, or when we lose a large part of our body.[382]

Unlike all the other entities, only ponderable matter is imperishable, capable of neither increase nor decrease.[383] Using a word that itself hints at his belief in a preterphysical realm, Autenrieth described ponderable matter as "unexpellable" (*unverdrängbar*): "But if ponderable matter, too, has only at one time emerged out of that which is beyond all space [*aus dem Jenseits alles Raumes hervorgetreten*] in order to fill it, so is it at least now on our earth without a trace of being able to return there; it has emerged from there completely and permanently, without preserving any connection with that Beyond."[384] Ponderable matter is thus quite unlike the weightless fluids, and one must not suppose that the former can be transmuted into the latter merely by becoming, say, rarefied, or that the imponderables are nothing more than a particular state of ponderable matter.[385] In support of his contention that "the vital force is something autonomous added to the organic plant and animal form and again separable from it, not merely the property [*Prädicat*] of a particular arrangement of the matter making up the organic body," Autenrieth cited the analogous behavior of heat, "an imponderable fluid, which can be expressed unchanged by presssure out of the air, a ponderable body, like water out of a wet sponge, and thus must be assumed to be something that exists by itself autonomously and not merely as an alteration in the activity of the air or a mere property of a particular state thereof."[386] Recall that Mayer, too, was at pains to deny that heat was a state of ponderable matter.

The imponderables represent a particular mode of being, between imperishable heavy matter on the one side and the merely transitory motion of matter on the other.[387] Against Autenrieth's former (and exceptional) inclusion of motion and the imponderables within the same class of forces, he here distinguished them: motion is merely a property of ponderable matter, whereas the imponderables are independent entities *sui generis*. Such was Autenrieth's more usually stated view. The invariable motion-producing force of attraction belonging to ponderable matter occupies a kind of middle conceptual ground between the pure motion of quantitatively invariable ponderable masses and the independent but quantitatively variable imponderables.[388] Unlike purely mechanical motion, whose existence is wholly dependent on ponderable matter, the motion-producing force of universal gravitation shares with the imponderables the quality of autonomy, that is, the ability to exist or act in empty space independently of matter.[389] Although Autenrieth did not emphasize the constancy of motion in the world, let alone venture to define a measure of the quantity of motion, he did at least once distinguish the imponderables from motion in terms of the capacity of motion, once created, to continue indefinitely if unopposed, whereas the imponderables require the motion of ponderable matter both for their creation and their continued existence.[390] Autenrieth insisted that *all* changes in ponderable and imponderable substances, including the vital force, take place only as a result of some kind of mechanical motion.[391] In an amazing analogy he even compared the ability of the spinning earth to call forth the "cosmic vital force" from the "dynamic source of life" to the ability of a spinning glass sphere to "call into appearance" an "inexhaustible stream of electricity that previously slept in an existence imperceptible to us."[392] And just as ponderable matter and the imponderables are not transformable one into the other, so, too, is each imponderable wholly distinct from all the others.[393] Electricity may call forth magnetism, but the process involves no conversion of one to the other, and each imponderable continues to exist with its own distinctive characteristics. Similarly, violet light can magnetize an iron bar (as he believed), but it does not thereby cease being light, nor does it acquire the directional property of magnetism. Autenrieth viewed this as another expression of the imponderables' autonomy.[394]

The issues prominent in Autenrieth's book would thus have provided an ideal context for Mayer's early reflections on the nature of force. Within this context a relatively small change in the way the imponderables and motion were conceptualized—now as imperishable "forces" wholly independent of ponderable matter for their existence and capable of being transformed one into the other—could have brought about a radical reorganization of his entire worldview. It is, to repeat, the *issues* that are most significant, more so than Autenrieth's own views on this or that point. What is a force? An imponderable? A property of matter? Recall Mayer's terminological wavering between forces and imponderables, and his insistence that forces must not be confused with the properties of matter. Repeatedly discussed in Autenrieth's book was the issue of the creation and destruction of the vital force and the imponderables: his idiosyncratic allowance of their passage into and out of the visible world was in fact premised on the absurdity of their creation *de nihilo* and destruction *ad nihilum*. He also made a point of the imperishability of ponderable matter—a point, as we shall see in chapter 4, that virtually no physics or chemistry texts of the period explicitly mentioned. Autenrieth was also unusual in at least occasionally considering motion and the imponderables to belong to the same class, that of forces; and even where he did not, he repeatedly raised the issue of their relationship. He insisted on the necessity of motion for the appearance of the imponderables and the vital force, even as he denied any form of transformation of one species into another. He considered the force of gravitation as a source of motion and suggested that the sun, the preeminent source of gravitational force in our world, is somehow the focus of the calling into this world of light and gravitation from the same preterphysical realm. In one place he implicitly contrasted mechanical and organic systems insofar as the latter—which include the solar system—are capable of generating imponderable fluids and vital force, that is, of enticing them out of their hidden realm of existence. And, finally, Autenrieth's use of the term "indifference" recalls Mayer's reasonably similar usage. If there remains no hard evidence that Mayer read Autenrieth's *Views on the World of Nature and the Life of the Soul*, then the least that one can say is that Mayer's *problématique* probably derived from cognate (if less rich) sources, which included Autenrieth's own earlier work and the writings of Müller and Elsässer.

That Autenrieth's ideas on these matters enjoyed some contemporary visibility is shown, finally, by the fact that Lotze regarded them as worth citing and opposing in his *Lebenskraft* article of 1843. Of the first essay in Autenrieth's book, which Lotze characterized as a "treatise on the vital force," he wrote:

> He borrows his abstractions from an already manifestly incorrect consideration of the imponderables. Although a healthy physics can only imagine two kinds of things under imponderables—either really existing imponderable substances or distinctive changes and motions of the common ponderable bodies—so is the doubt nevertheless

> not uncommon (as otherwise also—to mention this parenthetically—in physiology, especially apropos of the nervous principle) as to whether it is substances or motions or merely forces that exist here. Now the last is certainly never possible, for abstract relationships cannot run around in the world without something they belong to. Now, it is upon such misguided analogies that Autenrieth has erected his theory of the vital force as an autonomously existing force separable from matter; he even thinks to find empirical verification for this wandering of the vital force when the blood previously driven away as a result of the freezing of the limbs flows back into the parts upon heating. It is upon this simple error that his otherwise careful and learned work founders.[395]

For Lotze, as for many of his contemporaries, the imponderables are either really existing weightless substances (*Stoffe* or *Materien*) or, as forces, properties of ponderable matter. For him, the concept of force remained tied to its Newtonian roots, hence tied to matter; any other conception was a nonexistent abstraction. For Griesinger, writing in 1845, comparisons between the soul and the imponderables in terms of their similar relationship to matter—both appear to be immaterial, and to both effect and be affected by material transformations—are of little use. In his view, there is "nothing really analogous" to the "psychic or nervous agency."[396] The rich analogies of his immediate predecessors meant nothing to him.

## *3.2 Organisms as Machines*

The metaphor of the organism as a machine was widespread in the German scientific literature of the early nineteenth century. In some cases the metaphor was little more than a passing turn of phrase, not indicative of any particular commitment to mechanistic modes of thought—as, for example, when Burdach asked, "Where [is] the seat of life, and where the spring which sets the wheels of the machine in motion?"[397] Following a favored usage of Berzelius, Friedrich Tiedemann frequently likened the body to a laboratory (*Werkstätte*) for the carrying out of chemical processes: "A living body, considered as an object of chemical investigation, is, as Berzelius expresses it, a laboratory in which many chemical processes take place whose end result is to produce all of the phenomena which in their entirety we call life."[398] More charged with literal significance was Autenrieth's repeated characterization of the body as a machine or mechanical arrangement (*Maschine, Maschineneinrichtung*): for Autenrieth, as for Descartes, except for the operation of the soul, the physical and chemical functions of the body take place in a strictly lawbound, mechanical, fashion. Animals' involuntary movements are, he thought, "merely a mechanical arrangement set in motion by the vital force." Only the freely acting soul introduces a physically undetermined factor into organic processes.[399]

Müller, too, while not a defender of a thoroughgoing mechanistic reductionism of the phenomena of life, nevertheless often employed graphically mechanical imagery in characterizing organic functions. Recall his likening of the execution of the organic functions to the harmonious interconnection of the wheels of the bodily machine, whereby the respired oxygen was the cause of motion of the mechanism, and the lungs, heart, and brain were referred to as the mainsprings of that mechanism. Recall his reference to external stimuli as the source of motion of the gearworks of the whole machine. Recall, finally, his later clarification of the quasi-idealistic import of his mechanical imagery: if, to be sure, the body can be likened to a purposeful mechanical contrivance, then it also remains true that mechanical contrivances do not create themselves, but rather represent the realization of a creative idea. In another place Müller asserted that the maintenance of a mature organism, considered apart from the processes of development and growth that characterized its earlier life, can be compared with the maintenance of a machine: "If one compares the embryo of an organic being with its state in advanced age, then the whole (which according to Kant determines the existence of the individual parts) consists in advanced age almost only in the interaction of the individual parts and their forces, in a manner similar to a mechanism which is maintained merely by the interaction of its parts."[400] The tradition of natural theology, as represented by Henry Peter Brougham, paid particular attention to mechanical contrivances and compared organisms to purposefully constructed machines and instruments with essentially the same intention as Müller: a mechanism designed to accomplish a clear purpose cannot be imagined to have come into being by itself, independently of a guiding idea (whether divine or 'Platonic').[401] Brougham repeatedly referred to the special "mechanisms" by which animals accomplish their vital functions, such as the system of bones and joints or the paddlelike webbed feet of water birds. Writing in the same tradition, William Prout followed Paley in aiming to give readers "some notion of the 'concealed and internal operations of the machine'"—by which he meant the human body.[402] Prout, too, used the argument that from the existence of "a machine admirably fitted for the office it performs" we infer the existence of an intelligent creator.[403] The important point is that in the pre-1840s climate, the metaphor of organism-as-machine was not only widespread but theologically safe, not yet the polemical property of an assertive and spirit-denying materialism.

Kremer has pointed out the similarity between Müller's mechanistic language and a passage in Liebig's *Animal Chemistry* taken over from earlier publications:

> [R]espiration is the falling weight, the tensed spring, that keeps the clockwork [of the body] in motion; the breaths are the strokes of the pendulum that regulate it. We know with mathematical precision for our ordinary clocks the alterations exerted on their regular motion [*Gang*] by the length of the pendulum or by external temperature, but only a few clearly recognize the influence that the air and temperature exert on the state of health of the human body. Yet discovering the conditions for maintaining it in the normal state is not more difficult than for an ordinary clock.[404]

Valentin, in reporting Liebig's views on animal heat, asked the reader to consider the organism as a machine in which the amount of heat produced depends on the combustion of fuel.[405] His elaboration of this metaphor two years later in the introductory section of his *Textbook of Physiology* (1844) nicely signals the way in which broad-based general changes in physiological thinking had complicated its significance: "Organisms have already often—and indeed from very diverse points of view—been parallelized [*parallelisirt*] with machines or other artificial contrivances that human beings are capable of manufacturing, without, however, the friends of this view having been able to completely sustain the comparison, but without its opponents, who often thought to see in such endeavors only the products of a crude materialism, having been able to expose it as completely inappropriate."[406] Although Valentin did not so put it, the difference was that now more people were asserting, or were seen to be asserting, that organisms *are* machines, hence the greater need to specify the range of validity of the likeness. The similarity he stressed was the fact that for both machines and organisms the manifestation (or exertion) of force (*Kraftäußerung*) depends on the consumption of fuel, for which the steam engine provides an especially trenchent analogy.[407] The relative coarseness of humanly constructed machines vis-à-vis the microscopic fineness of organisms need not weaken the analogy, but rather can be taken as evidence of "the infinite skill of their creator."[408]

Valentin's use of the steam-engine analogy probably reflected Liebig's influence. Alluding to its ingenious purposive activity, Liebig had been one of the earliest to invoke the image of the organism as a steam engine: "A not entirely inappropriate image for what goes on in the animal body is provided by self-regulating steam engines, for which the human mind has shown the most amazing ingenuity for the production of uniform motion."[409] Significantly, a later, more elaborate analogy between organisms and steamships suggested in addition the indispensability of rational, purposive guidance: both have a motion-producing source of heat and force in the food or fuel consumed, and just as the ship is kept in repair through the activities of the skilled crew, so, too, is the living body maintained by its own carpenters, blacksmiths, and locksmiths.[410] As we have seen, the idea of an active vital force had a strong hold on Liebig's imagination.

The first person to insist upon this kind of imagery seems to have been Liebig's rival, Jean-Baptiste Dumas, who in a frequently reprinted lecture of 1841 spoke of animals as "producers of heat and of force, veritable combustion devices."[411] As already noted, Dumas suggested that "the poetic assimilation of the railroad locomotive to an animal rests on a more serious foundation than one perhaps thought."[412] Indeed, several authors have argued that technological change in Western Europe, in particular the advent of railroads, stimulated the development of a new way of thinking, of regarding nature as (in Herbert Breger's words) a "working machine."[413] Many came to think of machines in terms of their capacity to do work. For Breger, the mechanical tradition, which increasingly had come to measure performance in terms of the height to which a weight could be lifted, was an important element in Mayer's conceptual background: Mayer, he noted, referred repeatedly to the mechanical effect (*Nutzeffekt*) of machines, horses, and people, and measured the work done by the lifting of a weight.[414] Breger sought to date this conceptual break in people's view of the world to the years after 1830, when there was growing awareness of the radical impact of the burgeoning industrial revolution.[415] Indeed, an anonymous article from 1842 on the founding of physiological institutes characterized the times in terms of the restless whirl of daily events and the "roaring of steam engines."[416] Hermann Schlüter's study of nineteenth-century biology paid particular attention to the metaphor of organism as machine; he, too, singled out the railroad as the most important transformer of consciousness. Schlüter cited Henrich Steffens's impressions of the profound effect of the railroad on people's imagination after around 1830, and quoted physiologist Carl Ludwig's Dumas-like recollection of the imaginative impact the railroad locomotive had had on his science: "For the first time there was constructed with this machine a *self-acting* mechanism in which the interplay of forces took shape transparently enough to discern the connection between the heat generated and the motion produced. The great puzzle of the vital force was also immediately solved for the physiologist in that it became evident that it is more than a mere poetic comparison when one conceives of the coal as the food of the locomotive and the combustion as the basis for its life."[417]

Given this complex of images and attitudes, one can imagine a kind of grafting onto a more or less traditional post-Cartesian picture of the organism as an exquisitely contrived machine, of energetic considerations derived principally from steam engines and, in particular, locomotives. What was the balance between fuel, waste products, and the heat and work produced? If, to be sure, such general changes in worldview and their influence on any given individual are notoriously difficult to pin down, the least one can say is that Mayer, growing up during those times, showed a deep and early interest in the work performed by waterpowered machines, and he applied quite naturally and heuristically to organisms energetic considerations that derived from a general understanding of steam engines.[418] In the 1830s natural theology and industrial technology both reinforced the traditional physiological metaphor of the organism as a machine—and Robert Mayer was well situated to be affected from all three directions.

### 3.3 *The Solar System as a Living Organism*

Even more widespread and multifaceted in the German scientific literature of the 1830s and early 1840s than the metaphor of organisms as machines was the analogy between organisms and the solar system, an analogy that harked back to the traditional comparison of microcosm to macrocosm and drew force from the explanatory success of the Newtonian world system. An appreciation of the pervasiveness of that image and of the issues it entailed will go a long way in reducing the apparent strangeness of Mayer's understanding of the dynamics of central-force motions and his early belief, held even after the publication of his first paper, that force can be created in organic systems, including the solar system. Once again the development of Mayer's ideas depended strongly on the analogies he regarded as valid.

Recalling Autenrieth's extended analogies between processes whereby the soul, the vital force, the imponderables, or the force of gravitation are called into existence from a common preterphysical realm, one is not surprised to find him among those who drew an analogy between living organisms and the solar system as containing within themselves perpetual sources of motion:

> But that perpetual internal change in the vital force presupposes . . . in this force itself a not further explicable basis [*Grund*] that sustains such continual disturbance in the balance between the essentially interrelated antitheses and does not allow them to come to equilibrium. In the same way only a continuing source of all motion, not an impulse imparted once in remote antiquity, most likely also explains the continuation of the motion of all the stars, the sun, and our earth, since space does not appear to be any kind of absolute vacuum.[419]

Johann Carl Passavant, a practicing physician in Frankfurt am Main who published extensively on animal magnetism, advanced a similar claim in his desire to establish a link between the forces of organic and inorganic nature:

> Organic bodies are principally distinguished from inorganic in that they are determined by an autonomous, purposefully acting principle. By means of this autonomous principle they then also have the capability to develop [*sich zu entwickeln*]. Nevertheless, this distinction is not absolute. For nature in the large—e.g., our solar system— also has such an autonomy. The motive and conservative forces are reproduced in it; they are not given to it from the outside. Light and gravity hold sway in the entire visible world. . . . The universal powers of nature [*allgemeine Naturpotenzen*] thus appear to us as universal processes of nature, and these latter as functions of our planetary system, and probably of other solar systems, too. In this view physics is no longer essentially divided from physiology. Just as every individual organism is a totality in and of itself, and at the same time repeats in a distinctive manner that of a higher universal whole, so, too, does it repeat, though with modifications, the universal functions of the planetary system that generates and nourishes it. It follows already in general from this . . . that the organic forces can only be modifications of those universal forces of nature.[420]

Lotze's *Lebenskraft* article also broached this issue, though in a somewhat ambivalent fashion. In one place Lotze denied the justness of an analogy between organisms and the solar system, arguing that whereas in such natural mechanical systems the fundamental forces of nature "call forth that self-preserving interplay of motions of a perpetuum mobile," organic bodies are like artificial machines "since they are in perpetual need of a new replenishment and impulse for their motion."[421] Later on, however, Lotze maintained that organisms and the solar system are *alike* in both exhibiting a divinely ordered purposefulness. Even so, as he now concluded from a consideration of the relationship between the soul (*die Seele, das Ideale*) and the body, "the organism not only pursues the purposes [or ends (*Zwecke*)] predestined by the combination of its masses, but can set itself new purposes and is itself capable of procuring the means conducive to their realization with an absolutely new initiation of mechanical motion."[422]

The design—or, in the more Kantian terms used by Lotze, the purposefulness (*Zweckmäßigkeit*)—exhibited by both living things and the solar system was probably the most commonly insisted-upon similarity between the two manifestations of organization. Burdach, for example, wrote (in his "anthropology" of 1837) of the diverse bodies making up the solar system: "There is purposefulness in their placement such that each moves in its orbit without disturbing the others, and insofar as the influence of the sun over all parts of the earth is apportioned with a certain uniformity due to the obliquity of the ecliptic, it resembles an organic arrangement. The planetary system thus bears in itself the essential traits of the organic."[423] For the Romantically tinged Burdach, the force of this analogy was seemingly to make the world system into a kind of cosmic organism; indeed, he played up the Neoplatonic parallelism between macrocosm and microcosm.[424] For Volkmann, too, who also employed the language of macrocosm and microcosm, the likeness of the solar system to an organism implies that the universe is also a living thing (*ein Belebtes*).[425] For the more reductionist-minded Schwann, on the other hand, very nearly the same analogy served to make living things into finely crafted machines. The purposefulness Schwann admitted is not the expression of a purposefully acting organic force, but derives from "the creation of matter with its blind forces by a rational being, wherein also lies the basis of the purposefulness in inorganic nature. We know, for example, the forces that act in our planetary system. They are forces that act according to blind laws of necessity, like all physical forces, and yet the planetary system is extremely purposeful. The basis for the purposefulness does not lie in these forces, but in Him who so created matter with its forces that in obeying their blind laws they nevertheless bring forth a purposeful whole."[426] Although the language and intention were different, Schwann's physicalist position was very close to that of Brougham's natural theology: after describing, as evidence of design, a number of animals' contrivances for performing certain tasks, Brougham cited the stability and regularity of the solar system as likewise evincing design.[427] Here again variants of the same basic analogy could be employed to very different ends.

Schleiden's adaptation of that analogy echoed the spirit of Schwann's deorganicizing and despiritualizing understanding of scientific explanation. Schleiden rejected the notion that organic and inorganic beings are radically and fundamentally distinct, preferring instead to find only gradual differences:

> The actual riddle of life breaks down upon closer consideration into two problems:
>
> 1) the construction of a system of motive forces that maintain themselves in regular periodicity;
>
> 2) the construction of the process of configuration [*Gestaltungsprocess*].
>
> Thereby is embraced everything which even the unhandiest abstraction can designate with the word life, for such people who mix along the psychic [*das Geistige*] into pure natural science warrant no consideration here; their wrongheadedness is demonstrated to them in the realm of philosophy. But the solution to the one as to the other of the just-described tasks does not at all fall within the limits of the organic. The first has already been solved through the construction of the solar system, which is only the simplest form of such a vital process.[428]

Schleiden distinguished three orders of such organized systems: solar systems (the simplest and most independent), the individual planets, and individual terrestrial organisms, whereby he apparently followed the lead of his mentor, Jacob Friedrich Fries, who had specified a parallel division.[429] If Schleiden's and Fries's tripartite hierarchies go beyond the simpler analogies of other writers, they nevertheless effectively juxtaposed microcosmic organisms and macrocosmic solar systems as belonging to parallel conceptual classes. They, too, challenged the contemporary reader to reflect on the essential similarities and differences between those two all-important classes of beings.

In the view of several authors, just as we can know with great precision the laws of planetary motion without knowing anything about the origin of the solar system, so too can we aspire to know the laws of organic life even if its origin remains forever unknown to us. It belongs to science to investigate laws, not origins. In his review of Karl Wilhelm Stark's *General Pathology*, Lotze applauded Reil's desire to explain vital phenomena in purely physical terms, but criticized him for failing to recognize that no explanation of actually existing systems is possible solely on the basis of knowledge of the laws of action, without knowledge also of the particular circumstances in which the system finds itself, of what we would call the initial conditions. He drove home his point by referring to the analogous situation that we are in with respect to our knowledge of the solar system:

> For the action of the simple fundamental forces is always only possible; undecided as to different directions, it always acquires the form of its activity only from the more precise conditions of the phenomenon to be fashioned, conditions which do not lie in *it*. The force of attraction . . . would tend toward the collapse of the gravitating masses into a center; but in truth we only see this process occur in nature in a subordinate way on the surface of the individual heavenly bodies, whereas its realization in the totality of the system is impeded by a tangential velocity of that which is attracted, whereby a circular motion is generated as the true form of the solar system. That velocity, however, and the corresponding arrangement of the distances are determinations according to which the mechanism comes into operation but which are not given in the mechanism itself.[430]

Lotze reiterated this point three years later in his own *General Pathology*, and again the next year in his *Lebenskraft* article. In the latter work he recognized that although science can only deal with the general laws of existing systems, we nevertheless still desire to know the origins of things. Confronted with the choice between chance or special creation, scientists have come increasingly to reject the latter alternative as intrinsically unscientific: "The curiosity which wonders how in fact the solar system or the germs of the organic first came into being is, rather, accustomed always to presuppose that it happened through some chance mechanical occurrence." But, argued Lotze, we ultimately cannot avoid assuming some kind of divine creation in order to explain the complex and purposeful disposition of cosmic and organic systems—organized macrocosm and organic microcosm once again linked in terms of their organization.[431] His position, too, came very close to that of natural theology.

Lehmann invoked the same analogy between solar system and organism in defending precisely the same views in his *Textbook of Physiological Chemistry*

(1842). We have been able to discover the invariable laws of motion of the heavenly bodies—laws that allow the precise prediction of the system's future behavior—even though we must remain forever ignorant of how the inscrutable *primum movens* first imparted motion to it. In the same way we shall never know how a seed or an egg or a germ first came into being, "but we are not thereby prevented from investigating the laws of the organic motions once initiated just as well as the regularity in the continued motion of the heavenly bodies."[432] We thus expect to find organic processes to proceed according to invariable organic laws. In a similar vein, Valentin likened the periodically recurring processes of self-preservation of the "organic machine" to the law-bound organization of the solar system.[433] If the purpose of Lotze's and Lehmann's extended analogies was to defend a mechanistic view of life, nevertheless their significance here is in the fact that the metaphoric assimilation of organisms to the solar system was everywhere to be encountered in the German literature of the 1830s and 1840s; others, like Mayer, exploited that analogy for different ends.

Still another analogy between the solar system and living things had to do with the association of straight lines, flat surfaces (such as the faces of crystals), and straight-line motion with inorganic systems, curved lines and surfaces and curvilinear motion (such as that of the planets) with organisms. Wrote Karl Wilhelm Gottlob Kastner, professor of chemistry and physics at Erlangen: "Curvilinear motions that, proceeding out from a point return to it and thus endlessly renew themselves, we call *organic motions* insofar as they did not arise as artificial productions of willing [*wollend*] beings, but came into existence in the necessary course of the activities of nature, and either exist only within certain intervals of time or continue to exist beyond all determinable time. To the first belongs the *circulation of humors* of living higher animals and human beings, to the latter the visible *motion of the heavenly bodies*."[434] Although he did not explicitly invoke the analogy with planetary motion, Liebig expressed essentially the same views in an important speech of 1856: "All of the configurations of inorganic bodies are bounded by plane surfaces and straight lines, all of the configurations of the bearers of organic activity are bounded by curved surfaces and curved lines; in organic bodies a cause must be active which bends the straight line into a curve."[435] In the solar system, too, there acts a force which produces curvilinear motion by constantly deflecting bodies from straight-line motion.

Mayer and his contemporaries were thus confronted with a richly elaborated set of analogies that conceptually united the solar system and living organisms. Thus Mayer's readiness to apply to living systems conclusions derived from his analysis of the gravitationally controlled motion of the planets around the sun—in particular the creation of force—would have appeared to be entirely in harmony with generally accepted views. Is the organism like a machine, and hence incapable of creating force out of nothing, or is the organism like the divinely disposed solar system, capable (as some believe) of calling force continually into existence? At different times Mayer drew inspiration from both possibilities. The progress of his thinking depended crucially on the analogies he took to be significant at different times.

## 4 Physiology as an Opponent of Materialism

It has already become clear that a large number of the most important scientists among Mayer's contemporaries—men like Autenrieth, Müller, Schwann, Liebig, and Lotze—all gave place to theological considerations in their physiological writings. Although the self-styled genre of natural theology was more characteristic of English than of German-speaking writers, the latter nevertheless evinced widespread allegiance to its central line of argument: design—or, as the Germans more often put it, purposefulness—proves the need to assume the existence of a rational Creator. Also important in Germany were arguments of a more general 'spiritual' nature; that is, if science, and in particular physiology, can demonstrate the existence of nonmaterial agencies such as mind, soul, or vital force, then atheistic materialism is proven invalid. The issue was sometimes put in the more general terms of whether forces are merely properties of matter, or whether they, like the soul or the vital force, can exist independently of matter as a distinct kind of entity. As Volkmann put the issue in 1837 in an elementary handbook of anatomy and physiology (in words in part already quoted):

> If we designate the ultimate cause [*Grund*] of a phenomenon by the name force, then a force must also underlie life [*dem Leben zum Grunde liegen*], one that can appropriately be called vital force. But opinions already divide here insofar as one disputes what a force is. Materialists conflate force and matter in that for them forces are nothing but properties of matter itself. . . . Spiritualists proceed in the opposite fashion: either they allow matter itself to be produced out of force, or they at least regard force as in any case independent of matter, and assume an actual existence even without the filling of space. According to them the vital force is earlier than the organism and is related to it as cause is to effect. The spiritualist therefore also distinguishes the soul from the body, and regards it indeed as the cause of all vital phenomena, and as the architect of its body, as Stahl taught.[436]

And like many others, Volkmann identified God as the original creator of organic forms:

> The original force [*Urkraft*] which called organisms into existence continues to act, albeit now on only a reduced scale. We are thus in a position to follow the conditions under which the original creation [*Urschöpfung*] takes place [*sic*]; they are air, water, heat, and light, but the essence of the force itself does not thereby become clearer to us, for the determining elements are not the creating force itself. The original force, however, is in God, to whom the natural scientist returns just like the theologian, to be sure in a different way, but with the same necessity.[437]

The perceived materialism of science was of such concern to many of the religiously minded scientists of the day that a minor genre existed which addressed itself specifically to just that issue. Volkmann, for example, published a short work in 1838 entitled *Physiology as Opponent of the Doctrine of the Materialism of the Identity of the Body and the Soul*. The central issue for Volkmann was the immortality of the soul, belief in which found its most serious opponent in the materialistic assumption that all the phenomena of life are a function of the construction and composition of the body.[438] He began by citing some of the arguments in favor of the belief that the soul is nothing more than a manifestation of the physical organization of the brain: the soul is undeniably connected to the body, and we find a close correlation between the intellectual capacities of an animal and the complexity of its nervous system. Moreover, the divisibility of the soul, which would accord with its material nature, is suggested by the ability of organisms to regenerate lost parts and by the intellectual correspondence between parents and children. He then restated the materialists' argument against the existence of a corporeally independent soul based on an analysis of the relationship between force and matter:

> According to this system the word force is something entirely idle if one designates by it something other than activity. Force in contrast to matter, and without matter, is a conception without content, and just because force is nothing in and of itself it can also not be the cause of an activity. Every material substance must, insofar as it exists, also be active, for an existence which did not manifest itself would not be existence. Activity thus attaches immediately to matter, which without being active would not even exist. Every activity is therefore the property originally pertaining to and inseparable from matter.[439]

Thus once again the issue of the existence of the soul was inextricably connected with the question of the general nature of forces: are they entities *sui generis*, or properties of matter? Only the first alternative lends support to belief in an independent (and immortal) soul.

Against the materialist conception of the soul as simply a function of the brain, Volkmann argued that the level of an animal's intellectual development does not in fact correlate directly with the development of the brain—witness, he said, stupid fish compared with clever bees. Nor does severe damage to the brain necessarily impair a person's mental faculties. As for the purported divisibility of the soul, he countered that whereas the division of matter is always accompanied by a corresponding division of all its properties, the psychic force (*Seelenkraft*) of each regenerated half of a bisected worm is not less than that of the original creature. Volkmann's examples were no more than one is accustomed to find in the German physiological literature of his day. What decisively demonstrates the difference between the soul and the physical forces of nature, and hence supports the independence of the soul from the body, is the soul's spontaneous activity, its ability to initiate organic motions without being bound by the immutability that characterizes those forces, an immutability attested by the constancy of the movements of the planets. Nor did the fact that bodies' active chemical forces are capable of coming to rest in chemical compounds weaken his general conception of the nature of forces—forces in our sense of the word: "Matter acts by means of forces that it had from the beginning, by means of constant forces that cannot decrease, much less cease and begin."[440] Thus Volkmann's physiologically based antimaterialistic argument for the existence of the soul finds him asserting the indestructibility and uncreatability of the physicochemical forces of nature—without, to be sure, his having had a clearly elaborated general concept of force, let alone of its measure.

Volkmann's essay repeated in large measure the main points of an article entitled "Reasons against Materialism" published by Autenrieth in 1816 in a scientific journal he coedited and reprinted in his posthumously published work of 1836. The basic standpoint was that materialism could be refuted by demonstrating the existence of a soul independent of the body. Autenrieth, like Volkmann, based his argument on the existence of free will and on the nondivisibility of the soul during processes of generation and regeneration. Still another work in this genre was the *Attempt to Prove the Immortality of the Soul from the Standpoint of Physiology* (1830) by the professor of medicine at Bonn, Moritz Ernst Adolph Naumann, which based its argument on the manifestations of consciousness. These works, along with related discusssions in most of the more exclusively physiological works of German scientists published in the decades before around 1840, not only reinforced the theological connections of physiology, but also thrust into prominence the issue of the dependence or independence from matter not only of the soul but also of forces. At least one kind of force, the *Seelenkraft*, is not merely a property of matter but is an independently existing entity *sui generis*. Within such a context, Mayer's vindication of force as such an entity constituted another powerful argument against such a conception of materialism.

## 5 Homeopathy

The advent of homeopathy, an alternative to the drug-based treatments of traditional medicine, can be identified with the appearance of Samuel Hahnemann's *Organon of Rational Medicine* in 1810. Hahnemann's system won many devoted followers, both lay and professional, and aroused the equally passionate opposition of many "allopathic" physicians, as Hahnemann termed the practitioners of traditional medicine. In 1834 the new system found a new defender in the medically trained professor of "practical philosophy" at Tübingen, Carl August von Eschenmayer, who published a modest work entitled *Allopathy and Homeopathy Compared in Their Principles*. That work elicited the critical response of another Tübingen professor, Ferdinand Gottlob Gmelin—at the time one of Mayer's medical professors—who published a *Critique of the Principles of Homeopathy* in 1835. Homeopathy was thus a controversial part of the medical landscape during the years of Mayer's schooling, and one can only assume he had some knowledge of the major issues. Homeopathy was part of the broad common context which involved physiology, medical practice, ontology, theology, and the spiritualist phenomena of animal magnetism and somnambulism. Wunderlich, in his *History of Medicine*, noted that homeopathy had found favor with the opponents of materialism.[441] In part this reflected the materialistically counterintuitive basis of homeopathic practice; in part it had to do with the fact that homeopathy tended to find support among those who also believed in such scientifically suspect phenomena as animal magnetism and the like.[442]

Although there is no evidence that Mayer was attracted to or in any special way influenced by homeopathy, nevertheless several of the central issues were precisely those already identified as important aspects of his conceptual background, namely the relationship between cause and effect, the nature of force, and the refutation of materialism. The debate concerning homeopathy thus may well have contributed something to Mayer's awareness of such conceptually loaded topics. Homeopathy was a part of the stage-setting context of Mayer's intellectual milieu, even if one cannot say with any assurance that he attended to that component of his surroundings.

For Liebig, the plausibility of homeopathy reflected people's confused notions of force, cause, and effect. As he wrote in the first of his new series of "Chemical Letters" in 1844: "Without correct conceptions of force, cause, and effect, without practical insight into the essence of the phenomena of nature, without a thorough physiological and chemical education, is it a wonder that otherwise reasonable people defend the most absurd views, that Hahnemann's theory could arise in Germany, that it could find pupils in all countries?"[443] The acuteness of these issues stemmed from the fundamental tenet of homeopathy that properly prepared dosages of extreme dilution are medically more powerful than the relatively large doses customarily administered. (The homeopathic potentiation [*Potenzirung*] of the substance was achieved by prescribed shaking of the progressively diluted solutions.) Eschenmayer, for one, argued that *das Quantitative* (represented by gravitation) and *das Qualitative* (representated by heat) oppose each other in an inverse relationship: gravitation seeks to bind, heat to release, and as the homeopathically active substance is progressively divided—its bulk decreased—so must its qualitative (medical) action increase.[444] This phenomenon is an example of the general law "that the quantitative and qualitative element stand in an inverse ratio."[445] A remark of Eschenmayer's to his close friend Justinus Kerner in 1838 suggests not only the centrality of that contested doctrine, but the extent to which the Tübingen medical establishment regarded homeopathy as worthy only of ridicule: "At the university (Tübingen), where wigs have not yet long been put aside, the victory still belongs to the pill boxes, herb presses, and heavy-bodied mixtures. To want to cure a robust machine with a millionth of a grain these men call a piece of foolishness, and they also permit themselves the sarcasm that homeopaths will eventually succeed in teaching people to bathe in the thousandth part of a drop of water."[446] In countering what an anonymous author recognized is "the inconceivability of the effectiveness of such small doses," defenders of homeopathy typically appealed to experience.[447] To defenders of the system, homeopathic physicians are the true empiricists, who refuse to let theoretical preconceptions blind them to the truths of experience. In an attempt to render the inconceivable a little more plausible, analogies were enlisted from well-known phenomena about the ability of small physical and chemical causes to produce large effects:

Even in so-called dead nature the smallest, most invisible substances often produce the greatest effects. Whereas a glass is not damaged by the loudest sound, an insignificant tone having a particular harmony with it often causes it to explode. Steel and stone lie peacefully on gunpowder, but a stroke of the first two against each other melts small steel spheres (sparks) with the heat, which [then] produce a huge explosion or a devastating fire. In the most insignificant quantity of silver fulminate there slumbers the most prodigious production of force, which no one suspects there and which comes into being by means of an insignificant friction![448]

As for the vanishingly small weight of homeopathic dosages, Hahnemann pointed out that physicists accept the existence of "prodigeous force entities (powers) [*ungeheure Kraftdinge* (*Potenzen*)] . . . which are entirely without weight—for example the substances of heat, light, etc.—and which are thus still infinitely lighter than the medicinal content of the smallest dosages of homeopathy."[449] Nor was that all. Recalling that the potentiated homeopathic dosages are prepared not just by progressive dilution but under the accompaniment of a prescribed amount of shaking, a defender of homeopathy invoked the further physical analogy of the ability of friction to evoke large quantities of force:

For the rest, the effectiveness of the small homeopathic dosages becomes all the more *explicable* when one considers their preparation and the conditions under which they act. Infinite forces are produced in nature and transferred from one body to another through friction.* Therefore if (say) a magnet makes iron magnetic through friction, or a contagion communicates itself to other bodies, why shouldn't medicinal forces also communicate themselves through friction equally well to unmedicinal milk sugar or spirits of wine? The homeopathic *dilutions* are thus also improperly so termed, since they are at the same time a *generation of force*, a communication of force. Homeopaths therefore also prefer to use the expression (for example) third or thirtieth *development of force* or potentiation.

. . .

* . . . Count Rumford heated rooms by means of the rapid friction of metal plates, etc.[450]

In its innocence of any intuitive notion of the conservation of quantity of motion, such a defense of the plausibility of homeopathic remedies was neither more nor less unreasonable than Liebig's explanation of putrefaction and fermentation in terms of the propagation of molecular motions.[451]

Gmelin was especially critical of the homeopathic principle that the medically effective strength of a substance increases with decreasing concentration, especially since the dilution is typically so extreme. Our usual experience, he countered, is that the quality expressed by a material substance decreases in proportion to the amount present, at least up to a point. Below a certain concentration, contagions cease to be able to cause disease and frog semen ceases to be able to fertilize eggs, and there are limits below which small quantities of electricity, sound, and light can no longer be perceived:[452] "According to the foregoing there are thus extraordinarily fine motions of the imponderables, and otherwise imperceptible minima of ponderable matter, which still produce a reaction in dead and living bodies, but even with these it is evident that the reaction is stronger and the effect greater with a greater quantity than with a lesser, and that every kind of reaction ceases at a certain limit of division and fineness, a limit which occurs much earlier than the Hahnemannian dilution."[453] Alongside the issue of the quantitative relationship between cause and effect, Gmelin has raised the possibility of the disappearance of an effect once the cause has become sufficiently small.

## 6 Summary

However much a variety of factors influenced the course of Mayer's intellectual development, that he was trained as a physician was the central important fact of his background with regard to understanding his route to something like the conservation of energy. The crucial influence of that medical background was roughly speaking twofold: it prepared him to observe and react to the anomalous color of the blood he let in the tropics, the event that set off his unstoppable train of reflections, and it provided him with a rich context of issues concerning the ontological status of forcelike entities, in particular their possible creation and destruction. It was those issues, I believe, that led him to transform a limited conviction concerning the quantitative equivalence of heat and motion into a global theory about a new conceptual entity, force. Since it has been the purpose of this long chapter to lay out the full richness of that medical and physiological context in order to demonstrate just how profoundly Mayer was connected with important and highly visible issues of the day, it may be useful to summarize its principal findings in a few pages. As the theme of this book has emphasized, that context defined several fields of meaning with respect to which Mayer defined the issues and solved the problems of central concern to him.

It is important to recognize that although *we* see the issues of the color of the blood, the nature of respiration, and the sources of animal heat as intimately connected, in Mayer's day they were by no means all regularly considered *in the same context*. At the same time, each issue alone was more complicated and unsettled than would be the case after a theoretical consensus had allowed the separation of essential from nonessential factors. Thus in the 1830s the color and physiological function of the blood were not clearly or universally connected in the first instance with respiration and the generation of heat: the presence of saline matter was widely held to influence the blood's color, and its essential functions included the removal of excess carbonaceous matter from the body in an excretory process unconnected with the problem of animal heat. Indeed, it was that excretory function that probably shaped Mayer's initial expectations. According to a widely held view, the diminished oxygen content of warm air means that less carbon will be removed from the blood via respiration, implying that such blood will be *darker* than normal. Such seems to have been Mayer's expectation. At the same time, the body was supposed to compensate by increasing the production and secretion of bile produced in the liver from the venous blood passing through it—a complementary process that underscored the excretory function of the oxidation of carbon in the blood. To be sure, the production of animal heat was mostly ascribed to an oxidation process involving the blood, but discussions of that issue were typically independent of the foregoing kinds of considerations—that is, they were in different works or in different parts of the same work. All through the 1830s factors other than oxidation were also thought to contribute to the production of animal heat, most notably the influence of the nervous system. Only gradually in the course of the decade did the increasingly taken-for-granted conviction that all physiological processes must be understood in terms of underlying processes of material exchange begin to undercut the validity of such supposed additional sources of heat; and even at that, it was still only gradually, during the late 1830s and 1840s, that German scientists became generally sensitive to issues of what one might call 'energetic causality.' For some, the organism might well be a kind of perpetuum mobile. Dulong's and Despretz's universally known findings—that respiratory oxidation cannot account for all of the heat produced—seemed to give experimental warrant to the nonacceptance of the sufficiency of Lavoisier's theory. Thus it was a radical and decisive step for Dumas and Liebig to come out wholeheartedly in favor of the oxidation theory of animal heat in 1841 and 1842, a step not motivated by new experimental findings but by a new theoretical conception of physiological processes. Thus, too, in 1840 it is highly unlikely that Mayer made his initial bloodletting observations from the standpoint of that theory; rather, it seems more likely that he came quickly to accept that theory in part because it helped him make sense of his anomalous observations. The oxygen theory of animal heat was not Mayer's starting point, but an early conclusion from his developing train of reflections.

A few generalizations can be made about the status of the vital force as an issue relevant to the emergence of the concept of the conservation of energy during the 1840s in Germany. Always open to criticism as being merely a verbal cover for one's ignorance of the real causes of organic phenomena, from the mid-1830s on the vital force was in clear retreat from the advancing front of organic chemistry as the special agency responsible for the synthesis of complex organic compounds. Up until around 1839–40, though, less widespread than a principled general rejection of the vital force was the growing consensus that the task of physiology is to explain vital phenomena in terms of processes of material exchange. (Tiedemann and Müller are good examples of this position.) And as physiology proper came more and more to exclude issues of development—especially involving cross-species comparisons—in favor of a narrower science of the organic functions of the (tacitly) already developed single organism, physiologists no longer needed to invoke a special organic (or as Müller put it, *organizing*) force to explain a class of phenomena for which they no longer accepted responsibility. Freed, as it were, of that intractable explanatory burden, physiologists now had only to deal with the more plausibly physicochemical functions of the erstwhile vital force. Having long been a *topic* of discussion and debate, after around 1839 the vital force became a contentious *issue* among German scientists in ways it had not been during the preceding decades, even as its defenders became fewer and fewer. In many ways it was Liebig who brought the issue to a head by claiming for the vital force an unprecedentedly large and explicit role in the production of organic motions, a position more glaringly at odds with the general tenor of scientific thinking than the more traditional reliance on the vital force to explain processes of generation and development. At the same time, Lotze spoke out in favor of limiting physiological laws and explanations to those sanctioned by physics and chemistry, a position which would win an increasing number of adherents as the decade wore on and which left little room for any kind of special vital force.

Ironically, the otherwise opposing positions of Liebig and Lotze provided equally fertile ground for the germination of an awareness of the implications of the *absence* of some kind of energy conservation principle. Liebig's work-performing vital force and Lotze's variable organic force both grossly violated the feeling many scientists had long held, in appropriately circumscribed realms, that nothing can be created from nothing, that a physical effect must be assignable to a bona fide physical cause. If Helmholtz was misleading in claiming that he had been reacting against the "vitalists" of his youth, nevertheless the context surrounding the issue of vital force had never been more conducive to energetic reflections than it was during the first half of the decade of the 1840s.[454] Mayer's 1845 book, *Organic Motion in Its Connection with the Exchange of Matter*, reflected this new state of affairs.

Significantly, the vital force had not been a problematic issue for Mayer in the progress of his thinking, from mid-1840 to early 1842, leading up to his first published paper, the "Remarks on the Forces of Inanimate Nature." In part this reflects the relatively uncharged atmosphere surrounding the issue as of around 1840; in part it reflects Mayer's decision, in his earliest papers, to exclude the organic realm from consideration; and in part it has to do with the fact that Mayer followed other routes to his theory of force, as will be laid out in chapter 6. Nevertheless, although principled criticism of the vital force did not, I believe, play a significant role in Mayer's earliest reflections, a number of the issues regularly raised in discussions of the vital force, as part of the broader common context indentified here, were directly relevant to the kinds of issues that did occupy Mayer's thoughts. For example, discussions frequently touched on the general issue of what characterizes a scientifically legitimate concept of force, and whether or not the vital force meets those criteria. Are forces properties of matter, or can they have some kind of independent existence? With the significant and idiosyncratic exception of Autenrieth, almost everyone denied the latter possibility; but the *issue* was engaged. Similarly broached was the question of whether or not the vital force can be created (at birth) or destroyed (at death), of whether anything can be created out of nothing, of whether (as some maintained) the vital force is capable of indefinite self-multiplication. Such issues were precisely those that engaged Mayer's early attention as he struggled to forge a coherent conception of a new physical entity, force.

Of perhaps secondary importance here were attempts to liken the vital force to chemistry's catalytic force, an assimilation that depended on the suppression of energetic considerations as it suggested that the vital force, too, might be interpretable in terms of the properties of appropriately arranged matter. (The catalytic force will be dealt with at greater length in chapter 4.) Of great significance to Mayer was the fact that the growing tendency to reinterpret the increasingly disreputable vital force in terms of the arrangement of organic matter raised the specter of materialism, and made his relationship with his self-assertively progressive contemporaries somewhat problematic. Indeed, many German physiologists had, up till the end of the 1830s, seen physiology as an opponent of materialism, and it is clear that Mayer saw his concept of force as part of that tradition.

These discussions often involved the soul in ways entirely parallel to those of the vital force: Is the soul an autonomous entity? Can it (or the psychic force) be created? Does it generate itself? Where does it go upon the death of the individual? What is its relationship, substantive or analogical, to the general forces of nature and to the vital force? A further important issue which concerned the soul in particular was whether it (or the nervous system) can by itself initiate (and perhaps sustain?) animal motions. What is the relationship between cause and effect here? What is the relationship between the magnitude of the cause and the magnitude of the effect, and how is either to be estimated? Questions of this nature, widely discussed in the physiological literature—or at least strongly implied by the issues that were discussed—were, I believe, an important part of the context within which Mayer shaped his own conception of force as a distinct entity capable of neither creation nor destruction. As I have shown, the physiological literature contained a rich consideration of analogies among the imponderables, the vital force, and the soul: How are they similar, and how are they distinct? Müller, for example, defended a conception of the vital force as an entity *sui generis*, not reducible to chemical and physical forces, but which like heat, light, and electricity can exist in matter in a latent form. Significantly, both Müller's notion of the latency of the vital force and Autenrieth's image of its passage to and from a preterphysical realm are tacitly premised on the absurdity of assuming the creation *de nihilo* or the destruction *ad nihilum* of (vital) force. Recall, too, that for a period of time Mayer wavered terminologically between "force" and "imponderable": he created his conception of force less by rejecting the imponderables than by reinterpreting the nature of their "substantiality" and by extending their scope to include motion.

Much of Mayer's own thinking was guided by his search for valid analogies—for example, between force and ponderable matter in terms of roughly the same issues others had applied to the relationship among the imponderables, the vital force, and the soul. A second important guiding analogy, also widespread in the physiological literature of the day, was that of the organism as a machine. Indeed, that image was also dear to authors writing in the tradition of natural theology, for which Mayer may have had an ear. Although Mayer's antimaterialism kept him from believing that the organism *is* nothing but a machine, his willingness to treat it *as* a machine with regard to its ability to perform work and produce heat as a result of physiological processes that are themselves essentially chemical was a necessary component of his initial deduction of the equivalence between heat and motion. (Part of that deduction also drew upon his childhood failure to construct a perpetuum mobile.) Later on, a third common metaphor—that of the solar system as a living organism—encouraged Mayer to accept the possibility that force might well be created in nonmechanical "organisms," a conclusion reinforced by a common general distinction between mechanical and dynamic (or organic) systems and rendered all the more plausible insofar as Mayer did not begin with a principled rejection of the notion of a vital force, a concept only then just coming under increasing critical scrutiny. Mayer's context supplied him with rich if mutually incompatible analogies; an essential part of his creative process was to identify for himself the *valid* analogies, a difficult task when one keeps in mind that he was not somehow destined to proceed monotonically toward the conservation of energy. His ultimately decisive rejection of the vital force was an integral part of an extended process of clarification: it was neither simply the starting point nor simply the end point of his intellectual journey, part of which involved deciding what kinds of things his new theory of force would apply to.

Without insisting on its importance, this chapter has also looked at homeopathy as another potential component of Mayer's medical context that may have been relevant to the course of his thinking. That is, the controversies around this doctrine—which were active in Tübingen at just the time Mayer was a medical student there—brought to the fore issues of the relationship between cause and effect, the nature of force, and materialism that might conceivably have further stimulated his thinking on these important general issues.

• CHAPTER FOUR •

# Physics and Chemistry

ALTHOUGH the immediate context of Mayer's initial reflections was physiological—in the broad sense of the term as then understood—he carried with him a residue of his childhood fascination with machines and their capacity to do work, he had been well trained in chemistry, he had taken one course in physics at the university, and he early on turned his attention to physics in an attempt to secure a physical foundation for his developing conception of force. Central to Mayer's thinking, moreover, were an analogy between matter and force and a parallel conception of chemistry as the science of matter and physics as the science of force. It thus behooves us to have a clear sense of how his contemporaries dealt with the relevant concepts of force, matter, and the imponderables. Alas, Mayer encountered mostly confusion and contradiction in contemporary texts and compendia concerning the conceptual and ontological status of forces and imponderables, and he spent several years trying to work his way to consistent clarity. In what follows I have paid primary attention to German works of the 1830s, though I have extended the survey to include a few French and English works, in particular those Mayer consulted, such as Lamé, Biot, Whewell, and Herschel. Although, with the exception of Geiger's *Handbook of Pharmacy*, there is no direct evidence until several months after the appearance of his first published paper that Mayer had read any *particular* text, for the most part the works then generally available were in substantial agreement about the most important relevant issues. It is, therefore, for the most part of no great consequence that we do not know in more detail what he read when, although in chapter 6 I will argue that Mayer had Fries's *Textbook of Physics* on board the *Java* with him, a work that *was* different from the norm in certain important regards.

Among the many features common to German works of the period was a conceptual division of the world of physics into matter and force. Although Mayer offered a profound reinterpretation of the meaning of that distinction, the matter-force dichotomy itself was a commonplace. Muncke's *Handbook of Physics* was entirely typical: "As the *object* of our investigation, the complex of everything affecting our senses is called nature. We thereby distinguish matter and force, and call matter that which is knowable through the senses and which fills space, whereby we distinguish it . . . from an effect produced and from the forces producing the effects."[1] In his articles on "Physics" and "Matter" in *Gehler's Physical Dictionary*, Muncke essentially repeated this formula, adding that "if one admits that, with the exception of everything pertaining to the world of spirits, no force can exist autonomously, then all of nature, or the corporeal world, offers only matter with different forces peculiar to it."[2] Fischer's *Textbook of Mechanical Physics* added an epistemological twist to this standard distinction: "In the inner being of every body there lie inseparable motive forces, nor can we in reality withdraw any body from the influence of external motive force. But since we are compelled by a necessity lying in the inner being of our conceptual faculty to distinguish *matter* and *force*, we can thus also separate them in the idea, and we must do so for the sake of theory."[3] Eisenlohr's characterization of the subject matter of physics was entirely typical of what was to be found in most texts of the day:

> Everything we perceive through our senses is called *body* or *matter*; everything that causes a change, *force*.
>
> . . .
>
> *Physics* is thus *the science of the causes or forces that determine the phenomena and changes that take place in inorganic nature.*
>
> . . .
>
> The phenomena of nature take place according to determinate rules that we call *laws of nature*. The ultimate causes of these phenomena are called *fundamental forces*, and those laws of nature that, according to experience, express the simplest known effects of these fundamental forces are called *fundamental laws*.[4]

Even this brief presentation of matter and force as the two fundamental conceptual elements of physics has touched upon two of the most general and important characteristics of force: forces are the *causes* of change in the physical world (especially of motion); and forces (unlike *das Geistige*) have no independent existence in nature, but are *properties* of matter. The goal of physics is to discover the laws that embody the effects of the most basic forces. Those forces, however, did *not* usually include the so-called imponderables—heat, light, electricity, and magnetism. The issues are complex and important; let us consider in turn the three basic classes of force, the imponderables, and matter.

## 1 Force

Considering that Mayer believed he could prove the truth of his doctrine of force by invoking the axioms "forces are causes" and "cause equals effect," it is of great significance that statements to the effect that "forces are causes"—in particular the causes of *motion*—were part of virtually every treatment of force Mayer might have read.[5] Indeed, Mayer complained to his friend Baur that physics texts treated force as an "obscure cause of motion."[6] He cited Baumgartner, who had written:

> Often the cause [*Ursache*] of a phenomenon is itself in turn a phenomenon, and thus requires a new cause [*Grund*]. If it is present in experience, the latter presupposes in turn a new cause, such that, via a series of phenomena of which each is simultaneously cause and effect, one ultimately comes to a final supersensible cause that has its roots in the interior of nature. One calls it *force*, without wishing to designate with this expression more than a cause of a phenomenon that is in its essence completely unknown to us.[7]

Mayer also quoted Lamé, in French: "One gives the name *force* to every cause that is capable of taking a body from the state of rest to one of motion, or of producing the inverse effect."[8] And he cited Fechner's edition of Biot for likewise stating that "every cause that is capable of transposing matter from the state of rest to that of motion is called force."[9] Eight or ten months later, in another letter to Baur, Mayer broached the subject again, this time citing from the German edition of Whewell's *Elementary Treatise on Mechanics*. One passage he referred to read, in the original: "*Any cause which moves or tends to move a body, or which changes or tends to change its motion, is called* FORCE."[10] In another place cited by Mayer, Whewell had written that "*no change can take place without a cause*, and that *causes are measured by their effects*."[11]

In another English work Mayer owned in German translation, Herschel's *Treatise on Astronomy*, the author characterized astronomy as "a science of cause and effect."[12] In his chapter on gravitation he attacked the "attempts . . . made by metaphysical writers to reason away the connection of cause and effect, and fritter it down into the unsatisfactory relation of habitual sequence."[13]

Lorenz Geiger's *Handbook of Pharmacy*, which Mayer knew (and probably owned), contained an appreciable section discussing the general concepts of physics and chemistry. It followed the norm in characterizing physics as the science that investigates the causes of change in the material world, causes that are in general called *forces*.[14] Geiger, too, further identified forces more particularly as the causes of *motion*, but in a somewhat more quantitative fashion than most: "Where motion is to arise or again cease there must be a force that occasions or hinders it (see inertia). But the force experiences each time from the body upon which it acts a counter action proportional to its mass or to the magnitude of its motion, whereby the acting force is diminished or entirely annihilated."[15]

Mayer thus had ample license from the well-known and respected authorities he consulted for regarding forces in general as causes, and in particular as causes of motion—though to be sure, as Mayer complained, that was not the *only* way in which the word "force" was used. Conceptual and terminological inconsistencies abounded. Fischer included under the "motive forces of nature" the force of the will, momentum, elasticity, and the causes of gravity, magnetism, heat, and electricity.[16] In the event, however, he did not maintain consistent usage with respect to "forces" versus "imponderables." Among the effects for which Biot assigned various forces as their cause were the expansion and contraction of bodies by heat and cold, the mutual attraction of magnet and iron and of electrified bodies, the fall of bodies toward the center of the earth, chemical affinities, the endeavor of the planets toward the sun, and organic forces subject to some extent to voluntary control (though such forces played no role in his physics).[17] But Biot also referred to gravity, attraction, and light as "properties" of material bodies and urged the probability of their materiality, and he spoke of the "principles" of electricity and magnetism, both of which he likewise took to be "fluids," just as he leaned toward the materiality of heat.[18] Biot's physics did not have a consistently applied conception of force, nor was the term usually applied to the class of entities more commonly referred to by contemporaries as imponderables.

The most obvious and indeed paradigmatic example of a physical force was the attractive force of gravity. The law of universal gravitation admirably fulfilled the goal of physics of exlaining an ordered class of phenomena in terms of an underlying cause, a force not reducible to anything else. As Kämtz put it, in tracing back the chain of effects and causes we finally come to a cause that is inseparable from the essence of matter: "This ultimate cause of the [phenomenal] changes is called *force*."[19] The force of gravitation appeared to most to be inextricably associated with matter. For Fries all the fundamental forces (*Grundkräfte*) are either attractive or repulsive, and are the causes of increase or decrease in motion or of deviation from straight-line motion. He, too, attached his forces firmly to matter: "Every fundamental force or *original force* is an immediate invariable property of its mass."[20]

Muncke likewise repeatedly insisted that forces are necessarily bound to matter, that one cannot conceive of forces having an existence independent of matter.[21] In his article on "Force" in *Gehler's Physical Dictionary* he considered and rejected as scientifically valueless the dynamical theory of fundamental attractive and repulsive forces—that is, Kant's theory (or one of its later variants) whereby matter is composed of ontologically primitive opposing forces. He then added: "Empty in almost the same way are the hypotheses of those who regarded the imponderable powers (imponderables, incoercibles) as merely forces or activities, or fancied that they had explained their real essence by introducing such a name."[22] He rejected the notion of a latent force as "unthinkable."[23] Although for most writers matter and force were ostensibly (and formulaically) the two parallel constituents of the physical world, only matter was granted an independent existence.

Friedrich Tiedemann gave an unusually extended consideration of the nature of force in his highly regarded *Human Physiology*. In large measure his views were close to the consensus of his physicist contemporaries: forces are the causes of physical changes, and cannot be conceived to exist independently of matter. However, he also entertained the idea, rejected by Herschel, that the concept of force expresses the (purely conceptual) *relationship* between causes and effects, although he himself seems to have preferred giving the concept of causality a Kantian epistemological grounding. His treatment well epitomizes the context within which forces were discussed by German scientists in the 1830s:

> We call the internal conditions or causes of the manifestations of activity [*Thätigkeits-Aeusserungen*] of living bodies that lie in the constitution of the living bodies themselves their forces. The word force has long since been used with different meanings, which has thus caused confusion in the explanation of the phenomena of nature. One understands by it sometimes the property, or the cause lying in the constitution of a body, that produces certain phenomena; on the other hand one sometimes imagines by it a something which only (as it were) adheres to bodies, penetrates them, but which is objectively different from their matter. . . . Where we perceive certain phenomena in the sensory world, there we are obliged by the nature of our cognitive faculty to relate them as results or effects to a thing that we call the

cause or the basis in reality [*Realgrund*] of the phenomena. We attribute to this thing, as the cause of the production of certain phenomena, a force, and we call the phenomena its effects, or manifestations of force [*Kraft-Aeusserungen*]. Thus do we perceive the phenomenon of gravity in bodies. Physicists bestow the name force of gravity [*Schwerkraft*] on the cause of this phenomenon that lies in the bodies. As far as our experience with it reaches, this force is grounded in the material constitution of bodies, and we see the phenomena of gravity to be different according to their differences; we call this their specific gravity. Physicists regard the cause of gravity that lies in bodies [*die in den Körpern liegende Ursache der Schwere*], which they designate by the term force of gravity, not as something existing by itself, objectively different from the matter [of the bodies] and only just attached to them; rather they regard it as a property grounded in the constitution of the bodies themselves. . . .

> Force is thus a subjective concept by which we express the relationship obtaining between cause and effect or between the properties of bodies and their phenomena . . . . If we ask further: are the phenomena that bodies or material substances present really only just effects of their material properties, or are they not perhaps grounded in a something that is not matter? Here we must openly confess that the answer to this question lies wholly outside the realm of our knowledge; because such a something, which is not matter and consequently supersensible, can in no way be an object [*Gegenstand*] of our knowledge insofar as it has no access to our imaginative faculty [*Vorstellungs-Vermögen*].

> There is only one force known to us through experience, the psychic force [*Geisteskraft*] or soul, that knows itself, investigates the causes of the phenomena of nature, and conceives of a supreme or highest cause of nature and itself, something Absolute or Divine. We regard this force as grounded in a supersensible substrate insofar as it does not perceive itself, as it does other things, by means of the external senses, but knows itself simultaneously as object [*Object*] and subject, or is the object [*Gegenstand*] of its own internal perception. Even this force manifests itself, as far as experience goes, in and through a body. We have no experience and no real conception [*reeller Begriff*] of an immediate action of the psychic force, one not manifesting itself in a body and hence entirely independent of it. Whether and how this self-knowing force might be able to manifest itself as active even independently of and not bound to a body is not a matter for physics and physiology, which only deal with the phenomena and properties of bodies in general, and of living bodies in particular, as far as they can be known through sensible perception. The solution to that problem thus belongs in the domain of metaphysics.[24]

For Mayer to have advanced a concept of forces as independently existing physical entities was thus a severe challenge to those, the overwhelming majority, for whom such an idea was unthinkable and unimaginable.[25]

Let us look more closely at just what forces were invoked as causes of various classes of phenomena. According to Leopold Gmelin, "all changes in the material world can be reduced to the following causes, or *forces*": forces of repulsion and attraction (gravitation, cohesion, adhesion, and chemical affinity) and the vital force.[26] Baumgartner noted that the diversity of the phenomena prevents us from tracing them all to a common original force, hence "we therefore assume a particular force for every related series of phenomena that we cannot further explain, and name it according to the last phenomenon to be explained by it. One thus speaks of a force of gravity, or a force of adhesion, in order thereby to designate the ultimate cause [*Grund*] of gravity, of adhesion."[27] Baumgartner identified the *Grundkräfte* as (otherwise unspecified) attractive and repulsive forces. He did not, however, include the imponderables among the *forces* of nature. They constituted a separate class.[28] Fries, too, identified the *Grundkräfte*, the motive forces that cause bodies to change their state, as forces of attraction and repulsion, such as those responsible for expansion and cohesion.[29] He classified sound, light, and heat as "phenomena of immediate sensuous perception," magnetism, electricity, and organic processes as "phenomena of mediate observation."[30] Fries thus applied a phenomenological criterion to break up the conventional class of imponderables. For Geiger the *Urkräfte* were forces of attraction and repulsion whose mutual relationship determines the characteristic properties of matter (*Schwere*, cohesion, adhesion, and affinity) through which bodies interact.[31]

For the most part, then, Mayer's contemporaries meant by *Kraft* the attractive and repulsive forces of matter that were conceived to be the ultimate causes of the properties of matter and of phenomena of physical change (in particular, of physical motions), though to be sure there was appreciable latitude in the application of the term. That *motion* played such a prominent role as the general effect of a force may have encouraged Mayer to include motion within the new class of entities he was coming to identify as forces, entities of a radically different nature from those previously designated by that term.

In developing his own conception of force, Mayer realized at an eary stage that he needed a quantitative measure of its magnitude, in particular of the quantity of motion. To his great frustration and confusion, when he turned to contemporary physics texts for enlightenment he found himself at sea in a mass of inconsistencies and contradictions. Mayer never had a sure command of even the rudiments of mathematical physics, and was thus poorly equipped to correct his sources. Fortunately his friend Baur gave him valuable help in this area, though even he expressed exasperation at Mayer's sometimes dogged attachment to incorrect notions.

Writing as if the vis viva controversies of the eighteenth century had never taken place, writers of physics texts universally took the measure of the quantity of motion to be mass times velocity, even though such a definition could not be rhymed with the definition of force as cause of motion in the case of the motion produced by the force of gravity on a falling body.[32] Wrote Lamé, in a passage cited by Mayer: "In all cases the ratios of the masses of bodies are defined and measured by the ratios of the quantities of force of the same nature that are capable of impressing on these bodies motions with the same velocity. One is only repeating this definition, the only exact one, of the word *mass* when one says that *the instantaneous forces are proportional to the masses that they move with the same velocity*."[33] Mayer further quoted Lamé's statement that "*forces are proportional to the velocities that they would impress on the same mass*."[34] Biot and Whewell gave the same measure (*mv*) of quantity of motion and of force, respectively.[35] Fischer introduced his conclusion "that in all cases the quantity of motion is proportional to the product of the mass and the velocity" with a brief philosophical analysis.[36] When one considers both the "motive forces of nature"—that is, the force of the will, momentum, elasticity, and the causes of gravity, magnetism, heat, and electricity—and the force of inertia (*Beharrungs-Kraft*), then one easily perceives

> that *the conception* [*Vorstellung*] *of a force* is never something *perceptible* [*Anschauliches*], thus also in no case something in and of itself *measurable*. But since mathematicians have nevertheless long since represented motive forces by means of lines, numbers, and formulas, it is necessary that the beginner be clearly instructed right from the beginning in what sense such expressions for forces are to be understood.
>
> Now what can be said about this in general is first of all *that what can be measured is never the force itself, but only the effect it has produced or would produce*—insofar as it is again a necessary law of our cognitive faculty to recognize that the effect is always proportional to the force (or part thereof) that is actually used for the effect. Now since the effect of motive forces is nothing but *motion of a body*, we must first concern ourselves with a *measure for the motion of a body*.[37]

Fischer went on to set up what he took to be a more general law than Newton's third law, according to which a body loses the same quantity of motion it communicates to another. His gloss on his new law reveals how the inclusiveness of the definition of force as cause of motion, coupled with an appeal to experience as final arbiter, worked against any intuitive sense of the conservation of energy:

> That even this law is not a rational one, but only one known through experience, is evident because it does not apply at all to the production of motion by psychic force (the force of the will). One cannot even attribute motion to the will itself when it acts as a motive force; but it can produce and communicate motion without thereby suffering any loss. Indeed, one can even doubt whether this law is applicable to imperceptible substances, e.g., to the motions produced in perceptible bodies by heat, light, magnetism, etc.[38]

Muncke's article on "Force" in *Gehler's Physical Dictionary* provides a good example of the extent to which authors sometimes advanced several mutually incompatible definitions of force. In one paragraph Muncke applied the word *Kraft* to weight, momentum, and vis viva:

> The motive force of any mass can clearly be measured by the weight with which it presses against its support and sets in actual motion a body capable of being moved by it. If one thus calls the absolute weight of a body $P$, its motive force is $x = P$. But at the same time the intensity of a force must be all the greater, the greater the space is through which a given body [*Last*] is moved by it in a unit of time, because both the exertion of force [*Kraftanstrengung*] as well as the mechanical effect [*Nutzeffect*] are so much greater. Thus if the mass of a body expressed by its weight is called $M$, its velocity $C$, and if it exerts a force proportional to its motion against any other body,

then the measure of the latter [force] is clearly $k' = MC$. But one must consider here that one obtains a different measure of force when a movable body is kept in constant motion by another one moving with a given velocity, than when the latter hits a body at rest and communicates to it all at once the entire force of its motion; as is well known, the former is also called the mechanical moment of the motion, the latter the moment of inertia of the motion. . . . Huygens was the first . . . to point out that the effect of a moving body on one at rest must be equal to the product of the mass and the square of the velocity.[39]

A few writers called attention to this chaotic situation. An anonymous reviewer of Kämtz's *Textbook of Experimental Physics* chided the author for speaking of the motion of a force (*das Bewegtwerden der Kraft*) instead of the motion of the body moved by the force, and called this "a confusion of concepts which stems from the fact that quantity of motion and accelerative force are designated without distinction by the same word force. For that reason communication of motion and acceleration are also regarded as the result of the same kind of force."[40] The author of a textbook of mechanics, Johann Andreas Schubert, complained that the word *Kraft* had been given several different meanings: "One does not always designate the same concept with the word force; sometimes one understands by it the cause of the actual motion of a mass, sometimes the tendency for the production of motion, even (finally) a pressure, whereby no account is taken of any motion."[41] The Heidelberg professor of physics, Philipp Gustav Jolly, who reviewed the book, objected (with somewhat quibbling justice) to Schubert's charge, saying that everyone agrees on the *definition* of force, although they may give different measures for the *moment* of the force. Jolly's own understanding of force was quite the norm:

As the author himself remarks, and as is universally assumed in all scientific works, by force one understands the cause of motion of a mass or the cause of the tendency of a mass toward motion. A force itself cannot be measured, only its effect is measured, i.e., the motion that it produces or would produce when no obstacle to the motion opposes it. The effects of two forces are in the same ratio as the masses multiplied by the velocities which they possess or would possess if there were no obstacle to the motion.[42]

Jolly, the reader will recall, was one of the professors to whom Mayer turned in the fall of 1841.

Several authors noted the tendency of the forces of nature to produce a state of rest or "indifference." Berzelius, for example, coupled that belief (somewhat oddly and ambivalently, given our acknowledged understanding of the solar system) with our inabililty to understand the purposeful contrivances of living organisms, an inability that should instill in us a humble respect for the wisdom of the creator:

All effects [or actions (*Wirkungen*)] originate out of what we call forces; these in turn (like the will) strive to be carried out or satisfied in order to come to rest after having been satisfied, to a rest which cannot be disturbed and in which nothing can arise that

> accords with the notion of chance. We do not understand how it is precisely this striving of inorganic matter to come to an indifferent and resting state through the yearning for satiation of opposing forces [*Sättigungsbegierde wechselseitiger Kräfte*] that is employed to keep it in ceaseless activity; but we see this calculated regularity in the motions of the planets [*Welten*], our investigations discover every day more and more about the structure [*Gebäude*] of organic bodies constructed in so marvelous a fashion for certain final purposes, and it will always do us more credit to admire the wisdom that we cannot imitate than in philosophical arrogance to reason ourselves speciously to the supposed knowledge of that which is perhaps not given to us ever to understand.[43]

Davy, in his *Consolations in Travel*, considered more particularly the tendency of gravitation, if unopposed by other forces, to reduce the world to a featureless sphere:

> The property, which . . . universally belongs to matter, *gravitation*, is the first and most general cause of change in our terrestrial system . . . . The forms upon the surface of the globe are preserved from the influence of gravitation by the attraction of cohesion, or by chemical attraction; but, if their parts had freedom of motion, they would all be levelled by this power, gravitation, and the globe would appear as a plain and smooth oblate spheroid, flattened at the poles. The attraction of cohesion or chemical attraction in its most energetic state, is not liable to be destroyed by gravitation; this power only assists the agencies of other causes of degradation; attraction, of whatever kind, tends, as it were, to produce rest, a sort of eternal sleep in nature. The great antagonist power is *heat*. By the influence of the sun, the globe is exposed to great varieties of temperature.[44]

Such pronouncements help make intelligible Mayer's still odd statement, toward the beginning of his first (unpublished) paper, that "we can derive all phenomena from a primitive force [*Urkraft*] that tends to annihilate the existing differences, to unite everything that exists into a homogeneous mass in a mathematical point."[45]

### *1.1 The Parallelogram of Forces and Central-Force Motion*

Virtually all physics texts of the period followed the introduction of the concept of force with a discussion of the parallelogram of forces. In a straightforward and unproblematic fashion, they laid out the graphical technique of representing forces and motions by means of directed line segments, and of then combining and decomposing them so that, for example, two equal and oppositely directed forces (or "motions") cancel each other out (*sich einander aufheben*, or words to that effect).[46] Such common treatment may have suggested to Mayer a deeper commonality between force and motion. Be that as it may, other than reinforcing his inclination to take *mv* as the measure of force and providing him with a misleading instance of the annihilation of force, the parallelogram of forces was important insofar as it also provided a widely used schema for the analysis of central-force motions, in particular of the motion of the planets around the sun. Most German texts offered a close variant of a standard analysis of the production of constrained circular motion resulting from the quasi-vectorial composition of a central force and what was typically referred to as a "tangential force"—that is, *not* as resulting from the action of a central force on a body with a particular initial *velocity*. Even a text like Geiger's, which offered no detailed analysis of central-force motions per se, still employed the same language:

> Every curvilinear motion is a composite motion where at least two forces must operate.
>
> If one force hereby always seeks to drive the body toward a fixed point, while the other drives it steadily away from there, and both forces act equally, then the body describes a circle. One of the forces, the one that drives the body toward the midpoint, is called *centripetal force*, the other, which drives it away from there, *centrifugal force* (*tangential force*); both together are also called *central forces*.[47]

Having reflected on such language, and having compared the magnitudes of the two composing "forces" to that of the resultant "motion" in the typical diagrams of central-force motion, Mayer was undoubtedly led to the erroneous conclusion that force can be destroyed *and* created—since the solar system is a kind of perpetuum mobile—in gravitationally controlled central-force motions. Certain considerations led him to make a fundamental distinction between such "natural" physical systems and "artificially" constrained mechanical systems, in which force can only be destroyed. Let us look at some typical examples of the treatment of central-force motion.

One of the books in Mayer's library, one that his father had given him when he was a boy, was Tübingen professor Johann Heinrich Moritz von Poppe's multi-volume "physics for children," *Der physikalische Jugendfreund*. In a chapter on central forces, Poppe analyzed the motion of the moon around the sun as the resultant of two separate forces:

> Gravity causes it [the moon] to remain always near our earth, but gravity *united with still another force* is the cause of its always going *around* our earth.
>
> One can imagine this second force as an impulse which the moon received right at its creation in a direction going off sideways from our earth, and whose effect now continues undiminished. If S indicates the earth and the point *a* the moon (Figure 2, Plate IV), and if *a* is driven in such a way toward the earth and at the same time in such a way away from the earth that by means of these two forces the parallelogram *abcd* can be constructed, then *a* must travel along the diagonal *ad*. But now *a*, by virtue of its inertia, would continue on without end in the direction *ad* if the force of gravity of the earth S did not act on it anew. It therefore now runs along the diagonal *dg* of the parallelogram *edfg*; thereafter the diagonal *gi*, etc. Both forces, the force driving [bodies] toward the earth and the force driving [them] away from the earth, act on the moon without letup and in every moment; it must consequently go around the earth in infinitely many small diagonals, which, when composed, form a single curved line (the *ellipse*) . . . .

Those two forces are found wherever bodies move around a certain point. They are called *central forces*, and in particular the one which draws [bodies] toward the point (*S* in Figure 2, Plate IV) is called *centripetal force*, whereas the one that drives [them] away from the point is called *centrifugal force* or *fleeing force* [*Fliehkraft*].[48]

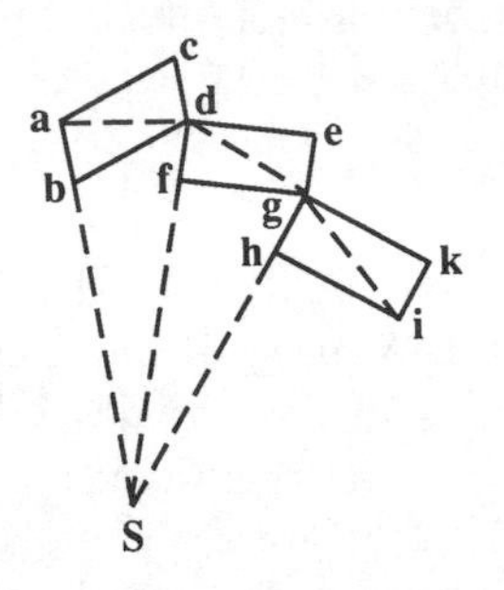


FIG. 4.1 Poppe's figure 2, plate IV:
The Motion of the Moon around the Earth.

Note that in Poppe's figure the initial impulse given the moon is not tangent to its orbit but goes off at an angle in a direction away from the earth. Later versions of his analysis of central-force motions were essentially the same, except that his diagrams showed the direction of the sideways "tangential force" to be along the tangent to the body's orbit.[49] Even when, as was usually the case, other authors also located the *Tangentialkraft* along the tangent to the orbit, they still regularly referred to it as a "centrifugal" force that holds in check the endeavor of the gravitational centripetal force to draw the revolving body toward the center.[50] Virtually every German physics text of the period contained some such representation of circular motion.[51] Muncke gave a rather confused account in his *Handbook of Physics*, in one place calling the "tangential force" a centrifugal force (*Centrifugal-Kraft*), in another identifying an apparently different *Schwungkraft* as the force with which the body strives to distance itself from the center. Although he remarked in passing that the *Schwungkraft* can be constructed as the diagonal of two other forces, it is not clear what kind of parallelogram of forces he imagined, and his accompanying diagram is both unenlightening and different from the standard form.[52]

Since I will argue in chapter 6 that it was probably Fries's *Textbook of Physics* that Mayer read at a crucial early stage, it is significant not only that its handling of central forces was entirely in the mainstream tradition, but also that some of the ways in which it differed from the norm made it an even more suggestive source of the kinds of constructions Mayer toyed with for more than a year. No other work depicted a parallelogram of forces similar to that of Fries's "Figure 21" (figure 4.2, below), which ostensibly produced the body's orbital motion via the neutralization of its centripetal and centrifugal forces:

> Point A (Figure 21) is attracted from C. If point A's own motion falls here in the direction AC, it will then approach C only with acceleration; but if A has by itself a

> motion in the direction *AB* oblique with respect to *AC*, one whose velocity is measured by *AB*, while in A there occurs a pull toward C measured by *AD*, then the point must now travel in composite motion along *AE*, in that *AE* is the diagonal of the parallelogram *BD*. Huyghens represented this in this way. In the extension of CA let *AM* = *AD*, and now consider *AB* as a motion composed of *AM* and *AE*. *AD* and *AM* are here equal and opposite, they cancel each other out, and there remains only *AE* as the motion of the point. Huyghens called *AD* the *centripetal force* and *AM* the always equal but opposing *centrifugal force*, which is otherwise also called *Fliehkraft*, or *Schwungkraft*. But here only the centripetal force is actually a force, and besides that it is only a question of point A's own motion, which we call its *tangential motion*.[53]

Even though Fries was careful to speak in terms of tangential *motion* and not tangential *force*, his analysis was scarcely more cogent than everyone else's. He, too, provided a parallelogram-of-force type drawing of continuous circular motion that was entirely in the spirit of Poppe et al.[54] From his analysis he concluded "*that every free central motion takes place with uniform speed in a circle when the [gravitational] pull and the [centrifugal] impulsion cancel each other out [indem Zug und Schwung sich einander aufheben]*."[55]

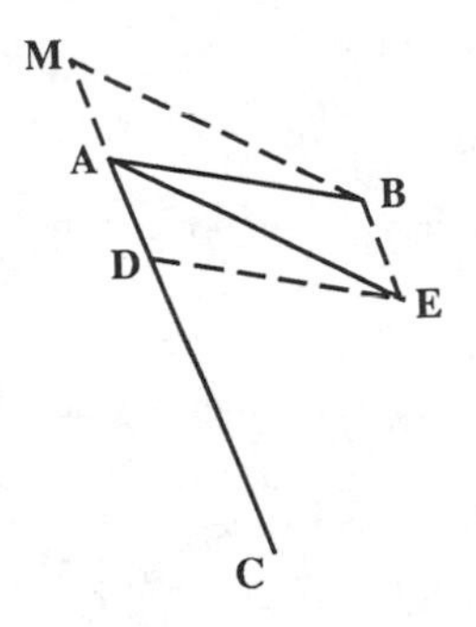


FIG. 4.2 Fries's figure 21, plate I:
The Action of a Central Force on a Moving Body.

On all sides, if not always in a consistent fashion, the reader of German physics texts in the 1830s was encouraged to apply a parallelogram of forces and motions to the several variously interpreted components whose resultant was the continuous circular motion of a gravitationally constrained body. If Mayer's speculations were confused and in some ways idiosyncratic, he was nevertheless seeking to employ what he can only have thought was an acceptable and entirely standard mode of analysis. It is easy to see why he thought he had to wrestle with the problem of the apparent disappearance and creation of force.

Mayer's distinction between "natural" physical systems (such as the solar system), in which force can be created, and "artificial" mechanical systems (such as a body constrained by a string to move in a circle about a fixed point), in which force can only be destroyed, reflected similar distinctions made by several contemporary authors. In a way reminiscent of Schwann's and Lotze's distinction between forces and initial conditions, Fries noted that the physical process of central motion is not simply the effect of a *Grundkraft*, the force of gravitation, but also of particular "purely geometrical relationships" such as the position and velocity of the affected body.[56] The same holds true for all physical processes, and Fries insisted that their effective cause is not just a force, but what he termed a *Naturtrieb*, which might be translated "natural disposition" and which expressed the combination of basic forces and given material circumstances:

> The process of central motion is not explained solely by the force of attraction, rather there is only a natural disposition which gives rise to a central motion when the directions of the originally present tangential motions of the mutually attracting masses at one time form an oblique angle with respect to each other. Thus we should not look for the causes of crystallization, plant life, or animal life in a force of crystallization or a vital force, but only in a natural disposition toward crystallization or in a vital disposition [*Lebenstrieb*] toward plant and animal formation.[57]

Fries distinguished in this context between "mechanical" and "organic" systems, whereby the latter include the solar system:

> A principal difference among these natural dispositions is that in some an interplay of motions is brought gradually to rest in equilibrium, whereas in others an interplay of motions preserves itself in a certain circular course [*Kreislauf*] of the same recurring phenomena. I call the former *mechanical*, the latter *organic natural dispositions*, the former a *striving toward equilibrium*, the latter a *striving toward a circular course*.
>
> In general those whose process is predominantly determined by forces acting in contact are mechanical. Thus the chemical process ends with an equilibrium of the mixture, the process of crystallization with the quiescence of solidified forms; thus do friction, cohesion, and the resistance of media gradually reduce all motions aroused on earth to equilibrium, and allow us to discern a perpetuum mobile only in the heavens. On the other hand, the processes that are predominantly determined by action-at-a-distance forces, by universal attraction from a distance, and those modified by radiant motion, show themselves in this general signification to be organic processes. But in nature as a whole, as in the central motions of the solar system, the action-at-a-distance forces must predominate and hence must secure for the *organism* dominance in the corporeal world.[58]

Fries had earlier expressed this distinction between mechanical and organic processes in terms of the difference between mechanical and "dynamical" interactions:

> [T]he mutual action between two bodies is *dynamical* when it is directly determined by the fundamental forces of their masses, whereas it is *mechanical* when it is only brought about by means of other mediating bodies. Thus we presuppose as dynamical the mutual action between the earth and the falling stone, between sun and planet, when we explain it by means of an original force of attraction, whereas the mutual action between force and load in machines is mechanical when it is mediated by levers, ropes, and gearworks.[59]

These distinctions had implications, not clearly or consistently worked out, for the measure of force. In mechanics, said Fries, one calls the quantity of motion a motive force and defines it as the product of mass and either velocity or acceleration. In mechanical systems we in fact measure force either as the product of mass and velocity or as weight. "Dynamical" action-at-a-distance forces, which are capable of continuously increasing the quantity of motion, must be measured by the acceleration they produce.[60] Fries did not seriously address, let alone resolve, the contradiction between, on the one hand, such a creation of motion or its destruction when mechanical systems reach equilibrium, and, on the other, his assertion that the law of equality between action and reaction implies that the quantity of motion in the world can neither increase nor decrease when opposing "directions" (*sic*) cancel each other out.[61]

Muncke, who subjected Fries's book—in particular its treatment of mechanical and organic *Naturtriebe*—to fifteen pages of critical scrutiny,[62] elsewhere elaborated a parallel distinction between what he termed physical and mechanical systems. Once again the solar system, governed by the continuously acting natural force of gravitation, occupies a special position. At issue was the distinction between a *perpetuum mobile physicum* and a *perpetuum mobile mechanicum*: "If we speak first of the *perpetuum mobile physicum*, then there can be no doubt that such can exist, since the circular course of things in nature is one which ever endures and uninterruptedly renews itself. . . . Our planetary system, among others, is thus a true *perpetuum mobile*, as is no less the earth rotating on its axis."[63] What distinguishes such systems is that "the motive force or cause [of motion] is supplied by nature itself."[64] Humanly constructed machines, on the other hand, have no intrinsic source of motion and are incapable of moving themselves out of a state of rest, since "a force arising by itself out of the mere arrangement of the parts of the machine presupposes the creation [*Entstehung*] of something from nothing, which is inadmissible in the domain of physics, where the principle *ex nihilo nil fit* is accepted as an irrefutable axiom."[65]

However, if such machines are once set in motion, then the law of inertia says that the motion initially imparted will persist. The *perpetuum mobile physicum* of our solar system is thus also a kind of *perpetuum mobile mechanicum*. Indeed, Muncke eventually seemed to imply that the distinction between the two does not amount to anything very important as long as one appreciates that no machine is capable of producing a greater *Nutzeffect* than the *Kraft* originally employed to set it in motion.[66] It is significant that, despite the effective evaporation of the initial distinction between physical and mechanical systems, Muncke's terminology continued to reflect an abiding, almost intuitively felt distinction between natural and artificial systems, according to which the solar system and the force of gravity are natural. Just such a distinction was a noteworthy feature of Mayer's thinking in 1841.

### *1.2 Catalytic, Contact, and Electrochemical Forces*

Several important chemical issues having to do with the nature of force and causality occupied Mayer's scientific contemporaries, and although there is no evidence that he paid particular attention to the debates going on in the more technical literature, his thorough grounding in chemistry as medical student and as apothecary's assistant suggests that a brief examination of them is in order. They formed part of Mayer's scientific context and *became* issues for him as he elaborated and defended his ideas on force, even if it is difficult to tell how much his earliest reflections owed to his possible awareness of them. Since the path of Mayer's life and work intersected with Liebig's in many important ways, I will pay special attention to Liebig's views on these issues, especially since a number of his important pronouncements appeared in the popular press in the fall of 1841, when Mayer was attempting to forge a coherent theory of force.

Alwin Mittasch suggested that Mayer's knowledge of Johann Wolfgang Döbereiner's discovery in 1823 of the catalytic transformation of hydrogen and oxygen into water by means of platinum made a lasting impression on him.[67] Unfortunately, the passage he cited attesting to Mayer's knowledge of this phenomenon was from 1876, in Mayer's last essay.[68] Yet, as we have seen, the transformation of *Knallgas* into water was Mayer's earliest and favorite example of the neutralization of chemical differences, and Mittasch's hunch may well be correct even in the absence of good evidence. Mayer's earliest surviving mention of catalysis comes in a letter to Baur, written two months after the publication of his first paper, in which he criticized the concept as an example of the fact that, as Liebig had written him, "such confused notions prevail in general over force, cause, and effect. Just as the concept 'force of gravity' is a nonsensical one, so is also that of many others—in particular of the catalytic force, etc.—equally foolish and pernicious."[69] The notion that a substance (or force) is capable of bringing about indefinite chemical change without itself undergoing any change or diminution was a challenge to Mayer's ideas on force and the necessary equivalence between cause and effect.

Eilhard Mitscherlich had introduced the concept of "decomposition and combination by contact" and the term "contact substance" in 1833. The latter is a substance whose presence brings about a chemical transformation without itself undergoing any chemical change.[70] Mitscherlich applied the concept to cases such as the decomposition of alcohol into ether and water in the presence of sulfuric acid, of hydrogen peroxide into water and oxygen in the presence of manganese dioxide and several other substances, and of sugar into alcohol and carbonic acid (in fermentation), as well as to the oxidation of alcohol to form acetic acid. He seems not to have spoken in terms of a contact *force*.

The terms "catalytic force" and "catalysis" were introduced by Berzelius in 1836 in such a way as to stress the connection between the newly named force and electrochemical affinities. It was not his intention to hypostasize a new causal agency:

> It has thus been proven that many bodies—both simple and compound, both in solid and in dissolved form—possess the property of exerting on compound bodies an influence entirely different from the usual chemical affinity, in that they thereby bring about a transposition [*Umsetzung*] of the constituents of the body into other relationships without them and their constituents thereby themselves necessarily taking part, although that can sometimes also be the case.

> This is a new force for the production [*Hervorrufung*] of chemical activity, one pertaining to both inorganic and organic nature, which may certainly be more widespread than one hitherto believed, and whose nature is still hidden from us. If I call it a new force, it is in no way thereby my intention to declare it to be a power [or capacity (*Vermögen*)] independent of the electrochemical relations of matter; on the contrary, I can only suppose it to be a peculiar mode of their expression. However, as long as its reciprocal connections remain hidden to us, it facilitates our investigations to regard it for the time being as a distinct force, just as it also facilitates our dealings with it when we have a particular name for it. Therefore, to make use of a derivation well-known in chemistry, I will call it the *catalytic force* of bodies, and decomposition by means of it *catalysis*, just as with the word analysis we understand the separation of the constituents of bodies by means of the usual chemical affinity.[71]

One of those to speak swiftly against the new concept was the ever-contentious Liebig, who countered that the fact that a number of chemical reactions cannot be explained according to the normal laws of decomposition "affords not the slightest reason for the creation of a new force by means of a new word, which likewise does not explain the phenomenon. The assumption of this new force is detrimental to the development of science in that it ostensibly satisfies the human mind, and in that way imposes a limit to further investigations."[72] Although Wöhler suggested to Liebig that Berzelius intended nothing more by the new term than the designation of a group of phenomena, and that he no more than they believed in a distinct new force, Liebig responded that he thought the whole idea of the catalytic force is false.[73] Although he was still ridiculing it to Wöhler as late as 1839,[74] two years thereafter, in the second of his serialized and anonymously published "Chemical Letters," Liebig effectively accepted the operation of such a force, though without deigning to use Berzelius's term. Under discussion was the ability of certain forms of platinum to absorb a considerable volume of oxygen as it mediates a number of chemical reactions: "In this platinum black, even in the platinum sponge, one indeed has a perpetuum mobile, a clock that, having run down, winds itself up again, a force that never exhausts itself, effects of the most powerful kind that renew themselves ad infinitum."[75] Liebig's letter appeared in the widely read *Allgemeine Zeitung* of 16 September 1841, and may well have come to Mayer's attention at an important stage in the development of his ideas. Although Liebig did not choose to reprint the letter in the book version of his *Chemical Letters*, his *Animal Chemistry* of 1842 again effectively accepted the action of the again-unnamed catalytic force in effecting certain chemical transformations. As we have already seen, he was there prepared explicitly to accept the existence of chemical forces capable of producing motions and transformations without any diminution in their strength. The context of those passages suggests that he may have been motivated by a desire to render his conception of the vital force more plausible by likening its action to that of a bona fide *chemical* force. That he consistently declined to use the *term* catalytic force has more to do with polemics than with (in the event) any de facto objection to such an agency: his own explanation of such chemical transformations was based on an inconsistently maintained distinction without any real difference. Bischoff, for one, believed he was following Liebig's lead in appealing to contact effects and catalytic forces—interpreted in terms of molecular motions—as analogies for the explanation of the alleged fertilizing action of sperm-free seminal fluid.[76]

Liebig's attitude toward the concept of a catalytic force was influenced by his understanding of processes of fermentation, putrefaction, and decay, topics of great importance in the development of his scientific views that he addressed in 1839 and 1840. Those processes are representative of the phenomena "that accompany the decompositions and transformations [*Veränderungen*] of organic substances that take place by themselves, i.e., which are produced by unknown causes."[77] In Liebig's view, complex organic compounds tend to be unstable as a result of the different degrees of affinity (or attraction) between the compound's radical and the other elements in it or among those elements themselves. He termed "organic metamorphoses [*Metamorphosen*]" transformations of organic molecules characterized by a rearrangement of the already present constituents.[78] Such disturbances in the equilibrium between a given radical and the elements (or other radicals) bound to it are brought about by one or more of the following causes: change in the state of cohesion of two or more elements due to the influence of heat; "contact with a third body, which thereby enters into no [chemical] combination" with the original substance; and addition of water.[79] He noted that there are many compounds in which the slightest temperature difference or the slightest change in electrical state produces decomposition. Some compounds, such as hydrogen peroxide, decompose upon contact with finely divided carbon, platinum, or other solids, without the latter themselves undergoing any change. Liebig rejected the proposal—he did not say whose—to explain such decompositions in terms of a supposed catalytic force: if we (correctly) do not wish to suppose that friction and concussion decompose fulminates (or that a slightly elevated temperature decomposes perchloric acid) by producing a catalytic force in them—this was the way Liebig put it—then we should not invoke a catalytic force to explain the decomposition of (for example) hydrogen peroxide, since there are no essential differences among all these cases of decomposition. Here, as elsewhere, Liebig was groping to find appropriate analogies to either justify or reject particular forces—catalytic, vital, or otherwise—in the explanation of different classes of phenomena.

Liebig did not, however, further elaborate his own conception of such decompositions except to allude to his earlier explanation of organic metamorphoses. Instead, he went on to identify "a heretofore unnoticed cause" of his own which produces such decompositions as fermentation and putrefaction: "*This cause is the capability that a body possesses while undergoing decomposition or composition (i.e., a chemical action) to produce [hervorrufen] the same activity in another body in contact with it or to make the other body capable of undergoing the same change that it itself experiences.*"[80] Liebig did not explain this capability in terms of, say, molecular motions, but spoke of it as "a distinctive power [*eigenthümliches Vermögen*], a particular expression of the affinity, acting like a distinctive force."[81]

Substances that produce such decompositions are "*carriers of an activity which extends beyond the sphere of the decomposing body.*"[82]

In a supplementary note to this paper, however, Liebig explained that several of his friends had criticized his theory as unsatisfactory because it gave no explanation of *how* the new cause acts. He defended himself by arguing that we are at best capable only of imagining various hypotheses to explain the mode of action of a physical cause: "as far as I could I have avoided hypotheses and kept only to the phenomena."[83] But he did not long remain so strict with himself, and we find him a few pages later explaining several new classes of phenomena in terms of the communication of motion and the subsequent rearrangement of the molecules of the affected substances.[84] That conception was in short order then extended to explain fermentation: the molecules of the ferment are in a state of continual transformation (*Umsetzung*), of incessant motion.[85]

A remarkable letter of Liebig's to Wöhler exhibits the transformation in Liebig's thinking as it was happening.[86] Wöhler was clearly the unnamed friend who had criticized his account of fermentation for not explaining how the putrefying body actually acts (*sic*: Liebig went back and forth between the various classes of phenomena at issue). In his reply, Liebig first took refuge behind our analogous inability to explain the destruction of the properties of an acid by a base. Especially unacceptable to Liebig must have been Wöhler's assertion that "catalysis is the same thing," since Liebig had been at pains to distinguish his views from Berzelius's.[87] Liebig asserted that he regarded fermentation and putrefaction as metamorphoses brought about by contact with a body itself in a state of decomposition: "the cause [of the metamorphosis] is the decomposition that is transferred . . . . Berzelius says it is a metamorphosis that is brought about through contact with another body, which *thereby* plays *no role*."[88] In a postscript written the next day, Liebig went on to quote Berzelius's characterization of catalytic force in his *Textbook of Chemistry*—where the action was said to be due to the mere presence of the body and not to its affinity—and asked rhetorically: "What is that supposed to mean?"[89] Liebig argued that he wished to distinguish metamorphoses that are properly so called from decompositions—processes that Berzelius lumped together—and to explain fermentation and putrefaction in terms of the effect on an unstable complex molecule of "the action [*Action*] in which another body finds itself."[90] It was in seeking to clarify his thinking by means of an example that Liebig, in the next-to-last paragraph of the letter, first interpreted the heretofore vague "activity" and "action" of the causative agent in terms of *motion*: "A ball in motion comes into contact with another at rest; what will happen? Either the moving ball communicates its motion to the one at rest, such that both move, or the one in motion comes to rest and the one at rest comes into motion, or both lose their motion. This, now, is fermentation."[91] He immediately applied this imagery to putrefying bodies and ferments, the molecules of which are (he said) in incessant motion: when a ferment comes into contact with sugar, it produces motion in the sugar molecules; when it comes into contact with corrosive sublimate, its internal motion is annihilated (*aufgehoben*). (Note that there is decidedly no conservation of motion implied in any of this; quite the contrary.)[92]

Liebig re-presented these ideas the next year in his semipopular *Organic Chemistry in Its Application to Agriculture and Physiology*, whereby he glossed over an important difference between the (unacceptable) catalytic force and his explanation in terms of the "state of decomposition or combination" of the reaction-producing body: namely, the fact that the usual examples of catalytic action involve bodies (such as platinum) not themselves undergoing any chemical change, hence not obviously characterized by forceful internal motions.[93] Nor was Liebig any closer here to any conservation-of-motion type considerations: "the most remarkable thing" about these phenomena is "that extraordinarily small quantities of those substances which have gone over into the state of fermentation and putrefaction possess the ability to produce the same kind of decomposition in unlimited quantities of the same substances."[94] In considering processes of fermentation and putrefaction (e.g., of grape juice and blood) as purely chemical transformations triggered by a specific external agency, Liebig broached the issue of causality:

> There is hardly an error more widespread than the notion that organic substances left to themselves, *without external cause*, are capable of transforming themselves. If they are not themselves already in a state of transformation, then it always requires a disturbance in the state of equilibrium of its elements, and the most general inducement to such disturbances, the most widespread cause, is incontestably the atmosphere, which surrounds all bodies.[95]

Liebig pictured the chemical transformation itself as consisting of the motion of the smallest particles of matter. Although he couched his considerations in terms of external causality and the propagation of an initial disturbance, the analogies he appealed to in support of his views suggest that he had no inclination to insist on any quantitative proportionality between cause and effect:

> Oxygen acts here like friction, impact, or motion, which bring about the mutual decomposition of two salts, the crystallization of a saturated salt solution, and the explosion of silver fulminate; it occasions the suspension [*Aufhebung*] of the state of rest and mediates the transition into the state of motion.
>
> If this state has once been attained, then the presence of oxygen is no longer necessary. The smallest particle of decomposing, transforming nitrogenous body acts in its stead, propagating the motion to the particle lying next to itself. The air can be excluded, and the fermentation or putrefaction proceeds uninterruptedly to completion. One has observed in the case of many fruits that only contact with carbonic acid is required to produce the fermentation of the juice.[96]

Indeed, Liebig's attachment to this mode of explanation gave him a kind of vested interest in *avoiding* such considerations. Even though he here interpreted the relevant chemical transformations explicitly in terms of molecular motions, he still occasionally fell back on his earlier vague language of an "activity" that "acts like a distinctive force, . . . a force that reaches beyond the sphere of the attractions" of the body undergoing combination or decomposition.[97]

Perversely enough, the anonymous reviewer of Lehmann's *Textbook of Physical Chemistry* also accepted the reality of the catalytic force—or rather of those *reactions* termed catalytic—but as part of his argument *against* the acceptability of the vital force.[98] By implication, if (as it appears) the phenomena of life share important similarities with those Berzelius explained in terms of a catalytic force, then there is no need to assume other than chemical forces to explain them. According to the reviewer—who I suspect may have been Lotze—Lehmann sought to explain catalytic reactions by applying Liebig's ideas about the transmission of molecular motions, but the reviewer objected that the concepts of mechanics cannot be applied in a precise way to chemical reactions: "In any event, however, this conception [*Fassung*] of the matter, which in the end anyway just reduces the result thereby attained to the given disposition of certain circumstances, is more highly to be esteemed than the assumption of the catalytic force, which subsumes the same phenomena rather carelessly under the reflective concept [*Reflexionsform*] of a simple force, without being able to demonstrate its law of action."[99] In the end, an explanation which appealed to a mechanical disposition of atoms was preferable to one which invoked a force whose law of action was unknown, though the reviewer neither rejected as a matter of principle all explanations in terms of forces nor criticized the catalytic force on energetic grounds. His silence with respect to the latter and Liebig's positive insistence on the inexhaustibility of the work-performing catalytic force show once again how unobvious conservation-of-energy-type considerations were, even to those who addressed themselves to notably energetic issues. Mayer's critical response to these issues helped him to refine *his* understanding of causality and conservation.

One of the most bitterly contested scientific controversies of the first half of the nineteenth century was between defenders of the contact and chemical theories of the pile.[100] The former followed Volta's lead in assigning the cause of the activity of the pile to a contact force acting between the two different metals typically used in its construction; the accompanying chemical changes were thus seen as the *result* of that primitive force. Chemical theorists maintained that, even if there were such a force, it could not be the cause of the pile's sustained activity, which is due to ongoing chemical transformations. For the most part the debate revolved around an appeal to experiment—Does or does not the mere contact of metals produce a measurable electric tension? Can an electric tension be detected while the circuit is closed? and the like—though some, following a theory of Ampère's, insisted on a distinction in principle between tensional and current electricity. It was the exceptional writer like Peter Mark Roget who advanced energetic arguments on behalf of the chemical theory.[101] Only with Faraday's combined experimental and theoretical work of the 1830s—the "Experimental Researches in Electricity," begun in 1831—did a real resolution of the long-inconclusive debate become possible. From his assertion (in 1833) that there is a direct ratio between chemical decomposition and quantity of electricity, to his enunciation (in 1834) of the principle of electrochemical equivalence, to his epoch-making paper of 1840, "On the Source of Power in the Voltaic Pile," it was primarily Faraday who came to advance explicity energetic considerations in defense of the chemical theory of the pile. Their rarity elsewhere suggests the extent to which the notion of the impossibility of perpetual motion was restricted to purely mechanical systems.

Friedrich Mohr's progress report for Liebig's *Annalen der Pharmacie* in 1837 observed that "it is an important fact to report here at the outset that none of Faraday's more important principles, on which he based his view on the origin of the galvanic electricity and on its quantity in relation to the decomposed matter, has had to be retracted."[102] Later in the same report Mohr reviewed the status of the dispute between the contact and chemical theories of the pile and recorded his impression that a large majority of Europe's scientists followed Faraday in support of the chemical theory.[103] Surprisingly, Mohr did not advance any energetic arguments in favor of the theory that he, too, supported, despite the fact that it was only a few pages earlier in the same report that he published his "Views on the Nature of Heat," in which he considered the general nature of force and came close to a full enunciation of the principle of conservation of energy. Writing in *his* progress report of 1837, Gabriel Gustav Valentin found in the reinvigorated chemical theory of the pile both an analog for the organism's vital activities—both they and the electricity of the pile are the product of ongoing chemical transformations—and confirmation of the growing sense that the various agencies of nature are somehow intimately connected.[104] Via such routes did recent discoveries in physics and chemistry make themselves felt among physiologists, notably with regard to general conceptions of causality and the relationship among forces.

Faraday's "On the Source of Power in the Voltaic Pile" appeared in English in 1840 and in German translation in Poggendorff's *Annalen* roughly between January and August 1841. At the end of that paper Faraday devoted a separate section to "The Improbable Nature of the Assumed Contact Force," in which he rejected it for reasons of causality and energetic implausibility:

> 2071. The contact theory assumes, in fact, that a force which is able to overcome powerful resistance, as for instance that of the conductors, good or bad, through which the current passes, and that again of the electrolytic action where bodies are decomposed by it, can arise out of nothing. That, without any change in the acting matter or the consumption of any generating force, a current can be produced which shall go on for ever against a constant resistance, or only be stopped, as in the voltaic trough, by the ruins which its exertion has heaped up in its own course. This would indeed be *a creation of power*, and is like no other force in nature. We have many processes by which the form of the power may be so changed that an apparent *conversion* of one into another takes place. . . . But in no cases, not even those of the Gymnotus and Torpedo (1790.), is there a pure creation of force; a production of power without a corresponding exhaustion of something to supply it.
>
> . . .

2073. Were it otherwise than it is, and were the contact theory true, then, as it appears to me, the equality of cause and effect must be denied (2069.). Then would the perpetual motion also be true; and it would not be at all difficult, upon the first given case of an elecric current by contact alone, to produce an electro-magnetic arrangement, which, as to its principle, would go on producing mechanical effects for ever.[105]

Liebig likely drew upon Faraday's work—especially his concept of electrochemical equivalents—in his analysis of the relative effectiveness of battery-powered electric motors to do work vis-à-vis steam engines.[106] In his "Chemical Letter" of 30 September 1841 he addressed the exaggerated expectations many had had with respect to the economic utility of electromagnetism as a source of power. The chief question is which fuel, the zinc consumed in the battery or the coal in the steam engine, is cheaper in producing a given amount of motion:

In order to grasp this question in its true significance one must recall the chemists' equivalents. These are certain invariable, numerically expressible mutually proportional values for the effects produced [*Wirkungswerthe*]. In order to produce a certain effect, I need 8 pounds of oxygen, and if for the same effect I want to use not oxygen but chlorine, then I must take not more and [not] less than 35½ pounds. In the same way 6 pounds of carbon are an equivalent for 32 pounds of zinc. These numbers express entirely general *Wirkungswerthe*, which relate to all activities that are capable of manifesting them.[107]

Whereas in his "Chemical Letter" of two weeks earlier Liebig had been prepared to regard catalytic systems as self-winding clocks, true perpetuum mobiles capable of creating force indefinitely, he now confidently asserted the axiomatic uncreatability of force:

No force can come into existence out of nothing; in the case at hand we know that it is produced through the solution (through the oxidation) of the zinc; but if we abstract from the name that this force carries here, we know that its effect can be produced in another manner. That is, if we had burned the zinc under the boiler of a steam engine—i.e., in the oxygen of the air instead of in the galvanic pile—we would have produced steam and with that a certain quantity of force. We wish to assume now—something that has in no way been proven—that the quantity of force in the two cases is unequal, that for example one has obtained two or three times more force with the galvanic pile.[108]

The modern reader who expects Liebig to then develop an argument in terms of the impossibility of perpetual motion will be disappointed. Liebig's interest was rather in demonstrating the economic unfeasibility of using electric motors as sources of power.[109] His statement of the relationship between the heat, electricity, and magnetism produced by a galvanic cell and the chemical activity in it reveals how closely his thinking was tied to the concept of chemical equivalents:

Heat, electricity, and magnetism stand in a similar relationship to each other as that between the chemical equivalents of carbon, zinc, and oxygen. With a certain quantity of electricity we produce a corresponding proportion of heat or magnetic force, which are mutually equivalent. I purchase this electricity with chemical affinity, which expended in one form produces heat, in another electricity or magnetism. With a certain quantum of affinity we produce an equivalent of electricty, just as vice versa with a certain measure of electricity we decompose equivalents of chemical compounds. The expenditure [*Ausgabe*] for the magnetic force is then here the expenditure for the chemical affinity. Zinc and sulfuric acid provide us with the chemical affinity in one form, coal and a hefty draft of air in another.[110]

He went on to compare a magnet to a heavy rock pressing down on a surface and to the pent-up water of a lake which has no head (*Fall*): none of them is an effective source of motion. If here as in several other places Liebig appears to have come very close to an appreciation of the conservation of energy, in no case did he expound a general concept of *Kraft*, and he never made more than a feeble attempt to define its measure. Given his many contradictory pronouncements it is extremely difficult to penetrate his mind to understand what he 'really' believed was true, as opposed to what might be useful in this or that argumentative context. There is no question, however, that the contemporary reader of his writings—like Mayer—was confronted with a wealth of thought-provoking ideas relating to these issues.

In another "Chemical Letter," published two and a half years later, Liebig discussed in general terms the advances scientists had recently made in uncovering the secrets of electricity:

Certainly there was no state of matter more hidden and obscure to the corporeal and mental [*geistig*] eye than the one we call *electrical*. . . . As a result of tireless investigations, and unintimidated by difficulties without number, the natural scientist obtained its closest acquaintance and made it his maidservant; he now knows that it comes from one mother along with heat, light, and magnetism; with it he has subdued its siblings [*Geschwister*], they follow his call, with its help he indicates the way to lightening, with it he lures precious metals out of their poorest ores, with it he first succeeded in making out the true nature of the constituents of the earth, with its help he sets ships in motion and with it he reproduces art objects.

A force cannot be seen, we cannot grasp it with our hands; in order to know it in its essence and in its distinctiveness we must study its manifestations and investigate its effects.[111]

Even then, in 1844, Liebig still spoke of electricity variously as a "state of matter" and a "force" without apparent self-consciousness of any important distinction. As much as he recognized the close connections among what were termed imponderables, forces, or powers, so little did he himself have a clear conception of them *and motion* as a radically new physical entity—what *we* term energy. Given all that Liebig *did* say about forces, causality, *Stoffwechsel*, and the like, the fact that he did not take those last few steps makes even more impressive Mayer's clear sense that that was exactly what was called for. It also underscores the importance of understanding Mayer's work as coming out of a very particular problem context, and not just as a reflection of general and widely recognized ideas. With the all-important exception of the universally recognized transformation of heat into motion and vice versa, the many examples of what came to be seen as the interconversion of one force into another identified in the decades before 1840 appear to have played no significant role in the genesis of Mayer's ideas on force, and hence have not been made a special subject of study here.[112] Once he started looking for examples of such connections, they were obvious and abundant.

The first significant public acknowledgment of Mayer's "Remarks on the Forces of Inanimate Nature" came in 1845 in Christoff Heinrich Pfaff's last-ditch defense of the contact theory of the pile.[113] Pfaff cited Faraday's above-discussed criticism of the notion of an inexhaustible power, and quoted Leopold Gmelin's embellishment of Faraday's argument that the existence of such a contact force would allow the construction of a perpetuum mobile. As Gmelin put it, "One activity calls forth the other; frictional electricity comes into being as a result of mechanical motion; thermoelectricity is connected with the motion of heat, but contact electricity comes into being out of nothing, is a creation of force."[114] The chief reason for some people's mistaken conclusions, Pfaff said, was that they failed to recognize the true nature of a genuine physical force; that is, they did not distinguish force as a primitive cause from force as a derived cause. Derived causes may well be *indestructible* once they have been called into existence, but only primitive causes—such as gravitation, magnetism, the catalytic force, and the (unknown) cause of sunlight—are truly *inexhaustible* in their capacity to go on generating force indefinitely without diminution of their effect. The notion that a force only works by consuming something arose, he said, from people's experience with electric motors as a source of motion. One noted the consumption of zinc, and was led to compare the consumption of zinc in a voltaic cell to the consumption of coal in a steam engine: in both cases one obtains a given amount of motion by consuming a certain equivalent amount of substance. Pfaff continued his line of reasoning in a passage most of which Mayer quoted (anonymously) in what was essentially his last scientific publication, the *Remarks on the Mechanical Equivalent of Heat* of 1851:

> The same view of, as it were, the continuing nourishment of a force or the continuing new generation of it by means of a repeatedly renewed chemical process . . . has been brought to bear in the explanation of vital phenomena insofar as one rejected a vital force as a mere phantom and regarded its activity as in any case dependent only on the activity of this process. If we nevertheless pursue the chain of causes and effects to its first beginnings, in that way alone do we arrive at the true forces of nature, to the primitive causes [*primitive Ursachen*] which require for their activity no others that precede them, which demand no nourishment in the above sense, and which can, as it were, repeatedly rekindle motions out of an inexhaustible source [*Grund*] and maintain and accelerate existing motions. If it is completely true that no motion can be annihilated in nature, or that (as one expresses it) the quantum of once-existing motion remains unimpaired and undiminished, and if in this sense the attribute of indestructibility belongs even to every derivative cause, then in addition the charac-

> teristic of *inexhaustibility* is also among the attributes of a primitive cause, i.e., of a true physical force. These characteristics will best be able to be unraveled through the closer consideration of gravity, which is the most active and most widespread natural force (primitive cause)—as it were, the world soul—which indestructibly and inexhaustibly maintains the life of the large masses on whose motions the order of the universe depends, without its needing any nourishment from the outside that repeatedly rekindles its activity.[115]

From this standpoint Pfaff went on to criticize the misconception of force that underlay Mayer's 1842 paper and to claim for the electromotive contact force the status of an inexhaustible primitive cause. Mayer used his response to vindicate the justness of *his* understanding of the nature of physical forces. Ironically, Pfaff's notion of the indestructibility *but* indefinite creatability of force under certain circumstances is not far from Mayer's own early belief that forces, albeit indestructible—according to the axiom *causa aequat effectum*—may nevertheless still be created in certain noninanimate systems, including the solar system. As Mayer put it, what is at issue is not "what kind of a thing a 'force' is, but what thing we wish to *call* 'force.'"[116]

## 2 Imponderables and the Nature of Heat

With not many exceptions, few German scientists of the time spoke of heat, light, electricity, and magnetism as "forces"—even occasionally, let alone consistently.[117] For the most part, and especially up until around the mid 1830s, they were referred to as imponderables, as weightless substances or fluids. Thus for Leopold Gmelin they were "the unweighable, imponderable substances, imponderables, aetherial substances, radiant powers, incoercibles," for Christian Gottlob Gmelin "imponderable substances" and "imponderables."[118] Poppe, in a book Mayer owned as a child, regularly spoke of *Wärmematerie* and *Wärmestoff*, of *Lichtmaterie* and *Lichtstoff*, of *elektrische Materie*, *elektrischer Stoff*, and *elektrisches Fluidum*, and only rarely used the exceptional term *elektrische Kraft*.[119] Muncke called them "imponderable powers," "imponderables," and "incoercibles," but admitted that we do not know whether or not they are entirely weightless.[120] As late as 1842 Muncke wrote that "one is today pretty generally agreed that the incoercibles known up to now—*heat*, *light*, *electricity*, and *magnetism*—consist of fine aetherial substances [*Stoffe*] which have not been able to be weighed with the means hitherto applied, although it has not thereby been settled whether some of them—in particular heat and electricity—do not gravitate toward the earth, or are not attracted by it."[121] In his article on the "Imponderables" in *Gehler's Physical Dictionary*, Muncke discussed reasons why even in principle one cannot expect experimentally to be able to weigh light, magnetism, or electricity, as had in fact proven impossible. He emphasized that one was justified in calling those three *Potenzen* "weightless" only insofar as experiment warranted, not as a supposed intrinsic characteristic distinguishing them from ponderable matter. Indeed, in order to escape the detection of our finest balances, they would only have to be as light with respect to hydrogen as hydrogen is with respect to platinum. Muncke was clearly at pains to make plausible the materiality (*Materialität*) of the imponderables.[122] Fischer and August's textbook of 1837 called them imponderable and incoercible substances (*Stoffe*), and cautioned against confusing the terms *body* and *matter*: "We wish to call everything matter in which there lie active (but not psychic) forces, whether or not perceptible. Body, on the other hand, is only that which is perceptible. Everything corporeal is material, but not everything material is vice versa corporeal."[123] Thus both ponderable and imponderable matter may well be the *seat* of certain attractive and repulsive forces, but the imponderables themselves are not forces. For Fries, the imponderables were not only not forces, or even a separate class of weightless substances, they were considered to be highly tenuous and expansive states of aggregation of matter, on a dynamic continuum with ponderable matter.[124] Although Berzelius noted the disagreement over whether the imponderables are material substances or properties of matter, he himself spoke of them as if they were a kind of weightless matter, and speculated that one or another might be composed of the others, or that they might all be composed of common but as yet unknown elements.[125]

Alongside the dominant treatment of heat, light, electricity, and magnetism as imponderable substances, some scientists took a more positivistic approach to the interpretation of the imponderables. Biot, for example, did not usually refer to them by *any* general term, but chose rather to address the *phenomena* of heat, light, electricity, and magnetism. As he said with respect to the phenomena of magnetism:

> What is the nature of the principle [*Grundursache*] on which these phenomena depend? We do not know. Nevertheless, of whatever nature it may be we want always to designate it for brevity's sake by the name *magnetism*; just as we called the unknown principle of the electrical phenomena *electricity* and the no less unknown cause of the expansion of bodies *heat*. In order not to stray from the true path of science we will only have to guard against attributing to the magnetic principle any other properties and qualities than those indicated or required by the phenomena that it produces.[126]

Lamé played up the analogies between radiant heat and light and favored the undulatory over the caloric theory of heat, but for the most part maintained a positivistic neutrality with respect to the ultimate nature of the cause of these phenomena: we can treat the phenomena without subscribing to any particular theory about the nature of heat.[127] Wilhelm Friedrich Eisenlohr seems to have followed this largely French tradition: one considers in turn the several more-or-less distinct classes of phenomena without committing oneself to any underlying causal explanation. Magnetism and electricity are simply terms that refer to the otherwise unknown causes of particular classes of phenomena; heat is neither an imponderable *nor* a force.[128] And Thaddaeus Siber divided the otherwise untitled "Part Two" of his elementary physics text into separate sections entitled "Heat

Theory," "Extensible Form," "On Light," "Magnetic Phenomena," and "Electric Phenomena."[129] He said very little about the nature of the causes of these phenomena, and used no general term for them, neither imponderables nor forces nor anything else. As he said of heat,

> One usually designates the cause of this expansion by the name *principle of heat*, and one understands thereby those causes that produce in us the sensation of being warm. Much has been said and argued about the materiality or immateriality of this principle; but whereas the atomistic view (as in general, so also here) appeals more to sensible perception [*sinnliche Anschauung*], the dynamical view has by far more scientific value. We can here leave undecided the argument over the nature of this principle insofar as we are only concerned with its effects.
>
> This principle may, that is, be material or immaterial, [but] its effect is always the expansion of the matter on or in which it acts, either only as far as to increase its volume or even as far as to change its form of aggregation.[130]

Consistently, the principal phenomenological characteristic of heat was that it causes bodies to expand—and thermal expansion was to be the process through which Mayer conceptualized the transformation of heat into motion.

Since heat was the imponderable of greatest importance to Mayer, let us look more closely at how several major authors treated it. In 1827 Fischer indicated his preference for the assumption of a subtle fluid, a *Wärmestoff*:

> The cause of heat eludes all of our senses. Some mechanistic scientists are inclined to take it for an internal agitation of the smallest particles of bodies; chemical scientists unanimously assume a substance of its own, which they call *substance of heat* (caloricum), as the cause of these phenomena. In what follows we shall find if not decisive, yet important reasons for this way of conceiving things. In the meantime we shall use the expression *substance of heat* at least as a convenient means of representation.[131]

In 1837 Ernst Ferdinand August's revision of Fischer's text was more evenhanded, and concluded that the phenomena demonstrating the constancy of the quantity of heat are equally consonant with the assumption of the invariability of either motion *or* substance:

> All of the experiences reported in this and the previous chapter on latent and specific heat lead to the conception of an *invariability in the total effect of the heat*. No substance can lose heat unless another receives it. In this regard the effect [or action (*Wirkung*)] of heat accords entirely with Newton's third law of motion, and there is nothing absurd in regarding the essence of heat as a *distinctive motion* in substances, the total effect of which remains always the same. But since nothing ponderable in nature can disappear either, however much the sensibly perceptible constitution of substances is transformed by change in state of aggregation or by chemical action [*Einwirkung*]; so, too, on the other hand, can heat be regarded as an *imponderable substance* whose total quantity remains unchanged, however varied its distribution may be at any moment in the individual substances that it permeates as free or bound heat. The assumption of a substance of heat would thus not be refuted by any experi-

> ences up to now. On the contrary, the phenomena of bound heat find in it a simple explanation. The laws of disinterested investigation demand therefore that both hypotheses be tested in all phenomena and that, until the advantage of one has been completely decided, one understand by the expression *heat* . . . the unknown cause of the phenomena observed in this area.[132]

August had clearly not pressed his reasoning very far into the phenomena: the unbounded development of heat through friction remained inexplicable on the basis of the assumption that heat is an *unwägbarer Stoff*. Far from believing that there were two equally good theories of heat, Lamé underscored the inability of any existing theory of heat to render an explanation of the full range of thermal phenomena, but added that it did not appear possible to explain the production of heat by friction (as Rumford had demonstrated) other than in terms of the vibration of the solid body.[133]

Andreas Baumgartner noted that one is accustomed to explain the phenomena of heat in terms of an assumed *Wärmestoff*, but he advanced several (as he thought, decisive) objections to it.[134] It cannot explain how bodies radiate heat at temperatures below that of their surroundings or into empty space. It cannot explain the ceaseless production of heat of bodies subject to friction (as Rumford showed) or to the passage of electricity. The frictional generation of heat is especially significant: "If a particular substance is assumed for the explanation of the phenomena of heat, then an excitation of heat can only consist in a liberation of this substance or in the decrease in [heat] capacity; but the excitation of heat through friction cannot be explained in that fashion, and one is forced to assume that heat is really *created* there, not merely that already present heat is liberated."[135] Finally, one cannot thereby explain the relationship between light and heat, especially if one accepts the obviously superior vibrational theory of light.[136] In a similar vein, Baumgartner noted that one usually also assumed the existence of an "electric matter" to explain electrical phenomena, but *he* gave favorable consideration to Ampère's vibrational theory.[137] Nevertheless, in 1836 he still regarded magnetism as wholly distinct from the other agencies of nature, notwithstanding Ampère's electrodynamic explanation of it: "These [magnetic] phenomena are so very different from those of light and heat that one is incapable of deriving them from the same fundamental cause. One assumes almost universally for their explanation an uncommonly fine imponderable (aetherial) fluid which consists of two distinct parts or two particular fluids."[138] By 1842 he noted that some physicists saw magnetism as an electrical phenomenon and suggested that the assumption that the phenomena of the electric current are due to the vibrations of a subtle matter or of the smallest particles of ponderable matter had the advantage "that it reduced the four great powers [*Mächte*] of the physical world—light, heat, magnetism, and electricity—to a common source."[139] That was just the kind of aetherial-vibrations theory Ampère himself had hoped would unify the full range of physical phenomena.

Oddly, elsewhere in his text Baumgartner seems to have forgotten his favorable disposition toward the vibrational theory of heat. For example, he explained the heating of bodies through pounding in terms of their diminished volume and hence lessened capacity for heat. He believed that a body could be so heated only as long as its volume continued to diminish, and he cited Berthollet's evidence that successive blows with a hammer develop successively less heat in beaten metal plates. That liquids cannot be heated significantly through percussion was thus due to the fact that they cannot be significantly compressed. He then considered friction: "One knows the internal course of events for the excitation of heat through *friction* not at all as well as for percussion. The development of heat appears to be sure to keep step with the friction, and according to Morosi the heat increases with the duration of the friction, but one cannot assume that the bodies being rubbed together increase in density as long as the friction continues, although a densification may have taken place in the beginning."[140] As far as Baumgartner considered the matter here, the generation of heat through friction is an anomaly. There was certainly no hint of any energetic considerations, nor reference to any mode-of-motion theory of heat. In still another place Baumgartner followed the common practice of explicitly bracketing the question as to the nature of heat in favor of using the term to designate the otherwise undetermined cause of thermal phenomena.[141]

Geiger was largely either noncommittal or indecisive with respect to the nature of heat, light, electricity, and magnetism, treating their phenomena in a largely empirical and phenomenological fashion. Indeed, one is not justified in using any single general term (such as imponderables) to represent his views, despite the fact that the section of his pharmacy text dealing with heat, light, electricity, and magnetism was headed "On the Powers [*Potenzen*]."[142] His characterization of them left their ontological status unclear as it suggested their substantiality: "Through this interaction of bodies there reveal themselves to our senses distinctive phenomena . . . which we derive from particular causes that can be considered by themselves apart from the bodies that produce them, and which one designates by the names *powers*, *imponderables*, *aetherial substances*, *incoercibles*, *primitive substances*, etc."[143] Later language suggested both their materiality and his attachment to a vibrational theory of light, and in one passage dealing with the "transformation of light into heat" he implied not only some kind of proportionality between the light absorbed by a body and the heat thereby generated, but also vaguely suggested that heat is produced only to the extent that light loses its propagative motion: "When light falls upon opaque, especially dark bodies with a rough surface, not all of the light is thereby reflected, a portion disappears; but to the degree that the light ceases its progressive motion as such, it appears as *heat*."[144] How this process might rhyme with a conception of heat as a weightless substance was not explored. One notes, however, that the gist of the paragraph corresponds closely to Mayer's views: its treatment of light as some kind of motion, its finessing of the question of the essential nature of heat, and its image of the creation of heat as a result of the cessation of the motion which is light. As Mayer wrote with respect to the transformation of light, sound, and percussion into heat, "only the motion that is lost through resistance will become heat."[145]

Indicative of a broad if sometimes tentative movement in physical thought from the mid-1830s on, Halle professor of physics Ludwig Friedrich Kämtz used the analogy between radiant heat and light to suggest the possible unification of the erstwhile imponderables in terms of some theory of vibrations. He went on: "This view also shows, then, the intimate connection between heat and the corpuscular forces. Everything that agitates the parts of the body contributes to setting it into livelier vibration, and we therefore see that heat must necessarily be generated through friction, indeed that there lies in this operation an inexhaustible source of heat."[146] Rumford's sixty-year-old experiments acquired new significance once a vibrational theory of heat had become more plausible on other grounds. Kämtz concluded his section on heat with a paragraph (the last in the book before a brief historical sketch) on the "Sources of Heat Change," in which he gave an uncommonly suggestive representation of not just the connections between heat, light, electricity, and magnetism, but also of the effective *transformation* of one into another as manifestations of a common underlying but otherwise unspecified "process":

> If now at the end of our investigations we again look back at the whole series of phenomena, then the number of difficulties grows to which attention was already called at the end of the theory of electricity. Just as the phenomena of heat and light can in part be derived from particular substances [*besondere Stoffe*], so, too, with magnetism and electricity. Individually, light and heat can also be derived from particular vibrations, but it will be more difficult to demonstrate the connection between the two. But if we keep this fact in mind, then it will be little likely that particular substances underlie electrical and magnetic phenomena. Perhaps all of these four classes of phenomena are only modifications of the same set of circumstances [*desselben Vorganges*], as it were particular actions of the universal all-embracing process, in such a way that where one ends, the other begins. We need only consider that the moment the electric current begins, and the phenomena of heat, light, magnetism, and the chemical process [also] begin, every trace of electric tension disappears; that, however, the phenomena of the latter appear with magnets as the current in the closed circuit is interrupted or established; that with many crystals the electrical phenomena begin as the process of heating or cooling is begun; that light is capable of producing similar chemical effects to heat; that friction generates not only heat but also electricity, etc. In this connection it is the chemical process that is most especially to be considered. The fact that the elements only combine in definite ratios is completely analogous to the definite arrangement of parts with crystals . . . . The connection discovered by Dulong and Petit between heat capacity and atomic weight affords another connecting link in this chain, and to it is most intimately joined the fact discovered by Faraday that chemically equivalent substances are liberated by the decomposition in the voltaic pile. Someone who simultaneously keeps these phenomena in view and wishes to find out their inner connection stumbles upon a dark field where only individual unconnected points of light are visible; a large series of experimental investigations is still required in order to find their connection here; only in this way is it possible to reach the goal, for what the *Naturphilosophen* were able to disclose through speculation during the last decades has only served to spread greater darkness over this entire field.[147]

If, again, there is no evidence to suggest that Mayer was particularly attentive to such speculations *before* he set out on his own path of discovery, he may well have been aware of some of the issues—in particular those concerning the nature of heat—in a general way, and he must have encountered some of them as he began to read more widely in the textbook literature of the day. It is thus not unlikely that they affected the development of his thinking at least in the years after 1842, especially as leading up to his more broadly conceived work of 1845. That such unifications as there were tended to be in terms of some kind of theory of vibrations may also have (at least in part) lain behind his early attempt to define forces qualitatively in terms of a coefficient equal to unity for heat and infinity for light. On the other hand, Mayer quite pointedly *rejected* the explanation of heat in terms of molecular motions. His insistence on the ontological autonomy of forces may have represented a conscious attempt to get beyond the inconclusive debate over whether they are weightless fluids or properties of ponderable matter. It would seem not at all farfetched to regard physics as having been in something like a period of Kuhnian crisis during the 1830s with respect to its understanding of the traditional imponderables. Mayer's self-appointed task was to bring clarity into this confused situation.

Mohr's report of 1837 recorded his sense that, due in large measure to Macedonio Melloni's experiments on radiant heat, "the view that emerges from the experiments, that heat is a vibration of ponderable substances and not itself a substance, is gaining ever more acceptance, and is already even being introduced in textbooks as the leading view, for example, in Baumgartner's *Elements of Physics*, 1837, p. 132. The agreement between this view and all of the phenomena of heat gives it an incontestable superiority over the older view of the materiality of heat."[148] In the short section headed "Views on the Nature of Heat," which formed part of that report, Mohr repeated his belief that Melloni's discoveries "demand the assumption of vibrations after the manner of the vibration theory of light."[149] He elsewhere rejected out of hand the concept of a weightless substance.[150] Unlike Baumgartner and (later) Kämtz, however, Mohr did not stop with the suggestion that heat, light, electricity, magnetism, and (for Kämtz) chemical activity might receive a common explanation via some kind of theory of vibrations, but went on to redefine those five agencies *plus motion* as a discrete entity, *force*, which he introduced in terms of the generally accepted application of the term to the force of cohesion:

> Heat appears as force. It neutralizes [or annuls (*aufhebt*)] the cohesion of bodies, but the latter is a force: *but whatever is supposed to neutralize a force must itself be a force*. The expansion of bodies by heat is a phenomenon of force of the highest order. Whatever produces a motion or a manifestation of force must likewise be a force.
>
> The expansion of bodies by heat is accordingly an extended amplitude of vibration.[151]

Note the connection between heat—a force—and motion via the phenomenon of thermal expansion, not (explicitly) via assumed molecular motions; such would also be Mayer's route. In the version of this passage published in the *Zeitschrift für Physik,* Mohr further noted that cohesion can be neutralized (*aufgehoben*) not only by heat, but also by filing, sawing, and rubbing, and he asserted that "there is no single case in nature where one neutralizes a force other than through the opposition of another force."[152] In its own way, Mohr's way of thinking was closely tied to the notion of the qualitative and quantitative equality between cause and effect. In both essays he compared the expansion of water by heat to its heating by compression, and calculated that heating water by 1°C causes it to exert a pressure of 97 atmospheres.[153] The heat given off when a given amount of a substance solidifies *must* be exactly equal to the amount of heat absorbed when it liquefies, "since one can give an accounting of a force just as well as of a ponderable substance; one can divide it, subtract from it, and add to it without the original force becoming lost or changing its quantity."[154] Mohr thus exploited what was to be Mayer's central analogy, that between force and matter. He also insisted on their difference:

> The voltaic connecting wire is traversed by an electric current, it gets hot, it glows; how is this possible if light, heat, and electricity are different substances, like oxygen and hydrogen, like copper and zinc? From the countless transitions between these phenomena, however much remains in particular still to explain, I believe it possible to formulate the following general principle [*Satz*]: Aside from the 54 known chemical elements there is in the nature of things only one further agency [*Agens*], and this is called *force*; it can appear under appropriate circumstances as motion, chemical affinity, cohesion, electricity, light, heat, and magnetism, and from each of these classes of phenomena all the others can be produced. The same force that lifts the hammer can, if it is otherwise employed, produce each of the other phenomena.
>
> I must remark that this view is very different from the one which some time ago interpreted all of the imponderables as identical substances. Much is gained through the use of the common concept of a force.[155]

Mohr had sent a draft of his essay to fellow coeditor of the *Annalen der Pharmacie* Justus Liebig, who discouraged him from having it printed with the curious comment that "force without matter is inconceivable," and anyway, physicists already subscribe to the views you present, only in a different form.[156] For some unstated reason, Liebig believed Mohr's views were incompatible with Melloni's discoveries in radiant heat. One physicist who read and commented on Mohr's article was Muncke, who cited its support for the view that heat can be explained solely in terms of motion.[157] Although Muncke ostensibly urged caution with respect to making assumptions about the nature of heat, since both theories of heat—as motion and as a fluid—have difficulty explaining some of the phenomena, in fact he came out for his old favorite, the fluid theory: "By far the most and, one might well say, the most experienced physicists assume a material, albeit extremely fine, aetherial substance of heat [*Wärmestoff*]."[158] There is no evidence that Mayer read either of Mohr's two essays—nor any real necessity to assume so—yet the possibility of his having seen the briefer version in Liebig and Mohr's *Annalen* should not be totally discounted: part of Mohr's progress report was a short section on "Santonin," the subject of Mayer's doctoral dissertation the next year, and he may have consulted Mohr's report—to be sure without necessarily reading all of it—in his search of the literature on the drug.[159] At the least, this small fact symbolizes the complexity of the web of real and potential connections between Mayer and his plausibly reconstructible context. Insofar as their reflections were grounded in contemporary physics and chemistry, Mohr and Mayer arrived at strikingly similar conceptions of force.

Fries, who had an idiosyncratic conception of the imponderables as tenuous states of aggregation of ponderable matter, defended a kind of vibrational theory of light and heat. In his view, their phenomena demonstrate "that the radiations of light are different only in intensity from radiant heat, for wherever the latter has attained a certain degree of energy [*Heftigkeit*] it passes over, in glowing, into light, and where rays of the one or the other kind are absorbed they act to raise the temperature."[160] All bodies are subject to a "law of radiation" according to which "the higher degrees of this radiation are light rays, the lower degrees are radiant heat, and that which radiates [*das strahlende*] itself appears as heat when its radiating motion is annihilated [*aufgehoben*]."[161] Although Fries did not further explain what he meant by describing heat as neutralized motion, such language vividly recalls Mayer's conception of heat, according to which heat is not itself motion, but rather the result of the disappearance of motion. As he wrote to Baur, "Heat does not arise as long as the corresponding quantum of motion continues in waves; for I repeat: heat is motion which has become qualitatively 0."[162] So, too, does Fries's rudimentary attempt to distinguish light and heat according to a vaguely quantitative measure of their "degree" recall Mayer's coefficient $n$ in his formulas $\frac{M}{n}nC$ and $0\,\frac{2A}{n}nC$ from his first essay.

### *2.1 Thermal Expansion of Gases and Related Phenomena*

Mayer's status as discoverer—even if only as one of the discoverers—of the conservation of energy has come to rest in large measure on his published calculation of the numerical equivalence between heat and what we call work. As is well known, he performed that calculation by applying the numerical ratio of the specific heats of gases at constant pressure and constant volume to the work done by an expanding gas in raising a column of mercury. As Kuhn pointed out, the phenomena of adiabatic compression (and expansion) not only provided, qualitatively, an ideal demonstration of the conversion of work into heat (and vice versa), but, quantitatively, they also afforded the only means of computing a conversion coefficient with the data then at Mayer's disposal.[163] Since Mayer derived his knowledge of these phenomena—and some valuable clues as to how to carry out the requisite calculation—from contemporary textbooks, it will be useful to ascertain just what he might have found in some of the works we know he consulted, such as those of Lamé and Baumgartner.

Good if somewhat late evidence from Mayer's correspondence reveals that Lamé was probably his chief source of information.[164] Lamé called attention to the fact that the specific heat of gases at constant pressure is larger than that at constant volume by an amount corresponding to the heat disengaged when a gas is suddenly compressed by 1/267 of its volume, and he alluded to experiments that indicate that the amount of heat thus disengaged is the same for all gases.[165] Without telling the reader quite how, Lamé drew an important conclusion from those experiments:

> The measurement of specific heats and the identity of their values obtained by different experimental methods furnish the most direct proof of the existence of an agency [*agent*] whose quantity varies with temperature for the same body, and which can neither increase nor decrease in a ponderable medium without decreasing or increasing in other neighboring media in such a way that its total quantity remains constant. This general fact accords very well with the two hypotheses of §216: in the emissions theory the caloric, considered as a material fluid, cannot disappear at one point without being forced back [*refoulé*] at another, where one must be able to verify its accumulation; in the wave theory the vis viva of vibratory motions cannot diminish in one body without increasing in others. Thus the two hypotheses in question rest upon two incontestable principles, the indestructibility of matter and the constancy of the quantity of vis viva [*somme des forces vives*], which completely explain the results furnished by the measurements of specific heats. From this perspective one of the two theories has no essential advantage over the other.[166]

In a later section Lamé dealt with the heat or cold produced by the compression or expansion of a gas, and reported Gay-Lussac's experiment of 1807 in which he allowed a gas to expand into a previously evacuated chamber equal in volume to the first, and noted that the drop in temperature of the gas in the first chamber was equal to the rise in temperature of that in the second—that there had thus been no net change in the temperature of the gas overall.[167] Lamé underscored the physical significance of the ratio ("*K*") of the specific heats of a gas at constant pressure and constant volume, and reported its value of 1.421 as given by calculation and experiment relative to the speed of sound in air. Mayer would later quote Lamé's summary of Dulong's important findings with respect to the heat given off or absorbed by different gases while undergoing compression or dilation:

> Dulong, in a series of experiments that we shall have occasion to describe later when we speak of the propagation of sound in different gases, has found that this ratio *K* varied in a very noticeable way between simple and compound gases, and has deduced from the comparison of these different values this law remarkable in its simplicity: first, that equal volumes of all elastic fluids, taken at the same temperature and pressure, having been suddenly compressed or dilated by the same fraction of their volume, release or absorb the same absolute quantity of heat; second, that the resulting variations in temperature are in inverse ratio to their specific heats at constant volume.[168]

At that later place, where he restated Dulong's results in somewhat more detail, he explained that the decimal part of *K*—that is, 0.421—represents the heat disengaged when the gas is compressed by 1/267 of its original volume.[169]

Mayer thus had in Lamé's physics text all the data he needed for his calculation of the mechanical equivalent of heat, presented in such a manner as to all but lead him to the method of its calculation. That was not the only work he consulted, however, and other works—Baumgartner's physics text, in particular—also contained treatments of the relevant relationships that would have greatly facilitated his manipulation of the data. After mentioning the heat developed by percussion (as discussed above), Baumgartner added that one can even calculate the heat developed through compression of a gas:

> It is as large, namely, as that which the gas needs in order to reattain its previous volume. If we set the quantity of heat that it has for the volume decreased by 1/267 equal to 1, that for the original volume equal to $1 + x$, then clearly 1 is to $1 + x$ as the capacity C at constant pressure is to the capacity $c$ at constant volume. But for chemically simple gases $C/c = 1.421$, thus $1 + x : 1 = 1.421 : 1$ and therefore the quantity of heat $x$ developed at 0°C by the compression of 1/267 of the original volume is equal to 0.421.[170]

Kämtz briefly discussed the production of heat through compression in terms of both material and vibrational theories of heat, and seems to have followed Baumgartner very closely in running through a similar calculation of the relationship between the compression of a gas and the heat thereby released.[171] If Mayer's calculation of the mechanical equivalent of heat was a stroke of genius, it was nevertheless a step well prepared for the attentive reader of the contemporary literature.[172]

## 3 Matter

In an essay containing one of the most perceptive analyses of Mayer's work to date, Alois Riehl rejected the notion that Mayer had simply deduced the principle of the conservation of energy from the principle of causality or had otherwise discovered it by means of some a priori thought process. Instead (Riehl insisted at length and with references to Mayer's earliest writings), Mayer had "transferred Lavoisier's analogous principle of the conservation of matter from chemistry to physics and expanded it into the principle of the 'quantitative invariability of the given.' "[173] Indeed, as we have seen, Mayer himself insisted on the centrality of that analogy to his entire system of thought. There remains, however, a fundamental problem with Riehl's explanation. The problem is that, with very few exceptions, the scientific literature of the time—at least the German—contained no explicit statements of the conservation of matter. The principle was, to be sure, tacitly assumed as a working principle by chemists, but it had no visibility as a fundamental principle, let alone *the* fundamental principle upon which the science of chemistry rested.[174] Thus until Mayer forged for himself the dual principles of the indestructibility and uncreatability of force *and* matter, there *was* effectively no principle of the conservation of matter to be exploited as the basis for a fundamental analogy. Moreover, that analogy itself only crystallized into awareness as part of the process by which Mayer gradually defined for himself the meaning of force, whereby he ultimately *and with great conceptual difficulty* decided (for example) that force, too, cannot be created under *any* circumstances. He never doubted the indestructibility *and* uncreatability of matter, but whether that 'principle' applies also to force was precisely the difficult question to answer, and it was not answered by invoking an analogy. Which is not to say that the example of matter was of no importance to Mayer; quite the contrary. It is only to say that this example *by itself* was insufficient to convince him of the conservation of energy even after he had a rough conception of the new entity, force. Once he had attained clarity on these issues, of course, the analogy between matter and force could be advanced as a strong *argument* in favor of his new theory.

The overwhelming majority of German physics and chemistry texts of the period contain no explicit mention of anything like the conservation, indestructibility, or uncreatability of mass, matter, or substance.[175] Most list as the general properties of matter things like extension, impenetrability, divisibility, porosity, elasticity, compressibility, inertia, and motility, rarely also weight or gravity (*Schwere*). Some do not even mention mass as an important concept, let alone its conservation. Even Brandes's article on "Mass" in *Gehler's Physical Dictionary*, albeit short, made no mention of its conservation or indestructibility.[176] Nor did Muncke's seventy-five-page article on "Matter" in the same work specifically mention its conservation; the closest he came was in noting that chemists recover the original quantities of qualitatively unchanged gold and silver from any mixture of the two, and in rejecting the possibility that plants can manufacture carbon out of pure water, because that would presuppose "a creation out of nothing in self-contradiction with the forces of nature."[177] In a similar fashion Muncke's *Textbook of Physics* reported the empirical finding that simple substances—that is, the chemical elements—can be neither created nor transformed one into the other, and that thus all natural and artificial processes consist solely in the combination and separation of those simple substances, but he never ventured to assert as a general principle the conservation of matter.[178]

Mention of such a principle does occur, to be sure, here and there in the course of particular discussions—for example, Lamé's reference (cited above) to "the indestructibility of matter and the constancy of the quantity of vis viva" inserted into his discussion of the constancy of the quantity of heat in certain reactions. But Lamé otherwise passed in complete silence over the conservation of matter. Interestingly enough, the two other incidental references that I have encountered to something like the conservation of matter have to do with reactions involving oxygen—perhaps a resonance of the historical association of that principle with Lavoisier. In Christian Gottlob Gmelin's *Introduction to Chemistry* one reads: "But if the new body arising out of the combination of the combustible body with the oxygen of the air is weighed, then it is always found *that its weight is precisely equal to the sum of the weights of the vanished combustible body and the vanished oxygen gas*. The development of imponderable heat and imponderable light is therefore merely a phenomenon accompanying the combination of ponderable substances that neither increases nor decreases the weight of the combining substances."[179] But that is as far as he ever went; one does not find any statement of the conservation of matter as a general principle. Fischer and August drew a more general conclusion from the discussion of the combination of oxygen and hydrogen (as *Knall-Luft*) to form water: "If this experiment is performed with an apparatus that allows one to determine in the most precise fashion the weight of the constituents first separately before the experiment, then after their union into oxyhydrogen gas, and finally after the transformation into water, then it can be shown that for these processes the absolute weight undergoes not the slightest change, that therefore ponderable matter neither comes into being nor is annihilated; a fact confirmed without exception by all chemical experiments."[180] Even at that, the conclusion was expressed simply in terms of the general results of experiment.

Muncke was one of the very few physicists to venture a more principled assertion of the logical impossibility of either the creation or the destruction of substance. That occurred not in one of his several physics texts or in his article on "Matter," but in his article on "Physics" in the same handbook:

> For physics, whose essence consists in deducing general laws from given facts and in applying the former in turn to the phenomena that occur, the principle of sufficient reason [*der Satz vom zureichenden Grunde*] is in any case irrefutably certain; physics would destroy itself and lose its whole existence if it wished to deviate from this principle, and the axiom *ex nihilo nil fit* must accordingly also be assumed in it . . . . In the same way, in accordance with the mathematical mode of argumentation peculiar to it, it finds an internal contradiction in the assumption that something that exists should in turn disappear, because in the transition from something to nothing there is an infinite distance.[181]

Such reflections were entirely in accord with Mayer's, though the specific influence of Muncke's essay would be difficult to ascertain; in any event, it is exceedingly unlikely that Mayer would have read it until after his return from the Indies.

Fries—more philosopher than physicist, though the author of a respected physics text—sometimes skated close to a statement of something like the conservation of matter. In his *Mathematical Natural Philosophy* he laid down a "law of the permanence [*Beharrlichkeit*] of mass and force": "In all changes in the corporeal world the quantity of matter remains on the whole the same, unincreased and undiminished. Its fundamental forces also remain invariable and only the states of its motion can be changed."[182] His later *Textbook of Physics*, however, expressed this basic idea in a much more obscure fashion, without any direct mention of matter or mass. Fries there defined natural philosophy (*Naturphilosophie*) as that part of natural science in which we investigate truth by means of reflection without recourse to perception (*Anschauung*) or experience.[183] One of the central tasks of natural philosophy is to demonstrate "that our knowledge of the most general necessary laws of nature (for the corporeal world the laws of the permanence of substance [*Beharrlichkeit des Wesens*], of inertia, and of the equality of action and reaction) are of philosophical, in particular of metaphysical origin."[184] The metaphysical laws that apply to every natural object are that "every phenomenal condition of things is underlain by *substance*; the properties of these substances are *forces*, by means of which they become causes of changes. The substances and their forces are invariable, absolutely permanent in nature, only their states can be changed; every change of a state has a cause by means of which it takes place."[185] That Fries sought a metaphysical foundation for the fundamental laws of physics is not surprising given his close affiliation with Kantian philosophy. It is also significant that it was Kant and those who stood in the Kantian tradition who more than anyone else urged some form of the conservation of matter. As Kant wrote in the *Metaphysical Foundations of Natural Science*—in words Fries silently appropriated—the first law of mechanics is that "in all changes in the corporeal world the quantity of matter remains on the whole the same, unincreased and undiminished."[186] In his proof of this law Kant invoked the general metaphysical law "that in all changes in nature no substance either comes into being or passes out of existence."[187] Schmid quoted Kant's words directly in his *Physiology Treated Philosophically*, and Eschenmayer wrote in his dissertation that "metaphysics teaches that in all changes of material nature the quantity of matter is always the same."[188]

Perhaps the most widely influential aspect of Kant's natural philosophy on nineteenth-century physical science was his attribution to matter of fundamental attractive and repulsive motive forces (*bewegende Kräfte*).[189] Those primitive forces were the dynamical archetypes of the *Grundkräfte* and *Urkräfte* that played such a prominent role in Schelling's *Naturphilosophie* and elsewhere. They came to be treated as ontologically more basic than matter, with matter reduced to a construction based on forces.[190] Schmid enthusiastically took over the construction of matter out of opposing forces *without* eliminating matter itself from his worldview.[191] Autenrieth, not himself a follower of Kant in any meaningful sense, singled out Schmid's work for special praise in this regard in his own *Handbook of Empirical Human Physiology*: Schmid had reinstated the rightful claims of empiricism in science alongside theory, and had asserted the need to assume not only primitive forces in nature but also the independent existence of the physical world.[192] Mayer's world, too, was based on the assumption of the equally valid existence of both matter and force, if to be sure the word *force* had a different meaning for him than it did for others. But for the possible exception of his treatment of central-force motion, there appears to be little trace of the Kantian conception of fundamental attractive and repulsive forces in Mayer's writings. Indeed, his enterprise revolved around the *independence* of force from matter, not the post-Kantian reduction of matter to force.

The nearly universal absence of an explicit principle of the conservation of matter should not be taken simply as evidence that is was obvious and thus tacitly assumed. Although that surely was the case with some, especially as regards the conservation of matter as a working principle of chemists, it should be emphasized that for others such a principle was highly problematic because of the lack of precision concerning the conception of matter as a distinct entity. In particular, those dynamicists who, unlike Kant himself, completely reduced matter to the interplay of attractive and repulsive forces so denied the materiality of matter as to make it difficult to see what an assertion of its conservation could mean. When Fries interpreted ponderable matter and the imponderables as merely different states of aggregation of the same underlying substance—a notion which held out the implicit possibility of the effective disappearance of ponderable matter—he undercut some of the ground from under the utility of a principle of the conservation of *matter*. As we have seen, the closest he came in his physics text of 1826 was a vague statement about the *Beharrlichkeit des Wesens*. Similarly, when Muncke insisted that the alleged weightlessness of the imponderables had not been proven empirically, he held open the possibility that they are only tenuous states of matter, again blurring the concept of ponderable matter and rendering its conservation less than obvious. Autenrieth concluded his discussion of phenomena of electricity, galvanism, magnetism, chemical reactions, heat, and light with the judgment that "imponderable substances" differ only in degree from "ordinary heavy bodies": "They do not constitute a class of entities [*Wesen*] wholly different from the other material substances [*materielle Stoffe*], and between the magnetic fluid and rigid flint there is an almost continuous transition."[193]

Fries's discussion of the general characteristics of matter—as far as I can tell alone among physics texts of the period—included a fundamental distinction between its quantitative and qualitative properties.[194] Matter is (in principle) quantitatively divisible to infinity, into ever smaller particles; it is qualitatively divisible insofar as it can be separated into qualitatively distinct chemical components. Such language recalls Mayer's attempt at a quantitative and qualitative definition of forces, and in chapter 6 I will argue the likelihood that Mayer did indeed know Fries's text.

## 4 Metamorphosis, Neutralization, and Indifference: The Chemical and Physical Contexts

As discussed in chapter 2, Mayer's thinking was characterized by the sometimes peculiar usage of phrases like "the neutralization of differences" and the origination of force out of a state of "indifference." At first blush such terms might suggest that his thinking owed something to *Naturphilosophie* or some other cognate variant of post-Kantian Romantic philosophy. Chapter 7 will further explore those possible connections. Similarly suggestive was his frequent usage of the word "metamorphosis" in his *Organic Motion* of 1845, as in the following passage: "What chemistry has to accomplish with respect to matter, physics has to accomplish with respect to force. To become acquainted with force in its different *forms*, to investigate the conditions of its *metamorphoses*—this is the only task of physics, for the *creation* or *destruction* of a force lies beyond the realm of human thought and action."[195] A specific instance of such a transformation was "the metamorphosis of a chemical difference [*Differenz*] into mechanical effect."[196] But Mayer spoke not only of the metamorphosis of force, but also of "the chemical metamorphosis of the blood," and such a usage was commonplace in the chemical literature of the day, particularly with respect to organic reactions.[197] Berzelius, Liebig, Wöhler, and Lehmann, for example, all regularly used the word in that sense.[198] Thus what one has here is another instance of Mayer's exploitation of a chemical analogy in elaborating his concept of a physical force.

Erich Pietsch pointed out long ago the likely chemical origins of Mayer's characteristic usage of "difference"—both spatial and chemical—and "neutralization" in the neutralization of acids and bases to form "indifferent" salts.[199] Where Mayer spoke of "chemically indifferent bodies," his contemporaries spoke of chemically nonreactive substances like nitrogen as "indifferent," and defined as "indifferent substances" all those which are neither acids nor bases.[200] They matter-of-factly used the general phrase "chemical difference"—which derived its meaning at least in part from the notion of the full or partial saturation of chemical affinities—while Berzelius spoke more particularly of "the degree of electrochemical indifference."[201] Geiger employed similar language, in some ways even closer to Mayer's, in his explication of the general role of the *Urkräfte* in governing material interactions:

> In order to comprehend the possibility of material activity, the leveling out [*Ausgleichung*] of the mutually interacting primitive forces [*Urkräfte*] . . . must be thought of as only relative . . . . In this way the possibility exists of a new counter action and leveling out, which in turn continues acting in the same fashion, and in this must be sought the cause of the eternal interaction of matter (bodies) against each other. It is a continuous striving of the manifold toward a leveling out (of the different toward indifferentiation), which again and again creates new diversity. The eternal activity, the life of nature, can be thought of in this way; for absolute rest would be absolute death, and for that reason alone could not be an object of sensible perception.[202]

Such language vividly recalls that of Mayer's first (unpublished) essay.

Similarly reminiscent of several of the themes of Mayer's work is the discussion of equilibrium and neutralization in Prout's Bridgewater Treatise. In Prout's eyes, the phenomena of nature—which evince providential design—depend principally upon "a certain due adjustment to each other of the *qualities* and *quantities* of the different substances . . . of which our globe is composed," for example the relative amounts and respective properties of water and air.[203] Among the "questions utterly beyond our comprehension" are how salt can be composed of two poisonous elements, or why water should contain the extremely combustible hydrogen.[204] Prout noted a general tendency in nature toward the neutralization of qualitative differences leading to a state of rest:

> Amidst all that endless diversity of property, and all the changes constantly going on in the world around us, we cannot avoid being struck with the general tendency of the whole to a *state of repose* or *equilibrium*. Moreover, this tendency to equilibrium is not confined to the ponderable elements, but prevails also in the same striking degree among the imponderable agencies, heat and light . . . . Now, the formation of this state of equilibrium, and its preservation, may be considered as the results of those wonderful adjustments among the qualities and quantities of bodies above alluded

> to—the qualities being such as to neutralize each other's activity, while the quantities are so apportioned as to leave one or two only predominant.[205]

Such motifs were definitely not the sole property of *Naturphilosophie*.

In addition to its chemical meaning, the concept of neutralization was also regularly used with respect to the *Neutralisierung* or *Aufhebung* of opposing forces in the parallelogram of forces, just as Mayer, too, used it. Here his parallel treatment of matter (as in chemistry) and force (in his new physics) could exploit an existing linguistic commonality. The concept of indifference, too, was also of regular use in physics in at least one important context, that of magnetism. An unmagnetized iron bar was said to be magnetically "indifferent," and the point at the center of a magnetized bar where the opposing magnetic forces neutralize each other was termed the "indifference point."[206]

In a manner characteristic of his whole way of thinking, Autenrieth exploited his analysis of the behavior of what he called the magnet's *Indifferenz* or *Indifferenzstelle* or *Kraftindifferenz* in order to construct an analogy with the properties of *Geist*. For Autenrieth the indifference point could be regarded as the source of the bar's magnetism:

> The polarity of the magnet behaves to be sure in a way contrary to the forces of attraction or explosion or expansion; whereas the latter become weaker the greater the distance from their source of origination [*Entstehungsquelle*], it appears stronger the farther it is from its indifference—in that regard similar to the galvanic pile, whose polar force on both ends increases with its length. But we can nevertheless also think of the indifference point of the magnet as the source of a force that increases in its out-flowing course, just as an avalanche, beginning from its place of origination, gets larger and larger in its course. In every imaginably smallest part of space . . . there begins on one side of the magnetic indifference an infinitely weak northern current, on the other side an infinitely small southern. The indifference of the magnet thus has, to be sure, a place, but it itself, as absolute indifference, has no extension in space, just as a geometrical point can have a position in space but not fill it, or as a line has no breadth and a plane in and of itself no thickness.[207]

Autenrieth's writing teems with analogies large and small, analogies we would regard as mostly invalid or ineffective. In his opinion, the sudden appearance and disappearance of magnetism in iron and a few other metals demonstrates that magnetism is "not merely a variable predicate of the iron," but something that exists independently.[208] He imagined that the indifference point, too, exists independently of matter, and that its entrance into the material world (to use his language) is accompanied by the appearance of magnetism in the iron bar. One thus has the image of something real having a definite location in space without occupying space. If one breaks off the end of a bar magnet, the indifference point, seen again as "the source of the magnetic force," moves instantaneously to the middle of the remaining piece.[209] That phenomenon is sufficient for Autenrieth to conclude that the indifference point—now also called a *Krafteinheit*, as he edged his terminology toward his goal—is capable of moving instantaneously over any imaginable distance, in which case it becomes possible to consider the indifference points of two magnets as the same point occupying the two magnets successively with a rapidity that appears to us as simultaneity. That placed Autenrieth within striking distance of his goal:

> We even obtain in this way a notion [*Vorstellung*], which would have nothing contradictory about it, of how the dynamic unity [*dynamische Einheit*] of our ego, or the soul which feels in us, may really be a unity, see by means of our eye, and apparently simultaneously even feel in our foot, and nevertheless still have no spatial extension. Nothing prevents us, either, from mentally putting in the place of a dynamic substance that essentially has the capability of generating magnetic currents any other likewise nonspatial entity [*unräumliches Wesen*] whose existence manifested itself by another kind of capability, thus for example by emitting light that is differently colored in accordance with the condition of its entrance into the spatial world.
>
> . . .
>
> We are now also free to imagine the nonspatiality even of that dynamic unity [*Krafteinheit*] that underlies the visible personality of the living organic body, and to ascribe to this dynamic indifference [*Kraftindifferenz*] as an essential property the capability of assembling, upon its entrance into the spatial world, appropriate ponderable corporeal substances . . . into the most diverse coherent organizations, roughly as the magnetic force orders iron filings little by little into determinate vortex figures, or as electricity attracts light bodies.
>
> We can, finally, raise ourselves in our thoughts from these not yet self-determining but being-determined dynamic indifferences—which, although in and of themselves nonspatial, are nevertheless autonomous entities—to free spatial entities, to the dynamic unity [*dynamische Einheit*] which determines itself, and which is capable in the animal and in the human being of combining with the organic vital force.[210]

The existence of such "nonspatial sources of activity," of such "nonspatial substances," demonstrates that "even when one imagines away all the spatial relationships of a substance it by no means becomes merely an empty concept."[211]

The parallels between Autenrieth's train of conclusions, originating in his analysis of magnetism and its all-important indifference point, and some of Mayer's ideas are striking. Of paramount methodological importance to both was the appeal to meaningful analogies as a basis for their nexus of ideas, culminating for Autenrieth in an important analogy between physical forces and the soul in terms of their "dynamic indifference" and "dynamic unity." Autenrieth's identification of the indifference point as the source of magnetic force recalls Mayer's creation of opposing forces out of nothing at a point, just as Autenrieth's insistence that forces are entities with an existence independent of matter recalls Mayer's vindication of the independence and substantiality of force. If none of this proves that Mayer necessarily read Autenrieth's book, it at least demonstrates that his speculations had a strong affinity with the kinds of ideas prominently discussed in the contemporary literature.

## 5 Summary

If, to be sure, physiology and medicine constituted the initially most important aspects of Mayer's context, contemporary physics and chemistry nevertheless provided him on numerous counts with essential points of reference and rich fields of meaning—most centrally, perhaps, with regard to Mayer's understanding of the commonalities and differences between force and matter: both are indestructible and uncreatable, but whereas chemistry deals with material substances, physics is the science of the immaterial substance, force. The distinction between matter and force was indeed a textbook commonplace in Mayer's day, although what those books meant by force were things like elasticity, chemical affinities, the various forces of attraction and repulsion that had been hypothesized to explain the phenomena (e.g., gravity and cohesion), and (even) the force of the will. Mayer kept the fundamental terminological distinction between matter and force but radically altered its meaning by reconceptualizing the nature of force.

Likewise standard was the interpretation of forces as causes of change, especially as causes of motion. Although contemporary writers did not elaborate this notion with any great cogency, Mayer seized upon it and used it as an essential characterization of his new (and evolving) concept. For example, contemporary authors did not regularly include motion among the forces, even as they spoke of motion as the effect of the action of a force; but if forces produce motions and *causa aequat effectum*, then motion must also be a force, especially since motions can obviously produce other motions. English writers in particular, like Whewell and Herschel, likened force, as a cause of motion, to the effect of the will, that is, to a spiritual agency. A major aspect of Mayer's creative process was in deciding what kinds of things, which kinds of causes, one should properly call forces. If the will is a bona fide force, then that implies that force, as motion, can be created out of nothing; whether or not Mayer wished to allow for that possibility was something that took him several years to clarify in his own mind. Once he had shifted his attention from the sequence of considerations that had led him initially to deduce the quantitative equivalence of heat and motion, he did not simply assume as a universal principle the quantitative uncreatability of force.

Part of the general lack of consensus over the application of the term "force" had to do with whether or not forces are merely properties of matter, or whether they enjoy an autonomous existence, the first alternative being far and away the more common. One asked whether the fundamental forces of nature are variable or constant, whether they can ever cease to act or to exist. Some maintained that force is merely a subjective concept of the understanding by which we connect an effect to its supposed cause. As Mayer worked his way to a coherent conception of force—that is, what we would term energy—he would have to come to terms with issues such as these that pertained to a like-named concept having a different field of application.

One of the major confusions Mayer encountered in his sources was with respect to the measure of force, since most authors defined its measure as the product of mass and velocity even as they might also equate it with the elevation of a weight or with pressure. Such inconsistencies were scarcely addressed, let alone resolved. Physics texts did not typically discuss the concept of vis viva; some, such as Fischer's, denied that a force itself could ever be measured, insisting that what one could measure was only the effect of a force.

Similar confusion confronted Mayer when he attempted to assimilate contemporary accounts of the parallelogram of forces, especially as that figure was also applied to the combination and decomposition of motions. If on the one hand that commonality of treatment may have reinforced his understanding of motions as forces, the problems that it threw up for him were daunting indeed. For one, the motions so dealt with were measured by the product of mass and velocity; for another, comparison of the lengths of the sides representing the component forces (or motions) to the length of the diagonal implied to Mayer that force disappears or comes into existence in certain circumstances. That conclusion was reinforced by the standard application of the parallelogram of forces to central-force motions, in particular to the gravitationally controlled revolution of planets around a central sun. Most authors derived that motion by combining the force of gravitation with a "tangential force"—that is, an initial impulse—in such a way that Mayer readily convinced himself that somehow force must be continuously created in the solar system. Such a conclusion was in turn reinforced both by the widespread analogy between the solar system and an organism—since organisms were widely regarded as capable of producing some kind of force in some unknown but relatively unproblematical fashion—and by a common distinction between 'natural' physical systems (such as the solar system) governed by the fundamental forces of nature and 'artificial' mechanical systems (such as the constrained motion of a body attached to a central point by means of a string). Here again Mayer was able to exploit and adapt a common terminological distinction as he attempted to come to terms (literally!) with his particular nexus of problems.

Debates surrounding catalytic, contact, and electrochemical forces also involved difficult issues about the nature of force, about causality, and about whether something like a perpetuum mobile is possible in (inorganic) nature. Is the catalytic force a legitimate force? Is it distinctive? Its legitimacy and distinctiveness depended in large part on the further analogical implications one wished to draw, as for example in defense of an inexhaustible work-producing vital force. Can it be explained, say, in terms of molecular motions? Liebig's interpretation of ferments and catalysts in terms of the transfer of molecular motions implicitly raised energetic questions, as it demonstrates how unobvious the conservation of energy was to most of Mayer's contemporaries. A similar conclusion attends the issue of whether the electrical action in the voltaic pile is due to the mere contact of heterogeneous substances or to the presence of ongoing chemical transformations. Even if the argument that the continued production of electrical power must depend on an antecedent chemical reaction would seem to be entirely in accord with the growing sense in physiology that all organic effects must be underlain by processes of material exchange, nevertheless the simple historical fact is that most scientists were extremely slow to appreciate energetic arguments or their principled significance.

Among the things that were *not* generally included under the rubric of force were the so-called imponderables—that is, heat, light, electricity, and magnetism. As imponderables they were generally taken to be weightless fluids, usually distinct from one another. Some (like Muncke) held open the question of their absolute weightlessness, and some (like Fries) regarded them as particular states of aggregation of ponderable matter. But anomalies and problems abounded, most significantly with regard to heat. It was widely recognized that the two contending theories of heat—as a weightless fluid (caloric) and as the motion of the smallest parts of corporeal bodies—were each favored by a select set of phenomena that the other theory had difficulty accounting for. A common way out of this dilemma, especially favored by the French, was the positivistic solution of simply using the term "heat" to designate the unknown cause of a certain class of phenomena (as similarly with the other imponderables). Each theory allowed for its own version of a conservation principle, at least with respect to a circumscribed set of phenomena. Perhaps the principal anomaly was the apparent impossiblity of accounting for the unlimited development of heat through friction on the basis of the otherwise generally favored caloric theory of heat, since (obviously) no substance could just be created out of nothing. Fries, for one, tried to make coherent sense of the relationship between light, radiant heat, and the heat of ponderable bodies by interpreting the difference between the first two in terms of a vague notion of their different intensity, while explaining the heat of a body in terms of the annihilation of its radiant motion. None of this was either well worked out or widely accepted, but it reflects the extent to which there was, below a rough textbook consensus that repeated the traditional line about imponderable substances, a profound general uncertainty as to the nature and possible interconnections of heat, light, electricity, and magnetism. The lack of an adequate theory was widely felt. Mayer, of course, believed that his concept of force and its transformations brought intelligibility into this confused situation. Mayer clarified to himself the meaning of force in significant part by removing all traces of anything material from the autonomous "substance" force, whereby he also had to reject the common notion that forces are inseparable from ponderable matter. That the latter forces are forces in our sense of the word, not in Mayer's new sense, underscores the extent to which Mayer was both stimulated and stymied by the conceptual and terminological confusion surrounding "forces."

Mayer's historical claim to the status of (co)discoverer of the principle of the conservation of energy depends heavily on his 1842 calculation of the mechanical equivalent of heat—in the only way, as others have pointed out, then possible with existing data. My reconstruction of Mayer's thinking in chapter 6 will demonstrate how difficult it was for him to get from a general sense of the equivalence of motion and heat to an appreciation of the fact that the thermal expansion and compression of gases was even relevant to his problem. However, once he had reached that conclusion, the textbooks he used (Lamé and Baumgartner) provided him with not only the necessary data but also with strongly suggestive leads as to how to perform that computation.

I have already pointed out the significance of the usual conceptual pairing of matter and force, and have noted the issue of the materiality of the imponderables. One's understanding of the origin and significance of Mayer's central analogy between the indestructibility and uncreatability of force and the indestructibility and uncreatability of matter depends on recognizing that contemporary works of physics and chemistry contained very few explicit statements concerning the conservation of matter. There simply *was* no such explicit *principle* in the literature Mayer read, and he effectively forged *both* conservation principles for himself as he reflected on the similarities and differences between matter and force.

The absence of such a principle was surely due in part to uncertainties surrounding the very conception of matter and the imponderables, for if, as some thought, the so-called imponderables are somehow of the same substance as ponderable matter, if some transition is possible between imponderable and ponderable matter, then it would be difficult for the latter to be rigorously conserved. And for dynamicists, who constructed matter out of opposing forces, there was no 'matter' to be conserved. Mayer's redefining the imponderables as a clearly nonmaterial substance, force, created (as it were) a clear conceptual space around ponderable matter such as to allow it to be something which *could* be conserved conceptually and in principle and not merely tacitly or as a chemist's empirical law for the behavior of chemical substances in the laboratory.[212]

Mayer's training in chemistry, both as student and as apothecary's assistant, provided him with another rich source of concepts. Thus the notion of chemical equivalents provided him with an analog for the quantitative equivalence of different manifestations of force, and he transferred the concept of metamorphosis from its original application to organic chemical transformations to the qualitative transformation of forces from one phenomenal form to another. The concepts of neutralization and "indifference" also had solid chemical credentials, as in the neutralization of acids and bases to form indifferent salts or in the elimination of a "chemical difference" when hydrogen and oxygen combine to form water. Part of that imagery saw the neutralization of (chemical) differences as producing a state of indifference or equilibrium or rest, just as Berzelius and Davy described the forces of nature—in particular, the force of gravity—as tending to reduce nature to a state of rest or indifference, a state opposed by the action of heat. Recall, too, that the concept of the neutralization of opposing forces (or motions) was also a feature of the standard parallelogram of forces, just as the concept of indifference also had an analog in physics in the description of the midpoint of a magnetized iron bar. The concept of neutralization, in its several applications, acted as a kind of analogical connection between chemistry as the science of matter and physics as the science of forces. Thus many of the significant concepts and terms Mayer used which might suggest to some an influence of *Naturphilosophie* can be interpreted quite adequately on the basis of concepts and terms widely employed in the mainstream physics and chemistry of the day. Whether there nevertheless remains any explanatory role for *Naturphilosophie* in understanding Mayer's work is the subject of chapter 7.

• C H A P T E R   F I V E •

# Science Circumscribed

HAVING LOOKED in some detail at the relevant substantive content, language, and philosopical associations of physiology, medicine, physics, and chemistry as they pertained to the issues touched upon by Mayer's work, we turn now to consider certain other topics that circumscribed those core areas of concern. Of particular importance were philosophical and metaphysical issues concerning the nature and scope of science that influenced Mayer's conception of scientific knowledge and, as the character of the life sciences changed—profoundly—in the years around 1840, gave his work a significance quite different from what he had originally imagined. At the same time, Mayer's immediate geographical and intellectual environment was strongly marked by nominally nonscientific issues and controversies that bore directly on certain issues of central concern to him. Of particular importance in this regard were the connections between science and religion, the controversies that swirled around David Friedrich Strauss, and the mystical-religious aspects of the highly visible debates over the reality and significance of spiritualist phenomena. These two aspects of the context of German science during the 1830s and 1840s, different as they were, contributed substantially to the shape and meaning of Mayer's theory of force. The period of Mayer's principal creative activity, from around 1840 to 1845, coincided with a major historical turning point in the self-conception of the life sciences in Germany: from a time when physiologists routinely addressed problems of mind, soul, and vital force to one in which the field's pacesetters excluded such issues on both methodological and ontological grounds. If defenders of psychic phenomena had earlier enjoyed at least minimal space at the outer reaches of the scientific community, they would soon find themselves quite outside the pale. And if Mayer had originally regarded his conception of force as proof of the existence of nonmaterial entities, he would live to see the concept of energy assimilated comfortably into the minimally modified ontology of an ever more powerful and strident scientific materialism. The increasingly general exclusion of the soul and the vital force from the competency of the natural sciences represented a range of reinforcing considerations: the denial of their ontological reality; their methodological exclusion from empirical science; a contraction in the subject matter of physiology; a negative association (whether correct or not) with *Naturphilosophie* and spiritualist excesses; and a weakening (it would seem) of traditional religious belief in an immaterial and immortal soul.

## 1 The Nature and Scope of Science

The form that Mayer's work took was the result not only of his specific problems and their solution, but also in some measure of his perception of what science in general was. Not for nothing did he begin his first (unpublished) essay by affirming that "it is the task of natural science [*Naturlehre*] to explicate the phenomena of both the inanimate and the living world in terms of their causes and effects."[1] By so doing he clearly sought to identify one of *his* major themes with the general task of science. Similarly indicative of his sensitivity to what he perceived to be the character of scientific explanation was a prominently placed sentence in a later essay on medical physiology: "Physiology, if it wishes to uphold the name and status of an autonomous science, not only has to enumerate descriptively the natural phenomena belonging to its domain, but it must also *explain* them, that is, it must, in accordance with the rules of an inductive science, reduce the individual facts to general natural laws and connect them together by means of a common bond."[2] Not surprisingly, Mayer had in mind here the mechanical theory of heat in its relationship to the body's production of mechanical work. Appealing to the proper scope of science, Mayer insisted that it was no weakness of his theory that it could not explain *how* heat is transformed into motion: genuine science, the exact sciences, deal with positive knowledge of the phenomena, and leave speculation over the unknowable essence of things to speculative *Naturphilosophen* and their mystical ilk.[3] It was thus possible for him to invoke a principled defense, based on a somewhat rhetorical but widely repeated conception of science, of what might otherwise have been seen as a defect of his theory.

It is likely that Mayer's sensitivity to such issues was enhanced by his recognition that he, a medical doctor in a small German city, was an outsider to science, as well as by his decision to submit his ideas to the judgment of the community of professional scientists. It thus behooves us to look at the kinds of pronouncements relative to the nature and scope of science that Mayer would have encountered in the kinds of scientific literature he likely read. It should be kept in mind that here, as elsewhere, the scope of this study has been defined by its goal of explaining Mayer in terms of *his* scientific context—in the first instance textbooks, compendia, and broadly pitched works of physiology, chemistry, and physics—and it thus makes no claim to provide an in-depth assessment of contemporary scientific opinion *in general*, especially as might be revealed from a study of the vast body of more technical literature.

Attention has already been called to the extent to which Mayer's contemporaries identified causes as forces and defined physics as the science that investigates the causes of change in the physical world. The overall goal of science was to discover the laws that govern the behavior of the phenomena, ideally in terms of one of the fundamental forces of nature. Muncke's article on "Physics" in *Gehler's Physical Dictionary* expressed all the common views. A "philosophical treatment" of a science, he said, demonstrates "the necessary connection [of phenomena] according to causes and effects."[4] The object of physics is "to discover those forces on which depend the origin, the continuation, and the constant changes of the entire phenomenal world [*Sinnenwelt*]."[5] Its essential task is "to recognize the causes of things (*causas rerum cognoscere*), or rather to investigate the laws that underlie the phenomena of nature and their uninterruptedly succeeding changes."[6] However, as all commentators agreed, there is a limit beyond which human understanding cannot go: we can never hope to learn the *essence* of force or matter or the *origins* of things.[7] Fries underscored our inability to know the internal essences of things by quoting Albrecht von Haller's famous line, "Ins innre der Natur dringt kein erschafner Geist" ("No created spirit penetrates into the interior of nature").[8] Liebig urged that the task of the scientist is to trace phenomena back to their "ultimate causes," though without hope of ever being able to attain that goal:

> The investigation of nature has a definite limit that it may not overstep; it must always remember that . . . it is not possible to learn what kind of thing light, electricity, and magnetism are, precisely because the human intellect only has conceptions [*Vorstellungen*] for things that possess materiality. But we can investigate the laws of their state of rest and motion precisely because they manifest themselves in phenomena. Thus the laws of life and everything that disturbs, promotes, or changes it can doubtless be investigated without our ever knowing what life is.[9]

This general attitude toward the unknowability of essences and ultimate causes was, of course, acutely applicable to the situation surrounding the so-called imponderables, whose essence *really* was not known, thus the widespread decision to speak of heat (and the like) as the otherwise unknown cause of a particular range of phenomena. For some physicists like Lamé, in the tradition of Lavoisier and Laplace, it was one of the advantages of a physics based on experimentally corroborated mathematical laws that its results were compatible with either of the two chief theories of heat. For Mayer, the general acceptance of such notions meant that he could cover his ignorance of the true nature of (in particular) heat with an appeal to the intrinsic limitation of all scientific explanation.

Mayer would have encountered similar views on the nature and scope of science among the major physiological authors of the day such as Tiedemann, Autenrieth, and Müller. Wrote Tiedemann:

> Although it is the task of physiological theory to investigate the phenomena of life and its laws, to discover its causes, and to present the results of the investigations as a scientific whole possessing internal coherence, we must nevertheless recognize a limitation in the accomplishment of this task. Human reason, by its nature striving toward the highest perfection [or consummation (*Vollendung*)] of knowledge, seeks to be sure to raise itself to a clear grasp of the ultimate and supreme causes of nature and its diverse phenomena, and there expresses itself in the human spirit an untamable yearning to investigate the whole connection and chain of causes and effects up to its last link. But the satisfaction of this yearning has up to now shown itself to be beyond the limit of our understanding [*Fassungs-Vermögen*]. If one considers the phenomena of nature with any attention at all, then one soon perceives that the

> causes first recognized are only effects of higher causes, that one therefore proceeds from one cause to another, and finally comes upon still higher and incomprehensible causes where the thread of the investigation breaks off.[10]

Tiedemann assigned our inability to know the origin or first cause of organisms and organic matter—as well as of inorganic bodies and matter, and of the various forces of nature—both to the limitations of experience and to the arrangement of our cognitive faculty (*Erkenntniss-Vermögen*).[11] The latter, in particular, obliges us to assume the existence of certain forces in order to explain the phenomena, without our being able to explain *their* existence or origin any further. He quoted from Swammerdam's *Bible of Nature* to the effect that all our scientific knowledge is either of the phenomenal effects of unknowable first causes or of the effects produced by them in turn as new causes.[12]

Autenrieth, too, referred the character of scientific explanation to the nature of our cognitive faculty. It is a "requirement of our understanding" to investigate the cause of every perceived effect, even though our pursuit of the causal chain sooner or later ends in the dark, where we can do no more than assume a cause (*Grund*) that can be investigated no further.[13] In understanding a phenomenon like instinct we cannot rest content with teleological explanations, but must seek out the real, the physiological causes that act in accordance with necessary laws. A true physiological explanation must be in terms of the physical cause in the organic structure of the organism.[14] Autenrieth believed that our intellectual capacities, such as our desire for causal explanations, stand in a kind of preestablished harmony with the real nature of the external world: both stem from the same Creator.[15] We are not the creators of the laws of logic and mathematics, whose cause lies deeper: "The undeceptiveness of this discipline surely rests ultimately on the fact that one and the same fundamental cause has created the principle of contradiction for human thought in just as compelling a fashion as it also makes it impossible in the external world that a magnitude could simultaneously be and not be."[16] Thus, while maintaining our inability to fathom ultimate causes, Autenrieth allowed that the human mind is still capable of recognizing some fundamental truths. Mayer would not have needed to be a student of philosophy in order to come away from his reading with the conviction that "what is subjectively correctly thought is also objectively true."[17] For his part, Autenrieth went on to urge the danger of imposing on nature humanly derived concepts (such as an a priori dualism) and thus the necessity of empirical knowledge.

Müller's *Handbook of Human Physiology* likewise addressed general epistemological questions. The goal of natural science is to discover facts from which others can be derived as from a (unifying) concept, ideally after the manner of mathematics. From the law of gravitation we can derive the laws of planetary motion, "but the essence of the gravitation of matter remains hidden."[18] Similarly, Ampère discovered the laws of electric currents, "but the nature of electricity is still hidden," and psychology studies the phenomena of the soul, "but the essence of the soul still remains hidden."[19] Müller placed the burdon of science on the abstraction of concepts from experience:

One can see from this which method must be the most fruitful in the natural sciences. The most important truths in them have been found neither solely through analysis of the concepts of philosophy nor solely through mere experiencing [*Erfahren*], but through a reflective experience [*eine denkende Erfahrung*] that distinguishes the essential from the accidental in the experiences and thereby discovers principles from which many experiences are derived. This is more than mere experiencing; it is, if you will, philosophical experience.

Concepts are found in all sciences, for they are the actually present universal which is no longer experienced by the senses themselves, but which is abstracted by the mind. Concepts come to us only from the analysis of experiences.[20]

Müller's "reflective experience" is a reasonably accurate portrayal of the method followed by most scientists, including Mayer. Among the physics texts of the day it was perhaps Fries's that most clearly explicated the three-part relationship of scientific knowledge to a conceptually preparative natural philosophy, to mathematics (for fundamental concepts and methods of representation and calculation), and to experience (which has the last word).[21]

It has appeared, from the evidence presented in earlier sections of this book, that German physiology was in a period of profound flux during the late 1830s and early 1840s.[22] The writings of Schwann, Lotze, Schleiden, and Liebig—ambivalent as the latter was with regard to so many important issues—signaled a growing push for the thoroughgoing application to physiology and biology of the concepts and methods of the physical sciences.[23] Schleiden's 1842 review of Liebig's *Agricultural Chemistry* was a kind of manifesto of the new movement. Schleiden imputed to Liebig a fundamental philosophical principle that should guide all natural science: "It is the exclusive possibility of a natural science from a materialistic worldview [*aus hylologischer Weltansicht*]. The subordination of all of nature under strict mathematical natural laws that allow of no exceptions, laws that must ultimately all be reducible to the laws of motion of matter—motion that itself derives from the fundamental forces of nature—is the sole principle of our science and holds without exception just as well for organic as for inorganic nature."[24] In Schleiden's view the progress of science has been marked by the steady exclusion of the mystical, spiritual, and unlawlike from the natural world—a step that has been all the harder to make with respect to living things because our own experience confronts us with a mysterious parallelism between the spiritual (*das Geistige*) and the corporeal:

But here, too, the awakened science goes [*fortschreitet*] its assured way, only momentarily led astray by Schelling's and his school's poetic fantasies, and our whole science of organisms in physiology and medicine presses irresistibly in the endeavor of all its distinguished cultivators toward the goal of clearly declaring, proving, and scientifically grounding the complete independence of our corporeal worldview from all grounds of explanation involving spirits [*alle geistige Erklärungsgründe*]. This is now precisely the bond that invisibly entwines all the distinguished investigators of our time, this is the goal which all perceive ahead of them more or less obscurely, clearly, or distinctly, and which, however great the deviations and differences of opinion,

> unites all individuals into one great school; it is this, finally, that impresses upon the natural science of our day the character that it will triumphantly maintain against the quickly passing reveries of every philosophical mysticism.[25]

The new science was thus to be marked by an ontology of matter subject to invariable laws, by the exclusion of anything having to do with soul, spirit, or free will, and by the severing of all connections with religion and metaphysics. Of course, Schleiden and like-minded scientists did not solve, or even try to solve, the traditional problems of life, mind, and soul that their immediate predecessors still worried about: they simply defined them out of positive science. Indeed, although Schleiden and other propagandists presented their program as if it were directed primarily against *Naturphilosophie* and its baleful influence, in fact the mainstream physiology of the 1830s routinely addressed such issues. They were important to Mayer, too, and he never wholeheartedly reconciled himself to the new dispensation, however much part of him applauded the explanatory goals of a physiology based solidly on chemistry and physics.

The materialistic tendencies developing during the 1840s found prominent polemical expression in Carl Vogt's *Physiological Letters* of 1845–47, in which he identified psychic activities (*Seelenthätigkeiten*) as "functions of the brain substance" and, in a famous analogy, compared the relationship of thought to the brain to that of bile to the liver and urine to the kidneys.[26] Vogt developed these issues most fully in his twenty-seventh letter, in which he attacked the notion of an individualized immaterial and immortal soul that somehow uses the brain as its material instrument.[27] Similarly inadmissible was the explanation of embryological development in terms of a directive idea (*leitende Idee*).[28] All explanations must in principle be in terms of the form and composition of material parts: "physiology knows only functions of material organs."[29] Vogt's position stood in stark contrast to Volkmann's decade-earlier defense of physiology as an opponent of materialism; wrote Vogt:

> Physiology accordingly declares itself determinately and categorically against an individual immortality, as in general against all notions connected with that of the special existence of a soul. It is not only completely entitled to have some say in these matters, but it can even be reproached for not having raised its voice earlier in order to point out the sole correct path by which they can at all be resolved. . . .
>
> People have tried to save themselves from the (as they say) disconsolate materialism of the physiological way of seeing things by saying that what is immortal is not the special functions but the idea that underlies their development. The fundamental cause that gives rise to the formation of organs and their function is supposed to be imperishable, and thereby also the function participating in this immortality of its cause. I must confess that I cannot make sense of this reasoning. Matter is the only imperishable thing we know.[30]

In an anonymously published essay of 1850 Vogt repeated his standpoint in many of the same words, and appealed more explicitly to the changing character of physiology: "The whole direction of the science is a purely material one, and there is no longer any important physiologist who assumes an immaterial soul dwelling only in the body and treating it like a mere instrument. Common to all is the view that the so-called psychic activities are only products of the substance of the brain."[31] Mayer, of course, had hoped that his concept of an immaterial and imperishable force would undercut the ground from Vogt's brand of materialism, which treated forces only as properties of matter.[32]

The central issue was indeed materialism. During the mid-to-late 1830s it was still safe for Mayer and others to treat the organism for certain purposes *as* a machine, especially since he long entertained the possibility that organisms, unlike artificial mechanical systems, might indeed be able to create force, and since he regarded his concept of force, a nonmaterial but still really existing entity, to be an answer to traditional materialism. In the face of the more aggressive science-based materialism of the 1840s, however, which denied or excluded from scientific consideration everything that smacked of soul or spirit and for which the organism *is* a machine, Mayer's position had become irrelevant, especially since he had by then come to abandon the special status of organisms with respect to the creation of force. Mayer thus had to reconsider the *meaning* of his scientific work as its context changed around him.

Several other contemporary authors remarked upon these changes as they were getting under way. In 1836 August Vetter noted a movement in contemporary physiology to adopt the methods of precise experimentation of the physical sciences, whereby the theory of life took on more the character of a positive science, "unconcerned with certain ultimate grounds and causes that one knows for heat as little as for life."[33] For Valentin, writing in 1839, the conflicting diversity of the directions represented by the physiology texts published the year before was proof that the preceding few years had been a transitional period in that science.[34] The anonymous reviewer of Lehmann's *Textbook of Physiological Chemistry*—who I suspect may have been Lotze—opened his review by calling attention to the heightened opposition between teleological-organic and mathematical-mechanical approaches to natural science: "Two completely opposed and thoroughgoing endeavors having consequences in often remote connections divide now as before—but perhaps more decisively in our time—the works of natural scientists, where they extend over the entirety of organic life and inorganic existence, as well as over their mutual relationship to each other."[35]

At the same time a parallel movement was underway in Germany that sought a self-consciously "scientific" reform of medicine. Its chief protagonists included Wilhelm Roser, Carl August Wunderlich, and Wilhelm Griesinger—"the three Swabian reformers of medicine"—all of whom had been friends of Mayer's at Tübingen.[36] In 1843 Roser and Wunderlich wrote in an essay "On the Current Situation in Physiological Medicine" that "one would have to be little acquainted with the spirit and course of present-day science not to notice that it is in a critical stage, and that it is about to enter into a new period of development and to attain—after many a wrong turn and with, to be sure, also new dangers—a more exact and positive direction."[37] Roser and Wunderlich had founded the self-assertively progressive *Archiv für physiologische Heilkunde* in 1842, and they reprinted as an introduction to the first volume a polemical essay published in August of the previous year entitled "On the Deficiencies of Today's German Medicine and on the Necessity of a Decisively Scientific Direction in It." They issued a rallying cry to like-minded individuals desirous of transforming German medicine into a "positive science": "Medicine, as an empirical and inductive science, must also appear in the corresponding garb, and one can demand for it the same method as for the exact physical sciences. Nothing dogmatic may here be tolerated; rather every law which is proposed must bring with it the tests of its justification, it must appear in the company of the facts, observations, and experiments from which it is supposed to be derived."[38] They also expressed the need to overcome widespread skepticism about the possibility of treating medicine as a science, as opposed to an empirical art, and they were especially critical of the so-called ontological approach to medicine, of the physician whose chief concern is to classify diseases according to their supposed species and who neglects to investigate the *causes* of disease. They likewise ridiculed as an "ontological abstraction" the traditional concept of the "healing power of nature" as "a particular cause [*besonderes Motiv*] to which the healing is supposed to be referred."[39] According to Sticker, Wunderlich was among those who wished to get rid of the concept of vital force, and to follow La Mettrie in totally despiritualizing our conception of the human organism: he saw in the sick person only the disturbed machine, and the physician's task was that of the watchmaker, to repair the machine.[40]

Griesinger's special concern was to make psychology a part of scientific medicine, in particular a part of neurology: "Mental diseases are diseases of the brain."[41] As he wrote to Mayer in 1842, "the development and achievement of a purely physical view of vital processes I regard as the task of the physiology of our day."[42] He echoed Roser and Wunderlich's support for a new, scientifically based medicine—"the progressive part of present-day medicine"—in his first (and apparently only) "Medical Letter."[43] He, too, rejected both teleology and "ontology," insisting that "physiology and pathology have to do not with abstractions, but with material processes."[44] Griesinger pushed for a radical physicalism that excludes all speculative abstractions, ideal powers, and special organic forces.[45] He allied his *Pathology and Therapy of Psychic Diseases* of 1845 with the movement known as "physiological medicine," and explicitly accepted its characterization as "materialistic."[46] Mental diseases are, he wrote, strongly localized in the brain; nevertheless, he left open the scientifically unanswerable questions about the nature of the soul and its relationship to the body, and warned against an "abstract and shallow materialism" that denies the facts of human consciousness. Methodological rather than ontological commitments guided Griesinger's conception of a scientific psychology.[47] Similarly, although Lotze had gained fame as an opponent of the vital force in physiology and as an advocate for a physiology based on physical and chemical principles, he refused to accept what he saw as the dominant ontological materialism of the day that denied the existence and scientific legitimacy of the soul.[48] Significantly, he saw himself as defending what had become a minority position. The majority viewpoint was increasingly that expressed in assertive terms by Rudolf Virchow in an address of 1845, "On the Need for and the Correctness of a Medicine from a Mechanical Standpoint":

> Modern medicine has defined its point of view as mechanical, its goal as the establishment of a physics of organisms. It has demonstrated that life is only an expression for an aggregate of phenomena, each of which takes place according to the customary physical and chemical (i.e., mechanical) laws. It denies the existence of an autocratic vital force and healing power of nature. Indeed it is only for convenience that the natural scientist imposes the name 'forces' at all on the law according to which the properties of bodies manifest themselves in their relations to each other. Force is always only the resultant of the properties of two bodies, a concept, but not a real thing. Knowledge of the law is also completely sufficient; to search after the basis of the law is transcendental impertinence.[49]

Just as Mayer was trying to vindicate the substantial reality of a new concept of force, many of his more prominent contemporaries were trying to eliminate abstractions like force from a positive science of life.

The formative period of Mayer's ideas and the years of their publication were thus situated within a profoundly changing context, and if at first that context encouraged his thinking in certain important directions, it later caused him to pull back from some of the apparent implications and associations of his ideas. As Rümelin recalled, Mayer had been averse to the materialism widespread among medical students of the day, and he must have greeted Griesinger's words with mixed feelings.[50] Such a regimen was moreover theoretically too spare for Mayer, since it threatened to judge as unscientific his admittedly abstract conception of force. Indeed, Griesinger had earlier leveled precisely that charge against it, likening Mayer's "force" to the inadmissible ontological abstraction "disease":

> Your principle, that motion is *transformed* into heat and heat into motion, appears to me clearly too abstract. Motion in and of itself is a pure abstraction [*Abstraktum*], a mere representation [*Vorstellung*], or a concept; we can have empirical knowledge only of *matter in motion*, and it is the same thing with heat, the same with color, etc. All these are words that our unphilosophical language employs for something general [*ein Allgemeines*] in the phenomena of matter, as it does for example when it speaks of disease, while this "disease" itself is nowhere objectively present, rather there are only *diseased organisms* in the world. Matter furnished with the relevant properties is itself precisely the concrete content of those abstractions [*Abstraktionen*], and that itself is what science first has to deal with. But if you now say instead of "motion is transformed into heat" more concretely: matter in motion is transformed into hot matter!—then, insofar as this is [already] known from frictional phenomena, you have only enunciated either a platitude, or at most the general expression for a known fact, but you are still every bit as far from the explanation of the matter [*Sache*] as before.

I would thus have wished that you had operated more with material substances [*Materien*] than with concepts.[51]

In his response to Griesinger, Mayer could only repeat his refusal to speculate as to *how* motion becomes heat, and he insisted that an abstraction such as his concept of force is the "common bond" necessary to embrace the diversity of phenomena.[52] In 1844 he professed to Griesinger his allegiance to the new physicalist tendency in physiology, to the "striving for positive knowledge" that, "in allegiance to the spirit of the times," they both embraced.[53] Then, of course, he was still looking for his friend's approval of his work—he had sent him an early draft of what became his *Organic Motion*. In a later, unpublished, essay he evidenced his growing disaffection from the tendency of contemporary science: "For the rest, the force of custom and fashion must ultimately yield to the facts, and the time will surely come when every more profound view of the vital process is not driven out by microscopical and chemical subtleties."[54]

Ironically, Griesinger's objections proved Mayer right in one regard: a hard-nosed materialist of the traditional variety Mayer had in mind, especially one hostile to all abstract concepts, *would* reject Mayer's conception of force. Of course, a decade or so later, after the general acceptance of the conservation of energy, latter-day materialists would learn to embrace energy along with matter in a new synthesis, though I suspect that that synthesis was facilitated by widespread attachment to refurbished *aetherial* interpretations of the erstwhile imponderables, which obscured the need for a radically new view of the kinds of entities in the world.[55] Mayer's conception of energy was not to come into its own until publicly recognized by Josef Popper in 1876 and embraced by the energeticist movement of the 1880s.[56]

Liebig, too, felt the threat of the new materialism, and he spoke out againt it as he tempered his once more confident reductionist-sounding language. Hall concluded that Liebig's reluctance in later years to cite Helmholtz et al. was due to his realization, after around 1845, "of the totality of their commitment to the reductionist programme, a totality he did not share."[57] As Hall put it, "Liebig was conscious of having to steer a middle course between . . . the Scylla of vitalism (and of *Naturphilosophie*) and the Charybdis of outright reductionism."[58] We have already noted Liebig's ambivalence toward the vital force, as well as his programmatic exclusion of *das Geistige* from science. Marking a subtle but important focusing of his position, in his "Chemical Letters" of 1844 he vigorously defended a conception of the vital force as able to direct the chemical forces of the organism, and granted to the human will the ability to control all the otherwise invariable laws of nature:

Nowhere outside of themselves do human beings observe a will having attained consciousness, they see everything in the fetters of immutable, unchangeable, fixed laws of nature, only in themselves do they recognize a something that can control all of these effects, a will that can control all of the laws of nature, a mind [or spirit (*Geist*)] which in its manifestations [or expressions (*Aeußerungen*)] is independent of these natural powers, which in its complete perfection issues laws only to itself.[59]

Even to recognize the efficacy of such a 'spiritual' power, if not yet to bring it within the purview of science, was to go against the reductionist program of many of his contemporaries.

Two years later, in an essay published (rather oddly) in the largely nontechnical *Deutsche Vierteljahrs Schrift*, Liebig assumed a more aggressive stance against those who sought to eliminate the vital force from science. Having insisted that anatomy alone, without the assistance of chemistry, is incapable of answering physiological questions, he argued further, in a section headed "Chemistry Alone Not Sufficient," that

> [a]nother fundamental error entertained by other physiologists is that in order to explain the vital phenomena one can make do with chemical and physical forces alone or in combination with anatomy; it is indeed difficult to comprehend that the chemist, who is closely acquainted with chemical forces, recognizes in the living body the existence of new laws, of new causes, whereas the physiologist, to whom the knowledge of the effect and the nature of chemical and physical forces is remote or lacking, wants to explain the same processes with the help of the laws of inorganic nature.
>
> In its true significance the latter view is the extreme consequence of a reaction against one which preceded it. In the era of philosophical physiology not yet long past one explained *everything* by means of the vital force. The reaction entirely rejects the vital force and believes in the possibility of being able to reduce all vital processes to physical and chemical causes. In the living animal body—so one said forty years ago—*other* laws hold sway than in inorganic nature, all processes are of *another* nature. Many of today's physiologists, on the contrary, consider them to be of the *same* nature. The unprofitability of both views for us lies in the fact that neither then nor now has one attempted to determine or to discover the deviations in the effects of the vital force and in the actions of the inorganic forces, or their similarity or identity. The conclusions one came to were not based on an acquaintance with the similarity or dissimilarity of their mutual relations, but on a lack of acquaintance with them.[60]

Thus Liebig at his polemical best (or worst). It was certainly not the case that *Naturphilosophie* was the locus of belief in the vital force in previous decades, and it was a slanderous disparagement of his opponents to assert that it was primarily their ignorance of the facts that underlay their denial of the vital force.

In a popular address ten years later, in 1856, Liebig associated his defense of the vital force with his opposition to materialism. Echoing the natural theologians of an earlier day, Liebig argued that the existence of a house implies the agency of a builder, that it would be absurd to think that it "came into existence by itself through the interplay of natural forces."[61] And just as the existence of a house implies the prior existence of an *idea* and a *cause* capable of directing and controlling the forces of inorganic nature, so, too, does the existence of a plant imply the existence of an idea realized by a purposefully acting cause. Perhaps counting on his audience to supply the obvious theological conclusion which even he—a Christian scientist but a scientist nonetheless—declined to draw, Liebig invoked the necessity of assuming a vital force to explain the special character of vital phenomena, and he belittled the scientific credentials of those who denied it:

> If you look at the people who champion those opinions, then you immediately see that they are strangers in the domains that have as their task the investigation of chemical and physical forces; no competent physicist or chemist has ever agreed with them. And if you ask our great physiologists, to whom we owe the discovery of the facts on which the deniers of the vital force base their claims, then you will obtain the answer that these masters of science regard such claims and conclusions as neither well-founded nor justified. They are the opinions of dilettantes.[62]

According to Liebig's typically distorted history, the vital force only *seemed* to be denied by those scientists who for good methodological reasons search first for explanations of vital processes in terms of known chemical and physical forces; for the rest, the vital force had come into disrepute because of its misuse by unscientific *Naturphilosophen*, and its denial by materialists represented an excessive reaction against that misuse. His characterization of an earlier state of affairs in physiology was the kind of caricature that by then passed for history and has continued to distort many people's perception of early nineteenth-century German science:

> In order not to be unjust to the apostles of materialism, one must take into consideration that their views are essentially nothing but the extreme consequence of a reaction against the doctrines that still held sway some years ago. The *naturphilosophisch* physiology lacked the basis of exact investigation, of experience; it explained all processes of nutrition, respiration, and motion by means of a single imaginary cause that it called vital force; one believed that the chemical and physical forces play no part in the organic body, it creates for itself in its own way the iron it needs, as it does heat. The exact investigation of nature has demonstrated that all forces of matter really do play a part in the organic process, and the extreme reaction now maintains, in opposition to the earlier view, that only the chemical and physical forces determine the vital phenomena, that no other force at all acts in the body. But just as little as the *Naturphilosophen* of that time could provide proof that their vital force did everything, so little can the materialists of yesterday adduce proof that the inorganic forces do it and suffice by themselves to produce the organism, indeed the mind [*Geist*]. All their claims rest, as they did then, not on the acquaintance but on the lack of acquaintance with the processes. Truth lies in the middle, which raises itself above the one-sided views of things and recognizes for organic life a formative principle [*formbildendes Princip*], a dominant idea in and with the chemical and physical forces.[63]

Few of Liebig's scientific colleagues in 1856 would have accepted his claim that the middle road of truth leads to the rejection of materialism and the acceptance of a creative organic force, a quasi-Platonic idea that also accounts for mind. Even Mayer, who likewise abhorred the antireligious tenor of contemporary scientific materialism and its denial of reality to the mind or soul, did not seek support for his views in a vital force that by then had been scientifically pretty well discredited.

In some ways Liebig's relationship to the principle of the conservation of energy is even more enigmatic than Mayer's. Although he was thrust into early contact with Mohr's and Mayer's work, and worked within a relatively similar context of problems, and although he delivered himself of many an utterance that, in isolation, looks like an understanding of that principle, in the event he was extremely slow to appreciate its significance or even to grasp the need for a new concept. As far as I can tell, it was not until 1858, in an address "On the Transformation of Forces," that he either acknowledged the uncreatability and indestructibility of force or even mentioned Mayer's name.[64] Insofar as he there considered living things, the views he expressed scarcely went beyond those advanced by Dumas seventeen years earlier; of any complicating *Lebenskraft* there was nary a mention. Other than as evidence for Liebig's confused thinking or reluctance to acknowledge the work of others, it seems to me that he was rendered incapable of following Mohr's or Mayer's lead by his attachment to a physicochemically irreducible vital force and—ironically, considering his opposition to materialism—by his denial that forces could exist without matter. Just as the course of Mayer's career cannot be separated from Liebig's, so too does a comparison between them highlight that combination of similarity and difference that makes an appreciation of Liebig's work an unusually powerful foil against which to view the peculiarities and commonplaces of Mayer's, and against which to assess the distinctive components of Mayer's creativity.

## 2 Religion and Spiritualism

It has been one of my contentions in this book that for Mayer the meaning of force and the question of its indestructibility and uncreatability were intimately connected with religious and metaphysical issues, and that this constellation of issues occupied a prominent place within the widely ranging medical and physiological literature of the day. The existence and nature of the soul/mind/spirit (and its relationship to the vital force and the imponderables), materialism (and the atheism usually associated with it), causality and creation out of nothing (of which God's original Creation was the archetype)—such issues helped circumscribe the context of Mayer's thoughts on force. This section will examine in more detail the strong natural-theological strain in the writings of many of the leading scientists of the day—such as Tiedemann, Liebig, Autenrieth, and Berzelius—and will explore other aspects of that very popular program. For such scientists the study of nature leads ultimately to an acknowledgment of God as the purposeful First Cause. Timothy Lenoir has maintained that the post-Kantian teleomechanist tradition he identified "never made use of the design argument or the notion of a purposeful divine architect."[65] Although he was correct in identifying teleology à la William Paley and Samuel Wilberforce as characteristically British, his desire to decouple German teleology from religion and attach it to "good science"—his words—went too far, since from among his group both Tiedemann and Liebig argued explicitly from the structures of the natural world to the existence of a Divine Architect, and even Müller, though noncommittal, gave serious consideration to the cosmological and theological implications of different conceptions of the world.

To compare the churchmen Paley and Wilberforce to professional scientists like Tiedemann and Liebig is, moreover, to obscure their radically different enterprises, and one would hardly expect the latter's scientific works to double as full-blown tracts on natural theology. Yet they still allowed themselves to extrapolate from their science to God. As Tiedemann wrote in a long excursus on the meanings of the word "nature,"

> Finally, one designates by the word nature the first or highest fundamental or final cause of all things and phenomena in the world, out of which everything has arisen and through which everything exists. The notion of a highest cause in the universe is the work of reason, which sees everything in nature determined by eternal and invariable laws whose purposefulness it recognizes as in accord with reason. The unity and harmony that prevail in the cosmos, and the relationships among the countless heavenly bodies recognized there by reason . . . [as working] toward one purpose, the preservation of the whole, prove that there can be only one final cause in it. Reason, which is compelled by itself to acknowledge nature as a self-contained whole that is cause and effect of itself, imagines that which preserves, that which creates in the universe as the absolute, as the world soul, as God.[66]

For the most part Liebig did not throw lines from his science to religion until the mid-1840s, when he began to be sensitive to the antireligious implications of the materialist reductionism that was beginning to dominate much of the general scientific worldview. In what was to be republished as the first of his *Chemical Letters* in book form, he asserted that "without knowledge of the laws and phenomena of nature the human mind fails in the attempt to create for itself a conception of the goodness and unfathomable wisdom of the Creator; for everything that the richest fantasy, the highest cultivation of the mind is ever capable of inventing by way of pictures appears, compared to reality, like a multicolored, sparkling, empty soap bubble."[67] Most of his second letter, which specifically invoked the connection between science and Christianity, went on in the following spirit: "The investigation of nature teaches us to recognize the history of the omnipotence, the perfection, the unfathomable wisdom of an infinitely higher being in its works and deeds; unacquainted with this history, one cannot conceive of the perfecting of the human spirit, without it its immortal soul does not attain consciousness of its dignity and the status it occupies in the universe."[68] Thus was science to be harnessed to belief in response to those who were coming to see it ever more inextricably associated with an antireligious standpoint. We have seen how Liebig employed the traditional a-building-must-have-a-builder argument of natural theology in his antimaterialistic address of 1856. In his analogy, a house, an individual plant, and a plant species all proclaim their need for an author (*Urheber*)—whether a builder, the vital force, or the Divine Architect—purposefully following an idea: we recognize a purpose and a guiding idea in the ordered lawful development of a plant, and our reason concludes that, as the idea of the species must have an author, so, too, must a cause, a force, exist in the living organism that controls the chemical and physical forces of matter in order to construct out of it forms never otherwise encountered.[69]

Authenrieth's writings, in particular his *Views on the World of Nature and the Life of the Soul*, were full of arguments in the spirit of natural theology; indeed, in the preface to that posthumously published work his son wrote that if his father had not died he would doubtless have composed a classical work "in the spirit of the Bridgewater books."[70] The third essay reprinted in it, "Natural History of Man," developed the Bridgewater-like theme of the appropriateness of our terrestrial environment to our survival. And just as Autenrieth's *Physiology* of 1801 spoke casually of the "first creation" of the original members of a species, so this work invoked the beneficent control of "the wisdom which not only shines forth from the arrangement of the great heavenly bodies, [but] which allows so many wonderfully ingenious [provisions] to arise continuously even in the smallest creature."[71]

For Berzelius, too, the beauty and wonder of the world were grounded explicitly in God, just as the existence of life (i.e., the vital force) in an otherwise dead universe cannot be explained except through the assumption of a purposefully acting Wisdom.[72] Berzelius was especially hostile to the notion that the world as we know it could have arisen by chance:

> The purposeful quality [*das Zweckmäßige*] in everything that belongs to organic nature, which distinguishes the productions of a lofty understanding, gave human beings occasion—on comparing their calculations toward the attainment of final purposes with those they found in the edifice of living nature—to regard their capacity to think and to calculate as a likeness [*Abbild*] of the being to whom they owe their origin. Nevertheless it belonged more than once to the fancied profundity of a short-sighted philosophy to let everything be the work of chance, in which only that fraction of the products could endure that by chance had acquired the capacity to maintain and propagate themselves. But this philosophy did not perceive that what it assumed in dead matter under the name chance is a physical impossibility.[73]

On into the 1830s it was thus quite usual for leading scientists to express religious views of a natural-theological character even in their most important scientific works. In works pertaining more directly to the genre, such as Brougham's *God and Immortality from the Standpoint of Natural Theology* of 1835—to quote from the title of its German translation of the same year—one finds additional arguments pertaining to causality and creation that suggest an even greater and more particular relevance to the kinds of issues central to Mayer. Basic to the entire enterprise was the argument for the necessity of a rational and providential "First Cause" of the wonderfully contrived universe.[74] Brougham was, accordingly, also concerned to defend the philosophical legitimacy of the general concept of causality from the Humean challenge. He, like Herschel and Whewell, traced our belief in causality to our consciousness of the power of the will to move our bodies, and concluded that "I cannot understand how, but for the consciousness of power, we could ever have been led to the belief in the existence of a First Cause."[75]

In arguing for the independence of the willing mind from the body and for the former's continued existence after the latter's death, Brougham broached the issue of the indestructibility of matter (and mind) and of the uncreatability of matter (but not of mind):

> All our experience shows us no one instance of annihilation. Matter is perpetually changing—never destroyed; the form and manner of its existence is endlessly and ceaselessly varying—its existence never terminates. The body decays, and is said to perish; that is, it is resolved into its elements, and becomes the material of new combinations, animate and inanimate, but not a single particle of it is annihilated; nothing of us or around us ever ceases to exist. If the mind perishes, or ceases to exist at death, it is the only example of annihilation which we know.
>
> . . .
>
> It may be further observed that the material world affords no example of creation, any more than of annihilation. Such as it was in point of quantity since its existence began, such it still is, not a single particle of matter having been either added to it or taken from it. Change—unceasing change—in all its parts, at every instant of time, it is for ever undergoing; but though the combinations or relations of these parts are unremittingly varying, there has not been a single one of them created or a single one destroyed. Of mind, this cannot be said; it is called into existence perpetually, before our eyes. In one respect this may weaken the argument for the continued existence of the soul, because it may lead to the conclusion, that as we see mind created, so may it be destroyed; while matter, which suffers no addition, is liable to no loss. Yet the argument seems to gain in another direction more force than it loses in this; for nothing can more strongly illustrate the diversity between mind and matter, or more strikingly show that the one is independent of the other.[76]

If the calling into existence of mind sounds like Autenrieth and recalls the like-spirited reflections of Tiedemann and Müller, that only means that the same issues of mind and body, of creation and annihilation, served to link together into a common context both physiology and natural theology. Given the general silence of contemporary chemists and physicists with regard to the conservation of matter, it is noteworthy that the issue of its indestructibility and uncreatability should occupy a prominent place in theological and metaphysical contexts. In that connection Brougham cited the opinions of a number of ancient Greek philosophers with respect to the impossibility of conceiving of the act of creation, "of calling existences out of nothing": "Upon the uncreated nature of things,—for the doctrine extended to mind as well as to matter,—the ancient philosophers founded another tenet of great importance. Matter and soul were reckoned not only uncreated, but indestructible; their existence was eternal in every sense of the word, without end as without beginning: . . . '*Nothing can be produced out of that which has no existence, nor can anything be reduced to nonentity*' "—quoting from Diogenes Laertius's account of Democritus's philosophy.[77] He also quoted six lines from Lucretius and one from the first-century poet, Persius: "De nihilo nihil, in nihilum nil posse reverti."[78] From Brougham's perspective, what the ancients lacked was a conception of the Christian Creator-God.[79] He also attacked the Godless materialism of Holbach—which he took to be the assertion that "nothing [exists] beyond or different from the material world"—by insisting that mind, too, exists and is wholly different from matter.[80] Such was the brand of materialism Mayer believed was undone by *his* conception of force.

Such was the general context in which a contemporary of Mayer's might have encountered such deliberations and such texts, such concepts and such language. Strauss, too, discussed the theological issue of "creation out of nothing," which the ancient philosophers had objected to on the grounds that "*ex nihilo nihil fit*."[81] So did Schelling, in a work reprinted in 1834.[82] Not that there was any novelty to the subject, which had exercised Christian theologians and philosophers at least since Augustine. An eighteenth-century work of natural theology in Mayer's library, whose centerpiece was the divinely regulated equal number of male and female births, used similar language: "The wisest creator and ruler of the world lets the vast multitude of humankind arise by means of generation out of his Nothing, as many of them as he has ordained to life."[83] Mayer's own strong religious sentiments and the fact that he, a student of medicine, drew his closest associates from among students of theology greatly enhance the likelihood that he had been confronted by such issues. So, too, does Mayer's attachment to the Latin tags *ex nihilo nil fit* and the like suggest their possible source in the theological-philosophical literature concerning divine creation and the immortality of the soul. Such language was not common in the physical and chemical literature of the day.[84]

Contemporaneous with Brougham's and Prout's natural theology and with a theologically attuned mainstream science, the decade of the 1830s in Germany was marked by lively controversy over interconnected issues of spiritualism and religion, fueled most notably by Justinus Kerner's *Seeress of Prevorst* (1829) and David Friedrich Strauss's *Life of Jesus* (1835–36).[85] Significantly, Kerner was from Weinsberg, a town just over the hill from Heilbronn, and Strauss was in Tübingen when Mayer was a student there. Significantly, too, those on the spiritualist side were often medical doctors, and the discussion of their works often took place within a quasi-medical context. Thus Kerner had been a student of Autenrieth's at Tübingen, got his medical degree there in 1808, and practiced medicine in Weinsberg from 1819 on.[86] In May 1842 Kerner, as president of a regional physicians' society, opened the group's meeting with a talk on sympathetic cures; Mayer reproduced the full text in his published report on the proceedings.[87] Another major author, Johann Carl Passavant, had also studied with Autenrieth at Tübingen, gotten his medical degree there in 1810, and become a practicing physician in Frankfurt am Main in 1826; he met Kerner at the *Naturforscherversammlung* in Heidelberg in 1829.[88] The second edition of Passavant's book on animal magnetism and clairvoyance was reviewed in 1838 in the *Jenaische allgemeine Literatur-Zeitung* under the rubric "Medicine," and Friedrich Fischer's book on somnambulism was reviewed there the following year under "Physiology."[89] Strauss, moreover, had been a friend of Kerner's at least until he repeatedly criticized Kerner's and others' work on these topics during the 1830s. One of Kerner's more permanent friends and supporters was the Romantic enthusiast for all things mystical, Carl August von Eschenmayer, with whom Strauss also exchanged a number of acerbic writings. Eschenmayer had received his medical degree from Tübingen in 1796 and had practiced medicine for around ten years before becoming extraordinary professor of medicine and philosophy at Tübingen in 1811 and ordinary professor of practical philosophy there in 1818.[90] In other words, someone like Mayer studying medicine at Tübingen in the 1830s would have found himself immersed within a rich context of theological and metaphysical debate with strong medical associations. To be sure, there is no direct evidence of Mayer's close familiarity with this spiritualist literature, or that he shared its more particular concerns, but he could hardly have been unaware of the issues, some of which were closely relevant to those touched upon by his own work: the nature of the mind/soul/spirit and its relation to the body; the defense of Christianity via the demonstration of the reality of a nonmaterial world; the nature of scientific explanation; and, implicitly, where the force comes from to rattle tables in the next room. As with homeopathy, all the historian can do in this case is to describe what appear to be relevant props on Mayer's historical stage, without being certain how much attention Mayer in fact paid to them.

I have used the term "spiritualism" here to cover the broad range of phenomena—or, as some would have it, alleged phenomena—of animal magnetism, somnambulism, clairvoyance, possession by spirits, psychokinesis, and the like. They were all 'spiritual' insofar as they were usually interpreted by their defenders as evidence for the existence of a spiritual realm, of psychic forces different from the more commonly recognized forces of nature. They were of sufficient contemporary significance to physiologists that Müller took the trouble to dismiss animal magnetism as deception and superstition, as Autenrieth similarly rejected possession and *Geistererscheinungen*.[91] Yet Autenrieth accepted animal magnetism as demonstrating the influence of one person on another independently of the transference of any ponderable matter.[92] Even Burdach, who was close in spirit to those philosophical tendencies (such as *Naturphilosophie*) that eagerly embraced such phenomena, went no further in 1837 than to leave open the possibility that some of the phenomena of somnambulism and clairvoyance might be valid and hence require a scientific explanation. At the same time he was acutely aware that larger philosophical-ontological issues were at stake: "The rarity of these phenomena, their deviation from the customary course of life, and especially the frequency of their deceitful imitation as well as their pernicious misuse, has led to a situation where their reality has been entirely denied, particularly by those who fancy to have comprehended all of nature as a mechanism and who therefore also wish to acknowledge as true only that which is mechanically comprehensible."[93]

Passavant's strategy for rationalizing the acceptability of animal magnetism and clairvoyance was to present evidence that organisms are capable of producing electricity and magnetism and of acting independently of the mediation of organs or direct physical contact. Examples of the latter included the influence of the mother on the fetus, twins who always got sick at the same time, and pouters (*Kropftauben*) who, given a young trumpeter (*Trommeltaube*) to raise, subsequently gave birth to young pigeons who resembled not the parents but the adoptee.[94] Such phenomena attest to the power of *Geist*. Passavant's purpose included an appreciable and explicit religious component:

> Finally, the human spirit—mostly in its highest powers—can become a free organ of the eternal spirit. The immediate contemplation [*Schauen*] of the divine spirit [*gotterfüllter Geist*] reveals itself to us in the glimpses of light that anticipate and thereby prophesy a higher existence, glimpses that, in their highest perfection at the final limit of the development of intelligent creatures, must become a pure contemplation of the divine essence and of the creation as a divine system of thought.[95]

For Friedrich Fischer, professor of philosophy at Basel, the general significance of somnambulism was that it affords new insights into the nature of the soul and its connection with the body. Noting (in 1838) that the subject had recently fallen into discredit, he insisted that it was of potentially great significance to physiology and medicine. The problem, he said, was that people had separated corporeal from psychic (*geistig*) life, the vital force from the soul, and had thus failed to recognize their identity: somnambulism was "nothing but an *awakening of the vital force to the soul*."[96] Fischer reinterpreted the usual distinction between the vital force as working according to blind necessity and the soul as the locus of free will in terms of their differing degree of connection with the body. He solved the problem of whence the somnambulist derives his or her greater physical strength and intellectual capacities by appealing to the vital force as an unproblematically inexhaustible source of force:

> This intensification [*Steigerung*] of the animal and psychic operations is easily explained if it is the vital force that takes charge of them by means of an awakening to the soul. The vital force is always equally powerful in its operations, equally fresh and alive. Out of its inexhaustible source the soul—which every night, wearied by the day's work, sinks back into its womb [*Schooß*]—draws new force and new vitality and liveliness. In the womb of the vital force the fatigued voluntary muscles strengthen and refresh themselves anew. The nerves themselves become vegetative in sleep in order to replenish the lost force and excitability. Given such an inexhaustible store of force, elasticity, and liveliness, it is easy to understand that the motion of the limbs would turn out to be incomparably quicker, more elastic, and more untiring if the vital force instead of the soul takes charge of them.[97]

Thus on several counts somnambulism and related phenomena raised pointed questions about the sources of force mobilized by the human agent. Fischer's appeal to the vital force as an inexhaustible source of *work-producing* force ran counter to the tendency of most contemporary physiologists to look in such instances for underlying processes of material exchange, but it was quite in line with Lotze's later appeal to a quantitatively variable work-producing psychic force and with Liebig's sometime work-producing vital force.

One of the most persistent critics of Kerner, Passavant, et al. during the 1830s was David Friedrich Strauss. Strauss had developed an early interest in somnambulism and had read (without satisfaction) some of Eschenmayer's writings. He then read Kerner's book with great interest and enthusiasm, and developed close personal connections with Kerner, Eschenmayer, and Friederike Hauffe, the "seeress of Prevorst," but after her death in 1829 he became disenchanted with their brand of philosophizing, especially once he became a follower of Schleiermacher.[98] Strauss opened his "Critique of the Different Views on the Apparitions [*Geistererscheinungen*] of the Seeress of Prevorst" of 1830 by distinguishing those who granted the phenomena an objective existence from those who saw in them only the seeress's subjectivity. The latter view could not, however, account for the "visible, audible, and tangible manifestations of force" of the alleged spirits.[99] Strauss himself was willing to accept the objective reality of the phenomena, but he otherwise rejected Kerner's explanation of them and his attempt to use them as evidence for the existence of a spiritual realm independent of the material world. According to Strauss, in order to explain the action of spirits on physical objects Kerner had supposed that the soul, on passing into a realm beyond this world, takes with it its *Nervengeist*.[100] As far as Strauss was concerned, however, the supposed *Nervengeist* can only be a product of the nervous system, and can only exist insofar as it is constantly regenerated through vital processes. Ultimately he was left accepting the reality of the immediate phenomena without being able to explain them, but with the conviction that they were somehow due to the direct action of the subject and not to the mediation of disembodied spirits:

> The matter stands accordingly as follows: they accept the miracle of the somnambulist's action-at-a-distance, plus a second one into the bargain, the realm of spirits; we, however, accept only the first. But let us not hide the extent to which this capability ascribed to the seeress still lacks any proof in the laws of organic life . . . . In any case, however, it seems only logical to ascribe to an organism whose sphere of sensation extended so indefinitely far that it felt every metal and every plant in the room, some even at greater distances, a similarly extended sphere of action as well by means of which she was capable of producing (in part voluntarily in magnetic sleep) that rapping in distant houses and (in part involuntarily while awake) those sounds regarded by the seeress herself as her spirits' manifestations of force.[101]

After 1830 Strauss became increasingly dismissive of the whole business, and in another context he put down as ridiculous and misguided Kerner's and Eschenmayer's acceptance of possession by demons.[102]

One of the responses Kerner made to criticism of his inability to give a scientific explanation of these phenomena was to insist that one should be prepared to accept their validity even if they cannot be explained.[103] Strauss accepted this position in principle, and took the occasion of his review of a later work of Kerner's to discuss at some length the relationship between a theory and the data to be explained, from which he concluded that "theory is at no given point in time by itself already the full truth and reality: the excess of the factual, the empirical over it is also part of the picture, and the life of science depends precisely on the fact that experience [*Empirie*], just when theory thinks it has caught up with it, immediately gains the lead again, and the latter thereby gains a new task."[104] In its own way, this whole debate again brought up for discussion general issues on the nature and scope of scientific explanation.

The extent to which our inability to explain a phenomenon should influence our belief or not in its reality was also an important issue for Strauss in a more straightforwardly theological context, namely with respect to his decision to deny miracles and their traditional role as underpinnings of Christianity.[105] For his part Eschenmayer regarded Strauss as a traitor to Christianity, a person who denied the traditional role of faith by attempting to judge religious truths on the basis of "the laws of logical reason and physical nature."[106] He, on the other hand, had reaffirmed his faith in the truth of the Gospel by freeing himself from the captivity of said laws and from the erroneous notion that abstract concepts are decisive in questions of faith. In his view, "holy truths are of another nature than logical, aesthetic, and moral truths."[107] Eschenmayer ridiculed Strauss for (in effect) saying that because he could not imagine a miracle they were impossible, that because his critical understanding could not get beyond logical and physical laws there *are* no higher laws.[108] He regarded the biblical criticism of the day as a "disguised materialism" whose aim was "to make nothing out of something, and something out of nothing."[109] Such a position belonged, he alleged, to certain intellectual and social groups: "[Strauss's] book counts (or will count) among its admirers all the Enlightenment types who have long since concluded a cheap bargain with religion, and all men of the world and industrialists, since his morals are as hastily gotten together as his belief in miracles: according to the latter, 'everything which is not imaginable is impossible,' and according to the former, 'whatever is in agreement with human interests is allowed, as e.g. hypocrisy from the pulpit.' "[110] Eschenmayer saw ranged against his brand of mystical Christianity a rational theology, a pervasive materialism rooted both in science and in modern society, and a misguidedly restrictive concept of explanation. In that context it is not surprising that he would exploit, as Strauss would reject, alleged cases of demonic possession to prove the existence of a spirit world and hence to undercut the philosophical materialism which he felt threatened Christian faith.[111] Mayer's metaphysics of force reflected some of these same concerns.

Regarded as a locus for the consideration of issues relevant to Mayer's problem context, one work in particular from the spiritualist literature of the 1830s warrants further consideration, Passavant's *Investigations on Vital Magnetism and Clairvoyance* of 1837. Passavant wove together strands from *Naturphilosophie*, from the handling of the imponderables (and their increasingly prominent interconnections) by contemporary physicists, and from wider ontological reflections, in ways that intriguingly suggest a possibly fertile context for the encouragement of conservation-of-energy type considerations. I discuss them here as an example of the conceptual breadth and richness of the German medical/spiritualist literature of the day, a literature Mayer must have had *some* knowledge of. There is, however, no evidence that Mayer read Passavant, and little to be gained in assuming he knew this particular work even though some of its language is reminiscent of some of his own. The points of contact were of too general occurrence in a broad range of literature of the day to be of very specific diagnostic value. And although it is easy to imagine a work like Passavant's as having prompted someone else to pursue its leads in the direction of conservation of energy, I have no reason to suspect that it ever actually performed such a function: abstract intellectual affinities are not yet concrete historical connections.

Passavant opened his book with a section "On the Universal Forces of Nature." He presented an image of nature producing diversity out of simplicity, "das Differente aus dem Indifferenten," and urged that matter must not be assumed to be "dead" but rather to be the product of forces, of an internal activity.[112] He identified gravity (*Schwere*) as the one force to which all bodies are subject, "the *universal* unifying force" connecting all bodies with each other; at the same time, bodies are subject to a number of more powerful "qualitative forces" such as electric and magnetic attraction, the chemical process, and organic forces.[113] Light, heat, electricity, and magnetism—"the so-called imponderables"—are the "universal powers of nature," powers that exhibit great similarity to each other.[114] Once regarded as peculiar substances, in the present state of physics they are more justly interpreted as activities, as motions of bodies (like the vibrations of elastic bodies). Perhaps the most decisive phenomenon in the case of light, as also of radiant heat, is interference, since we can imagine how two motions can cancel each other out (*sich gegenseitig aufheben*), but not how two substances can destroy each other.[115] And just as the parallels between light and heat and their interconvertibility imply that they are both only different modifications of the same *Grundkraft*, so too can one bring electricity and magnetism into the circle of interconvertible forces: all the imponderables can be regarded as but modifications in the motion of an aether—perhaps also of the smallest parts of material bodies themselves—and they can be understood only in terms of their genetic connection.[116] Belying any intuitive sense of the conservation of motion, Passavant wrote that when the motion of the aether is suspended there arises, according to circumstances, darkness, cold, or the cessation of electrical and magnetic tension.[117]

Passavant opened the next section, "On the Organic Forces," by asserting that what principally distinguishes organic from inorganic bodies is that the former are governed by an autonomous and purposefully acting principle.[118] Nevertheless, that distinction was not absolute, since "nature in the large"—for example the solar system—also possesses such an autonomy: "The motive and conserving forces are reproduced in it, and not given to it from the outside. . . . As every individual organism is a totality unto itself and at the same time repeats that of a higher cosmic whole in a peculiar manner, so too does it repeat (albeit with modifications) the general functions of the planetary system which creates and nourishes it." From this he drew the conclusion "that the organic forces can only be modifications of those universal forces of nature." One must avoid, however, taking the forces of organic nature to be *identical* to those of inorganic nature, when they are merely *analogous*.[119] Thus even decidedly 'spiritualist' works used language and broached issues concerning the nature of forces and the imponderables, including the possibility of the former's generation, which might conceivably have contributed to Mayer's larger problem context.

## 3 Summary

That Mayer was of a mind to present his ideas to the community of physical scientists meant that he was likely to pay attention to contemporary norms concerning the characterization of a good scientific theory. Such attention would have reinforced certain strong tendencies in his way of thinking about force and would have provided him with a principled means of responding to certain potential weaknesses in his theory. At the same time, scientists' attitudes toward the status of forces and spirits, in the context of growing materialistic and reductionist sentiments during the 1840s, throw into relief some of the difficulties Mayer faced in getting others to accept his ideas.

Whatever Mayer might have read of the textbook and compendium literature of the day would have confronted him with the commonplace that scientific explanations consist in the linking of causes with their effects, expressed in terms of the laws of the (fundamental) forces of nature. That, of course, was precisely the nexus of his own theorizing. Common, too, was the idea that we can only know things (like forces) through their effects, that we can discover the laws of the phenomena even though we must remain ignorant of the essence of things. Such a line of defense was welcome to Mayer when he had to confess that he could not explain *how* motion is transformed into heat: it was enough, he could argue, that he had identified and measured their relationship. On these centrally important counts Mayer must have been convinced that his theory of force well answered the general characterization of a good scientific theory.

Reinforcing methodological and ontological considerations worked together from the late 1830s on to delegitimate any talk of spirits, souls, or minds in positive science. Not surprisingly, such an attitude had profound implications for an entity such as Mayer's force, nor were matters helped by the use of the same word to designate several very different concepts. For reform-minded physicians like Mayer's friends Griesinger, Wunderlich, and Roser, "force" meant such unacceptable things as the vital force or the healing power of nature (*Naturheilkraft*, *vis medicatrix naturae*). Griesinger's inability to understand, and refusal to accept, Mayer's concept of force as an autonomous ontological entity was of a piece with his and others' rejection of what they termed "ontology" in medicine, and signaled Mayer's predicament. Griesinger and others were undisposed to accept what they regarded as illegitimate abstractions from the phenomena, whether disease entities or forces-as-causes, let alone force-as-substance. Mayer was thus left little conceptual room by the mid-to-late 1840s, when the only entity many scientists would countenance was matter endowed with various attractive and repulsive forces, but not force in Mayer's sense of the word.

That Mayer was personally intensely interested in religious issues, and surrounded himself as a student with people themselves so interested, suggests that he may well have had an ear for contemporary discussions of religious and spiritualistic matters, especially since many of those discussions took place in a quasi-medical context, and took place furthermore in his home territory largely during the years he was a student at Tübingen. A broad natural-theological strain in physiology would seemingly also have appealed to him, as it might have led him to pay attention to some of the more prominent works of that genre, such as those of Brougham and Prout. Such discussions formed (as it were) the extreme limit of a broad common context involving physiology, medicine, metaphysics, and theology. In other words, it was a context within which Mayer as a metaphysically concerned physician would naturally have functioned, even as ongoing changes in the self-conception of science were in the process of transforming it virtually out of existence. Mayer's ultimately unresolvable dilemma was how to reconcile his personal concern for what his scientific contemporaries were increasingly putting down as philosophical and metaphysical questions with his desire to be a good scientist, when the bounds of good science were being drawn increasingly tighter.

The further significance of that context is that it involved the consideration of a number of specific questions of particular relevance to the issues Mayer wrestled with. Discussions of the existence of mind and soul touched upon some of the same issues of causality and ontology as attended Mayer's reflections on force and matter: Are mind and soul legitimate entities? Are they autonomous, and can they exist independently of matter? Are they ever created or destroyed? During the 1830s one finds a more explicit discussion of the conservation of matter in works of natural theology than of chemistry, and it is in the former, not the latter, that one is likely to encounter Latin phrases such as *ex nihilo nihil fit*. The more explicitly spiritualist literature of Kerner and Passavant, including that of its critics like Strauss, dealt centrally with the nature of mind/soul/spirit and its relationship to the body; with the reality of a nonmaterial world; and with the manifestation of force. It also discussed the nature of scientific explanation with some sophistication, arguing that empirical science must be prepared to accept the validity of well-attested phenomena even if it cannot explain them. It is my suspicion that familiarity with some of this literature contributed to the process by which Mayer defined both the content and the scope of his science.

## • PART III •

# Mayer's Work in Context

• C H A P T E R S I X •

# A Contextual Reconstruction of the Development of Mayer's Ideas

IN THIS CHAPTER I will attempt a reconstruction of the development of Mayer's ideas based on the analysis of the leading ideas and peculiarities of his work given in the second chapter, on the relevant context established in the subsequent three chapters, and on such insights as can be gleaned from Mayer's retrospective accounts. In tracing the progress of his thinking in the years after his return to Heilbronn in February 1841 one can, in addition, draw upon a wealth of contemporaneous documentation both for Mayer and for those with whom he interacted, such as Griesinger and Jolly (in private) and Liebig (largely through published works). If in many important regards the focus and scope of Mayer's ideas changed markedly over the years, in another his identification of his 'discovery' with the mechanical equivalence (later, *equivalent*) of heat is a red thread running from beginning to end. Even his all-important concept of force tended, in later years, to be surprisingly underplayed. That his present fame is as codiscoverer of the conservation of energy is due almost as much to the influence of the evolving scientific context as it is to anything Mayer himself actually said. In any event it is essential to appreciate that Mayer's acceptance of the indestructibility and uncreatability of force was the result of a difficult process of rethinking extending over at least two, and possibly four, years. He *did* have an important insight in response to his bloodletting experience in June 1840, but it was not the conservation of energy (*or* force), and even the concept of force itself took some time to crystallize out of his ill-formed and wide-ranging mix of ideas.[1]

As I see it in broad outline, Mayer went from his surprise at the bright color of the venous blood he let to an acceptance of Lavoisier's oxygen theory of respiration and animal heat. A logical deduction, one that required him to consider the organism as (in Breger's phrase) a working machine, then convinced him of the necessity of a quantitative equivalence between heat and "motion." That much is fairly certain. Although the existing evidence does not permit too fine-grained a reconstruction of the next steps, soon thereafter he must have begun to elaborate both a unifying concept of force and a general ontology of entities that included ponderable matter, the erstwhile imponderables, and the soul. An important aspect of the meaning of force for him was that if forces really exist and are independent of matter, then the ground has been cut from under an atheistic and despiritualizing materialism. Only some such development can explain why he was swept away by the universal and "metaphysical" significance of his ideas. It is also significant that Mayer's rejection of both imponderable fluid and mode-of-motion theories of heat created a need for a new conception of force. It seems that it was in conjunction with the guiding analogy between force and matter, within the context of widespread discusssion in the physiological and theologically flavored literature about the creation and destruction of the soul and the vital force, that Mayer began to ponder the creatability and destructibility of force.

At some early stage Mayer decided that the clarification of his developing concept of force required assistance from physics, especially since he accepted the conventional ranking of the sciences whereby physiology is beholden to the more fundamental physics for the certification of its most general concepts. From physics he learned to identify forces as causes, which probably encouraged him to try to ground his conclusions in the laws of thought: *causa aequat effectum*, etc. He would also have encountered the standard conceptual division of the physical world into matter and force, though he was to give that duality a radically new interpretation. Unfortunately for him, he also found in the physics texts he consulted tremendous confusion and inconsistency with respect to the measure and application of the term "force." Perhaps his most fateful entanglement involved the foredoomed attempt to apply the parallelogram of forces and *its* application to central-force motion to *his* inchoate concept of "force" in the sense of energy. Supplied on all sides with a powerful analogy between the solar system and organisms, Mayer found himself obliged to contend with the possibility of the ongoing creation of force in the solar system and thus also with the possibility that living things might also be effective creators of force, notwithstanding that an earlier stage in his thinking had crucially entailed his likening organisms to machines. Not that Mayer's having accepted the possibility of the creation of force in the solar system should be seen only as an obstacle on his path to the discovery of the conservation of energy: that belief *was* a source of great confusion to him, but for a while what excited him most was that he believed he had solved the age-old problem of the source of the sun's heat. Moving toward the conservation of force meant abandoning, for a while, the solution to *that* important problem. Mayer was thus slow to apply his evolving theory to living things or to assert the general *uncreatability* of force. That it was only in the course of the early 1840s that the vital force became an issue of significant contention meant that Mayer was not initially strongly prompted from that direction to deny the creation of force in living organisms.

Such, roughly, were the considerations that informed Mayer's early thinking and found expression in his earliest surviving essay, the justly unpublished "On the Quantitative and Qualitative Determination of Forces." How his thinking evolved after that—both as a result of its own internal dynamic and in response to his interaction with, for example, Jolly, Liebig, and friends Baur and Griesinger—I will not attempt to recapitulate here; that is the subject of section 2 of this chapter. It is, however, useful to recall that even as Mayer was publishing his calculation of the mechanical equivalence of heat, the conventional *sine qua non* of all pretensions to the discovery of the conservation of energy, he still believed only that force cannot be destroyed, not that it cannot be created. It was not until 1845 that Mayer came publicly to a position reasonably close to what one expects of a putative discoverer of the conservation principle.

## 1 Through the Publication of his 1842 Paper

Let us return to Mayer on board the *Java* off the coast of Batavia in June 1840. As noted in chapter 1, Mayer's several sketchy accounts of his voyage imply that his prebloodletting reflections had centered around problems of the physiology and pathology of the blood. And as I argued in chapter 3, there is strong reason to believe that Mayer's understanding of the carbon-purging function of respiration led him to expect to find the venous blood of people in warm climates to be *darker* than normal. In quoting a passage in which Mayer recounted his surprise at observing the venous blood to be brighter than expected, I isolated it from its context which—at first blush confusing and not obviously relevant—on closer examination provides an important insight into the very earliest stages of Mayer's thought processes:

> Observations that I made in the tropics taught me to recognize the role blood corpuscles play in the combustion process in the body. During a 100-day sea voyage there had occurred no appreciable incidence of disease among the 28-man crew; however, a few days after our arrival at the Batavian roads there spread in epidemic fashion an acute (catarrhal-inflammatory) affection of the lungs. In the copious bloodlettings I performed, *the blood let from the vein in the arm had an uncommon redness, so that from the color I could believe I had struck an artery*. At the same time the blood was very rich in fibrin,** the crassamentum stuck fast to the sides of the basin, and after 12–16 hours usually only a few spoonfuls of water-clear serum had separated out; never, however, did there appear a buffy coat. After three weeks, during which we had sailed to Surabaya, the chest ailments disappeared; but dysentery and acute liver diseases soon appeared, which, along with the contagion received from beautiful black women, accompanied our ship back as far as the Cape of Good Hope. Since the military doctors at the Simpang hospital on Surabaya had characterized venesections on acclimatized Europeans to me as risky, I therefore restricted myself almost entirely to local bleeding. In one of the copious bloodlettings I performed two months after our arrival in Java on a robust seaman afflicted by an inflammation of the liver, I found a normal black color of the blood.
>
> . . .
>
> **One sailor, Bornet, formed an exception. Spared the chest ailments, he was afflicted at the same time by a severe syphilitic iritis, against which, among other things, two venesections were applied. The blood was rich in dark black cruor, on the other hand poor in fibrin.[2]

The blood's color was thus only one of its properties to which he paid initial attention. Its richness in fibrin, which he would have judged from the blood's quickness to coagulate, would have agreed with the arterial character of the blood, since arterial blood normally coagulates faster than venous blood.[3] (The behavior of Bornet's blood was counteranomalous with respect to both color and coagulability.) Contemporary authors disagreed as to whether or not the blood of those suffering from inflammation contains more fibrin than usual, and even as to whether it coagulates faster or slower than normal blood: Schönlein and Schill taught that such blood contains more fibrin and coagulates faster, while Berzelius denied any quantitative increase in fibrin and John Davy maintained that a sign of inflammation was "an unusual slowness in coagulating."[4] All agreed, however, that the so-called buffy coat (*crusta phlogistica, Entzündungshaut, Speckhaut*) is usually a sign of inflammation, hence its absence would also have struck Mayer as anomalous.[5] Apparently anomalous, too, was the failure of the crassamentum (*Blutkuchen*) to contract—that is, to pull away from the sides of the bowl—in cases of inflammation.[6]

What Mayer faced, then, was a confusing and anomaly-filled set of observations on the state of the blood, where the indices of color, rate of coagulation, behavior of the crassamentum, and presence or absence of a buffy coat had to be related to the independent variables of climate (including both temperature and humidity), state of disease, length of time in the tropics, and the peculiarities of different individuals. That Mayer the physician paid attention to such factors is clear; how far he pursued them is impossssible to say. What *is* clear is that in the end only the color of the blood and the ambient temperature crystallized out as *physiologically* significant *for him*.

Mayer connected these two variables via Lavoisier's theory of respiration and animal heat, and the evidence suggests that it was only then that Mayer came to regard the physiology of the blood and respiration as related in the first instance to the production of animal heat, as opposed to the removal of excess carbon. As he put it in a late recounting of his bloodletting observation, where else should that diminished difference in color derive "than from the fact that, with the much diminished need for the organic production of heat, the arterial blood is substantially less deoxidized than in colder surroundings. The physiological doctrine that animal heat results solely from a combustion process thus obtains manifest confirmation from the phenomena cited."[7] Or more fully in his book of 1851:

> In the summer of 1840, in the course of the bloodlettings I undertook in Java on recently arrived Europeans, I made the observation that the blood taken from the vein in the arm showed almost without exception a startlingly bright red coloration.
>
> This phenomenon captured my complete attention. Taking as my starting point Lavoisier's theory, according to which animal heat is the result of a combustion process, I regarded the double change in color that the blood undergoes in the capillaries of the lesser and greater circulations as a sensibly perceptible sign, as the visible reflection of an oxidation going on with the blood. For the maintenance of a uniform temperature of the human body the heat *produced* in it must necessarily stand in a quantitative relationship to its heat *loss*, hence also to the temperature of the surrounding medium, and therefore both the heat production and the oxidation process, as well as the *difference in color of both kinds of blood*, must on the whole be less in the torrid zone than in colder regions.[8]

This account implies that at the time of his observations he had already accepted Lavoisier's theory, as may indeed have been the case. It may be significant that the physics text of the philosopher Fries, a work I suspect Mayer had with him, gave more unequivocal support to the combustion theory of animal heat than did most physiologists, albeit in only a single sentence.[9] On the other hand, the account from 1845 quoted above states that it was the observations he made that *taught* him the role of the blood in the oxidation process in the body. Moreover, as already shown, very few of Mayer's contemporaries subscribed to Lavoisier's theory without appreciable reservation—recall the widely accepted influence of the nervous system in the production of heat—and in the only relevant surviving pre-1840 expression of his views on the physiology of the blood he followed Hasper and Müller in regarding the function of respiration as the removal of waste carbon from the blood and not (also) the production of heat. I suspect, rather, that Mayer's unqualified acceptance of Lavoisier's theory was in effect another crystallization of meaning, one which took place in close conjunction with his search for the meaning of the phenomena he had observed. From having been one among several theories of animal heat, one beset by any number of serious experimental difficulties and alternative hypotheses, it now appeared to him as *the* correct theory, in terms of which his observations made sense. It should be noted that, as far as one can tell, Mayer never paid any attention to the anomalies the theory could not easily dispose of, such as the well-known shortfall in the results of Dulong's and Despretz's experiments: Mayer's only interest in the theory was as a stepping-stone to *his* conclusions.

An essential aspect of Lavoisier's theory, which Mayer seems to have appreciated, was that it presupposed that the oxidation of a given amount of fuel always produces the same amount of heat, no matter what specific steps the process follows inside or outside the body. He seems also to have assimilated one of the most important general ways of thinking characteristic of Lavoisier's chemistry, namely to view phenomena and reactions in terms of a balance between input and output, between consumption and product. As he wrote in 1851 in connection with Lavoisier's theory of respiration and animal heat:

> Now, if it follows from this that in general a balance is to be struck in the organism between intake and output, or between work performed [*Leistung*] and consumption, then it is unmistakably a principal task for the physiologist to become acquainted as closely as ever possible with the budget [*Budget*] of his object of investigation. Consumption consists of the combusted material, the work performed is the heat production. But the latter takes place in *two different* ways, in that the animal body in part produces heat directly in its insides and in turn throws it off through communication to its immediate surroundings, but in part also possesses, by means of its motor apparatus, the capability of generating heat mechanically through friction, etc., even at distant places. It is, now, necessary to know:
>
> *Whether the directly produced heat alone, or whether the sum of the directly and indirectly produced quantities of heat is to be charged to the account of the combustion process?*[10]

It is likely, too, that Mayer's thinking in these regards was reinforced by a general acceptance of the central role of *Stoffwechsel* in the animal economy, as Tiedemann, Müller, Valentin, and others had been arguing during the 1830s. Physiologists were increasingly taking for granted the idea that all vital processes are due to underlying processes of material exchange, and were becoming increasingly, if only implicitly, sensitive to energetic and causal considerations. Such a position was also basic to the physicochemical reductionism of Mayer's university friends, Griesinger, Wunderlich, and Roser. As Mayer wrote to Griesinger, "A logical instinct led physiologists a short while ago to the axiomatic principle 'no action without exchange of matter.'"[11] In the same vein he later wrote that "as a general law of nature, to which there is no exception, stands the principle that for the production of heat a certain *expenditure* [*Aufwand*] is necessary."[12] That expenditure, he added, can always be reduced to one of two chief categories, either a chemical substance or mechanical work. Significantly, for Mayer these principles had the force of logically necessary truths. Although such testimonies from his writings come appreciably after the period in question—there *is* no surviving contemporaneously recorded information of any kind regarding his thought processes—there seems little reason to doubt that he appreciated their implications at an earlier time, especially since some such assumptions appear necessary in order for him to have been able to draw the conclusions he did.

Armed with these insights and general views, Mayer was poised to make his portentous discovery of the equivalence between heat and motion. When Griesinger asked him how he had gotten involved with the whole business, he replied that "the simple answer is that, occupying myself during my voyage almost exclusively with the study of physiology, I found the new theory [*Lehre*] for the sufficient reason that I vividly recognized the need for it; following the light obtained, there unfolded more and more a new world of truths."[13] There is good reason to take Mayer at his word here—that is, that the next crucial step in the development of his new theory *was* the result of a chain of logical deductions. In a passage continuous with the one quoted above from his autobiographical sketch of 1872 he wrote:

> Connected with this quite simply and necessarily is, however, yet another question of great importance in principle. That is, the organism with a temperature constantly higher than its surroundings continuously generates in general not only a determinate quantity of directly perceptible heat, it also produces *mechanical work* [*mechanische Leistungen*], and the latter in turn, as everyone knows, also generates heat. Is, now, this indirectly produced heat also a product of an organic combustion process, or does it derive from another source? Already as a child Mayer had, from a failed attempt at constructing a perpetuum mobile, arrived at the insight that mechanical work [*mechanische Arbeit*] cannot be generated out of nothing. But if one once assumes this as a fundamental truth, then it necessarily follows that the heat generated by the organism indirectly through friction must also be charged to the account of the vital combustion. But one cannot stop there, for one thereby obtains at the same time the insight that there must exist in general an invariable quantitative relationship between consumption of work and generation of heat, the numerical determination of which is a physical task of fundamental importance.[14]

I believe that such a sequence of conclusions must have been close to Mayer's thinking in 1840.

Noteworthy in Mayer's account is the recollection of his childhood experiences with machines and of the lasting impression that one cannot generate motion out of nothing. Mayer's observations and reflections had thus brought together into interactive proximity a reinforcing set of ideas: that processes of material exchange underlie all organic functions; that the production of something requires the expenditure of something else; that production is proportional to consumption; and that machines cannot create force out of nothing. And he had no compunction against applying those considerations to the human organism, against considering the body as a machine in which fuel and oxygen are consumed and heat and motion are produced. Whether Mayer at the time specifically invoked to himself an analogy between the organism and a steam engine is impossible to say. The earliest mention of that analogy in his surviving writings occurred late in 1842 in his first letter to Griesinger. After recapitulating his theory to his friend he added:

> The heat thereby arising (namely through the leveling of chemical differences), or, more generally, the force that must thereby make its appearance, manifests itself partly as free heat, partly as animal motion. If there were nothing else than motion and heat to consider, then we could call a steam engine a warm-blooded animal, too; in it, too, the chemical difference that exists between its food and the oxygen of the atmosphere is transformed partly into heat, partly into motion; both taken together naturally again give the measure of the first. To be sure, if one prefers to explain heat and motion in the animal organism by means of *vital aether*, *nervous spirits*, or *muscle force*, then everything comes to a halt and things go as everyone well knows how.[15]

But nothing in Mayer's probable course of thought depended on his using that analogy: for his purposes the water-powered mills of his youth were sufficient. Nonetheless the analogy with steam engines was a particularly good one for driving home the cogency of his general way of thinking.

Herbert Breger has argued at some length that "the conception of the human being as a working machine was the fundamental thought of his [Mayer's] investigations."[16] For Mayer and others of his generation, "mechanics is the basis of medicine, for the human organism, like all of nature, is a machine."[17] In Breger's view the concept of nature as a machine that does work was crucial to the formulation of the principle of the conservation of energy in the nineteenth century: despite the prevalence of (mostly water-powered) mills in the seventeenth and eighteenth centuries, only the widespread introduction of steam engines in the nineteenth was capable of exerting a *weltbildprägende Kraft* on people's imaginations.[18] Although Breger was, I think, correct in identifying a major change in people's way of thinking about the workings of nature, and in wishing (somehow) to associate Mayer's work with it, he went too far in claiming Mayer as a wholehearted representative of the new worldview. Mayer never said that mechanics provides a foundation for medicine, and he was quite willing to allow the unmachinelike solar system to create force out of nothing. It is significant that in one of the key passages Breger cited—"If there were nothing else than motion and heat to consider, then we could call a steam engine a warm-blooded animal, too"—Mayer's qualification can be read as implying that the analogy breaks down as soon as one considers *more* than heat and motion.[19] Breger mistakenly interpreted Mayer's heuristic use of a powerful *analogy* as belief in an underlying *identity*. Mayer's unwillingness to take that step was one of the things that separated him from many of his contemporaries. As shown in chapter 3, the *metaphor* of the organism as a machine was a commonplace among the likes of Autenrieth and Müller who would have rejected the complete *reduction* of living things to machines. That metaphorical way of thinking was thus also available to Mayer without his having had to embrace the kind of materialistic reductionism that was anathema to him.

Mayer's last full exposition of his ideas, in the *Remarks on the Mechanical Equivalent of Heat* of 1851, was structured in a way that closely followed what I believe was his route of discovery. For him the logic of his discovery provided a compelling logic of justification:

> The physiological combustion theory takes as its starting point the fundamental principle that the amount of heat that arises from the combustion of a given substance is an *invariable* quantity—i.e., one *independent* of the circumstances accompanying the combustion—from which it is more specifically concluded that the chemical effect of the combustible materials undergoes no quantitative change even as a result of the vital process, or that the living organism, with all its mysteries and marvels, is not capable of generating heat out of nothing.
>
> If one holds fast to this physiological axiom, however, then the answer to the question posed is thereby already given. For unless one wishes to turn around and grant the organism the capability of creating heat that has just been denied it, then it cannot be assumed that the sum total of the heat produced by it could ever turn out to be *greater* than the chemical effect taking place. The combustion theory, unless it wishes to renounce itself from the beginning, has therefore no alternative but to assume that the *total* heat developed by the organism, partly directly and partly mechanically, agrees quantitatively with (or is equal to) the combustion effect.
>
> From that, however, there now follows with the same necessity *that the mechanical heat generated by the living body must stand in an invariable quantitative relationship with the work thereby expended*. For if, according to the different construction of the mechanical devices serving to produce heat (etc.), by means of the same work and with a constant organic combustion process, *different* amounts of heat could be attained, then indeed the heat produced for one and the same consumption of material would turn out to be sometimes smaller, sometimes greater, which goes against the assumption. But since, furthermore, no qualitative difference exists between the mechanical work performed by the animal body and other, inorganic forms of work,
>
> *an invariable quantitative relationship between heat and work is consequently a postulate of the physiological combustion theory*.
>
> Insofar as I generally maintained the direction indicated, I had therefore ultimately

to direct my main attention necessarily to the physical connection existing between motion and heat, where the existence of the mechanical equivalent of heat could then not remain hidden from me. But even if I do owe this discovery to a chance occurrence, it is nevertheless still my property, and I do not hesitate to claim the right of the one who gets there first.[20]

Thus did Mayer arrive at his pivotal discovery of the equivalence between motion and heat, of the existence of a fixed quantitative proportionality between them, of (as he quickly came to see it) the transformation of motion into heat and heat into motion. *That* was what he emphasized in all his later autobiographical sketches. The insight (*Einsicht*) that he came to on his voyage to the East Indies was, he said, "that heat and motion or mechanical work must be converted into each other according to an invariable ratio."[21] What he brought home with him were, he wrote, "the for the most part still confused ideas . . . *on the transformation of motion into heat and of heat into motion*."[22] That insight, variously expressed, was the core of his soon-to-be glimpsed general theory of force. It really did derive, as Mayer always insisted, from his observation of the unexpectedly bright color of the venous blood. What is crucial to realize is that Mayer had not yet thereby discovered the 'conservation of energy,' nor did he yet (it seems) even have a general conception of force, as opposed to specific ideas relating to heat and motion, let alone of force as an entity two of whose properties are uncreatability and indestructibility.

It was in shifting his speculative venue that Mayer achieved his next conceptual breakthroughs. All that he later reported was that, having begun his ruminations over problems of physiology and pathology, he sought clarification of his developing concepts from the more fundamental fields of chemistry and physics. That move opened up to him an expanding world of insights and connections. As he wrote in his first letter to Baur:

> Taking as my starting point physiological and pathological investigations—in that it seemed to me that I had here arrived at correct principles—I was, in pursuing these principles consequentially backwards, necessarily led over into the domain of chemistry and physics, whereby a view of nature [*Naturanschauung*] took shape which completely clarified for me an immense and truly infinite series of heretofore inexplicable phenomena, and which, outside of the domain of the natural sciences and of specialized medicine, also solved for me the most important questions of metaphysics. You'll cry out, laughing: a lot at once! but I confess to you, too, that these investigations formed the exclusive subject matter of the most strenuous activity during my entire journey, and now also since my stay here, and that this alone and only this compensates me tenfold for all the hardships, etc., of so long a route.[23]

He reported a similarly motivated progression three years later in a letter to Griesinger:

> In no way have I hatched out the theory at my desk. After having occupied myself zealously and unremittingly with the physiology of the blood on my journey to the East Indies, the observation of the altered somatic conditions of our crew in the

> tropics, the process of acclimatization, again gave me manifold material for contemplation; the forms of disease and especially also the constitution of the blood steered my thoughts in the first instance unremittingly to the generation of animal heat by means of the process of respiration. Now, if one wishes to become clear about physiological issues, then knowledge of physical processes is indispensable, unless one prefers to treat the matter from a metaphysical standpoint, which disgusts me immeasurably; I therefore kept to physics and gave myself over to the subject with such predilection that I . . . inquired little into the distant part of the world, but preferred to stay on board where I could work uninterruptedly, and where for many an hour I felt as it were inspired, such that I can never recall anything similar before or after. A few sudden insights [*Gedankenblitze*] that shot through me—it was on the Surabayan roads—were immediately diligently pursued and led in turn to new subjects.[24]

Alas he was never more specific. Note that Mayer was off Surabaya during most of July and August 1840—that is, roughly two to ten weeks after his initial bloodletting experiences. His insights did not all come to him in a single flash, but followed over a period of some weeks or months as he pursued his deductive and associative chain of thoughts. It appears, too, that his attention had already shifted to physics before the *Java* had left the Indies. In this progression one finds further evidence of the persistent consequentiality that characterized Mayer's thinking in general. In his view it was logically necessary to move from physiology to physics if one wished to get to the bottom of what heat, motion, and force are.

Unfortunately, at this point in the development of Mayer's ideas the evidence does not permit one to define a clear path. Two things, however, seem to have taken place in close connection with each other: he elaborated a general conception of force as an ontologically primary entity on a par with matter, and he pursued the implications of regarding forces as causes. Since Mayer seems from an early period to have been caught up by the universal and "metaphysical" significance of his unfolding ideas, my hunch is that it was the former lead that he followed first, especially since the ideas it drew upon would have been available to him from things he presumably already knew, whereas the latter probably depended more on the subsequent following up of his ideas in the physics literature.

As shown in chapter 3, there was extensive discussion in the physiological literature of the nature and interrelationship of ponderable matter, the imponderables, vital force, and the soul. Mayer would have encountered this multifaceted topic in the writings of Autenrieth, Müller, and Elsässer, and it is hard to imagine he was not abundantly aware of such fundamental ontological issues, even if there is no direct evidence of that knowledge. In particular, he surely had encountered Autenrieth's and Müller's discussion of the possible independence of force from matter, as opposed to the general view of force as a *property* of matter. The concepts and issues that characterized those discussions would have provided an unusually rich context within which Mayer could have identified a rough new concept of force and then specified its nature and significance. Which properties might forces share with ponderable matter? With the imponderables? With the soul? Are forces like ponderable matter and hence both indestructible and uncreatable?

Are they like the imponderables and hence (à la Autenrieth) in a sense creatable? Are they like the soul and hence (à la Brougham) creatable but indestructible? Which are the valid analogies? It is precisely such a context that would have vastly facilitated Mayer's initial conceptualization of "force" and have immediately given it a worldview-transforming significance. Forces, he may have concluded, are nonmaterial weightless "substances," like matter in being quantitatively indestructible and qualitatively transmutable. His conception of the "substantiality" of force may thus have encouraged him to think in terms of its conservation. But forces share with souls the trait of immateriality; hence if forces really exist so, too, might souls. In any event, the existence of forces alone would be enough to refute the materialistic contention that the only thing that exists in the world is bare matter.

However much the foregoing scenario must remain hypothetical, even if eminently plausible, it *is* rather more certain that at an early stage Mayer latched on to a guiding analogy between force and matter as the two basic components of the physical world. Force was *defined* in terms of its possession or not of the basic properties of matter: materiality, ponderability, uncreatability, and transmutability. And physics he now defined as the science of force just as chemistry is the science of matter. As he wrote to Baur in 1841:

> It is the same with the theory of forces (physics) as with the theory of material substances (chemistry); both must be based on the same principles. . . . My first endeavor is now to secure the axis about which rotates the theory of material substances for the theory of forces as well; *from thence is dated the axiom of the invariable quantity of forces*. Physical laws become very simple by virtue of the fact that their objects, the different forces, can be reduced to one another—something one longs for in vain in chemistry; *how overjoyed I was when I gradually found out this result, the isomerism of forces!*[25]

Mayer's words strongly imply that he transferred to forces the principle of the quantitative invariability of matter—although, as we have seen, he was in fact long in doubt as to the uncreatability of force—just as he found a counterpart for chemical isomerism (the fact that radically different substances can have the same elemental components) in the more general interconvertibility of physical forces: the several phenomenally distinct *forces* can be interpreted as merely different manifestations of the same elementary (and unitary) *force*.[26] His concept of force was beginning to take clearer shape.

At some point during Mayer's voyage he turned for help to physics texts. It is certain that he had at least one with him—very possibly Fries's *Textbook of Physics* (1826), as I will argue below. It seems quite possible that he might also have brought along the fifth edition of Baumgartner's popular text: he later cited from that edition, published in 1836, and since the sixth edition came out in 1839, one would think he would have obtained the more recent edition if in fact he only acquired it after his return. Of course he still might have owned it and not taken it with him, but since the only physics course he took at the university was in 1832—theoretical and experimental physics with the *Privatdozent* Ludwig Felix

Ofterdinger—it looks like he might have purchased the book before leaving Germany in June 1839 expressly to have it with him on his voyage.[27] In any event, the only direct evidence relating to this next stage in the development of his thinking came in one of his late autobiographical sketches. He was describing the situation in the spring of 1841 shortly after his return to Heilbronn:

> At that time it was still two principal errors that disturbed the course of my thoughts and would not let me come to a clear view [*Anschauung*] of things. That is, in the physics compendia of the time, alongside the doctrine of the parallelogram of forces, *mc* was often given as the measure of motion, and this, combined with a remnant of concepts of a centripetal and a centrifugal force deriving from the Kantian school of *Naturphilosophie*, led me into a labyrinth of hypotheses and contradictions. Thus, for example, I had already brought back with me from the voyage the whole mixed-up notion that the continuous solar radiation is to be regarded as the centrifugal activity of the central body corresponding to the centripetal action of the planetary and cometary motions.[28]

That implies that Mayer must have had access to some physics text(s) while still on board ship. As already noted in chapter 4, essentially all the available texts gave *mc* as the measure of force (or quantity of motion, or whatever) and applied the parallelogram of forces not only to motions but also to the combination of centripetal and tangential forces in the case of central-force motions. Mayer struggled mightily to make coherent sense out of these components and his initial insight of the equivalence of heat and motion.

Seeing force as (in our terms) momentum meant that he was confronted with the example not only of the conservation of momentum, but more importantly of the mutual annihilation of equal and opposite momenta in inelastic collisions. Such was probably the route via which Mayer embraced the concept of neutralization, which he was able to exploit by saying that heat arises when opposing motions neutralize or annihilate each other. More problematic were cases where motion (or force) seems to be produced when one decomposes the hypotenuse of a triangle into sides whose total length is greater than the original magnitude. Nor did either of those situations seem to jibe with the texts' general assertion of the constancy of the quantity of motion. Such problems occupied a perplexed and frustrated Mayer for several years. No one could have made coherent sense out of contemporary textbooks' treatment of force and motion, and Mayer was not well equipped to do anything other than take what he found at face value.

Mayer's entanglement with central-force motions had serious consequences for him. In applying the principle of the parallelogram of forces to the diagrams of central-force motion typically given in German physics texts of the day, it seemed to Mayer that a certain fraction of the two original forces—the centripetal force and the "tangential force" (i.e., the tangential component of velocity, sometimes also called the centrifugal force)—is being constantly annihilated. Yet the planets keep going around the sun with eternally undiminished motion. Mayer solved this apparent anomaly by supposing that the divinely organized solar system, unlike humanly constructed mechanical systems, is capable of continuously generating force out of nothing, of transforming 0 into + and –. The sun's heat and light are the outwardly directed counterparts of the continuously annihilated centripetal motion of the planets and comets. I suspect that Mayer's ability to entertain such a conclusion was facilitated by the widespread availability of metaphors likening the solar system to living organisms. Even though there is no evidence that at this stage Mayer had reflected deeply one way or the other on the issue of the vital force, that he, too, specifically used the word organism with reference to the solar system strongly suggests that he regarded the ability to produce force out of nothing to belong in some way more unproblematically to organic than to inorganic systems—thus the restriction of Mayer's first papers to the forces of inanimate nature. If, as seems likely, the analogy between force and matter had already raised the issue of the possible uncreatability of force (as of matter), the conclusions he drew from his analysis of gravitationally controlled central-force motions led him for a time not only to limit his theory to inanimate nature, but also to insist only on the indestructibility of force. To be sure, giving up its uncreatability gave him an explanation for the origin of the sun's heat and light—hardly a mean accomplishment. As he wrote to Baur:

> The absolutely most magnificent, important, and splendid phenomenon in all of nature is the constant development of light in our solar body. . . . I assure you, dear friend, that since I began to occupy myself with matters having to do with nature, it was my most ardent desire that science might be in a position to solve this question, and hence the solution of the problem is even now of inestimable value to me. The solution is, in a few words, as follows: the sun's emission of light is the central motion in the planetary system corresponding to the peripheral motons.[29]

Baur was not wholly impressed.

While Mayer was pursuing his thoughts on heat, motion, and force, he had at his disposal an appreciable knowledge of chemistry from which he could draw appropriate analogies as he hammered out the full significance of his developing concepts. The indestructibility and uncreatability of matter, although not insisted upon explicitly in either the chemical or physical literature of the period, was perhaps *the* guiding analogy behind the elaboration of his ideas on force. Of great importance, too, was the ready availability of chemical analogs to the neutralization of forces and motions in physics: the neutralization not only of acids and bases to form chemically indifferent substances, but also (a favorite example) the qualitative "neutralization" of hydrogen and oxygen to form "indifferent" water. As chemistry is the science of matter, so is physics to be the science of force. In 1841 he spoke of his discovery of the "isomerism of forces," and in later years he even identified his theory as "physical stoichiometry," underscoring its parallels with chemistry.[30]

Although virtually all German physics texts of the day contained similar treatments of central-force motion in terms of the vectorial composition of centripetal and centrifugal (or tangential) forces, it may be significant that Mayer asssociated those concepts with the "Kantian" school of natural philosophy. That is, *he* may have derived his knowledge of them from a work in that tradition, in which case there is, I think, no more likely candidate than Fries's *Textbook of Physics*. Muncke called it one of the most complete and thorough (*gründlich*) of recent texts, harder but richer than those of Baumgartner and Fischer, and an anonymous reviewer praised it in 1828 as the best available.[31] It would thus not be surprising for Mayer to have used the work in connection with the physics course he took in 1832.

A number of aspects of Fries's work were not common among German physics texts of the period but were very close to motifs characteristic of Mayer's thinking.[32] Even in his handling of central-force motions, Fries's diagram of the production of a planet's orbital motion via the neutralization of centripetal and centrifugal forces was more suggestive than others of Mayer's confused understanding of the subject. Fries made a fundamental distinction between mechanical and organic (or dynamical) interactions, according to which the circular motion of a body attached to a fixed point by means of a string is "mechanical," while that of a planet around the sun is "organic" (or "dynamical")—and "organic" interactions like the latter, which are controlled by a gravitational accelerative force, are accompanied by an increase in the quantity of motion. He elaborated a quasi-vibrational theory of light, radiant heat, and heat in terms of their different degrees of intensity, and spoke of heat as neutralized motion. And he applied a quantitative-qualitative distinction to the properties of matter: matter is quantitatively divisible into infinitely small particles and qualitatively divisible into distinct chemical components. Fries even made a distinction between "dynamics" and "stoichiology" that assigned the study of forces and motion to physics and of matter to chemistry:

> By dynamics (δύναμις)—which in the narrower sense is also called *physics*, also the *general theory of nature*—I understand in particular the theory of the laws governing the forces in corporeal interactions in general (i.e., the theory of the different forms of *physical processes*); it is in essence the science of the laws of motion. Alongside *stoichiology* (στοιχεῖον, elementum, matter), the *theory of matter*, there belongs the theory of the heterogeneity of material substances; it is the proper task of *chemistry*.[33]

Most of the apparent idiosyncrasies of Mayer's work dissolve the more one knows about the kinds of scientific sources he likely drew from. Although he was professionally an outsider to science, his work had its roots in and drew most of its concepts and vocabulary from contemporary scientific sources of unimpeachable credentials.

The two major divisions (*Theile*) of Fries's book dealt with gravitation, contact forces, crystallization, capillary action, and the chemical process on the one hand, and sound, light, heat, magnetism, and electricity on the other. In the last section of the book Fries reflected on an important difference between the topics so dealt with:

> After we have now run through the circle of phenomena that we found defined as the task of experimental physics, and in conclusion take an overview with an eye toward a unification of the whole under the highest grounds of explanation, a striking

contrast is evident between the theories of our first and second parts. In the first theories, everything came together under the mathematically precisely defined law of gravity, and each circle of phenomena found its sufficient explanation. In the second part, a great wealth of observations remains in separate groups standing alongside each other, without our having known how further to unify them in an explanatory fashion. Nevertheless, even here all of the principal classes of phenomena cluster together in such a way that they demand a common highest ground of explanation, which we have, however, not yet found.[34]

After cataloging a number of the striking connections between light, heat, electricity, magnetism, and chemical substances (with their neutralizing forces), Fries noted that although summary laws have been found for some of these phenomena, none has been explained in terms of the fundamental concepts of dynamics—and recently discovered laws for double refraction, the polarization and interference of light, and electromagnetism argue against the adequacy of a dynamical construction in terms of attractive and repulsive forces acting along the straight line connecting the interacting bodies. His concluding paragraph looked to the future:

In none of these aids to [our] overview or to calculation can the mathematical philosophy of nature recognize a true explanation of the phenomena, but refers the demand for the latter entirely to the future; nevertheless, the stage of development of these theories now appears very similar to that at which Newton found the development of astronomy and mechanics, and we are thus given hope that . . . the time is not far away when these theoretical discoveries will give greater perfection to the theory as a whole.[35]

If Mayer did read Fries, one can imagine his fancying himself to be the new Newton who had succeeded in discovering in his conception of force a unifying theory of heat, light, electricity, and magnetism—*plus motion and gravitation* (the latter as fallforce). The challenge was before him.

One of the most striking aspects of Mayer's thinking was his continuing search for valid analogies; indeed, his entire theory was in some ways nothing more than an elaboration of an analogy between force and matter. Again and again he expounded their characteristics in terms of fundamental similarities: quantitative invariability, qualitative transformability, "stoichiometry" of force (as of matter), neutralization of differences, independent existence as "objects" (or "substances"), plus a host of specific examples. Entirely typical is the following passage from a letter of 1842 to Griesinger:

Such an object which is not matter—[i.e., an] imponderable—is motion; it does not arise out of nought [*Null*] insofar as it must always have its cause, but, having once arisen, it no longer becomes nought because no cause can be conceived with the effect nought. We thus know: motion is *one* manifestation [*Erscheinungsform*] of an object that is not matter; it arises out of another manifestation and becomes, insofar as it ceases to be motion, another manifestation of the same imponderable object. In other words, the cause of motion, the motion itself, and its effect are nothing but different manifestations of one and the same object, just as the same thing can be said about ice, liquid water, and steam. But just as steam can again become water, and water ice, so, too, with motion and its causes and effects; *cause and effect designates in general nothing but different manifestations of one and the same object.*[36]

Likewise, the analogy between organisms and machines and that of the solar system as an organism played a crucial role in the development of his ideas.

At some point before the composition of his first essay—perhaps while still at sea, perhaps upon his return after he had had a chance to consult several more physics texts—Mayer seized upon the identification of forces as causes common to all the German texts of the day. Such a conception not only underscored the role of forces as agencies of change, but also reinforced a strong logicodeductive strain in Mayer's thinking. If forces are causes, if every effect must have a sufficient cause, and if *causa aequat effectum*, then the quantitative indestructibility of force is a logical necessity, a truth confirmed by the laws of human reason. Mayer certainly convinced himself (ultimately!) that it was, and he clearly expected others to accept the ineluctable logic of his argument.

Still another aspect of Mayer's search for the proper circumscription of his theory involved decisions as to which phenomena are real and in need of explanation, which the appropriate conceptual tools to apply toward their understanding. What was the scope of his theory to be? Should he address issues of the soul and the vital force? Did he need to restrict his theory to inanimate nature? What are the general limits of scientific explanation with respect to both range of phenomena and depth of reduction to simpler entities? *Are* there situations in which force is produced? *Nihil fit ad nihilum* may be true, but what about *de nihilo nil fit*? Which objects are to be included in the class of forces? What is the measure of force? Why does the parallelogram of forces yield such confusing conclusions? Mayer's attention was first drawn to the transformation of motion into heat and then of heat into motion: their quantitative equivalence was one of his earliest insights. As far as the evidence shows, he did not begin with a grand notion of the unity of force or anything remotely like it. On the contrary, Mayer paid little attention early on to forces other than heat and motion, and only gradually extended his theorizing to embrace chemical difference, fallforce (or spatial difference), electricity, light, and magnetism.[37] Finally, he had to decide who his audience would be, and where he would try to have his work published. It is a testimony both to his intellectual distance from the science of his day and to his conviction that his ideas were of profound and fundamental importance for physical science in general that he sent his first crude essay to the editor of Germany's leading journal of physics and chemistry.

Mayer's longtime friend Gustav Rümelin gave a valuable description of the state of Mayer's thinking in 1841, a description that beautifully captures Mayer's obsessions, uncertainties, and probing of the boundaries of his developing theory. The two spent many hours discussing his ideas:

I was together with him a lot in the fall of 1841, and it was difficult then to talk with him about something other than this business. . . . He himself had up to that time never read a philosophical book, and, as far as I know, never did so even later. When I once at his request gave him Hegel's *Logic* and the volume of the *Encyclopedia* that contains the philosophy of nature to take home, he brought both back a few days later with the remark that he had not understood a single word, and would not understand anything even if he read in them for a hundred years. *Ex nihilo nihil fit; nihil fit ad nihilum. Causa aequat effectum.* Those were the three slogans he then always had on his lips, which he several times called at me as I came and then after me as I went. I was supposed to tell him what might be objected to these principles. Since later we still spoke often and much about these things, so, too, do I still remember much from our conversations then. Against the first principle, that nothing is created out of nothing, I had no objection; it was only another expression for the general law of causality. But the second, that nothing becomes nothing, is not contained in the first and is not a principle of general experience. The latter offers many examples of annihilation [*Vernichtung*]. . . . All forgetting is an annihilation [*Zunichtewerden*] of ideas. Intellectual forces can simply become lost. All organisms are annihilated as such through external destruction or internal death. The continuation of species changes nothing about that and does not belong here, as little as does the question of immortality. Indestructibility is not among the characteristics of the organic and psychic [*psychisch*] phenomena of life, and for that reason the principle *nihil fit ad nihilum* has no general validity. *Causa aequat effectum* in the first place only means that nothing can be contained in the effect for which the cause is lacking, and vice versa, and thereby only repeats the first principle, the general law of causality, in another form.

He responded to this more or less as follows: How it was with the organic and the psychic he would leave aside for now; he was talking about the physical, about what one called forces in physics. Now it was entirely inconceivable to him, and his imagination failed him if he was supposed to imagine that something real and active could somehow ever become nought and nothing. All causality would come to an end if from the cause something could also get lost underway and not pass over into the effect. If forces were causes—and that they must be, surely no one would deny—then they must appear in the effect and remain preserved in it until the effect itself again became a cause. Just as little as a positive magnitude could become nought in the course of a calculation—this was an image and example he often liked to repeat—so little could a force disappear in the effect and vanish into nothing.[38]

Rümelin recounted further how the occasion of their encountering a horse-drawn wagon on the road prompted a discussion of what the physical effect was of the horses' muscular force:

The heating up of the horses, the accelerated internal combustion process of the food they had eaten, the frictional heat that the moving wheels leave behind in the visible blue streaks on the road, the consumption of fat in the wagon grease on the axles—all those were not accidental epiphenomena, which I appeared to take them for; rather, the motion of the horses, their mechanical work, was converted into those heat phenomena, in particular according to a constant quantitative relationship, in the discovery and formulation of which he saw the most important part of his task, while he had not the slightest doubt any more over the correctness of the principle.[39]

Such, roughly, were the kinds of ideas and considerations that occupied Mayer up to the time of composition of his first literary endeavor and during the succeeding few months. Having incubated his ideas in effective isolation, upon his return to Heilbronn he found an audience first in his elder brother Fritz (an apothecary), then among friends such as Georg Kehrer (a *Gymnasium* instructor in Heilbronn) and—in July 1841—Carl Baur (then pursuing private studies in Tübingen).[40] Sympathetic, critical, and competent, Baur in particular would prove to be a great help to Mayer in coming to terms with his serious weaknesses in mathematics and mechanics. Wishing to obtain the judgment of professional physicists (*Männer vom Fach*) as to whether his ideas were either well known or untenable, Mayer also sought contact with the professors of physics at the universities of Tübingen and Heidelberg, visiting Nörrenberg in early September and Jolly around November 1841.[41] How little real progress he had made by then is revealed by an unpublished essay from around September of that year, untitled but bearing the motto (from Horace) "sapere aude"—dare to know.[42]

The major themes of that essay do not go beyond those of the one he submitted to Poggendorff three months or so earlier. It opened with an analysis of the relationship between the necessarily equivalent links in an indefinite chain of causes and effects: "No matter how far a chain of causes and effects may continue, according to what has been said its originally given magnitude can neither increase nor decrease."[43] It emphasized the compelling logic of its argument, asserting that "a logical cataloging of results is what one calls science."[44] It drew upon chemical analogies, in this case the (idealized) addition of oxygen to iron to produce ferrous oxide and heat. He took the occasion to point out the errors of the phlogiston theory, adding that "the conclusion of antiphlogistic chemistry that $\mathrm{Fe} + \mathrm{O} = \dot{\mathrm{F}}\mathrm{e}$ contains the emancipation of ponderable from imponderable matter."[45] It spoke rather indifferently of "imponderables or forces," and called the "spatial difference" of two masses a "cause of motion" and a "force."[46] It measured the quantity of motion by mass times velocity, and confused the motion-producing "spatial difference" of two gravitating masses with the distance apart of two colliding bodies. It sought to pair the measure of the quantity of motion with a specification of its qualitative determinations (*qualitative Bestimmungen*), such as direction. Thus in the formula 0 2AC—more idiosyncratically symbolic than conventionally mathematical—the symbol (*Zeichen*) "0" represented the "direction" of the combined motion of two oppositely directed bodies. In other cases the "0" might represent paired motions in indefinitely many directions, as in vibratory or wave motion, or in thermal expansion: "The expression one makes use of here: heat expands bodies, is nothing but a paraphrase of the principle: heat is the cause of opposing motions (positive and negative motions, motions with the symbol 0) as a necessary corollary to the principle: opposing motions are a cause of heat."[47] Thus did Mayer make the crucial step from considering the transformation of motion into heat (via inelastic collision) to that of heat into motion (via thermal expansion): thermal expansion is radial motion in infinitely many opposing directions.

Absolutely central to the essay's line of thought was the relationship between heat and motion: heat (the effect) arises when motion (the cause) disappears. There was talk of motion being annihilated not only by an opposite motion but also by means of a "presumed fixed point."[48] It appealed repeatedly to experiment in a general way, but revealed no appreciation of which experiments might actually be doable, of which might even in principle provide decisive information. It contained simple mathematical mistakes. It combined in mathematically meaningless fashion forces and masses, and heat and the apparatus used; at times it lost hold entirely of the concept of force and even ignored the heat of fusion in considering the effect of rubbing two steel disks together, looking only at the change in their "state of aggregation," as in the production of filings.[49]

In one slightly new development, Mayer asked whether the heat produced upon the disappearance of motion was due to a change in the volume of the substances. A consideration of solid bodies convinced him that a diminution in volume could play only a small role in the evolution of heat; his further conclusion hinted, perhaps, that he had begun to consider the relationship between heat and volume changes in gases, too: "[B]ut since this [i.e., the decrease in volume] is often very significant in the phenomena that one has tried to explain in this fashion, we thus see from this that this mode of explanation—which is still the only encouraging one, and which proceeds from a correct idea—is of little good even where it most applies."[50] It would appear that Mayer had temporarily convinced himself, on the basis of specious reasoning and dubious symbolism applied to solids, that change in volume could not be the cause of the heat produced when gases are compressed, either. He also seems to have made a distinction between elastic bodies as those that restore the momentarily disappeared and stored force of motion, and inelastic bodies as those that transform motion into heat: "Alongside the active expansion of elastic bodies, by means of which they again give off the force applied to them, stands accordingly the capability of other bodies, which are not capable of this kind of expansion, to give back the force received in the form of heat."[51] Since gases are the preeminently elastic substances, they would thus be unsuitable objects for the investigation of the transformation of motion into heat. Mayer clearly had to work his way past several confusing issues before he would be in a position to exploit gas phenomena to his own end. At least he had now started to think about them.

At the end of the last paragraph before his summing-up Mayer added his only reference to an actually performed experiment: "In conclusion let it be mentioned here that heat can be developed not only by means of the motion of solid bodies, but also liquids—by shaking water—as the author found through many careful experiments, which is also connected with the observation of all seafarers that agitated seawater is warmer than water that is calm. cf. Rumford's experiment."[52] Mayer had cited the example of seawater being warmer when the waves are rough in a letter to Baur that August, before he met with Nörrenberg and without any mention of his having performed any experiments.[53]

The records of Mayer's encounter with Johann Gottlieb Nörrenberg, the entirely undistinguished professor of physics at Tübingen, are few but of one voice. As Mayer wrote to Griesinger fourteen months after the event:

I first went for a talk with Nörrenberg, who said to me: "Those are basically nothing but new views of things that one can look at equally well in other ways; but if you can base a new experiment on your theories, then, then you've got it made." He himself reckoned here whether I could demonstrate that liquids get warmer through shaking; I performed this experiment in the most careful fashion immediately after my arrival home, and it succeeded completely.[54]

His letter directly after the event to Baur—who, a student of Nörrenberg's, had arranged the meeting—confirms that account, and contains Mayer's earliest mention of the problem of calculating a numerical value for the mechanical equivalent of heat:

Since our meeting I have zealously further cultivated my theories and in particular found that Nörrenberg's objections are completely untenable. The proof Nörrenberg demanded, that water can be warmed by shaking, I have supplied by means of many always successful careful experiments. A vital question for my theories, which themselves can be developed with mathematical certainty, remains now the solution to the question: How high must a weight—say 100 pounds—have been raised over the ground in order that the quantity of motion (quantité de mouvement) which corresponds to this raising, and which can be obtained by lowering the weight, be equal to the quantity of heat necessary to transform 1 pound of ice at 0° into water at 0°. By means of this solution the equation that exists between spatial difference of masses, motion of masses, and heating of water is determined in concreto.[55]

Several writers have rightly insisted on the importance of Nörrenberg's having driven home to Mayer the importance of experimental confirmation and the determination of a numerical value for the equivalence between heat and motion.[56] It is less clear whether it was Nörrenberg's suggestions that put Mayer on the track of determining that number in the way indicated in his letter. Baur, who had studied mechanics—and not just physics—might more likely have introduced his friend to the idea of measuring the vis viva of a body in terms of the distance it falls under gravity. It would then have been a small step for Mayer to have made the equation between the descent of a weight and the production of heat, the two phenomena linked conceptually via motion and its measure, the quantity of motion.

What Mayer succeeded in doing (as of 12 September) was to bring together that conception of the effect of letting a *weight* fall through a given distance with his heretofore abortive reflections on the expansion and contraction of *gases*: if the thermal *expansion* of a *solid* had been his original example of the transformation of heat into motion, now the *compression* of a *gas* represented the production of heat via the descent of a compressing column of mercury—that is, via the disappearance of a "spatial difference." Thus did Mayer continue his train of reasoning in this all-important letter:

If a spatial difference $= P$ is given, then, according to Newton's law of gravitation, this is equal to a definite quantum of motion $= Q$; $P$ can be converted into $Q$, but never into nought; this is necessarily contained in the equation $P = Q$. The solution of this formula is the business of mechanics. . . . To determine the quantity of heat equal to Q is in turn a matter for experimental physics, it is contained in the solution to the above-set task [i.e., to determine the equivalence between the motion generated by the fall of a weight and the rise in temperature of a quantity of water].

We can achieve this solution in the following way. Let *ACB* (figure 6.1) be a precisely graduated, completely closed glass tube, at C a stopcock; let the tube be filled with mercury from A to C, from C to *B* with some kind of gas. Let the column of mercury be tall enough so that after opening the stopcock the gas is somewhat compressed. Let one then open it and note precisely how much the mercury falls and how much absolute heat the compressed gas gives off until the original temperature has been reestablished. *The fall of the column of mercury and the absolute heat obtained are equal to each other.* Upon closer inspection this apparently so simple sentence arouses many doubts, and it was only through long study and unremitting reflection that I came to complete clarity about it.[57]

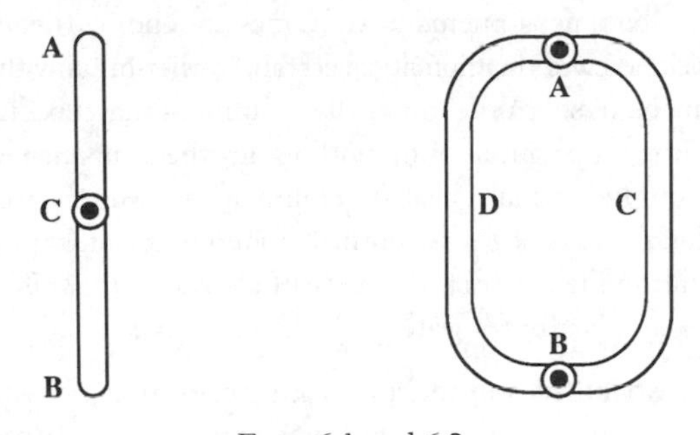


FIGS. 6.1 and 6.2

Just how long Mayer had spent pondering the matter is impossible to say, nor did he spell out just what his reservations were about the cogency of this example. But one can venture a guess along the following lines: Mayer had probably regarded the effect of the falling column of mercury not only as the production of heat but also as the compression of the gas, which is itself capable of producing additional effects (i.e., motion). (Recall Mayer's earlier treatment, as effects of motion, of both the generation of heat and the production of filings and other gross physical changes.) What clarified the matter for him was to imagine a way of eliminating that nonenergetic effect by means of a procedure in which the gas returns to its original state and the only appreciable net change is the fall of a certain quantity of mercury, which then *must* be the cause of the only remaining effect, the heat released. Crucial to his reasoning was the empirical fact that the expansion of a gas into an empty space—or, as he may have put it to himself, the *Aufhebung* of a prior compression—neither releases nor absorbs heat, that is, produces no additional effect. Such a reconstruction is consistent with the proof he then gave:

The proof lies basically in the following. Figure [6.2] is a glass tube that runs back into itself, at A and *B* [furnished] with stopcocks, *ADB* filled with mercury, *BCA* with a gas. After opening the stopcock at *B* the mercury falls from *AB* to *DC*, the gas is compressed, thereby giving off heat, in *AD* however there occurs *no decrease in temperature*! After the reestablishment of thermal equilibrium one closes *B* but opens A. The gas now again expands as far as *CAD*; there occurs thereby in *CA* a decrease in temperature, in *AD* an increase. But we know from precise experiments by Gay-Lussac that the increase is equal to the decrease in such a way that after opening the stopcock at A the temperature of *DAC* taken as a whole remains unchanged, i.e., that the gas in *DA* gives off precisely as much absolute heat as the one in *AC* has to take up in order to return to the temperature of the surroundings. As is well known, Gay-Lussac took two [glass] balloons, one evacuated, both [furnished] with very precise thermometers. After the reestablishment of communication between the two the thermometer in the previously empty one rose precisely as high as the one in the one filled with air fell. . . . The experiment teaches that after opening stopcock *B*, as a result of the falling column of mercury the gas in *AC* is both compressed as well as heat is released from it (if one collects this latter heat, then one can melt ice with it). After opening A the compression produced is again undone [*aufgehoben*], *but no heat is thereby consumed.* We are therefore left with the earlier-obtained quantity of heat; it must necessarily be equal to the fall of the column of mercury. This result, which can be obtained in an a priori fashion, has long ago been confirmed in advance by experiments on compression and heat development in gases by means of pressure; they teach that for all kinds of gas the heat obtained is precisely proportional to the decrease in volume, and the latter to the pressure.[58]

(The last sentence is probably an allusion to the results obtained by Dulong in deducing from the speed of sound in various gases that, in Kuhn's words, "the heat liberated by a given compression was the same in all simple gases.")[59]

By tipping over the apparatus of figure 6.2 and returning the mercury to its original height, one could transform mechanical force into heat indefinitely. His conclusion was that "spatial difference *P* is to motion *Q* as *Q* is to the quantity of heat *x*, or more precisely $P = Q = x$."[60] Motion was thus the conceptual connecting link between heat and what he would soon call fallforce. An important crystallization had also taken place with respect to the meaning of "spatial difference"—now applied only to the height of a body above the surface of the earth—which only days before he was still also applying to the distance between colliding bodies. Once again Mayer's success lay in a brilliant application of logic, supported by a few important facts, to a complex situation, one whose resulting clarity belies the nonobvious character of the deductions that led to it. Mayer felt himself poised, if not yet quite prepared, to make the calculation on which his claim to greatness would later rest:

The already so variously, carefully, and ingeniously performed experiments on the specific heats of gases, their temperature changes under compression, and their changes in volume give us all the data necessary for the calculation of the fall of the column of mercury and the quantity of heat thereby produced, and one can thereby again solve the task first set. I am now at work on this task, but I continually encounter difficulties. On Kehrer's advice I now summon up my courage to apply myself to you, whether you would not be so good as to look into the matter when you have the chance; . . . with regard to the science at hand I am almost as isolated here as on the ship.[61]

Unfortunately, no reply from Baur has survived, and we do not know how much he might have contributed to the solution of Mayer's problem.

Of Mayer's encounter a few months later with Jolly, somewhat conflicting accounts have come down, those of Mayer's and those of Jolly's. As Mayer described it to Griesinger a year after the event (after mentioning his visit with Nörrenberg),

> A few months later I went with substantial improvements to Professor Jolly in Heidelberg, who soon expressed himself to the effect that the matter pleased him very much; the theory of heat (which was the principal topic of conversation) is, he said, necessarily in need of such improvements; but I should develop the matter further. This was a natural bit of advice; but since the matter was expanding endlessly in front of me, I always had to be more mindful of concentrating myself than of expanding.[62]

In a later autobiographical sketch he was more explicit:

> Taking as a basis Gay-Lussac's well-known experiment according to which the temperature of a gas that expands without performing external work remains on the whole constant, I spoke with the already mentioned first-rate physicist about the method by which the amount of vis viva of motion uniformly proportional to a determinate quantity of heat can be calculated from the compression of atmospheric air. This exposition enjoyed in general Jolly's approval . . . and he encouraged me to pursue the subject further, to think it through, and then to publish.[63]

Except that Mayer here used language he was not much at home with earlier—*äussere Arbeitsleistung* and *lebendige Kraft*—there is every reason to believe that that in fact was the topic of their conversation. It is possible that Mayer had by then found out for himself, with or without Baur's further help, how to perform that calculation by exploiting information contained in Lamé's and possibly also Baumgartner's physics texts. It formed the centerpiece of the paper he sent to Liebig at the end of March 1842.

A radically different account of Mayer's meeting with Jolly has come down from an anecdotal version that Jolly apparently enjoyed telling. One variant was reported by Ernst Mach, who said he had first heard the story orally from Jolly and later had it confirmed in a letter (which he did not publish):

> During a hurried meeting with Mayer in Heidelberg once, Jolly remarked, with a rather dubious implication, that if Mayer's theory were correct water could be warmed by shaking. Mayer went away without a word of reply. Several weeks later, and now unrecognized by Jolly, he rushed into the latter's presence exclaiming: "Es ischt aso!" (It is so, it is so!) It was only after considerable explanation that Jolly found out what Mayer wanted to say.[64]

The incident was supposed to illustrate Mayer's utter preoccupation with his own ideas. In another of Mach's retellings Jolly had trouble understanding Mayer—whom he "hardly recognized"—on account of his "entanglement in the views of the school."[65] Early historian of science Edmund Lippmann, who apparently also had the story from Jolly, reported that "Jolly still recounted with pleasure in his old age" how Mayer, months [*sic*] after their first encounter, burst suddenly and unrecognized into his laboratory, shouting in his Swabian dialect "Es ischt aso."[66] According to the version Carl Voit had heard (apparently secondhand), Mayer arrived at Jolly's door in a state of "great agitation" swinging a water-filled flask.[67]

It is hard to know quite what to make of these stories, retold and embellished, except that they attest to the fact that crazyman Mayer had become a fit subject for 'humorous' and demeaning anecdotes. What is troubling historically is that they imply that Mayer had had more than one meeting with Jolly, whereas Mayer mentioned only one, and that as of when they met the idea that water could be heated by shaking was new to him, whereas there is abundant evidence that it was after his meeting with Nörrenberg that he settled that issue experimentally. I can only suggest that in later years Jolly enhanced his shaky memory of events he may not have thought about for ten or twenty years, by which time Mayer had gained a measure of recognition, in order to tell a good story. Perhaps Mayer had told him of Nörrenberg's challenge and his subsequent experiments, which demonstrated (he insisted) that "it *is* true!" Perhaps the difficulty Jolly recalled in recognizing him from their first "hurried" meeting was his way of subconsciously reprocessing the fact that he *hadn't* seen Mayer before. In any event, I prefer to go with Mayer's account of the matter: I have otherwise not found it necessary to correct any of his retrospective accounts.

Jolly was an appropriate person for Mayer to have sought out. He wrote his inaugural address that year "on the effect of machines," including things like the relationship between velocity and moving and resisting force, and may have been prepared to discuss ways to measure the effects Mayer was concerned with, or at least to appreciate Mayer's problem.[68] And although there is no indication that he and Mayer discussed such matters, Jolly was even intimately familiar with the relevant physiological issues, having assisted Theodor Bischoff (at Heidelberg since 1836) with the latter's study of respiration.[69] At issue was the presence of oxygen and carbon dioxide in the blood and where oxidation takes place; Bischoff identified arterial and venous blood as the respective carriers of those two gases, explained the blood's varying color in terms of its varying charge of gas, and concluded that carbon dioxide is not formed in the lungs.

It is possible, too, that Jolly might have discussed his meeting with Mayer with his close colleague, Liebig. It is not known when their acquaintance began, but their earliest surviving letter, of 29 May 1842, implies that they had already been in contact.[70] Perhaps they had met at the 1834 *Naturforscherversammlung* in Stuttgart, which both attended. In November 1842 Jolly responded at length to Liebig's *Animal Chemistry*, and elaborated an analogy between the organism's production of force or motion via the consumption of vital force attendant on the breakdown of organic matter, and the production of motion via electromagnetically induced currents when magnetism disappears from the iron particles of a steel magnet.[71] Liebig evidently urged Jolly to publish a review of his book, which Jolly declined to undertake.[72] Thus by 1842, at least, they were writing to each other on topics closely related to those touched on by Mayer's work. Kremer has suggested that Liebig's concern for mechanical concepts in the third section of his *Animal Chemistry* may have been prompted in part by Mayer's 1842 paper.[73]

Liebig himself had begun to consider the application of organic chemistry to agricultural and physiological problems at the beginning of 1840.[74] As we saw in chapter 3, by the end of 1841 Liebig had seemingly arrived at a powerful and coherent view of the organism as a device that, like the steam engine and the voltaic pile, generates a certain amount of heat and motion by means of processes of material exchange, that is, through the expenditure of a corresponding measure of chemical transformations. Mayer read the paper Liebig published in his *Annalen* on the vital process and the atmosphere and began to run scared when he encountered there a view of heat as transformed motion and assertions such as "No force, no activity can arise out of nothing" and "We have a substrate, something given, which assumes the form of another substrate, in every case a force and an effect." When Griesinger called Mayer's attention to a particular page of Liebig's *Animal Chemistry*—a book in which he had found many resonances (*manche Anklänge*) to his friend's ideas—Mayer wrote back:

> The place in Liebig's *Chemistry*, etc., p. 32, which you write me about, appeared first in his *Annalen*—I believe in the February or March issue—and directly decided me to give a provisional presentation of some of my principles in dogmatic form in a short article. Liebig wrote me among other things: "About what force, what cause or effect is, there prevail in general such confused ideas that an easily understandable analysis [*Auseinandersetzung*] must be regarded as a true service." One might be supposed to think from that that he knew himself to be elevated above the general confusion, but that this is in no way the case I could tell to my satisfaction from his "Phenomena of Motion in the Animal Organism." It is nothing but a new patch on an old coat; instead of proceeding with the necessary radicalism he mingles together new ideas with old errors and thus ends up making real mistakes (pp. 206 and 207).[75]

Thus was Mayer prompted to compose his "Remarks on the Forces of Inanimate Nature" and to send the manuscript to Liebig, who accepted it. Since he appears to have come across Liebig's paper in February or March, and mailed his own essay to Liebig at the end of March, he must have responded quickly. One recognizes in the published paper many of Mayer's characteristic motifs, in particular the centrality of his concept of force and its transformation, presented both in terms of an analogy with matter undergoing chemical transformation and as a corollary of the fundamental principle of causality. This analogy and this principle provided the basis for the assertion that both forces and matter are quantitatively indestructible and qualitatively transformable. Mayer still sometimes mixed matter and motion—both "causes"—in the same equation as he sought to identify what the effect was (for example) of the rubbing together of two metal plates: in a paragraph addressed primarily to his own past confusions, he argued that heat, not a change in state of aggregation, is the sole effect of the disappearing cause (i.e., the motion).[76] He now had a clearer grasp of the meaning of spatial difference as a *cause* of motion, that is, as a *force*—the "fallforce," represented by the distance a body could fall under gravity. He now somewhat shakily measured the quantity of motion by $mc^2$, and the fallforce by mass (not weight) times height (i.e., ignoring the factor $g$). He identified the law of the conservation of vis viva as a special case of "the general law of the indestructibility of causes."[77] Except for the barest passing mention of chemical difference, electricity, and light (as vibratory motion), his presentation was limited to heat, motion, and the new fallforce. And although he wisely left out any discussion of the parallelogram of forces and the creation of force in "organisms" like the solar system, that he explicitly restricted himself to "inanimate" nature and spoke only of the indestructibility and not also of the uncreatability of force implies that he still had not worked his way to clarity on those important but confusing issues.

In moving toward the culmination of his paper, the calculation of the as yet unnamed mechanical equivalent of heat, Mayer invoked a new and inappropriate analogy between the fall of a body and the compression of a gas:

> We can represent to ourselves the natural connection existing between fallforce, motion, and heat in the following fashion. We know that heat appears when the individual massy parts of a body move closer to each other; condensation [or densification (*Verdichtung*)] generates heat; now what holds true for the smallest massy parts and their smallest interstitial spaces must surely find application to large masses and measurable spaces. The descent of a weight is a real decrease in the volume of the earth, and hence must certainly be connected with the heat that thereby becomes manifest . . . .
>
> . . .
>
> If fallforce and motion are equal to heat, then heat must naturally also be equal to motion and fallforce. Just as heat arises as an effect with decrease in volume and cessation of motion, so does heat disappear as a cause upon the appearance of its effects—motion, increase in volume, lifting of a weight.
>
> In water-powered machines the motion that arises at the expense of the decrease in volume that the earth constantly undergoes through the fall of water, and which [then] disappears continuously, delivers a significant amount of heat; contrariwise, steam engines serve to decompose heat into motion or the lifting of a weight.[78]

Mayer was attached enough to this analogy not only to repeat it in 1845, but to expand it by associating the concept of heat becoming latent with the transformation of motion into fallforce attendant upon the rise of a heavy body against gravity: the fall of the body, representing a diminution in the volume of the earth, and a decrease in the volume of a gas are both accompanied by a becoming "free" of a previously "latent" force.[79] Even without quibbling over his continued use of the ambiguous word "force," it would be difficult to maintain that as of 1842 Mayer had discovered the conservation of energy, despite the fact that he had introduced a new conception of force and had calculated the mechanical equivalent of heat. If one wishes to maintain such a claim, then one must be prepared to say that a person can have discovered something he does not believe in, because Mayer had not yet convinced himself of the universal uncreatability of force.

## 2 Later Developments and Changing Emphases

More than a year after the composition of his first published paper, Mayer was still carrying on a dialog with Baur over mechanics, the definition of force as the cause of motion, and the proper measure of force.[80] Gravity, he insisted, is not a force: its measure is acceleration, not motion. Old motifs assumed new form: having earlier spoken of his theory of force as the axis of a reformed science of physics, he now argued that the reform of mechanics and physics requires a change in coordinate system, or rather "the introduction of such a system as the fixed axis for all physical sciences."[81] Moving from a more metaphoric to a relatively more concrete application of the term "axis," Mayer stipulated that the $x$ axis designate *force*—"for example the motion, or rather the vis viva of the motion"—the $y$ axis *time*.[82] Not surprisingly for Mayer, his new foray into mathematical physics was fundamentally misguided and of only suggestive significance:

> For uniformly accelerated velocities the ordinates representing time form a parabola whose parameter increases and decreases inversely with the magnitude of the so-called accelerative force, with the magnitude of $g$. [Cf. figure 6.3.] Under other circumstances the ordinates of time form other curves, whereas the x-axis remains fixed under all circumstances. For the oscillations of a pendulum the curve would have something like the following form [figure 6.4], which is perhaps a cycloid, etc. What the value is of this representation (which I do not wish to pursue further)? The great one, that the concept "force" has been worked out consequentially; that then in mechanics nothing else will be understood by force than in the rest of physics, "cause of motion"; that one at all knows what force is; the product of mass and its height, the product of mass and the square of its velocity, free heat—those are objects, not abstractions.[83]

Figure 6.3 is mine, while figure 6.4 is taken from Mayer's letter. In figure 6.3, $pp'$ represents the so-called parameter of the parabola through its focus $f$. This abortive attempt at symbolic reform is, even in its failure, characteristic of Mayer's continuing search for the proper form and language in which to express his ideas.

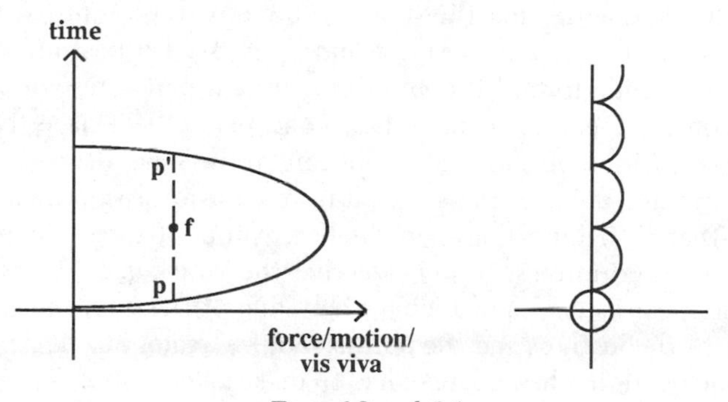


Figs. 6.3 and 6.4

By June of the next year (1844) Mayer had drafted a long essay, which, after repeated and extensive revision in response to suggestions from Griesinger and Baur, would be published toward the middle of 1845 as *Organic Motion in Its Connection with the Exchange of Matter*. The original draft contained a long "polemic" against certain usages in mechanics—an abiding concern of his—which he toned down at Baur's urging.[84] It likely contained a critique of the concept of "Schwer*kraft*" as a source of motion, criticism (again) of the customary definition in mechanics texts of force as the cause of motion coupled with their use of *mv* as its measure, and a misguided attack on the concept of *force instantanée* (as he put it, after Lamé).[85] When he sent the essay to Griesinger, he explained the small part physiology played in it—"the inorganic has become without qualification my principal concern"—in terms of the need first to secure his theory on physical grounds before seeking wider applications.[86] He pictured his work as an impregnable citadel over which flew a banner bearing the words "Heat can be transformed into motion."[87] In the same letter to Griesinger he added that "'Motion is transformed into heat': in these five words you have implicitly my whole theory," and to Baur he wrote that he considered it to be his "life's task" to prove the principle that "$mc^2$ = heat."[88] The centerpiece of Mayer's work was still, without question, the equivalence of heat and motion.

Although there is no evidence as to just how or when Mayer achieved clarification on the matter, sometime between 1842 and 1844 he must have discovered the fallaciousness of his reasoning about central-force motions, the apparent cause in the first place of his having entertained the possibility that force might be creatable in divinely contrived "organisms," and thus have convinced himself that forces can never be either destroyed *or* created. His published essay of 1845 could thus *lead* with just that confident assertion: "*Ex nihilo nil fit*."[89] Quite ignoring the fact that he himself had for years thought otherwise, he now boldly asserted that "the *creation* or the *destruction* of a force lies outside the range of human thought and action."[90] He now for the first time criticized the notion of an inexhaustible vital force in the same terms: "The creation of a physical force . . . [is] already in and of itself hardly conceivable."[91] The transformability of forces he likewise presented as a logical necessity: "It can be proven a priori and confirmed everywhere by experience that the different forces can be *transformed* into one another. *There is in truth only a single force*."[92] It is also possible that Mayer's newly critical attitude toward the vital force was in part a response to Liebig's forceful, if confused, employment of that concept. Earlier, Mayer had rendered acceptable the creation of force out of nothing in the solar system by identifying the solar system as a divine organism, as if it were obviously more acceptable that organisms—that is, living things—might be creators of force. Such an analogy gave Mayer a vested interest in *not* rejecting the vital force. His consciousness raised by the inconsistencies of Liebig's handling of the vital force, Mayer may have rethought the justness and the necessity of his analogy from the other direction, and concluded that he no longer had to make an exception for *any* organism.

The somewhat misleadingly named booklet that resulted from Mayer's interchanges with Griesinger and Baur contained his fullest statement of his theory of force. Even at that, the sentence just quoted was the closest he ever came to anything expressive of the notion of the "unity of force," a term he never used. Nor did he there or elsewhere ever settle upon a single phrase to express something like the notion of the conservation of energy. For him his theory was first and foremost "the mechanical theory of heat" whose central fact was "the equivalence of heat and motion." In the surviving manuscript from the series of drafts that preceded his 1845 work he once spoke of "the axiom of the indestructibility of force," but nothing like that ever became his standard way of designating his theory.[93] Intriguingly, instead of "axiom" Mayer had earlier written "principle," and instead of "indestructibility of force" he had first written "positive motions," then crossed that out in favor of "conservation of force," finally crossing out "conservation" and writing above it "indestructibility."[94] Mayer was indeed *almost* the discoverer of "the principle of the conservation of force"! But he consistently declined to adopt such language.

Even after Helmholtz's work had become generally known and its terminology standard, Mayer clung to his favored terms. In an essay from 1862 he spoke of the enrichment physics had recently experienced "through the discovery of the law 'of the indestructibility of force.'"[95] In another from 1870 he spoke of "the mechanical theory of heat, or as Helmholtz calls it, the law of the conservation of force."[96] Since Mayer himself had once weighed and rejected the phrase "conservation of force" years before Helmholtz arrived on the scene, his resistance to it cannnot have been solely a question of sour grapes and *amour propre*. Rather it had also to do with what he perceived to have been his prime accomplishment. In a review written in 1877, the year before his death, of a published speech by Helmholtz on medical thinking, Mayer recalled that the purpose of his first publication had been "to secure my priority rights to the mechanical theory of heat and to the calculation of the mechanical equivalent of heat, which I was the first to carry through."[97] Quoting a passage in which Helmholtz recounted some of the background to his discovery of the law of the conservation of force, Mayer countered that, as far as he knew, the law of the conservation of living force had been discovered two hundred years earlier by Huyghens and then defended by Leibniz against Descartes: "This law was thus known much earlier than the discovery in our day of the mechanical equivalent of heat with its connection to medicine."[98] Mayer resisted what he perceived to be the danger of having his discovery appear to be merely an extension or a special case of the law of the conservation of vis viva in mechanics: better to avoid any suggestive terminology. In his 1842 paper he had even sought to subsume the older conservation law under his new principle of the "indestructibility of causes"—forces being of course also causes.[99] An important aspect of the progressive crystallization of meaning that attended the evolution of Mayer's thinking was his choice of appropriate terms.

Although he did not accompany its statement with a handy phrase, Mayer nevertheless did give clear if somewhat awkward expression in 1845 to the essence of the conservation of energy: "*In all physical and chemical processes the given force remains a constant quantity*."[100] He identified what he took to be the five principal forms of physical force as fallforce, motion, heat, magnetism and electricty (considered together), and chemical force.[101] He then listed examples of each of the twenty-five possible forms of transformation among the different forms of force. For the first time either in print or in private, Mayer had significantly expanded his attention beyond heat, motion, and (as of late 1841) fallforce. In working up to this conclusion, he had devoted separate sections of his essay to each of the five principal forms of force, in which he anticipated the full network of interconnections. He analyzed, for example, the transformation of mechanical effect into electricity and magnetism in the electrophore and in the magnetization of previously "indifferent" steel rods.[102] And most importantly, he offered a full and clear calculation of the numerical equivalence between heat and fallforce.

Here again it would have been terminologically anachronistic to have spoken of Mayer's calculation of the mechanical equivalent of heat, since no such phrase had yet made its appearance in his writings. It was the theoretically central *equivalence* of heat and motion that was of greatest importance to him, less so the numerical *equivalent* connecting heat and fallforce: the latter was, to be sure, important, but its function was to confirm the former. Even in his next major work, in 1848, he again calculated the *number* without assigning it any particular descriptive *name*.[103] Only in 1851, in his *Remarks on the Mechanical Equivalent of Heat*, did he expressly introduce the term.[104] What had taken place in the meantime was his interchange with Joule, though even as Joule spoke in 1849 of "the mechanical equivalent of heat [*calorique*]," Mayer responded with a claim to his priority in discovering "the law of the equivalence of heat and vis viva and its numerical expression."[105] For me the fact or not of giving something a name is an important indicator of the self-consciousness a person has with respect to its importance, hence it would appear that Mayer required an external stimulus before he 'realized' just how important the number in question was: it had become the focus of an international priority dispute, a specific fact on which hard claims could be based. The meaning of Mayer's work experienced another small shift, another small but significant sharpening of its focus.

Although Mayer always considered the establishment of the mechanical equivalence of heat on physical grounds to be his foremost scientific undertaking, he nevertheless maintained an interest in the relevance of his ideas to physiological phenomena. Such is hardly surprising, since he was a physician and had begun his train of speculations by addressing issues of respiration, the function of the blood, and animal heat. Contemporary authors' use of a vital force might also have provided him with a rather obvious target of criticism against which he could argue in terms of cause and effect and the like, assuming that those works had caught his attention. Yet his desire to devote his energies first to physics, coupled with his lingering suspicion that organisms might be capable of creating force, effectively diverted his attention from physiological issues until (it appears) sometime between mid-1843 and mid-1844.[106] In May 1843 Griesinger inquired of him what the situation was with respect to the intention he had earlier expressed of writing a physiological essay for Wunderlich and Roser's journal.[107] Mayer replied in June 1844, sending his friend a draft of such an essay for criticism.[108] He mentioned that Carl Pfeufer, who in 1844 moved from Zürich to Heidelberg as professor of clinical medicine, had been informed of his work (by an unnamed third party) while he was in Heilbronn, and that Pfeufer had urged him strongly to publish something else, "since he thought the theory promised to do much for physiology, which decided me all the more to move the matter as close as possible to physiology."[109] In reply Griesinger offered Mayer two bits of advice: "In the first place you should spread people some critical butter and sprinkle them some polemical salt on the dry bread of mechanics and mathematics."[110] He suggested in particular targeting Liebig's theory of animal motion (in the controversial third section of his *Animal Chemistry*) and Lotze's handling of the concept of force (in his *General Pathology*). In the second place he recommended that Mayer make the physiological part of his essay longer and broader, addressing issues of general relevance to the understanding of vital phenomena. He suggested considering the views of others—for example, Valentin, who, he said, stood methodologically close to Mayer's approach to physiology. Mayer thereupon asked Griesinger to send him Lotze's book and something of Valentin's, and expressed his profound disagreement with Liebig's work, adding that it would be easy to refute Liebig's entire theory on the basis of internal contradictions—a task that, following his friend's advice, he would set about to do.[111] Griesinger responded by suggesting that Mayer separate the physiological from the physical part of his work and seek its publication in his friends' journal, and he again urged him to address general questions—"on force, vital force, etc."—and to write with sharp polemics against the already mentioned authors.[112] To Baur, Mayer wrote at the end of July that he had decided to include "a modest polemic against Liebig's 'Phenomena of Motion in the Animal Organism'; maybe that will succeed in bringing the matter into consideration."[113]

Mayer's response to Liebig's work was strange, given the obvious points of major disagreement between them on fundamental physiological matters. It is striking that the errors he alluded to, late in 1842, in the third section of the *Animal Chemistry* had to do not with the vital force but with Liebig's stillborn attempt to introduce physical concepts such as *Bewegungsgröße*, *Kraftmoment*, and *Bewegungsmoment*.[114] Even in 1844 the first thing he criticized about Liebig's use of the concept of vital force was that he associated it with "forces" of gravity and adhesion (Liebig had said cohesion): "Since I firmly combat these latter two [misapplications of the word force], there lies therein indirectly a polemic against Liebig, the contest can thus most easily be settled on physical grounds."[115] However 'natural' Mayer's eventual criticism of Liebig's physiological ideas may appear—and *conceptually* it does follow directly from important differences in the views they held—the evidence strongly suggests that it was primarily Griesinger's pressure that prompted him to compose the highly critical passages directed against Liebig's *Animal Chemistry* that occupy such a prominent place in his *Organic Motion in Its Connection with the Exchange of Matter*. That Mayer himself may have been uncomfortable with such a level of polemics is suggested by the fact that he omitted those nine pages from the collected edition of his works in 1867. It is hardly less odd that arch-polemicist Liebig let such an attack go unanswered.

I argued earlier that, although Mayer's pre-1845 reflections do not seem to have involved a *critique* of the vital force, nevertheless some of the *issues* raised in connection with it may well have stimulated his thinking about problems of the creation and destruction of substances like forces, the imponderables, and the soul, as well as of their mutual similarities and differences. The present evidence suggestive of a protracted reluctance to attack the concept of vital force head-on accords with the possibility that Mayer may have been attached to the concept because it lent support to his belief that organisms, including the solar system, are in fact capable of producing force even in the absence of antecedent physical or chemical transformations. It would certainly seem that his use of the term "organism" in that connection implies a tacit understanding that living beings are somehow more naturally acceptable as creators of force. It would appear, too, that it was only after he gave up the idea of the creation of force *ex nihilo* in the solar system that he in fact came out against the concept of a work-performing vital force. One must be cautious, however, in the absence of clear evidence, of making too strong and one-sided a causal claim, since Mayer may also have become increasingly sensitive to the increasingly negative judgment on the status of the vital force by German scientists during the half-decade in question.

Slightly less than a third of Mayer's 1845 book—the first part—was devoted to physical and chemical matters. Thereafter came a six-page depiction à la Dumas of the energetic and material interrelationships among sun, earth, plants, and animals. The first physiological controversy to engage him was whether the production of force in the muscles is due, as many (including Liebig) thought, to the consumption of muscle tissue or, as Mayer maintained, to the oxidation process in the capillaries of the muscles.[116] However, the burden of the polemics against Liebig was directed against his use of the concept of vital force in the third section of the *Animal Chemistry*. Mayer did not fail to call attention to the "logical inconsistency" in Liebig's explanation of organic motions, one of those "internal contradictions" not infrequently encountered in his writings.[117] He quoted with approval three paragraphs from the first section of Liebig's book in which Liebig traced the production of all heat and motion to the exchange of matter.[118] He then quoted with disapproval two paragraphs from the last section of the book in which the expenditure of vital force was identified as the cause of an animal's production of mechanical work.[119] Mayer concluded:

> The exchange of matter that was first declared to be the *cause* of the work performed by the muscles is here made into its *result*. To put it in terms of one of Liebig's examples, first the river drives the mill, and afterwards the mill drives the river.
>
> A specific capability of resistance of living tissue "against external causes of disturbances," against the influence of atmospheric oxygen, against putrescent decomposition, as the effect of a distinctive vital force in Liebig's sense, we cannot allow because we do not regard ourselves as justified in hypothetically introducing specific causes where no specific effects have been demonstrated; in other words, because we should not think anything without a reason [*Grund*].
>
> . . .

> Since we perceive in a chemical process, in the exchange of matter, an adequate basis [*vollwichtiger Grund*] for the continued existence of living organisms, we must therefore protest against the proposal of a vital force in the sense of Liebig, Autenrieth, Hunter, etc. But as regards Liebig's hypothesis of the *expenditure* of a vital force to [produce] mechanical effects, that appears to be still more venturesome than the positing of such a *vis occulta* in and of itself.[120]

Mayer quoted another four paragraphs from Liebig's book in which Liebig described the production of mechanical effects in the animal organism in terms of the transformation of animate into inanimate matter, the precondition for which was a reduction in the power of the vital force to resist oxidation, that reduction in turn having been caused either by a withdrawal of heat from the affected part of the animal or by the consumption of vital force to produce mechanical motions.[121] Mayer offered a chemical analogy to demonstrate the weakness of Liebig's argument: certain substances (such as gadolinite and five others he named) undergo a change in resistance to chemical action attendant upon the production of heat or some other (unspecified) force, but unlike the effect of the organism's loss of heat or its production of mechanical effects in *reducing* its resistance to chemical decomposition, the resistance of gadolinite (etc.) *increases*. The point Mayer drew from this line of reasoning was important, and recalls the kind of questions physiologists had traditionally asked with regard to an alleged vital force:

> But the causal connection between vital force and mechanical effect must also be declared inadmissible for another reason than that of analogy. Just as an effect does not arise by itself, so too no cause vanishes without a corresponding effect. The binary combination of carbon and oxygen cannot occur, nor can gadolinite (etc.) acquire that capability of resistance, without a corresponding development of heat (or an equivalent force). Liebig says (*op. cit.*, p. 249): "The sum total of force expendable for mechanical effects must be equal to the sum total of the vital force of all the organic structures [*Gebilde*] suitable for the exchange of matter." Now, what effect does this force manifest after death has occurred? According to Liebig's theory, the body of a beheaded person would have to produce a great sum of mechanical effects, or a corresponding amount of heat, before the onset of putrefaction; but experience confirms nothing of the kind.*
>
> . . .
>
> *Question: What becomes of the vital force after death? Answer: Nothing. Conclusion: The vital force is therefore = nothing.—*Nihil fit ad nihilum*. A force or cause that can vanish without manifesting an effect is not a force.[122]

Where Müller had posited the latency of the vital force in all matter, where Autenrieth had assumed the existence of a preterphysical realm to which the vital force could return after death, and where Lotze had assumed the existence of a variable force, Mayer tacitly invoked his identification of forces as causes and his principle of the indestructibility and uncreatability of force in order to deny legitimacy to the very notion of a vital force à la Liebig.

Mayer would not have been Mayer had he not elaborated still another analogy, this time between the irritability of organic tissues and the expansibility of gases. Irritability was, in his interpretation, the capacity of living matter to transform chemical force into mechanical effect, while expansibility was the capacity of gases to transform heat into mechanical effect.[123] He compared the disappearance of the elasticity of a gas at a low enough temperature (at which the gas condenses) to "the well-known gradual difference in the *permanence of irritability*, or tenacity of life, among different classes of animals"—namely, that cold-blooded animals remain capable of irritable muscular response under circumstances (such as reduced oxygen supply and death) when the muscles of warm-blooded animals cease to function.[124] Thus "the apparatuses of motion [i.e., the muscles] of cold-blooded animals appear in general analogous to permanent gases, those of warm-blooded animals to vapors."[125] He reinvoked the analogy a few pages later when considering the "mechanical quotient" of muscles, that is, their efficiency in transforming a given amount of chemical force into work. Although the necessary experimental data were lacking, he ventured two tentative conclusions: "1) The stronger the chemical process or the formation of carbonic acid is in an animal, the smaller the mechanical quotient and the less the mechanical work in relation to the production of heat. 2) By analogy with elastic fluids, the mechanical quotient is greatest for permanently irritable muscles."[126] In other words, cold-blooded animals may be slower and consume less oxygen than warm-blooded ones, but their muscles work more efficiently. This time, however, unlike most instances when he appealed to sometimes rather farfetched analogies, Mayer seems to have appreciated the intrinsic limitations of *all* analogies:

> If the parallel drawn between the elasticity of gases and the irritability of muscles is carried further, then it shares the fate of all analogies; the initially natural comparison soon becomes artificial and ultimately loses itself in paradoxes. Gases are formless substances, whereas muscles are organized, and their actions depend more or less on the influence of the motor nerves; we call this specific influence *innervation*, to which the expansible fluids in and of themselves reveal nothing corresponding.[127]

Few characteristics of Mayer's way of thinking are as striking as his attachment to analogies, which ran the gamut from the most essential component of his ideas to merely ephemeral usages, from the analogy between force and matter to that between the increasing rate of decomposition of dead animal tissue and the increasing rate of acceleration of bodies falling to earth from far out in space.[128] Undoubtedly in response to the perceived danger of a materialistic misinterpretation of what was only a heuristic analogy, in later years Mayer carefully qualified the all-important comparison of organisms to machines:

> The material necessary for this mechanical work [of animals] derives, however, from the plant kingdom, to which it has in any case earlier flowed from the sun; thus animals transform erstwhile sunlight into motion and heat. In *this* regard, I say in *this* regard the animal organism, with all the infinite diversity of its anatomy, can nevertheless be compared to a steam engine. That is, the steam engine also consumes for

> the production of its output [*Leistung*], for the production of work—and of heat—the sunlight stored up by the plant world, and we cannot help but . . . make frequent use of this comparison. But let it be stated right at the beginning that a comparison depends on the discovery of similarities, but that similarities are a long way from being an identity. The animal is by no means a mere machine; it stands far above even the plants, for it has a *will*.[129]

He was not always so fastidious in his distinctions.

We have already encountered Mayer's fascination with the perennial problem of explaining the source of the sun's unimaginably great output of heat and light. For a while he thought he had found it in the continuous annihilation of the centripetal component of the orbital motion of bodies circling the sun, even if that meant abandoning the (later!) logically necessary principle that nothing can be created out of nothing. Giving up that explanation was part of the cost of accepting the universal uncreatability of force. Yet the problem stayed with him of explaining what Herschel had termed (in the German translation Mayer read) "das grosse Geheimniss."[130] Mayer's new solution to the problem was a straightforward application of his mechanical theory of heat: the sun's radiation is the result of the transformation into heat of the vis viva of meteors falling at tremendous speed into the sun. As he put it with his usual rhetorical confidence in his announcement of his theory in 1846 to the Paris Academy: "The sun's radiation is produced by the fall of cosmic masses, because without that there would have to be effects without cause and causes without effect."[131] He developed his theory further in the *Contributions to Celestial Dynamics* of 1848, where he showed that the known effect can be accounted for on the basis of reasonably plausible assumptions—but for one problem. Even if the increase in the sun's diameter were unobservable over thousands of years, the increase in its mass should have produced a detectable effect on the periods of the planets. Mayer's solution revealed his profound ignorance of the nature of wave phenomena:

> An undulating motion proceeding outwards from a point or a surface in an unbounded medium cannot, however, be conceived without a simultaneous progressive motion, a being-pushed-forward [*Fortgedrängtwerden*] of the vibrating massy parts, and there thus lies—according to the vibrations theory no less than according to the emanations theory—in the sun's radiation a reason for a continuous decrease in mass of this fixed star. But why, despite that, the sun's mass does not suffer an actual decrease, for that a sufficient reason was already given.
>
> The radiation of the sun is the centrifugal action equivalent to a centripetal motion.[132]

The last sentence is a startling reincarnation of his earlier ideas: the physical mechanism may have changed, but the suggestive spirit of the language survived.[133] Ironically, at the beginning of the work he had invoked an analogy between light and sound that might have set him straight about the nature of wave motion: both consist of vibrations given off by glowing or resonating bodies to the surrounding medium, vibrations that must steadily diminish unless there is some source of continued motion. But when he noted that "people have frequently and appropriately compared the sun to a continually sounding bell," he failed to consider that the sound of a bell is not carried to our ears by air particles traveling out from the source of vibrations: the bell is not a source of air.[134]

Among the manuscripts in the Mayer archives in Heilbronn is an appreciable fragment on large sheets, numbered 57 to 67. It was apparently once part of a sizable work, and from the contents of what remains it would seem to have been a major summing-up of his ideas, presented in such a way as to give a good sense of the often unspoken considerations that attended his original reflections. It reveals a lot about his wrestling with the issue of the possible creation of force in nonliving systems and its implications for living organisms; about his concern for *das Geistige* and his conviction that science legitimates belief in the soul; and about the depth of his religious conception of the world in general. From internal evidence one can only say with some assurance that it dates from after the late 1840s, and probably after around 1860; on the basis of certain peculiarities of expression I have tentatively dated it to around 1866.[135] It thus provides an unusually valuable insight into both his mature thinking and the kinds of considerations that motivated him at earlier stages. The surviving manuscript picks up in the midst of a discussion of issues that might cause someone to question the uncreatability of force:

> . . . to spread light over what takes place in the microcosm; for if it could be proven that mechanical work were performed, that light or heat were developed somewhere in the macrocosm through the joint action of its parts, ordered with such great wisdom, then the possibility would be indubitably demonstrated that the same thing could also take place in the living creatures of the earth; in the other case, however, if the law valid for the inorganic world—no work without consumption—proves to be true everywhere, even in the planetary system itself, then we are [justified] by the strictest induction to claim the same law as well for muscles and the animal organism in general—a conclusion that henceforth could only be undone by assured facts, but not by scientific-vitalistic meditations.
>
> There are, however, in all three kinds of such macroscopic phenomena: 1) the motions of the heavenly bodies in their orbits around the sun, 2) the development of light and heat in the planetary system, and 3) the tides.[136]

Such had been his own concern in 1841 with respect to the second point; when he first perceived the tides to present a similar challenge is impossible to say. Whether or not he himself was ever troubled by the fact that the solar system is a kind of perpetuum mobile, as some contemporaries put it, is unclear, and he may have included it for expository purposes. Such as it might have been, that problem had been solved long ago by Newton's law of gravitation. Mayer continued:

> Now as regards 2) the development of light and heat on the sun, it would be entirely in accordance with the taste of the vitalists if one were to imagine this action as the effect of a vital force inherent in the planetary system produced by polarity, etc.; but

since, as everyone knows, the exact sciences will not content themselves with such phrases, I have therefore attempted to explain this natural phenomenon in the spirit of the latter, i.e., to represent it as a lawbound process.

The tides, finally, which must be mentioned here once again, are an interesting and important phenomenon for biology because it can be determined with mathematical precision whether the work produced by this macrosopic process is to be regarded as only the result of the combination of certain organ-parts of the planetary system, or whether here, too, work and consumption again go hand in hand. The question is, namely: can work be performed by gravitation in and of itself alone? If [the answer to] this question turns out to be affirmative for the tides, then on the basis of this example the possibility may no longer be denied that motion can also be generated by means of the irritability of the muscles, without any consumption. But insofar as I have demonstrated in the foregoing that the work of the tides can only be performed at the expense of the earth's rotational motion, then there remains no justification to claim for the organic arrangement of a system, or for the vital process, the capability in and of itself alone of being able to perform actions.[137]

Perhaps Mayer had been prompted to consider the problem of the tides in conjunction with his running battle against the common usage which identified gravity (*Schwerkraft*) as a force: since it was not a force in *his* sense of the word, it could not by itself be the cause of motion. Perhaps he had been bothered by the fact that the earth's rotation has apparently not slowed measurably since antiquity, despite the continuous friction of the tides, and hence continuous consumption of motion: his solution was to posit a precisely compensating contraction of the earth's radius. In his continuation, Mayer gave his most extensive surviving consideration of 'metaphysical' issues:

Applying this to psychic actions, it can thus no longer be a question of wishing to see these actions, in accordance with physical laws, as the result of the combination of the brain fibers or of the vital force of this organ; but there would still remain two assumptions. That is, if one makes an analogy between brain and muscle, then one can arrive at the conception that, just as the muscle performs its mechanical work with the help of the chemical process and with the consumption of the force supplied by the latter, so, too, the brain may be able to produce its psychic activity with the help of the same process.

At first glance such a view might have something seductive about it, [but] on closer examination it proves to be completely untenable. That is, the actions of the secretory and motor organs depend essentially on the fact that the work [*Leistung*] they produce—the product of secretion or the mechanical work [*Arbeit*]—is equivalent to the consumption taking place in these organs, an equivalence that can be and has been demonstrated by means of empirical laws. The carbonic acid exhaled by the animal is equivalent to the carbon and oxygen the animal consumes, its mechanical work is equivalent to the process of chemical combination taking place in its motor organs; but how is the psychic work produced by the brain suppposed to be taken as the equivalent of the chemical process, heat, electricity, etc.? And one would wish to undertake to adduce proof of this?

Now since the psychic activities can and should accordingly be regarded as neither the result of the animal's vital force nor the result of a chemical process in it, we are thus led necessarily back to the principle that the healthy reason of all humanity has at all times and in all places recognized as an axiomatic truth and [for which the expression is found in all languages], and we cannot hesitate to declare that the assumption of a metaphysicial [?] existence, the assumption of a feeling, willing, and thinking entity, a soul, asserts its authority as an irrecusable postulate for the natural scientist.[138]

After a one-and-a-half-page discussion (of the kind common in Autenrieth, Müller, et al.) of mental functions and the extent to which they do or do not depend on the integrity of the brain, Mayer concluded:

Now since there is no part of the brain, much less of the rest of the body, that necessarily determines a certain psychic disturbance by means of a change in its structure, it thus in turn follows that all corporeal organs are only capable of exciting disturbances in the psychic activities per consensum, or in other words that the organ of psychic activity is not corporeal.

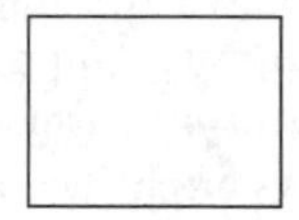

If I should now in addition speak out about the continued existence of the soul after the death of its material organs, about the human being's personal immortality, then I can repeat that which has very often been said and which is written in every breast.

The complete proof of this continued existence lies in the divine perfection [*Vollkommenheit*] of the entire arrangement of nature. Every plant and animal organism is constructed in such a way that it is capable of fulfilling the purpose for which it was created and of attaining the last stage of perfection, [and it would be a ridiculous presumption to regard an improvement in a single organ(ism?) in the long series of plant and animal species even as only possible]. And yet the fool complains rightly, as does the wise man, about the imperfection of this world; but the only thing imperfect [or incomplete (*unvollkommen*)] in it is the human spirit; it alone in this world is not permitted to attain its purpose, that of divine consummation [or perfection (*Vollendung*)], for what is the wisdom of a human being and the virtue of the most virtuous of this earth? The human spirit can be compared to the air enclosed within the embryonic membranes. If we take its future away from the infant organism at its separation from the umbilical cord, or from the spiritual organism at its separation from the body, then we are left in both cases with nothing but a miscarriage, incapable of bringing to maturity the germ placed in it.

There are thus only two assumptions possible: either nature must have neglected, in this world otherwise everywhere constructed with divine perfection, to impress the seal of consummation on the crown of its creation, the human being, or the human being's spiritual individuality must also have a future after its separation from the

> body, a future in which it will have the leisure [?] to attain its purpose, divine perfection.
>
> The means nature chooses for the attainment of its intentions are, to be sure, entirely hidden from us in very many cases, and in particular the most important epochs of human existence—generation, birth, and death—are enclosed in impenetrable mysteries. But since we know that the farthest fixed stars send each other light in an unknown fashion, and that a mysterious bond of attraction connects together all the masses of universal space, why should we wish to despair of the possibility of a connection between distant spaces and distant times in the moral world?[139]

In what remains of the manuscript, Mayer went on to discuss the production of heat in the sun from the fall of fast-moving "cosmic masses" and the transmission of that force to us as light; he addressed the contention of Georg Friedrich Parrot that projectiles cannot be heated by friction; and he argued against the existence of active vulcanism on the moon.[140] As usual, he sought to buttress his ideas by means of an analogy, this time between meteorites and blood corpuscles:

> One regards a meteorite as too inconsiderable to credit it with an important purpose in the economy of the planetary system. But what a small thing is a blood corpuscle, of which over 100 million float in a single drop of blood! And yet the organism can more easily do without whole limbs than these blood corpuscles, on which depend its manifestations of life, its heat, and its mechanical actions. In the microcosm as in the macrocosm it is the incalculable number [*unermeßliche*(?) *Zahl*] that is capable of endowing the tiniest objects with the highest significance.[141]

Mayer's attachment to his many analogies was such that it is hard to see them all as merely literary devices or heuristic illustrations; they seem rather to have constituted a wide-ranging web of connections that lent his ideas broad significance, that gave them meaning.

One of the most difficult issues Mayer faced at the beginning of his creative process was deciding what forces are and how they are to be measured. Is motion a force, and not simply, as was more usually maintained, the effect of a force? If so, is it to be measured by mass times velocity or mass times the square of velocity? (Mayer never made the conceptual transition from vis viva to kinetic energy—that is, by introducing the factor of one-half—nor was he fussy about the distinction between mass and weight.) Is gravity a proper force? If not, what *is* the force that causes bodies to fall? Similar uncertainties attended the concept of cause. If forces are causes, and *causa aequat effectum*, what about cases where the effect seems much greater than the cause, such as a spark setting off an explosion or a tiny (chemical or physical) stimulus eliciting a violent response in an organism?[142] Is the problem with the definition of cause, with the presumed equality between cause and effect, or with the application of those concepts to the organic realm? What did his basic terms *mean*, and what was his theory a theory *of*? The answers to such difficult questions evolved over time. As Mayer put the issue in 1851, as far as forces are concerned, the question is not "what kind of a thing a 'force' is, but what thing do we wish to *call* 'force.'"[143]

He achieved the desired circumscription of the term "force" within two to four years of his initial thoughts on heat and motion. How to deal with causes was a problem that occupied him till the very end of his active intellectual life. His last essay, in 1876, dealt with what he termed *Auslösung*, the "unloosing" or release of a large effect by a small cause. Not surprisingly, he began his essay with one of his oldest and favorite examples of chemical change:

> Very many natural processes only take place if they are initiated by means of an impulse, and this circumstance is what the science of the day calls "unloosing." One of the readiest examples is certainly oxyhydrogen gas [*Knallgas*]. As is well known, a mixture of oxygen and hydrogen gas in the ratio in which they yield water enters by itself into no chemical combination until the latter is initiated by means of heat or an electric spark or by platinum black. In the same way, when we ignite a match by means of a little frictional heat and by means of this burning match initiate a further, arbitrarily large combustion process, so, too, do we have here another simple example of "unloosing," and such examples confront us close at hand in infinite plenty. Light pressure with the finger produces a violent effect with firearms, etc., etc.[144]

Characteristic of processes of unloosing was that they cannot be defined in terms of a quantitative relationship between cause and effect. Organic phenomena present countless examples of such processes, for example the movement of a muscle in response to a nervous impulse under the direction (perhaps) of the will. Mayer implicitly likened such physiological processes to the function of a person controlling a machine: he recalled how, as a child, he enjoyed watching the operator of a water-powered sawmill set it in motion by simply pressing a lever and thereby withdrawing the sluice gate.

Mayer's concept of unloosing represented the culmination of long years of wrestling with issues of cause and effect. In a letter to Griesinger in 1844 he wrote that "with pedantic logic I cherish the pious wish that by cause and effect (in inanimate nature) one understand either things that stand in a quantitative relationship to each other *or* that are not in proportion to each other."[145] His example of the latter was a spark setting off an explosion. Forces are causes, but not all causes are forces. A principle of the 'indestructibility of causes' was not, ultimately, one he wished to defend.

Several adumbrations of his later distinction occurred in his *Organic Motion* of 1845. With respect to the phenomena of muscular motion he suggested a distinction between "conditions" and "causes," between "psychic" and "physical" factors:

> Two different things belong to a muscle's manifestation of activity: 1) the influence of a motor nerve as condition, and 2) the exchange of matter as cause of the work performed [*Leistung*].
>
> Like the whole organism, so, too, does the organ, the muscle, have its psychic [*psychisch*] and its physical sides; to the former we reckon the influence of the nerves, to the latter the chemical process.
>
> The movements of a steamboat obey the will of the helmsman and the machinist.

> But the psychic [*geistig*] influence, without which the ship would not set itself in motion . . .—it directs, but it does not move anything; continued motion requires a physical force, coal . . . .
>
> . . .
>
> The action of the muscle, the transformation of chemical force into mechanical effect, is conditioned [or determined (*bedingt*)] in a mysterious fashion by a contact influence that, as experience shows, pertains to the nervous system. Insofar as the motor apparatuses depend to a *different degree* on this influence, they are divided into voluntary and involuntary. If we imagine a steam engine of which one part works as soon as steam is generated, but another part is only set in motion as a result of the machinist's contrivance, then we have voluntary and involuntary motor apparatuses here as well. The machinist is only capable of bringing his influence to bear on the apparatus at his disposal by means of a certain expenditure of force; but the expenditure of force is *vanishingly small* in comparison with the mechanical work produced, and can in general, with the increasing perfection of the apparatus, be made smaller than any given magnitude. In the same way we must declare it to be not only possible, but even probable, that without [notable] expenditure of a physical force, without an electric current and without any chemical process at all, the innervation exerts its control over muscular action.[146]

But for the addition of a qualifying "notable" in 1867, Mayer came close to asserting a sharp distinction between the entirely nonphysical *direction* of organic phenomena and the physicochemical *sources of force* that produce the mechanical effect, especially considering his use of the phrase "contact influence." Such a distinction corresponds closely to the tacit distinction many earlier physiological writers made between the vital force as either the director of organic processes or the source of motive force. For Mayer the will belongs to the "realm of freedom," which is the business of philosophy and theology; science proper deals with the realms of necessity and purposefulness pertaining to the physical and organic worlds.[147] The relationship between force and the soul was a long-term concern of his.

Contact and catalytic forces could not be bona fide forces in Mayer's sense of the term, but they could nevertheless still be regarded as causes of certain phenomena; the terms could be retained as long as they were properly understood:

> Everyone knows that in numerous cases chemical actions are determined by the mere presence of certain substances that themselves take no part in the change taking place. If one wishes, without making any assumptions, to bestow a name on an established fact, then one can designate by means of the expression "contact influence" the role played in such circumstances by the matter that remains unchanged; as is well known, one is otherwise also accustomed to speak of "catalytic force" and "catalytic effect"; but if by force one is suppposed to understand only "the measurable cause proportional to a measurable effect,"* then for understandable reasons a specific force cannot be attributed to the phenomenon in question.
>
> . . .

> *A force is called "catalytic" insofar as it stands in no kind of quantitative relationship to the intended effect. An avalanche falls into the valley; a gust of wind or the beat of a bird's wing is the "catalytic force" that gives the signal for the fall and brings about the extensive destruction.—The "catalytic" or "paralytic" aspect of this force refers in the first place to logic, or the law of causality.[148]

In other words, "catalytic" properly relates to how we use the concept of cause, not how we use the concept of force. By "paralytic" he probably intended the word's etymological sense of "loosed from beside." He was still searching for concepts and terms corresponding to his evolving understanding of the intertwined issues, just as he continued to refine the scope of his ideas and the field of application of those concepts and terms. *Auslösung* stood at the end of a development of ideas that had begun with the particulars *heat* and *motion*, seen the creation of the general concept of *force*, and entailed a laborious struggle to come to terms with the incommensurable meanings of *cause*.

• C H A P T E R S E V E N •

# Mayer and *Naturphilosophie*

As I recounted in the Introduction, this study began as an investigation of the widely accepted belief that Mayer's formulation of what became known as the principle of the conservation of energy was due in some essential way to the influence of Schelling's *Naturphilosophie,* interpreted primarily as asserting a fundamental unity of the forces of nature. The omission of any evidence supportive of such a claim from the foregoing reconstruction of the context and development of Mayer's ideas implies that I have not found such a contention supported by the available evidence, nor does a satisfactory understanding of Mayer's work appear to require the assumption of such a philosophical influence. On the other hand, neither the inadequacy of past interpretations, nor the absence of compelling evidence, nor even the possible sufficiency of the present historical analysis necessarily entails the impossibility that Mayer's way of thinking in fact owed something to *Naturphilosophie*. There are, indeed, certain motifs in Mayer's writings that *are* reminiscent of topics addressed by some of the principal *Naturphilosophen*, notwithstanding the striking absence in his writings of many of the better-known and seemingly more characteristic aspects of *Naturphilosophie*. The issue does not, I think, ultimately lend itself to a categorical determination, although one can go quite far in circumscribing the possible extent and significance of such an influence. From a larger perspective, the now almost ritual invocation of *Naturphilosophie* to explain Mayer's discovery of the conservation of energy is worth addressing as an issue in its own right whose elucidation can help clarify the widely misunderstood and caricatured nature of that early nineteenth-century German philosophical movement and its relationship to the science of the day.

The first person to suggest, albeit guardedly, that Mayer's thinking may have owed something to *Naturphilosophie* seems to have been Salomo Friedlaender in an insightful study of Mayer's work published in 1905. Associating *Naturphilosophie* with belief in a vital force, Friedlaender (wrongly, as we shall see) took Mayer's opposition to the latter as evidence of his rejection of the former.[1] Nevertheless, he suggested that Mayer's early enthusiasm for the idea of the unity and indestructibility of forces betrayed atavistic traces of *Naturphilosophie.*[2] Friedlaender did not identify just what those traces were, but went on (without clear warrant) to identify as *naturphilosophisch* Mayer's lingering disposition to regard the solar system as a divinely ordered system in which force is produced. However, in previous chapters I have argued that the latter belief resulted from Mayer's attempt to make sense of the then-standard application of the parallelogram of forces to central-force motions in conjunction with the then-standard reference to the solar system as an organism.

In 1940 Alwin Mittasch asserted more forcefully that the characteristically Romantic notion of the "unity of force," coupled with empirical examples of the interconnectedness of all forces, had created a kind of general consciousness of the unity and transmutability of force that influenced Mayer during his student years. As evidence for the influence of *Naturphilosophie* he cited the ostensibly telling words from Mayer's unpublished 1841 paper, "We can derive all phenomena from a primitive force [*Urkraft*]." Two years later Paul Diepgen identified rather vaguely as *naturphilosophisch* Mayer's predilection for analogies and (somewhat inconsistently) both his frequently deductive style of presentation and the tendency he evinced even as a youth to leap over the connecting links of a logical argument. In his 1959 book on *Naturphilosophie in the Nineteenth Century*, Gerhard Hennemann asserted, without evidence or argument, that Mayer was unconsciously influenced by the Romantic notion of the unity of the forces of nature even though he claimed to oppose *Naturphilosophie*.[3]

As far as the current historiography of science is concerned, however, the largely unelaborated suggestions of Friedlaender et al. are obscure and forgotten forerunners of the argument more forcefully and visibly advanced by Thomas Kuhn in a seminal paper read in 1957 and published in 1959. Kuhn's now classic "Energy Conservation as an Example of Simultaneous Discovery" sought to isolate the factors behind the spate of closely contemporaneous formulations of concepts akin to energy conservation by comparing the ideas and backgrounds of the twelve most important scientists. He argued that there were "major conceptual lacunae" in the accounts of how six of his figures got to something like energy conservation, that "in the cases of Colding, Helmholtz, Liebig, Mayer, Mohr, and Séguin, the notion of an underlying imperishable metaphysical force seems prior to research and almost unrelated to it. Put bluntly, these pioneers seem to have held an idea capable of becoming conservation of energy for some time before they found evidence for it."[4]

Kuhn located "Mayer's leap" in the gap between his observation of the lighter-than-expected color of venous blood in the tropics and his conclusion "that internal oxidation must be balanced against *both* the body's heat loss *and* the manual labor the body performs. To this formulation, the light color of tropical venous blood is largely irrelevant. Mayer's extension of the theory calls for the discovery that lazy men, rather than hot men, have light venous blood."[5] Kuhn's error here was in trying to get energy conservation out of Mayer's physiological observations and reflections, whereas what Mayer thereby discovered (without any logical lacunae but only by invoking both a long-held conviction in the impossibility of mechanical perpetual motion and a heuristic energetic analogy between organisms and machines) was the equivalence between heat and "motion." Both the general concept of force and its indestructibility and uncreatability were ideas Mayer arrived at via a host of other considerations.

Kuhn found in *Naturphilosophie* a possible source for the kind of metaphysical conviction he identified in six of his codiscoverers of energy conservation: "Positing organism as the fundamental metaphor of their universal science, the *Naturphilosophen* constantly sought a single unifying principle for all natural phenomena."[6] Though Kuhn did not use the phrase "unity of forces," he did cite Schelling's belief in the interconnectedness of the forces of nature. Of apparently greater significance to him was the attention Schelling paid to conversion and transformation processes, though that was not the way Schelling conceptualized them. Kuhn did not substantiate the assertion that Schelling's teachings "dominated German and many neighboring universities during the first third of the nineteenth century," an overbroad generalization in need of specification, especially since it is not true of Mayer's university.

Kuhn further motivated the suggestion that *Naturphilosophie* might close the explanatory gap in the accounts of Mayer, Helmholtz, Liebig, Colding, and Hirn by citing what he admitted were only "biographical fragments" relative to each. Thus: "Mayer did not study *Naturphilosophie*, but he had close student friends who did."[7] Only further research, he recognized, could substantiate this hunch. But he argued that it was significant in this regard that five of his twelve pioneers were Germans, at a time when Germany "had not yet achieved the scientific eminence of either Britain or France," and that furthermore "a sixth, Colding, was a Danish disciple of Oersted's, and a seventh, Hirn, was a self-educated Alsatian who read the *Naturphilosophen*."[8] Kuhn thus concluded:

> Unless the *Naturphilosophie* indigenous to the educational environment of these seven men had a productive role in the researches of some, it is hard to see why more than fifty per cent of the pioneers should have been drawn from an area barely through its first generation of significant scientific productivity. Nor is this quite all. If proved, the influence of *Naturphilosophie* may also help to explain why this particular group of five Germans, a Dane, and an Alsatian includes five of the six pioneers in whose approaches to energy conservation we have previously noted such marked conceptual lacunae.[9]

Kuhn was cautious enough to do no more than urge the reasonableness of his conclusion as a stimulus to further research. Given the state of historians' understanding of both Mayer and *Naturphilosophie* when Kuhn wrote his paper, his hypothesis concerning Mayer's relationship to *Naturphilosophie* was not unreasonable. Nor was Kuhn alone in failing to distinguish *Naturphilosophie* from Kantianism, though that confusion weakened the cogency of his more general hypothesis, especially with regard to Helmholtz.

During the 1970s the historian of science who most insisted on Mayer's affiliation with *Naturphilosophie* was Armin Hermann. Such a connection was, he said, "demonstrable," though he himself consistently omitted that demonstration.[10] Hermann sometimes preferred to link Mayer's work not so much to Schelling's *Naturphilosophie* in particular as to the broader school of thought he termed "dynamism," which emphasized the ontological primacy of force and the interconnectedness of the various forces of nature.[11] Alas, there is no evidence Mayer ever entertained the central tenet of dynamism, the construction of matter out of oppposing attractive and repulsive forces, or that he ever rejected chemical atomism; nor do the terms "dynamic" and "dynamism" often appear in his writings. In Hermann's most elaborated account, however, he argued for Mayer's debt to specific ideas of Schelling's. Referring to Mayer's unpublished essay of 1841, Hermann wrote: "Mayer here adopted the Schellingian conception that everything that happens in nature depends on the separation and recombination of polar forces. But he followed Schelling not only with respect to general principles, but also with respect to specifics, in particular the explanation of the generation of light in the sun through the force of gravity, through the mutual attraction of planet and sun."[12] He noted (without citing any textual support) that Schelling had spoken of the separation of zero into two opposing forces of equal magnitude. Here, at least, Hermann identified some of the most *naturphilosophisch*-sounding aspects of Mayer's work, although polarities per se played scarcely any role in Mayer's thinking. Hermann's consistently brief discussions do not evince either a deep understanding of Mayer's work or a broad appreciation of the range of Mayer's likely sources, nor do they address the crucial question of how *specifically* Mayer might have been exposed to the putative influence of *Naturphilosophie*, nor of how its concerns related to problems Mayer might plausibly have been concerned with. Ideas are not just in the air, and one wants to know how a professedly unphilosophical medical student at the University of Tübingen would have encountered such ideas during the 1830s, decades after the heyday of *Naturphilosophie*.

During the last two decades it has been especially Dutch and German historians of science who have continued to assert the importance of *Naturphilosophie* to Mayer's conception of the conservation of energy, usually in terms of belief in the unity of forces, and always without anything approaching a cogent historical analysis.[13] None of these authors seems to have noticed that Mayer never made an issue of the unity of force(s). The focus of his attention was rather on the relationship between heat and motion, about which *Naturphilosophen* typically had little to say.[14]

Continuing to conflate *Naturphilosophie* and Kantianism, Norton Wise argued that early nineteenth-century German "Idealist traditions" taught the indestructibility of the primordial motive forces of nature, and that such ideas "form[ed] the basis . . . for attempts such as Mayer's to enunciate conservation more precisely."[15] But his only textual reference was to Kant's then-unpublished "Opus postumum," and we are left ignorant of the extent to which *Naturphilosophen* in fact made an issue of the indestructibility of force. Nor did Wise suggest what works Mayer might plausibly have gotten such ideas from. His analysis is typical of those that, like Heimann's five years earlier, prefer the philosophical analysis of ideas and isolated texts to the tracing of actual problems in specific, historically relevant sources.[16] Nevertheless, Wise did appreciate the broad historical significance of an aspect of Mayer's works related to his rejection of the mechanical (i.e., mode-of-motion) theory of heat:

> In refusing . . . to reduce heat (and electricity) to matter, motion, and the forces acting between parts of matter, Mayer maintained the *naturphilosophisch* notion of force as a sort of substance, now independent of matter but still having the same status as matter. . . . That the rejection of *Naturphilosophie* meant precisely the rejec-

> tion of such "metaphysical" force substances in favour of the mechanics of matter helps to explain why space-occupying force fields did not receive serious consideration in Germany much before 1880. A mechanical ether provided the only legitimate basis for unity.[17]

Wise had his finger on an important aspect of Mayer's thinking—the ontological ("substantial") independence of force from matter—but he was mistaken in associating such an idea with *Naturphilosophie,* which tended in part to regard force as ontologically more primitive than matter, in part to treat the agencies more generally known as imponderables as special states of matter. Mayer, on the other hand, adamantly insisted on the independence of *both* ontologically primitive substances, force and matter.[18] The historical background to those considerations was, as I have argued in earlier chapters, Autenrieth-like reflections on the nature of the soul, the vital force, the imponderables, and ponderable matter. As with the association of *Naturphilosophie* with belief in a vital force, so here its association with the notion of a metaphysical force substance represents a distorted picture created by its latter-day critics. As will be demonstrated below, even the association of *Naturphilosophie* with belief in the unity of the forces of nature represents something of a distortion of what *Naturphilosophen* actually said about forces, though this distortion has come primarily from those who find a positive historical role for *Naturphilosophie*.

Several historians have, to be sure, questioned the cogency of the putative connection between *Naturphilosophie* and the conservation of energy. Barry Gower concluded his 1973 paper on "Speculation in Physics: The History and Practice of *Naturphilosophie*"—still one of the best analyses of the subject—with the judgment that "[Oersted's] dynamical theories of physical action, with their emphasis upon the interaction of polar forces, contributed little to the creation of a conceptual framework within which the energy conservation principle could emerge and be understood. . . . For this principle involved the idea that some physical *quantity* is conserved under transformation."[19] Herbert Breger likewise emphasized that Schelling's concept of force was not amenable to quantification, and that Romantic scientists did not have the important concept of mechanical work.[20] Peter Heimann's insightful study of Mayer's concept of force and its possible philosophical connections led him to the conclusion that *Naturphilosophie* was unlikely to have played a significant role:

> The fundamental aim of Schelling's philosophy was to discover certain a priori principles that were inaccessible to empirical cognition. . . . Schelling's conception of physics stands in sharp contrast to that of Mayer, who sought to render the concept of force as empirically meaningful as the concept of matter. . . . Schelling's theory of the polarity of forces as an expression of the tension between productivity and its limitation has no parallel in Mayer's writings. The general thrust of the writings of the *Naturphilosophen* such as Ritter was to account for phenomenal changes by an appeal to inner essences, and their notions of the unity of nature and the polarity of forces bear only a remote analogy to Mayer's concept of the transformation of forces.[21]

Nevertheless, Heimann recognized—with notable ambivalence—that some of Mayer's concepts do bear some resemblance to those of *Naturphilosophie*, though that recognition did not ultimately move him far from his previously stated conclusion:

> It seems that he was familiar with their speculations, and it is possible that the influence of *Naturphilosophie* was so pervasive in this period that it was implicit in all scientific activity concerned with the unity and conversion of "forces"; indeed, Mayer's reference to an *Urkraft* and to the unity of forces suggests an indebtedness to the ideas of the *Naturphilosophen*. But while such an influence is possible, it must be emphasized that in other respects Mayer's conception of nature was fundamentally at variance with the ideas of the *Naturphilosophen*. His matter-force duality and stress on the concept of force as an empirical quantity suggest that there would be little reason to attribute to *Naturphilosophie* any substantive role in shaping Mayer's natural philosophy. His intentions were fundamentally opposed to those of the *Naturphilosophen*, and claims for the influence of their ideas on Mayer's work must therefore be regarded with some suspicion.[22]

In his last paragraph Heimann again kept open the possibility that *Naturphilosophie*'s notion of the unity of forces might have had limited influence on Mayer's otherwise markedly different natural philosophy.[23] Allowing for Heimann's contextually restricted picture of both *Naturphilosophie* and Mayer's work, for the common misidentification of the "unity of forces" as Mayer's central motif, and for the failure to spot the significance of concepts such as indifference and the creation of force out of nothing, Heimann's assessment was not unreasonable.

The attempt to separate Mayer from Schelling et al. in terms of a distinction between empiricism and speculation is too crude to have much force. Nor, on the other hand, does Mayer's predilection for analogies by itself imply any affinity with *Naturphilosophie*, especially since Mayer's particular analogies are largely without any Romantic resonances. Among the concepts most central to Mayer's work most notably absent from works of *Naturphilosophie* (though important to Kantians) was efficient causality. As Hermann Schlüter rightly emphasized, Schelling's system was chiefly concerned with "formal causes," with finding the place of a given phenomenon within a structure of interconnected concepts. He quoted from Schelling's *Ideas toward a Philosophy of Nature* (1797/1803):

> Explanations take place as little in the philosophy of nature as they do in mathematics; it proceeds from principles certain in themselves, without any direction prescribed to it (say) by the phenomena; its direction lies in itself, and the more faithful it remains to this, the more certainly do the phenomena appear by themselves in that place in which alone they can be seen as necessary, and this place in the system is the only explanation of them that there is.[24]

Among the concepts most central to *Naturphilosophie* most notably absent from Mayer's works were those of *Steigerung* and *Stufenfolge*, in particular the tripartite hierarchies of light, electricity, and magnetism (as in Schelling's system), of gravitation, heat, and light (as in Eschenmayer's), and of the forces of magnetism, electricity, and chemistry (as in Walther's)—typically in parallel with a host of other forces, realms, essences, or aspects. Such concepts formed the ideal structure within which individual phenomena acquired their significance. Nor, given their preference for the conceptual and the qualitative, were *Naturphilosophen* advocates of the kind of quantification exemplified by Mayer's all-important calculation of the mechanical equivalent of heat.

Any argument connecting Mayer with *Naturphilosophie* thus faces serious conceptual difficulties. Nor is such a case made easier by the lack of direct evidence connecting Mayer to *Naturphilosophen* or their works. His close friend Gustav Rümelin reported that, as of the fall of 1841, Mayer had never read a philosophical book—which, even if not literally correct, tells us something about Mayer's scanty background in philosophy. When, at Mayer's request, Rümelin lent him two works by Hegel, Mayer returned them after a few days, complaining that he had not understood a single word.[25] Nor do Mayer's few disparaging references to *Naturphilosophie* reveal that he had any real acquaintance with its doctrines: for him, as for many of his contemporaries, it had become a term of general abuse applied imprecisely to what were perceived to be mystically unscientific and speculative ideas.[26] That evidence, by itself, would not preclude the possibility that Mayer was influenced by some more properly 'scientific' work of *Naturphilosophie*. It is, however, hard to identify any such work that would seem to provide a likely context for understanding any of the leading ideas and peculiarities of Mayer's work, quite aside from the question of the likelihood of his having read it.

In an historical analysis such as this, one faces the problem of who to take as representative of *Naturphilosophie*. Although a more exhaustive analysis would have to deal with that issue in a more principled and sophisticated manner, I have tried to choose major individuals unarguably so identified such as Schelling, Lorenz Oken, Carl Gustav Carus, and Philipp Franz von Walther, plus a few others who evinced strong ties to that movement such as Carl August von Eschenmayer, Moritz Ernst Adolph Naumann, Georg Friedrich Pohl, and (with some qualifications) Karl Friedrich Burdach. I do not think my conclusions about *Naturphilosophie* have been distorted by idiosyncracies or unrepresentative ideas in the works I chose to review, nor do I think my selection will be controversial to historians who have some familiarity with the relevant primary sources. At the same time I have allowed myself to skate pragmatically past the otherwise difficult and important issue of the legitimacy of identifying an individual *as* a representative of a particular 'tradition,' although I will address that issue briefly in chapter 8.

After attempting briefly to identify the leading characteristics of *Naturphilosophie*, the remainder of this chapter will examine what specific leading *Naturphilosophen* had to say about such topics as force(s), the vital force, and the physiology of respiration and animal heat, thereby providing a basis for comparison with Mayer's thinking; and it will evaluate the significance of the *naturphilosophisch* flavor of certain of Mayer's ideas and usages such as *Urkraft*, *Indifferenz*, creation out of nothing, and the relationship between light and gravity in planetary systems. I hope the passages quoted from primary sources will give the reader an adequate sense of the style and tenor of *Naturphilosophie*, since *Naturphilosophie*, more than most philosophical systems, cannot be understood independently of its peculiar use of language, and since it is impossible to gain a faithful sense of the character of *Naturphilosophie* from summaries and paraphrases alone.

## 1 The Leading Characteristics of *Naturphilosophie*

It will be useful to begin with a rough sketch of the kinds of concepts and topics generally characteristic of works of *Naturphilosophie*, without claiming that what follows can be taken as exhaustive or canonical. Although other authors with different purposes would undoubtedly identify a slightly different set, the features I have singled out were quite general, if not all present in every single work.[27] The weighting I've given to their exposition here reflects both their relative significance to my purposes and their likely (un)familiarity to my readers.

The most renowned characteristic of *Naturphilosophie* is surely its speculative character, as exhibited by the aprioristic deductivism prevalent in (for example) Schelling's and Oken's writings. *Naturphilosophen* had a high regard for the ability of the human mind to fathom the deepest truths about the world. As Burdach put it, "the laws of our thought are realized in the external world, since our mind [*Geist*] is only the reflection of the world spirit [*Weltgeist*]. Only thus are we capable of knowing nature."[28] For *Naturphilosophen*, experience often served primarily to exemplify the fundamental principles supplied by philosophical analysis. Thus the polarities of electricity and magnetism illustrated the kind of fundamental antitheses (*Gegensätze*) found throughout nature, but the polar nature of the activities and forces of nature was an a priori truth independent of specific empirical verification. At the same time, however, *Naturphilosophen* subscribed to a quasi-inductivist ideology of scientific knowledge, according to which the basic truths of nature—as embodied, for example, in certain *Ur-* or *Centralphänomene*—could literally be *seen* (*angeschaut*). An essential feature of a good theory was thus its *Anschaulichkeit*, its capacity for being grasped via an immediate perception of its fundamental truths. Of similar importance to many of them was the direct experience of certain *Fundamentalversuche*.[29] Accordingly, I find the common charge that *Naturphilosophen* ignored experiment to be an oversimplification. More accurate is the general view of *Naturphilosophie* as opposed to mathematics and quantification; such mathematical notation as one finds was often used in an unconventional and purely symbolic, noncomputational manner. Real physical understanding was essentially qualitative.

Such a stance was in part a reaction to the mechanistic and reductionist hypotheses of mainstream physics and chemistry (such as atomism), especially since the latter were often highly abstract and mathematical, devoid of any *Anschaulichkeit*. As noted above from Schlüter, *Naturphilosophen* tended to downplay the efficient causality of conventional science in favor of acausal, merely formal explanations. It is in this context that *Naturphilosophie*'s penchant for analogies must be understood: the parallelisms identified in different realms were not merely suggestive, but expressed the essential and deep interconnectedness of things—without, however, necessarily implying any actual transformation between analogous entities. Thus Oken associated the processes of gravitation, heat, and light with the organic systems of bones, muscles, and nerves.[30] Thus for Yelin the identity of electricity and magnetism did not consist in one's being reducible to the other, but in their common polar nature and in other analogies in their behavior, whereby of special importance was, for example, the comparison between magnets and thermoelectric crystals such as tourmaline.[31]

*Naturphilosophen* did typically believe in the unity of nature, and in the unity of knowledge within a closed system. In some ways the systematic character of *Naturphilosophie* was its most striking characteristic, one which distinguished it from other strands of Romantic science: it is impossible to imagine Schelling assenting to Goethe's "Nature has no system."[32] *Naturphilosophen* such as Schelling, Eschenmayer, and Walther constructed elaborate tripartite and hierarchical schemas embracing all natural phenomena. Common, too, was a microcosm-macrocosm view of the formal relationship between human beings and the world. Two of the most characteristic terms of *Naturphilosophie* were the already mentioned *Steigerung* and *Stufenfolge*—roughly, "raising (or intensification) by degrees" and "graduated progression (or series)."[33] One thus often spoke of the relationship of one activity or force to another in terms of their being manifestations of different powers (*Potenzen*). Walther's conceptions were entirely typical of the genre:

> If on the other hand one also considers life in its relation, and thereby as a necessary power of the eternal substance of natural objects [*Dinge*], then it still expresses the unity of activity and existence, only no longer the original, but rather the synthetic unity of both, i.e., their indifference. On the one hand, therefore, the series of powers, as it is present in the real universe, proceeds in this way, that existence designates the first power (through matter), or the power with predominant reality, activity the second power (through light), or the power with predominant ideality, and finally life the third power, in which the first and second, the ideal and the real, are equal to each other.[34]

The passage also illustrates the kind of language that relentlessly characterized *Naturphilosophie*.

One of the chief tasks of Schelling's natural philosophy was "to derive a dynamical *Stufenfolge* in nature."[35] Thus for Schelling the relationship between magnetism, electricity, and the so-called chemical process lay in the degree of separation of the antitheses present in the original productivity of nature, not in any reducibility of one to another.[36] *Naturphilosophen* paid particular attention to processes of development and metamorphosis, or progressive qualitative change. Comparative anatomy and embryology were, not surprisingly, among the research areas most favored by those touched by *Naturphilosophie*. Thus, for example, the tendency to try to understand animal heat by comparing phenomena from across the animal kingdom, as opposed to analyzing the physicochemical relationships within a single organism in a cause-and-effect manner.

In addition to "productivity," Schelling also spoke of the "absolute activity" that forms the basis of all existence.[37] The necessary limitation of that productivity or activity produces "an original duality" in nature of opposing tendencies of expansion and contraction.[38] Thus the absolutely fundamental importance in all versions of *Naturphilosophie* of concepts of dualism and polarity. Polarities, polarities, and again polarities. One of the most fundamental was the opposition between the primitive attractive and repulsive forces that constitute matter. Indeed, matter as an ontologically independent entity tended to disappear via its construction out of opposing primitive forces (*Urkräfte*). *Naturphilosophen* were strongly attached to dynamism, and typically opposed what they saw as an unphilosophical and crudely reductionist atomism.

The original duality that Schelling et al. posited in nature arose from a prior state of "indifference": the process of development was thus one of differentiation of an originally "indifferent" state, of the creation of heterogeneity out of homogeneity.[39] As will be developed more fully in section 4 of this chapter, this set of concepts provides the most fruitful lead for establishing a connection between Mayer and *Naturphilosophie*, and hence warrants a few general examples here.

For Schelling, life is characterized by a continuously reproduced state of differentiation, unlike the general tendency toward indifference in the inorganic world:

> The product, as long as it is organic, can never sink into indifference. If it would succomb to the universal striving toward indifference, then it must first descend to [the level of] a product of the lower power. As an organic product it cannot perish, and when it has perished it is actually already no longer organic. Death is return to the universal indifference. For just that reason the organic product is absolute, immortal. For it is an *organic* product just because it can never come to indifference in it. Only after it has ceased to be organic does the product dissolve into the universal indifference.[40]

For Burdach, the organism comes into existence "through the vital activity in an indifferent primitive mass [*Urmasse*]," thereby introducing a basic polarity into its nature: "Organic formation is a differentiation of the indifferent, a development of antitheses that continue to reproduce themselves, that produce ever new antitheses in themselves."[41] The context of Walther's use of indifference in the following passage offers a rich harvest of many of the characteristic themes of *Naturphilosophie*:

> For the human organism forms not only the midpoint, the actual dynamical indifference point in the zoological series, but it represents at the same time the highest metamorphosis of organic nature in general, and the culmination point in the consummate organization of the earthly nature of our planet. All cosmic connections and configurations are therefore reproduced in it in the small, and even light and the perfection of the astral bodies of distant worlds have found their radiantly transfigured reflection in the human microcosm: for the divine essence of the universe shines forth in the same way out of them all; and they are only different with respect to that which

is accidental and alienated from the essence of things. Since the ideas of all sensible objects are eternally united and equal to each other in the absolute, so do the objects, into which those ideas have been sunk as (so to speak) souls into matter, stand in complete relation to each other, and the human microcosm can be compared not only with the other animals of the earth, but with everything that exists and moves in the universe. The laws of the planetary system, and all cosmic connections, must therefore again be found in the organism.[42]

Although Autenrieth was known in his day as a defender of empiricism in physiology—and his writings belonged overwhelmingly to the mainstream, non-Romantic tradition—he, too, occasionally invoked some of the central motifs of *Naturphilosophie*. In comparative-anatomical contexts he sometimes used the terms *Stufen der Entwicklung*, *Steigerung*, and *Stufenfolge*, though never as part of a grand tripartite schematization.[43] And in both organic and inorganic contexts he frequently spoke in terms of *Indifferenz*, *Indifferenzpunkt*, *Gegensätze*, and progressive differentiation.[44] Among those concepts, only indifference played a significant role in Mayer's thinking, and that was the concept with the most extensive connections to non-Romantic usages in physics and chemistry. To be sure, Mayer also spoke of the neutralization of differences, but he never made particular use of the terminology of antitheses.

Since tripartite schematization was such a characteristic aspect of *Naturphilosophie*, it will be useful to look at a few, more extended examples. The archetype of sorts was Schelling's "universal schema" of the "dynamic *Stufenfolge*," in which he associated with the three realms of organic, universal, and inorganic nature the parallel categories of formative force (*Bildungstrieb* or *Reproduktionskraft*), irritability, and sensibility (under organic), of light, electricity, and the cause of magnetism (under universal), and of chemical process, electrical process, and magnetism (under inorganic).[45] In the organic realm, at least, there was a kind of compensatory relationship among the three factors, so "*that through the whole of organic nature, as the* ***irritability*** *rises [steigt], the* ***sensibility*** *falls, and as the* ***sensibility*** *rises, the* ***irritability*** *falls*," and the like.[46] I have not found that Schelling so understood the relationship among the *physical* forces of nature, that is, those pertaining to his "universal" and "inorganic" realms.

Walther's *Human Physiology with Thoroughgoing Consideration of the Comparative Physiology of Animals* (1807–1808) divided medicine into three parts—physiology, pathology, and therapy—dealing respectively with the theories of life, disease, and cure: "These three parts express within the sphere of this science the three powers [*Potenzen*] of the infinite, the finite, and the eternal. For life is to be regarded as the infinite, disease as the finite, and cure as the synthesis of both (the third power)."[47] Walther elaborated an especially complex multileveled system of concepts and relationships as "potentiated series" of the "three-part division of the absolute unity of the idea."[48] The latter's "first reflection" was the division into nature, spirit, and art; nature's "second counterpart" was the division into being (matter), activity (light), and life (the organism). Being, activity, and life were in turn analyzed into series consisting of the unity of substance (*Wesen*), form, and totality; of the magnetic, electric, and chemical forces; and of the traditional favorites, reproduction, irritability, and sensibility.

A late example of this way of thinking is the *Outline of Nature Philosophy* published in 1832 by the then sixty-four-year-old Tübingen professor, Carl August von Eschenmayer. Invoking Schelling's name, though in fact introducing his own schema, Eschenmayer identified three realms of spirit, life, and nature, each divided into a particular series of three powers: thinking, feeling, and willing; reproduction, irritability, and sensibility; and gravity, heat, and light.[49] He further correlated gravity with the principles of necessity and the true, heat with the principles of life and the beautiful, and light with the principles of freedom and the good.[50] He also identified three "planetary powers" of galvanism, electricity, and magnetism, but I could not determine how they fit into his larger schema.[51]

None of the specific details from the foregoing samplings is of particular importance here. What is important is the consistent extent to which the general way of thinking characteristic of *Naturphilosophie* finds few resonances in any of Mayer's writings. He never employed tripartite conceptual schemas, nor did any of the central concepts of dynamism, polarities, activities, *Stufenfolgen*, and the like play a detectable role in his thinking. Mayer did, to be sure, employ mathematical notation in an idiosyncratic and primarily symbolic and noncomputational manner, but not, as far as I have been able to determine, in a way specifically identifiable with any particular work of *Naturphilosophie*, and in Mayer's day it was quite common for even mathematicians to multiply and divide with zero and infinity in ways we regard as illegitimate. Nor does Mayer's occasional (and in part strategic) appeal to the laws of thought bear much resemblance to the apriorism of *Naturphilosophie*: Mayer emphasized efficient causality and the notion (common in the physics texts of his day) of forces as causes; and the law of sufficient reason, which he especially invoked, was not one I have found any *Naturphilosoph* to have mentioned. Nor can it be too strongly emphasized that Mayer never made an issue of something like the "unity of forces": he began with a narrow focus on the relationship between heat and motion, and only gradually expanded the scope of his theory to include other forces (a.k.a. imponderables). One can slog through hundreds of pages of *Naturphilosophie* without encountering many passages that even remotely sound like anything Mayer wrote.

With the important exception of the concept of indifference and the attendant notion of the creation of antitheses out of a null state, the case for some kind of general influence of *Naturphilosophie* on Mayer's formulation of energy conservation is difficult to sustain. Nevertheless, there are a few specific topics dealt with by several *Naturphilosophen* having to do with planetary motion that do not allow this historiographical issue to be entirely disposed of. Before turning to them, however, we need to consider in more detail what *Naturphilosophen* had to say about two subjects of great importance to Mayer: the nature of force(s), and the explanation of respiration and animal heat.

## 2 Force and Forces in *Naturphilosophie*

Since concepts of force and the unity of forces would appear to constitute the strongest ostensible link between Mayer and *Naturphilosophie*—such, at least, has been the general argument—let us take a closer look at just what different *Naturphilosophen* had to say on the subject. In doing so it will become clear not only how unsettled such concepts were, but also the extent to which *Naturphilosophen* like Schelling and Oken offered largely materialistic interpretations of light, heat, electricity, galvanism, magnetism, and chemism. For the rest, they and others were much more likely to refer to these agencies as principles, antitheses, activities, and efficiencies than as forces. The widely held view that *Naturphilosophie* was centrally characterized by a preoccupation with forces and their unity has, to be sure, some foundation in historical reality, but it is a misleadingly oversimplified characterization quite at odds with the textual reality of many of the works themselves. It has owed its origin, I suspect, in large part to the subsequent centrality of a more precise concept of "force" and its retrospective imputation to a no-longer well understood philosophy of nature. And unless one pays careful attention to language, one may wrongly read a discussion of electricity (and so forth) as dealing with forces, when in fact no such general (let alone *clear*) concept may have underlain the particulars. The fact that *Naturphilosophie*'s "forces" included things like irritability, sensibility, and some kind of formative force did not bring that concept closer to the physics of the conservation of energy, either.

Schelling developed his protean 'system' of *Naturphilosophie* in four principal works originally published between 1797 and 1799: *Ideas toward a Philosophy of Nature* (1797/1803), *On the World Soul; an Hypothesis of Higher Physics toward an Explanation of the Universal Organism* (1798/1806/1809), *First Sketch of a System of Natural Philosophy* (1799), and *Introduction to His Sketch of a System of Natural Philosophy* (1799). The fact that none of these works was reprinted between (at the latest) 1809 and 1857/58 (in Schelling's collected works) adds to the unlikelihood that Mayer knew them. Nevertheless, they will help define a kind of baseline of views properly identified as *naturphilosophisch*.

In the first-named book—which, in Schelling's early estimation, had of all his works received the most attention from natural scientists[52]—the author revealed his debt to Kant in distinguishing between empirical and philosophical natural science and in characterizing force as a concept belonging essentially to the understanding, that is, one which is never an immediate object of *Anschauung* and which is hence out of place in empirical science:

> With respect to questions as yet still partly in dispute concerning the nature of heat and the phenomena of combustion, I followed the principle of admitting absolutely no hidden elemental substances in bodies, the reality of which can in no way be established by experience. One has recently mingled more or less philosophical principles into all these investigations concerning heat, light, electricity, etc., . . . principles

which in and of themselves are already foreign to experimental natural science and which are besides usually so indefinite that inevitable confusion arises as a result. Thus in physics nowadays the concept of force is played with more frequently than ever, especially since one began to doubt the materiality of light, etc.; one has indeed already asked several times whether electricity might not perhaps be *vital force*. All these vague concepts introduced illegitimately into physics I have had, in the first part of this work, to leave in their indefiniteness, since they can be rectified only philosophically.

. . .

Here is now the place to secure for the concept of fundamental forces of matter its reality, but also its bounds. *Force* in general is a mere concept of the understanding, hence something that cannot directly be any sort of object of perception [*Anschauung*]. . . . The fundamental forces of matter are thus nothing but the expression of those original activities *for the understanding*, for reflection, not the true [thing] in itself, which exists only in perception; and thus it will be easy for us to define them with perfect completeness.[53]

Following Kant, the fundamental forces of matter were the usual attractive and repulsive forces out of which matter is philosophically constructed.[54] Indeed, force and matter are in such intimate association that "neither forces without matter nor matter without forces is imaginable."[55] Such an understanding contrasts sharply, of course, with Mayer's central concern to vindicate for force an existence independent of matter. For Schelling, there is no force in nature that is not limited by an opposing force; the two are either in equilibrium or in continuous conflict (*Streit*).[56] Thus the dynamical polarities so prominent in *Naturphilosophie*. It should be kept in mind, too, that the attractive and repulsive *Grundkräfte*, the paradigmatic opposing forces, were forces in the traditional Newtonian sense. Such language represents no groping toward a more energetic conception of the term.

Reductive analysis into opposing forces was not, however, a feature of Schelling's treatment of light and heat, the phenomena of which do not readily lend themselves to a dualistic interpretation, especially before the discovery of the polarization of light. Instead, he argued repeatedly for their materiality in ways reminiscent of the imponderables language of later physicists (as discussed in chapter 4). He introduced the issue as follows:

The first question that must occupy us is this: How are light and heat connected? Are they both of a completely different nature? Is one perhaps cause, the other effect? Or do they differ only in degree? Or is one only the modification of the other, and in that case would the marvellously rapid, easily mobile element of light perhaps be a modification of heat, a substance [*Materie*] which, as it appears, diffuses with difficulty and only gradually in much smaller spaces?

They do not appear to be of a different *nature*; for the striving toward expansion and diffusion is common to both. But one diffuses infinitely faster than the other. Hence might they be different in degree? But the greatest heat is without light, while often far less heat is associated with great flame. These presuppositions therefore lead to no reliable result.[57]

In examining the relationship between light and heat, Schelling noted that light produces heat when it is absorbed by a black surface, such that the heat produced is proportional to the resistance the light experiences.[58] That transformation being (implicitly) paradigmatic of their relationship, he concluded not only that heat is a modification of light, but also—somehow—that light is matter (*Materie*) and that it becomes heat when its elasticity is diminished.[59] The crucial point here is that Schelling did not speak of light and heat as forces, but as material substances. "The most general assertion" we can make with respect to light is that "it is merely a modification of matter," though whether or not light is a *distinctive* substance one cannot say.[60] According to Schelling, the effects of the sun taught people that light and heat come from one source and are the only vivifying forces (*belebende Kräfte*) in the universe.[61] He likened the expansion of dead matter by heat to the expansion of a bud and to the development of a human embryo brought about by an "internal, living heat."[62]

Schelling introduced the chapter on electricity with words that suggest he might be about to adopt the language of force: "So far we have become acquainted with only one force of nature, light and heat, which could be checked in its action only by the opposing endeavor of dead matter; now an entirely new phenomenon arouses our attention, one in which activity appears to rise up against activity, force against force."[63] In the event, however, Schelling spoke consistently of the "electric matter," occasionally of the "electric fluid."[64] He even identified it as "decomposed vital air." Citing experiments that showed that electricity is capable of calcifying metals even in mephitic airs, he concluded: "Given so completely the same effects of both—of electricity and vital air—I do not know whether one can demand more evident proofs of their identity."[65] He later supported this interpretation by citing a contemporary chemist's quotation of Lavoisier's opinion to that effect.[66] Schelling tended to call magnetism a force, but he also sanctioned the "scientific fiction" of "two magnetic substances."[67] If terminologically Schelling sometimes applied the word "force" to light, heat, electricity, and magnetism, conceptually he had (in this early work) not broken with the ontology of subtle fluids. Nothing better reveals Schelling's imprecision, both terminological and conceptual, than his inclusion of *water* among the "forces": "only forces that penetrate bodies, such as heat and electricity, not such as only reach its surface, such as water (among others)—which are harmful to electricity—are capable of weakening this [magnetic] force."[68]

In a later chapter Schelling returned to the question of the nature of light and heat. Ridiculing the notion that light might be a *particular* substance (*eine besondere Materie*), he asserted that "everything we call matter is indeed only a modification of one and the same matter, which we admittedly do not know sensibly in its absolute state of equilibrium and which must enter into particular relationships to be knowable for us in this way."[69] Against calling light a force was (again) the inappropriateness of introducing philosophical terms into physics and the fact that everything we have direct knowledge of could be called a force. Nor is a *Lichtmaterie* any more hypothetical than any other kind of matter. Light can accordingly be regarded as matter characterized by a particular ratio between the underlying dynamic forces, one in which the expansive force most strongly predominates. A change in that ratio can cause the matter of light to enter into chemical combination. In the same vein, any substance can become light if its elasticity can be made to equal that of light.[70] Light was not a force; it was a species of dynamically constructed matter.

When it came to heat, Schelling now emphasized how it fundamentally *differs* from light:

> Things are quite otherwise with the *matter of heat* [*Wärmestoff*] than with light. Light itself *appears* as matter of a particular quality, whereas heat is itself *no* matter, but merely a quality—merely a modification of *every* (no matter which?) matter. Heat is a particular degree of expansion. This state of expansion is not peculiar to only one particular matter, but can belong to every possible matter.
>
> Thus the situation here is very different from that with light. For up to now we know of only one matter as such (vital air and a few that closely resemble it) that can pass over into the degree of elasticity that is accompanied by the phenomenon of light. We are therefore entitled to speak of a matter of light. But *every* matter can be *heated* in itself directly (through friction), and this not only by the *advent* of an unknown fluid, but by a simultaneous change that takes place in the body itself.
>
> If one now in addition considers that heat arises in very many *undoubted* cases through a mere change in capacity, then one will be inclined to regard heat in general as a mere *phenomenon* of the *transition* of a matter from the more elastic to the less elastic state (as from vaporous to liquid).[71]

Heat is thus neither a force nor a particular species of matter, but "merely a phenomenon," a "modification of matter," "a quality dependent on accidental conditions."[72] He now argued *against* Scherer's assertion of the identity of light and heat, and countered his arguments for their immateriality.[73] Misunderstanding the concept of heat capacity, and without considering the transformation of heat into anything else, Schelling denied that there is any measure of the absolute quantity of heat: "It is not only in general a *mere* modification of other matter, but also a modification for which there is no absolute measure (hence the concept of the capacity of bodies [for heat])."[74]

Schelling's next major work, *On the World Soul*, for the most part retained this materialistic interpretation of light and heat. The first edition of 1798 spoke explicitly of the "ponderability" of light; although the second and third editions of 1806 and 1809 did not use that word, and seemed otherwise to wish to tone down the earlier strongly materialistic language, the author left unchanged a statement of his belief in its "materiality."[75] He assumed the existence of a *Wärmematerie*, spoke of a *Wärmefluidum*, and postulated the existence of an *Urmaterie* present in bodies as either light or heat, depending on the degree to which the positive, expansive principle predominates.[76] He made no use of the word *Kraft* in this context. When he did use that word, in his discussion of electricity, he applied it to the positive and negative principles whose reciprocal relationship underlies and defines the electrical phenomena.[77] Even at that, he devoted a section to an "Investigation of the Ponderable Basis of the Electrical Matter," and wrote: "If the source of all light that we are capable of developing is to be sought in the vital air, so, too, must the *electrical matter* owe its origin to a decomposition of this air."[78] The electrical matter is, like light, a "*compound* fluid."[79] Significantly, in examining the relationship between heat, light, and electricity, Schelling looked not at transformations of one into another, but at separate classes of phenomena exhibiting analogous patterns of behavior. As he concluded, "If anything proves the *identity of the positive matter of light, of heat, and of electricity*, then it is this agreement in the laws according to which, in these different states of which it is capable, it is attracted or repelled by bodies."[80]

Schelling had earlier expressed his acceptance of what he took to be a fundamental principle of physics, namely "to assume a *material* principle as the vehicle of every force in nature."[81] What he opposed was recourse to "absolute forces," as opposed to the "primitive forces" introduced into science by the dynamical philosophy:

> They serve namely in no way as *explanations*, but only as *limiting concepts* of empirical science, whereby the freedom of the latter is not only not endangered but is even assured, because the concept of forces—since each one of them allows of an infinity of possible degrees of which none is absolute (the absolutely highest or lowest degree)—opens up for it an infinite space in which to manoeuvre, within which it can explain all phenomena *empirically*, i.e., in terms of the *mutual interactions of different substances*.[82]

Thus forces belong properly to the philosophical underpinning of natural science, whereas natural science itself bases its explanations on the assumption of material substances. Schelling applied this reasoning to magnetism, though he cautioned that one should not assume a distinctive magnetic matter.[83]

Such material principles cannot, however, explain the possibility of an ordered world since those principles are themselves only possible within such a universal system.[84] It is the ordered disposition of the universal forces of nature that appears as the various seemingly specifically different material substances. Although his language is vague, Schelling seemed to imply that only the force of gravity is invariable: "Only the general motion of the universe is dependent upon eternal and invariable causes; but variable causes betray material principles; thus the magnetic deviations, which one cannot explain without thereby assuming a material substance as active, one which is developed or brought to rest, decomposed and again compounded, and which (like atmospheric electricty) arises and disappears."[85] Very few passages in *On the World Soul* speak directly to the issue of the alleged unity of forces that is supposed to be such a hallmark of Schelling's *Naturphilosophie*. The following is the most striking:

Now, since in consequence of the preceding it is undeniable that a graduated series [*Stufenfolge*] of functions occurs in the living organism—since nature has opposed the animal process with irritability, irritability with sensibility, and thus has orchestrated an antagonism of forces that mutually hold each other in equilibrium, in that as one rises the other falls, and vice versa—one is thus led to the idea *that all these functions are only branches of one and the same force*, and *that perhaps the one principle of nature that we must assume as the cause of life appears in them only as in its individual phenomena*, just as without doubt it is only one and the same universally distributed principle that reveals itself in light, in electricity, etc., as in different phenomena.[86]

The forces of nature were thus seen as expressions of a common underlying dualism: it is that dualism that runs throughout nature, not a common and self-subsisting force somehow transformed in ongoing physical processes from one form into another.

Schelling's explicitly self-styled works of *Naturphilosophie*—the *First Sketch* and the subsequently published *Introduction* thereto of 1799—continued his unresolved attempt to come to terms with the proper ontology of nature, in particular the connection between force and matter. (Note that Schelling did not choose to reprint either of these two works as he did those published during the previous two years. Soon after 1800 his philosophizing took a different turn.) At times his "dynamical philosophy"—in its allowance of "no special material substances" to explain light, heat, electricity, and magnetism—appears to expound an ontology of pure force: "All original (i.e., all dynamical) phenomena of nature must be explained in terms of forces that inhere in *matter* even at rest (for even in nature's rest there is motion; this is the chief principle of the dynamical philosophy): those phenomena—e.g., the electrical—are thus not *phenomena* [*Erscheinungen*] or effects of particular individual material substances, but changes in the *existence* [*Bestehen*] of matter itself."[87] In his conception of things here, the essential properties of fluids provided him with an analogy applicable to nature's original construction: fluids are formless and indeterminate and thus close to the indeterminate source of all productivity; and in fluids, actions and attractions are the same in all directions. From this he concluded:

The most original and most absolute combination of opposing actions in nature must accordingly produce the *most original fluid*, which, since that combination takes place constantly (the act of organization is constantly in motion), will represent itself as a universally distributed entity [*Wesen*] which absolutely opposes the nonfluid (the rigidity) and which strives continuously to *fluidize* everything in nature.

(This principle is called *principle of heat*, which is accordingly not a simple substance [*Substanz*], no sort of matter at all, but always only a phenomenon of the constantly diminished capacity (of the original actions for each other), and it is therefore proof in nature of the ever persisting process of organization.—New theory of heat in accordance with these principles.)[88]

Similarly, Schelling reasoned that matter is not simply given in nature as a primitive entity, but is something that has become what it is—everything in nature is "ein *Gewordenes*"—as a combination of the primitive actions: "Thus there exists in nature absolutely no *primitive substance* [*Urstoff*] out of which everything might have *become*—roughly as the ancients imagined the elements. The only true primitive substance are the simple actions. Thus there also exist in nature no originally undecomposible, i.e., really *simple material substances* [*wirklich einfache Materien*]."[89]

Nevertheless, despite the apparent primacy of forces and actions—Schelling was nothing if not inconsistent in his use of terms—the original primitive fluid (and its manifestation as heat and electricity) came more and more to look like a conventional subtle fluid, allying Schelling with the plethora of aether philosophies developed throughout the eighteenth century.[90] Consider his introduction of heat:

> The matter of fire (or of heat) is familiar to us as the original phenomenon of the absolute fluid. The former appears to arise or to disappear where a mere *quantitative* decrease or increase in capacity (enlargement or diminution in volume) takes place. The matter of heat appears as *simple*, and as yet one has been able to perceive with it no duality or decompositon into opposing actions, as, e.g., with electricity. Just this is the proof that in this most original of all fluids the most complete combination appears still *undisturbed*.
>
> On the other hand, even the lightest contact of heterogeneous bodies (in galvanism and other recently performed experiments) produces phenomena of *electricity*, and since heat as well as electricity are excited by means of friction (constantly repeated and intensified contact), it thus appears that with every collision of different bodies the absolute fluid that permeates all of them (because it endeavors to fluidize everything) is displaced both mechanically out of equilibrium and dynamically out of its original combination. The former produces the phenomenon of heat becoming free, the latter the phenomenon of excited electricity. And there exists really almost no chemical process in which heat arises or disappears that does not also show traces of excited electricity; closer attention still has much to teach here. Not to mention the fact that in very many cases electricity manifests the same effects as heat, and that bodies behave the same way with respect to their conductive powers for both.[91]

Note that the apparent absence of duality shown by the matter of heat was here interpreted not as evidence for its nonpolar nature, but for the completeness of the combination of primitive forces in it. The difference between heat and electricity was attributed to the degree of disturbance of the equilibrium of the component forces. Note, too, that although this rendering of the sense of the passage required me to speak just now of "forces," the word does not appear in the original text. When it came to heat, Schelling still consistently spoke of *Wärmestoff* and *Wärmematerie*.[92]

There are, to be sure, a few passages ripe for quoting as evidence for Schelling's insistence on the unity of forces. The following is one of the best from the two works presently under discussion:

Thus instead of the *unity of the* **product** . . . we have now *a unity of the* **force** of production throughout all of organic nature. It is, to be sure, not one product but rather *one force* that we only perceive checked at different stages of the phenomenon. . . .

* [*] *

And it would thus then be about time to exhibit that graduated series in organic nature as well, and to justify the idea that the organic *forces—sensibility, irritability,* and *formative force—are all only branches of one force,* just as without doubt as in *light, electricity,* etc., *only one force appears as in its different phenomena.*[93]

Note that in this conceptualization, light, electricity, and so forth are in a sense not themselves different forces, but only the different manifestations of one common underlying force—the same force that, raised to a higher power, appears in the organic realm as the familiar sensibility, irritability, and so forth. In accordance with this conceptualization, Schelling more often spoke of the common fundamental dualism or antithesis that underlies the primary phenomena of nature. He had no clear or consistently deployed doctrine of force, let alone one that emphasized transformations, equivalences, or quantitative measures. Compare the following: "The antitheses occur in the *interior* of the universe, but all these antitheses are nevertheless only different forms into which is transformed the one primitive antithesis [*Ur-Gegensatz*] that extends itself in infinite ramifications throughout all of nature—and thus the universe in its absolute identity is nevertheless only the product of *one* absolute duplicity."[94] And again the following:

Thus the chemical phenomena, just like the organic, confront us with the question of the ultimate origin of all duplicity. . . .

It is thus **one** universal dualism that runs through all of nature, and the individual antitheses we perceive in the universe are only shoots . . . of that one primitive antithesis.

But what has that primitive antithesis itself produced, produced out of the universal identity of nature? . . .

But to *produce* heterogeneity means to create duplicity in identity.

. . .

But unity in dichotomy only exists where that which is heterogeneous attracts itself, and dichotomy in unity only exists where that which is homogeneous repels itself. . . . But this production of the heterogeneous out of the homogeneous and of the homogeneous out of the heterogeneous we perceive in the most primitive fashion in the phenomena of ***magnetism***. The cause of ***universal magnetism*** would thus also be the cause of the *universal heterogeneity in homogeneity and homogeneity in heterogeneity.*[95]

In Schelling's view of the idealized productivity of nature, it was through magnetism that an original antithesis entered nature; the question then was how, out of this original antithesis, all the other individual antitheses came into being.[96] His goal was to prove a priori "*that it is one and the same universal dualism which, from the magnetic polarity on through the electrical phenomena, ultimately loses itself in the chemical heterogeneities and which finally reappears in organic nature*."[97] The goal of Schelling's *Naturphilosophie* was to discover the abstract developmental schema according to which Nature produced (and produces) its full range of "products"—the historical echoes here are of Spinoza's *natura naturata* and *natura naturans*—by assuming some kind of check (*Hemmung*) to oppose the original expansive productivity of nature. As he summarized his overall conception of this process in his last major work of *Naturphilosophie*,

> The initial disposition [*erster Ansatz*] toward original production is the limitation of productivity by the original antithesis, which is still distinguished *as* an antithesis (and as the condition for all construction) only in *magnetism*; the second stage of production is the *alternation* of expansion and contraction, which is still visible *as* such only in *electricity*; the third stage, finally, is the transition of that alternation into indifference, which is still recognized as such only in *chemical* phenomena.
>
> ***Magnetism, electricity, and the chemical process*** are the *categories* of the original construction of nature—the latter withdraws itself from us and lies beyond perception [*Anschauung*], the former are that which is thereof residual, stable, and fixed—the universal schemata of the construction of matter.
>
> And—in order to again close the circle here at the point from which it began—as in organic nature the secret of the production of *all of organic nature* lies in the graduated series of sensibility, irritability, and formative force, so does the secret of the production of *nature out of itself* lie in the graduated series of magnetism, electricity, and the chemical process, as it can also be distinguished in the individual body.[98]

*Gegensatz, Dualismus, Duplicität, Polarität, Produktivität*: such were the central motifs of Schelling's *Naturphilosophie*. It is hard to escape the conclusion that historians' preoccupation with force as the supposedly central concept of *Naturphilosophie* represents a biased retrospective reading of the original texts.

Nor is that conclusion limited to Schelling's works. Landshut professor of medicine and surgery Philipp Franz von Walther's *Human Physiology* posited an absolute substance made up of activity and being: the word matter accordingly designates an excess of being over activity, light the excess of activity over being.[99] Force was not one of Walther's guiding concepts.

Zürich physiology professor Lorenz Oken's *Textbook of Nature Philosophy* posited an aether as the fundamental cosmic *Urmaterie* that, constituting time, space, and *das schwere Urwesen*, is nothing less than God's eternal position and first manifestation in material reality; it is the "*Ursubstanz* (= 0 + –)."[100] Through the separation of the aether into polar masses it becomes denser, heavier, and more material in a process conceived as taking place via the fixation (*Figierung*) of a determinate pole in a determinate mass of the aether.[101] Aside from that fixation, the aether itself exists in a state of indifference, and it strives to transform matter into an indifferent state.[102] Although I cannot follow the reasoning—such as it may have been—Oken asserted that it is through the activity of light that the aether congeals into matter, the light thereby "dying" and "becoming darkened."[103] The next "darkening" of light, its immediate transition into matter, produces oxygen, here also termed "corporeal light."[104] Heat was interpreted as an act of motion of the *Urmaterie*: "A material substrate does to be sure underlie heat, as it does light, but this substrate does not heat and illuminate; rather, the motion of the substrate heats, and only the tension of the substrate illuminates."[105] In accordance with Oken's ontological continuum, matter and heat are transformable one into the other: "The development of heat by a body is not a pressing out of a substance lying within it and foreign to it, but a dissolution [*Aufgehen*] of matter itself into heat. Matter does not develop or yield heat, it rather becomes heat."[106] The aether being so rare, the loss in weight experienced by a glowing body is negligible, though "philosophically" real. Note that such a conception makes problematic any principled notion of the conservation of (ponderable) matter. Thus Lorenz Oken, one of the best known and most influential of all the *Naturphilosophen*, defended in his *Textbook of Nature Philosophy* a worldview more in the spirit of materialistic aetherial natural philosophies than of force-based dynamism. Force was simply not one of his central concepts. Eschenmayer, for his part, approved of Oken's concept of aether as entailing the "substantiality" and "materiality" of the fundamental forces.[107]

Nor did a conception of force underlie Dresden professor of medicine Carl Gustav Carus's natural philosophy. In a review of Autenrieth's *Views on the World of Nature and the Life of the Soul*, published in 1837, Carus criticized the common notion of forces as something conceptually added to bodies in order to explain the phenomena, and he attacked the notion of force as an objective entity separate from the body.[108] He elaborated his views on force the next year in the first volume of his *System of Physiology*, which, like most physiological works of the period, paid particular attention to fundamental questions of ontology and the categories of explanation. He there criticized the concept of force as "a purely *subjective concept*" that, abstracted from the objects undergoing change, has wrongly been made into an object itself.[109] Echoing Oken, Carus wanted to replace the words *Materie* and *Substanz* with *Aether*, which, he argued etymologically ("ἀειθέω, to be in eternal motion"), contains in itself the idea of intrinsic motion:

> Thus the aether, or the aetherial substance, in this sense, is to be acknowledged as something that to be sure exists nowhere in a sensibly perceptible fashion, but as something that is nevertheless absolutely omnipresent, as that which, as something indifferent, underlies every however different and individual object, consequently as that out of which continuously proceed the differentiations that represent in every moment the order of the universe, and back into which the latter is continuously transformed.[110]

What he termed *das All-Leben* proceeds as a revelation of the divine being in the continuously repolarized aether.[111] Addressing himself to the issue of the supposed antithesis between force and matter, an antithesis that appears in physiology as that between activity and organ, Carus effectively dismissed the concept of force from science as a purely subjective representation. His words recall Schelling's similar sentiments:

> Now, if we wish to consider all of this properly, then upon closer examination it must also very soon become clear that, as soon as we speak of the activity or force of an organ, we thereby designate nothing but the series of certain temporally succeeding changes in the state of this organ itself, in that we separate off and distinguish this sequence just in the mind [*eben im Geist*]—i.e., artificially and merely intellectually—from the states themselves. . . .
>
> One should therefore at least never forget, when we speak of the force of a thing and imagine this force for the moment as something distinct from the thing, that this is *a completely subjective notion*, with which we may to be sure at least have the use of [a] more convenient and concise expression, just as with many other subjective notions, and which we may in particular indeed well employ *as a measure* of the minimum and maximum of the changes of a thing, . . . as long as we thereby take care not to lose sight of the actual state of affairs of the matter. Nevertheless we may in no way grant objective truth to this expression.[112]

Kant follower Heinrich Friedrich Link objected to Carus's dismissal of the concept of force: "The author rightly criticizes the antithesis of matter and force, but he ignores the simple conclusion that force is not distinct from matter in order to say that force is nothing."[113]

Bonn professor of medicine Moritz Ernst Adolph Naumann's heavily *naturphilosophisch* work *Pathogeny* of 1840 expounded another vaguely matter-bound conception of light, heat, electricity, and magnetism. Here again one looks in vain for a regularly employed general term for these four principal agencies. (Works of *Naturphilosophie* did not as a rule refer to them as imponderables, as did mainstream scientists and, following them, Mayer.) For Naumann they were "phenomena that indicate the state of matter passing or about to pass from one form into another."[114] They were to be regarded as "properties of matter" with respect to its change of form.[115] Even if, for the rest, Naumann tended to employ less materialistic and more idiosyncratically unintelligible language, it remains the case that force in other than a Newtonian sense was not one of his basic concepts, and he did not regularly so designate electricity et al.[116]

Nor did force (*Kraft*, *vis*, or whatever) play a central role in the natural philosophy of Georg Friedrich Pohl, the most prominent representative of *Naturphilosophie* in physics and chemistry during the 1820s and 1830s.[117] In a book devoted to magnetism, electricity, and chemism—the three "moments" of the efficiency (*Wirksamkeit*) of nature—Pohl for the most part used no generic term to characterize their common class membership; when he did, he preferred activity (*Thätigkeit*).[118] The word "force" appeared only once in his book: figuratively, as the "force of thought."[119] He expounded his natural philosophy at greater length in his 1837 inaugural address as professor of physics at Breslau, *On the Heretofore Sought-For Principles as Well of General Physics as of Its Chemical Part in Particular*. He began by criticizing the usual practice of assigning different classes of phenomena to "various kinds of efficient forces and powers as the causes of [the phenomena]."[120] Such a procedure leaves the phenomena without any fundamental unity, just as if one were to assume a separate force (*vis*) for seeing, hearing, etc., while ignoring the efficiency (or efficient power, *efficientia*) of the mind that performs those diverse actions.[121] A deeper understanding of nature thus requires that we have a firm mental and perceptual grasp of the "unity of the actions or *primary efficiency*" that remains one and the same through all the forms, degrees, and states in which it is manifest.[122] Pohl's central concept was just that *efficientia primaria*, what he might have termed *die ursprüngliche Naturwirksamkeit* had he been writing in German. His image, as vividly *naturphilosophisch* as it was vague, was of a "primary function" unfolding in continuous activity in which all phenomena are united and which unlooses one motion of the "universal efficency."[123] The "activity of the primary efficiency," the source of the unity of the phenomena of nature, is constrained between two opposing tendencies or moments, one urging forward (or positive), the other restraining (or negative).[124] The *efficientia universalis* (as he sometimes termed it) and its two active and opposing moments are joined in indissoluble unity. Although he agreed that the mind cannot conceive of force without matter or matter without force, he explicitly rejected the fundamentality of Kantian dynamism and asserted that force and matter are aspects of a third primitive entity, the *efficacitas* (or *efficientia*) *primaria*.[125] Matter, which is simply the lowest specific determination of that primary efficiency, can be brought into existence at any time and anywhere in space.[126]

Not force but a primitive active polarity—he also spoke of an "abstract polar impulse [*conatus*]"—was Pohl's central concept, and it was in terms of their fundamental polarity that the different realms and phenomena of nature are unified.[127] His language omitted all mention of force just where one might expect to find it: he spoke of the "phenomena" of light and colors, the galvanic "concussion," and chemical "activity."[128] The electrical "effect" or "impulsion" or "incitement" was now a *conatus*, now a *nisus*.[129] And although he noted certain similarities between electricity and magnetism, such as their relationship to metallic conductors, he warned strongly against assuming the "substantial" identity of electricity and the magnetic and chemical actions; such identity as they had lay in their common polar nature (as Yelin, too, had argued):

> One may therefore in no way approve of the common opinion according to which electricity is considered to be the identical and substantial cause of both the magnetic as well as the chemical action. For in truth the electrical and magnetic activities, although closely connected, nevertheless differ one from the other by a wide measure of opposition, and enjoy no other identity than that by which are produced, as of one and the same chemical activity, diametrical polar impulses as from a common center but tending in opposite directions, just as within the entire complex of earth and atmosphere, with chemical activity everywhere in force, the magnetic impulsion is only strong where the electrical is weak, and vice versa.[130]

As the rest of the paragraph made clear, what he meant by the last sentence was that electrical activity occurs primarily in the earth's crust and atmosphere, magnetic activity in the earth's metallic core. He was not talking about the transformation of one into the other or some kind of mutual limitation with respect to the magnitude of their joint effects.

It is very hard to imagine what kind of stimulation Mayer might have gotten from the treatment (or not) of force in works such as these. With the exception of Mayer's (for him) unusual use of the term *Urkraft* in his first (unpublished) essay of 1841 and in a letter of 1844, neither the general terms of discourse nor the specific problems addressed bear any striking affinity with Mayer's work. Nor did works of *Naturphilosophie,* unlike the works of mainstream science discussed in Part Two, treat the imponderables in ways suggestive of the issues of ontology and creation that I believe stimulated Mayer's reflections. As for the association of Mayer with *Naturphilosophie* via belief in the supposed unity of forces, such a concept did not strikingly characterize *either* party.[131]

## 2.1 *Vital Force*

Many critics, going back at least to the time of Liebig and Helmholtz, have associated *Naturphilosophie* with vitalism and the belief in a vital force.[132] As we saw, Friedlaender took Mayer's opposition to the vital force as evidence of his rejection of *Naturphilosophie.* Such an association is, however, based on a misunderstanding of *Naturphilosophie*'s fundamentally organicist worldview and its rejection of a mechanistic view of the physical world as essentially dead. Hermann Schlüter has given an excellent analysis of the issues in his book, *The Sciences of Life between Physics and Metaphysics* (1985), which deserves to be more widely known. As he pointed out, even if vitalists maintained that the explanation of life requires the assumption of a special vital force superadded to the otherwise dead mechanism of the physical world, they nevertheless agreed with mechanists that the material world is in a sense intrinsically inanimate, incapable of self-movement. *Naturphilosophen,* for whom the world is itself a kind of organism containing intrinsic powers of self-movement, had no need of such a "secondary animation [*Belebung*] by means of a vital force."[133] Although Schlüter's conclusion will require some qualification for Schelling, it is a reasonably accurate generalization for *Naturphilosophie* overall.

Already in the introduction to his first work of *Naturphilosophie* Schelling rejected the image of an organism as a machine or a chemical laboratory and argued against the notion of a vital force capable of somehow removing the organism from the normal laws of chemistry.[134] In Schelling's eyes, belief in a vital force seemed to hold out hope for an eventual physical explanation of life, whereas he believed in the higher directive agency of spirit. It was his *On the World Soul* of 1798, however, that contained Schelling's most extensive critique of the vital force, part of which repeated his earlier reasoning:

> How does nature bring it about that the chemical process active in the animal body never *oversteps the limits of organization?* Nature cannot (as one rightly maintains against the defenders of the vital force) suspend a universal law, and if chemical

processes take place in an organization they must proceed according to the same laws as in dead nature. How does it happen that these chemical processes always reproduce the same matter and form, or by what means does nature obtain the separation of the elements whose conflict is life and whose union is death?[135]

The context of these reflections was a consideration of "the opposing principles of animal life," principles that Schelling, as usual, sought to understand in terms of analogies drawn from other realms of phenomena:

To every positive principle in the world there stands . . . opposed a *negative* one: thus there corresponds to the positive principle of combustion a negative principle in the body, to the positive principle of magnetism a negative one in the magnet. The basis [*Grund*] of the magnetic phenomena lies neither in the magnet nor outside the magnet alone. Thus there must correspond to the *positive* principle of life *outside* animal matter a *negative* principle in this matter, and thus here, too, as elsewhere, truth lies in the union of both extremes.[136]

Not a unitary vital force, but a fundamental antithesis is the basis of life. What in the foregoing passage resembles an Aristotelian distinction between form (active, positive) and matter (passive, negative) represented for Schelling the antithesis of freedom and lawfulness, the union of which in *Naturphilosophie* avoided the unacceptable implications of both vitalism and mechanism, though Schelling did not use either of the latter two terms:

In general it appears to me that most scientists have up to now still missed the true sense of the problem *of the origin of organized bodies*.

When a portion of them assumes a special *vital force*, which like a magic power suspends the operations of the laws of nature in animated beings, they in just that way eliminate a priori the possibility of physically explaining organization.

On the other hand, when others explain the origin of all organization out of *dead chemical forces*, they in just that way eliminate all *freedom* of nature in formation and organization.

. . .

Nature should neither behave simply *lawlessly* (as the defenders of the vital force must maintain, if they are consistent), nor simply *lawfully* (as the chemical physiologists maintain), *but it should be lawless in its lawfulness and lawful in its lawlessness*.[137]

For this "union of freedom and lawfulness" Schelling employed the term *Trieb*; we can thus say that "a primitive *Bildungstrieb* acts in organic matter, by virtue of which it takes on, preserves, and continually restores a particular form."[138] The "first cause of organization" must lie outside the organized matter itself.[139] Schelling emphasized that *Bildungstrieb* is only an expression for that union, not an explanation of it. It is a synthetic concept, composed of two factors, "a *positive* (the *principle of nature*, by means of which the dead crystallization of animal matter is continually disturbed) and a *negative* (the chemical forces of animal matter). Out of these factors alone is the formative force constructible."[140] In the event, Schelling ended up with a conception of life seemingly not much different from the vital force he ostensibly rejected—that is, one based on the assumption of an "external principle *not subject to the chemical process*" that acts continually on the organic matter, disturbing the inanimate chemical laws and rekindling the vital processes.[141]

In typical fashion, Schelling abandoned the language of *Bildungstrieb* toward the end of the book in favor of what he once called "the common soul of nature" as the principle and cause of life.[142] That principle performed roughly the same two classes of functions—form-giving and energetic—commonly (if usually only implicitly) assigned to the vital force by most German physiologists of the 1830s: "It cannot of course be said that this principle *suspends* the dead forces of matter in the living body, but rather that it 1) gives these dead forces a *direction* that, left to themselves, they would not have taken; 2) that it repeatedly *rekindles* and continually sustains the *conflict* of these forces that, left to themselves, would soon have transposed themselves into equilibrium and rest."[143]

Schelling's *First Sketch of a System of Natural Philosophy*, published the next year, was again somewhat inconsistent in its handling of the vital force. He first gave it qualified acceptance as an immaterial principle controlling the chemical forces of organic matter:

> Now the cause that in part suspends and in part alters the chemical forces and laws of matter in the organism cannot in turn be a *material* cause, since every material substance [*jede Materie*] is itself subject to the chemical process—it is thus an *immaterial* principle, which is rightly called *vital force*.[1]
>
> . . .
>
> [1] . . . Nature can of course only suspend the chemical and physical laws through the opposing action of another force, and it is precisely this force that we call—because it is as yet still wholly unknown to us—*vital force*.
>
> There already lies in this deduction of the vital force the confession
>
> 1) that it has been invented solely as a makeshift of ignorance and is a true product of indolent reason;
>
> 2) that by means of this vital force we advance not a step further in either theory or practice.[144]

If this recalls the qualified use of the vital force by mainline German physiologists during the 1830s, all the more does Schelling's antagonistic way of thinking recall in particular Liebig's conception of the vital force as the internal creative force that opposes the external forces destructive of the organism's integrity—in particular, oxygen:

> α) The principle of life exhibits itself, *where* it manifests itself, as an activity that opposes every accumulation of matter from without, every press [*Andrang*] of external force . . . . External nature will thus attack life; most of the external influences that one considers to be conducive to life are actually destructive to life—e.g., the influence of the air, which is actually a process of consumption, a continuous attempt to subject living matter to chemical forces.

> β) But it is precisely this attacking by external nature that maintains life, because it continually reexcites the organic activity, rekindles the slackening conflict.[145]

Schelling did not rest with this qualified understanding of the vital force, but sought to go beyond it by incorporating Scottish physician John Brown's notion of excitability into his dualistic system as the expression of the alternation between receptivity and activity. Such a system would, he said, overcome the subject-object dichotomy of the traditional alternatives:

> The [Brunonian] system . . . takes this [position] between two opposing systems, of which one—the chemical—knows the organism merely as object, as product, and allows everything to act upon it only as object upon object (i.e., chemically), the other—that of the vital force—knows it only as subject, as absolute activity, and allows everything to act upon it only as activity. The third system posits the organism simultaneously as subject and object, activity and receptivity, and precisely this mutual determination of receptivity and activity comprised within one concept is nothing other that what Brown has called excitability.[146]

The beauty of Brown's system, as Schelling interpreted it, was that it was based on a "primitive duplicity":

> For even if we . . . wished to assume a *vital force* (although neither physics nor philosophy sanctions the assumption of a fiction), nothing at all is explained with this principle. . . . No force is limited except by an opposing force. If one now posits that there exists in nature a special vital force that would be a simple force, then indeed by means of this force it could never come to a determinate product, and if, in order to explain the determinateness of nature's production, one already posits something *negative* in this force, it ceases to be a *simple* force.[147]

If it would appear that what Schelling ultimately objected to about the vital force was that it did not fit his fundamentally dualistic system, one must nevertheless not forget that in his *World Soul* of the previous year he flirted with a matter-form dichotomy that readily accommodated the vital force. But the reader who hopes for either consistency or logic in Schelling's writings will be continually disappointed. For the rest, Schelling's aim was less to distinguish life sharply from nonlife than to interpret them as different manifestations of a hierarchy of forces: "life is nothing but an augmented [*gesteigert*] state of the common forces of nature."[148] Not a transformation or reduction of forces was intended, but a conceptualization of their ideal relationships.

Other *Naturphilosophen* than Schelling provide a more straightforward illustration of Schlüter's characterization of *Naturphilosophie*'s general rejection of the vital force.[149] Walther began the body of his book *Human Physiology* with a chapter "On Life" in which life was equated with self-activity and that activity identified as one attribute of the "idea of life."[150] In typical *naturphilosophisch* fashion, "the whole universe is a living nature, and all activity recognized in it must be predicated as an internal and necessary condition of its existence."[151] Walther explicitly rejected the notion of a vital force:

> The most erroneous of all modes of representation is, however, certainly that according to which one regards life as a particular condition, modification, or quality of individual material substances (of organized bodies) and on that basis (insofar as the active ground of any quality is called force) one falsely imputes to it a *vital force* of its own, which has always remained an occult quality—in such a manner that one assumes that dead matter would indeed become animated if one were able to impart vital force to it in an appropriate manner, and that living matter becomes dead when that force is destroyed in it.[152]

In a similar vein, Carus used his review of Autenrieth to criticize not only the common understanding of force in general, but in particular the illegitimate concept of vital force: "Nevertheless the notion of particular forces attached to material objects still causes the greatest mischief in every natural science, but especially in physiology the replacement of a healthy pure perception [*Anschauung*] *of life itself* as a primitive phenomenon of the world by the notion of a particular vital force which is supposed to penetrate and rule and then bring to life (one knows not *how*) intrinsically dead objects, has caused the greatest confusion."[153] In his *System of Physiology* Carus again criticized the view of life as due to the imposition of some force onto otherwise dead matter, as he criticized the concept of force in general as purely subjective.[154] Such a misuse of the concept of force he termed the first negative aspect (*Schattenseite*) of physiology.[155] The second concerned the notion of a soul:

> One assumed: 1) a number of particular material substances or elements, each furnished with manifold "forces" (although no one was able to say what a force is in and of itself and how it attaches to matter); 2) a particular force, "the vital force," which in part constructed from these material substances that which one called an organism, in part, like the weight in a clock, served to maintain this organism in vital motion (although again no one was able to say what the vital force is in and of itself and how this force might also be able to deal with the material substances in accordance with its purposes); 3) finally, after one had now imagined to have constructed an organism with these forces and material substances, one again took from somewhere a new force called "soul" which—without one knowing anything else about it, or being able to indicate *how* it ever enters into reciprocal interaction with the organism—was placed in the organism at a certain time and which again abandoned it at a certain time, whereupon the vital force likewise then escaped and the individual material substances were now again seized by other vital forces.[156]

A third crime (*Gebrechen*) was the tenaciously held distinction "between an organic and living and a so-called inorganic and dead nature."[157] Still another problem was that as scientists saw everywhere the same mechanical, chemical, and electromagnetic processes, and tried to apply them to living things, they began to regard organisms as machines. That was a fundamental mistake, however, since organisms arise via differentiation or separation within a single homogeneous entity, whereas machines are composed of many different heterogeneous parts. For Carus as for Schelling, the point was not to separate physical and chemical processes from physiological and organic ones, but to appreciate "that here as there, only *one* life holds sway, although to be sure the expressions of the one—of higher organic life—are much more complicated, manifold, and more difficult to follow in terms of their mental construction [*schwerer im Geiste durch Construction zu verfolgen*] than those of the other."[158] Schlüter was right: the problem *Naturphilosophen* had with the vital force was that the very need for it presupposes the intrinsic inactivity of inorganic (or "dead") nature. If on the one hand the discussion of the vital force in the mainstream physiological literature of the period broached topics of plausible relevance to Mayer, involving issues of creation and destruction and of the relationship of the generally accepted vital force to the other forces and imponderables of nature, on the other hand the widespread rejection of a specific vital force (as of force in general) in works of *Naturphilosophie* would appear to have offered a significantly poorer context for the stimulation of the kinds of ideas that led Mayer in the direction of conservation of energy.

## 3 Respiration and Animal Heat

As shown in chapter 3, mainstream German physiologists of the period discussed respiration not only in terms of its connection with animal heat, but also as one of the two chief means by which the body excretes excess carbon (along with the production of bile in the liver). Most did not unequivocally subscribe to Lavoisier's combustion theory of animal heat—the nervous system was typically assigned a role here—though that process was nevertheless well-nigh universally regarded as the *principal* source of animal heat. Mayer was well informed about both of those respiratory functions; indeed, his understanding of the blood's excretory function was what prepared him to be surprised at the color of the venous blood he let off the Java coast. Although there is no evidence regarding Mayer's views on animal heat before his voyage to the Dutch East Indies, those bloodletting observations led him very quickly to the acceptance of the complete sufficiency of the combustion theory, a crucial step in the progress of his thinking. In other words, Mayer's discovery is unintelligible outside the context of the dominant ideas regarding respiration and animal heat. Given the centrality of those topics, one might wonder whether their treatment by *Naturphilosophen* could have provided a similarly rich, if specifically different, context for the understanding of Mayer's thinking. In terms of the sources that I have read, the answer to that question is a decided "no." Indeed, *Naturphilosophen* had such bizarre and idiosyncratic views on the subject that they are difficult to paraphrase in language other than their own.[159] In the exposition that follows, we will encounter many of the leading characteristics of *Naturphilosophie*: analogies and parallelisms galore (notably between microcosm and macrocosm), *Potenzierung*, *Figierung*, antitheses, an appeal for *Anschauung*, and a rejection of both artificially contrived experiments and quantitative thinking. What we will not encounter is much that enlightens us with respect to Mayer's possible sources.

Schelling said little relevant to these topics in the works of his that I have read. In his *Ideas* of 1797 he asserted that animal heat flows from an internal source copresent with "the spark of life" itself; in that connection he noted that light and heat work against the laws of gravity that would otherwise rule an inert earth.[160] His *First Sketch* of 1799 related respiration to the now-central activity of irritability, but it said nothing about animal heat.[161] According to Burdach, Wilbrand's *Nature of the Respiration Process* (1827) taught that "there exist no oxygen and carbon at all, for one cannot see them; on the other hand the light nature [*Lichtnatur*] of the elements is a fact, since one sees the light in combustion, and respiration [*das Athmen*] consists in communicating to the organism the light nature inherent in the elements and thereby internal animation."[162] Pohl's *Views and Results Concerning Magnetism, Electricity, and Chemism* (1829) made passing mention of respiration, though it is harder to understand his positive views than his rejection of the conventional combustion theory:

> Respiration [*Respiration*] is the immediately victorious positive act of vanquishment against the chemism in the air that hostilely opposes the organism. The course and the products of this absolutely antichemical function must therefore no longer be judged in terms of the one-sided viewpoints in accordance with which the chemical experiment in and of itself is accustomed to being conceived, and the views that have become current regarding this subject—deriving, in accordance with a vague analogy with the process of combustion, from wholly unsuitable experiments with animals in enclosed, confined air—belong to the greatest aberrations into which a science that has deviated from the fundamental perception of natural life [*Grundanschauung des Naturlebens*] has ever been able to fall.
>
> Digestion, as the other polar moment of the vital process in the animal, is, vis-à-vis the positive act of respiration, only a negative, defensive side of the latter.[163]

For Pohl, respiration is an activity that *opposes* the fundamentally chemical action of the hostile atmosphere, and hence was not itself to be regarded simply as a chemical process.

Walther's *Human Physiology* provides a superabundance of characteristically *naturphilosophisch* ways of thinking. Consider his discussion of the heart:

> The heart, although representing in itself the identity of both types of vessels, is nevertheless according to its position venous, situated toward the left side. Positive, expansive motion has in general the direction from left to right. (The *dextrum* is therefore everywhere the good, the correct, the true, the *sinistrum* is everywhere the evil, the false, the baleful. The motion of the heavenly bodies is also governed accordingly.) For that reason, therefore, the arterial blood circulation goes from the left toward the right.[164]

Perhaps following Schelling's lead, Walther associated the production of heat with the irritable system. His section on animal heat is a tour de force of the genre:

> The relationship of cohesion to gravity is the same as that of heat to light. For just as gravity is innate in things in twofold fashion—as gravity, and then again, attaining

a distinctive life in them through the particularity of things, as cohesion—so, too, does the principle of light indwell in them in twofold fashion. For that reason heat is antithetical to cohesion, it is that which is anticoherent, that which liquefies and gasifies—the universal menstruum of things. . . . The process of heat is therefore also the inverse process of gravity. Just as gravity is the centripetal tendency of things, so, on the contrary, does heat act in the centrifugal direction. How can that which follows such a determinate, expansive, direction, and which is essentially completely antithetical, be taken as a manifestation of the third, synthetic, dimension? Heat is not the reconciliation of differences, the leveling out of antitheses, but it is that which ignites the antitheses: and all the affinities of things among themselves, which after all depend on the antithesis of qualities, appear only at a particular temperature of heat.

. . .

Heat has an immediate relationship only to light, not to gravity. It derives from light and belongs to those attributes of things that are an expression of their ensoulment by light. That which corresponds to gravity in the process of heat is cold: for just as there is a process of heating, there is also a process of cooling. Both are the opposing manifestations of one and the same cause [*Grund*]. But it is a highly objectionable view to regard cold as merely a withdrawal of heat, as a privation of the latter. Cold and heat are not merely quantitative differences of an arithmetical magnitude capable of increase and decrease; there is an antithesis of qualities between them: cold is a cohesion process, heat an anticohesion process.

. . .

Heat is the identity of light and gravity, but within the power [*Potenz*] of electricity: for that reason every process of the excitation of heat is the excitation of electricity. Heat conduction is an electrical conduction process. The sentence: the organic being generates heat in itself, means the same as the sentence: the organic being bears in itself the identity of gravity and light, reflected in the second dimension. For just as within the atmosphere of the earth heat only arises through the antithesis between sun and earth, in the same way the organic being, which bears in itself the center of its position and is simultaneously midpoint and circumference, has in itself the source of heat. The capacity to generate heat in itself and to bind it, to give itself its temperature through its own action, is the sign of the cosmic perfection of the organic being. But the irritable system is preeminently that of the generation of heat: for heat is the identity of light and gravity within the second power.[165]

It is hard to know how to evaluate a theory of animal heat that rests ultimately on an obscure cosmic analogy, on "the identity of gravity and light, reflected in the second dimension." On the other hand, although Walther's style of philosophizing was for the most part quite at odds with Mayer's, there are a few motifs reminiscent of some of the peculiarites of Mayer's earliest ruminations: a connection between light, heat, and gravity; centripetal/gravitational *versus* centrifugal/thermal tendencies; the neutralizing of differences; and the self-creation of heat by a cosmic organism. The significance of such connections will be explored in depth in section 4 of this chapter. For the present, I note simply that there is no evidence—and it seems to me little likelihood—that Walther's book of 1807–1808 was one of Mayer's sources.

Oken's brief consideration of animal heat in his *Textbook of Nature Philosophy* is as inscrutable as any extract from a work of true *Naturphilosophie*:

> 2575. The chief function of cell formation is the heat process. Cell process and heat process are one. Heat is the product of the condensation and rarefaction process.
>
> 2576. The temperature process is individualized in the skin.
>
> 2577. All temperature depends on the evaporation process. It is sometimes nerves, sometimes vessels, sometimes external influences that alter that process.
>
> Animal heat is produced like cosmic heat, through the alternation of fixation [*Wechsel der Figierung*]. But this alternation occurs principally in nutrition and transpiration.[166]

Naumann, in a work of 1835, concluded that the production of animal heat is principally "the result of the passage of the liquefied nerve-medulla [*fluidisirtes Nervenmark*] into the blood of the capillary vessels," a conclusion anticipated by the title of the book: *The Problems of Physiology, or the Antithesis between Nerve-Medulla and Blood*.[167] Naumann allowed that it was probable that the influence of air on the blood in the lungs might further increase the animal's temperature, but he denied that this could be the principal source of animal heat.[168]

Carus considered five possible causes of the production of heat in higher animals, two of which were relevant to physiology: "4) repetition of manifold telluric (or, if one wishes to follow customary linguistic usage, mechanical-physical-chemical) processes in the organism itself that are capable of developing heat; 5) repetition of an internal solar-planetary tensional relationship in the organism itself that, like the action of the sun on the earth, brings about the liberation of heat."[169] The processes included under (4) were in turn fourfold: "a) densification by means of pressure and b) the friction related to it, i.e., a variable pressure, c) metamorphosis of the phenomenal form of the substance in the sense of increasing densification, and d) the combination of diverse substances whereby new products are formed, and the electrical processes that accompany these [last] two processes."[170] Of these the most important physiologically were (c) and (d). Accordingly, Carus addressed the pros and cons of the (there unnamed) oxygen theory of animal heat, mentioning the well-known experiments of "Depretz." But Carus ultimately rejected the attempt to quantify the production of animal heat and to assign a percentage to any given chemical process. More "indicative of the dignity of the human organism" than the mere production of a certain degree of heat is, he maintained, the body's maintenance of the same temperature under widely varying circumstances.[171] That ability indicates the influence of another heat-producing relationship—the fifth possibility on his earlier list—namely, the opposition between the nervous system and the other tissues of the body: "Now as such an (as it were) regulative relationship . . . one can only recognize *the internal mutually opposing action of planetary and solar structures*, upon which depends in general the more strongly pronounced individuality of the human being, one that rises [*sich steigt*] as far as the concept of the person. It is the relationship between the nervous system, which represents the unity of the idea of the organism, and the other structures that it pervades."[172] Animal heat is thus not the result of a conventional chemical reaction, but the result of an antithesis that is itself the expression of a cosmic analogy. Such reasoning is, of course, pure *Naturphilosophie*, and demonstrates once again how far that movement was from the mainstream of German science by the 1830s. I see nothing here particularly conducive to conservation-of-energy-type reflections, and no explanatory gain from assuming that Mayer might have read Carus.

Burdach, in a book Mayer owned, offered his own variant of the explanation of animal heat in terms of some kind of polar opposition between nerves and (in this case) the vascular system: "As with an electrical antithesis, *heat* is generated through the interaction of the nerves with the vascular system on which they are distributed; it is developed in the blood, as its carrier and conductor, . . . everywhere that nerves and blood vessels are present, and it is the greater the more energetically (on the one hand) the nerve acts, the more complete (on the other hand) the blood and the more rapid its circulation."[173] He said nothing about oxidation in this connection. Against a purely mechanical explanation of the circulation of the blood, Burdach taught that the organs of the body attract the bright red (and oxygen-rich) arterial blood, just as the air in the lungs attracts the dark red (and carbon-rich) venous blood.[174] As one would expect, he was ready with a raft of analogies to render the phenomena intelligible:

> This action is not mechanical, but can only be compared to phenomena where bodies attract a fluid chemically related to them in order to enter into combination with it; or where, by virtue of the constitution of the substance, no change in composition comes about, but only adhesion—where, for example, a body attracts water out of the air, a narrow tube raises water, bodies floating on water are attracted or repelled by the rim of the vessel according as their substance and that of the vessel are different, etc.; or where, without regard to the substance, particular force relationships are at work, north- and south-polar magnetic bodies or positive and negative electric bodies attract each other, while the like-named repel each other. We recognize as a rule that two bodies which agree with each other in general, but which are opposed to each other in particular, attract each other, that therefore related but dissimilar bodies, or those in dissimilar states, strive toward unity and to level out their differences [*sich auszugleichen*], and we may be permitted to presuppose a like relationship between the blood and the organs.[175]

In the sixth volume of his well-known *Physiology as an Experiential Science* (1840), Burdach considered the problem of animal heat at greater length. He introduced the topic with words redolent of the conceptual apparatus of *Naturphilosophie*:

> Heat is accordingly generated wherever life stirs. But as we cannot, like the ancients, regard the capacity to generate it as identical with the vital principle, just as little could it suffice when one simply declared in general that it is generated by the vital

> force. For although its essence lies in the concept of life, it can nevertheless only be produced by life through the mediation of the universal forces of the world. A) Heat in general manifests itself as the expansive force that emerges free from the conflict with contractive force. In particular, it is produced by all kinds of conditions that presuppose contraction: a) mechanically, by means of pressure, friction, impact, and hitting; b) chemically, with the interaction of dissimilar related substances, whereby there takes place a leveling out of their difference, an exhaustion of their chemical force, and a densification of substance; c) dynamically, by means of the retarding effect of the earth on sunlight, and by means of the conflict of positive and negative electricities whose indifferentiation has been made more difficult.[176]

Burdach reviewed the experiments and conclusions of Lavoisier and Laplace; considered and rejected explanations based on friction, compression, or changes in specific heat; and weighed arguments in favor of the nervous system. He was of two minds as to whether or not respiration and the production of heat are directly proportional, though the comparative evidence from other animal species disposed him to deny such a connection. He concluded negatively that "respiration [*das Athmen*] does not produce heat directly, though to be sure it belongs to the conditions of its production."[177] He appealed to Johannes Müller in support of his positive conclusion that the source of animal heat lies principally in "the interaction of the nerves with the remaining tissues" of the body.[178] Burdach believed, moreover, in the qualitative difference between inorganic and organic electricity and heat: "Solar heat, the heat of a stove, and animal heat make an entirely different impression at the same degree [of heat]; Rumford found that his hand had a stronger effect on the thermometer than an inorganic or dead organic body of the same temperature; with someone suffering from putrid fever we feel a distinctive, disagreeable, biting heat that the thermometer does not indicate."[179] Such a view is not apt to dispose a person to accept a purely chemical explanation of animal heat, nor to look for a universal measure of the quantity of heat.

If the sources that I have looked at are representative, it would not appear likely that Mayer's understanding of physiological processes owed anything to *Naturphilosophie*. Its peculiar modes of thinking were in the main entirely foreign to the considerations that characterized Mayer's. Nevertheless, if the major themes, assumptions, and conclusions of *Naturphilosophie* do not appear to have been likely contributors to Mayer's theory of force, there are still a number of its recurring motifs that do have rough counterparts in Mayer's work. It is to those we now turn.

## 4 Echoes of *Naturphilosophie* in Mayer's Work?

From the material presented in the foregoing sections and chapters, I conclude that the traditional link between Mayer and *Naturphilosophie*, the doctrine of the unity of forces, cannot support such a connection: not only was the unity of forces not one of Mayer's guiding motifs, but it was not all that prominent among

*Naturphilosophen*, either. Neither did the latter speak of forces as causes, or consider the possibility of their quantitative measurement. Only Mayer's (for him) rather exceptional use of the term *Urkraft*—twice in the entire body of his surviving published and unpublished writings—remains to suggest a possible link between his doctrine of force and that of *Naturphilosophie*. It does not appear even remotely likely that Mayer's thinking owed anything to its peculiar physiological teachings. Nor, with the important exception of indifference, do the most prominent traits of *Naturphilosophie* find significant counterparts in Mayer's writings. Mayer never even made particular use of the words "dynamism" and "dynamic," despite the fact that his was a theory of force. There remain, however, certain peculiarities in Mayer's first (unpublished) essay and his correspondence from that early period that do have counterparts in the *Naturphilosophie* literature—most notably in Schelling's *World Soul, First Sketch*, and *Introduction*—that require us to leave open the question of influence: in particular, ideas such as creation out of nothing and the relationship between light and gravity in planetary systems, ideas that were themselves connected with the central concept of indifference.

The first paragraph of Mayer's earliest surviving essay, "On the Quantitative and Qualitative Determination of Forces," contained the striking and (for him) unusual sentence, "We can derive all phenomena from a primitive force that tends to annihilate the existing differences, to unite everything that exists into a homogeneous mass in a mathematical point."[180] As noted in chapter 2, Mayer's development of these ideas suggests that that *Urkraft* might be the force of gravity, which tends to eliminate the spatial separation, or "difference," between bodies. Even allowing for the fact that *Urkraft* might conceivably be here just another German word, formed according to quite regular principles—and thus not betraying any particular philosophical provenence—and despite the fact that *Naturphilosophen* tended to speak in terms of *opposing* forces (*Ur-* or otherwise, but in plural), the sentence still has a strongly *naturphilosophisch* flavor.

As far as I can tell, the only other use of the word *Urkraft* in Mayer's writings occurred in a letter to Baur from August 1844 in which he indicated he wanted the following passage added to the mechanical part of the long manuscript he had sent Baur nine days earlier:

> If one has once begun to ascribe the properties of things to particular forces, then logically one should also derive the property to exist from a distinctive force, a force not unjustly called the primitive force. Out of this "force of being" ["*Seikraft*"] there would emanate the composing principles of matter, the forces of repulsion and attraction, like the spectrum out of the sun's ray. The prism that accomplishes such a wonderful feat is the wisdom, rising above normal human understanding, of the philosophical investigator.[181]

This passage contains one of Mayer's very few references to the dynamical centrifugal and centripetal forces from which Schelling and the rest had constructed matter. His somewhat obscure analogy with the sun's spectrum also heightens the *naturphilosophisch* tone of the passage.

This is suggestive evidence, but it does not take one very far. The themes broached do not play a major role in Mayer's work, and the sense of the passage appears to be dismissive: Mayer did *not* wish "to ascribe the properties of things to particular forces"—he repeatedly insisted that forces are not mere properties of matter—and the alleged exceptional "wisdom" of the "philosophical investigator" was surely intended ironically. More impressive are certain similarities between Mayer's early attempts to generate the sun's light somehow out of its gravitational interaction with the earth, and some of Schelling's and others' ideas concerning planetary motions and the nature of the universe. Here again, only fairly extensive quotation from the primary sources can convey the quality of the resonances between their conceptions and Mayer's.

Schelling's earliest pertinent discussions occurred in the *World Soul* of 1798, in the first half of the book entitled "On the First Force of Nature." His opening paragraph would seem to be a clear allusion to the generation of cyclically recurring planetary motion out of two opposing forces, one acting centripetally, the other tangentially:

> Every motion that returns back into itself presupposes, as condition for its possibility, a *positive* force that (as *impulse*) *initiates* the motion (as it were, produces the initial tendency toward the line) and a *negative* force that (as *attraction*) *directs* the motion *back into itself* (or prevents it from striking out in a straight line).
>
> In nature everything strives continuously *forwards*; we must seek the reason that this is so in a principle that, being an inexhaustible source of *positive* force, continually initiates the motion anew and uninterruptedly sustains it. This *positive* principle is the first *force of nature*.[182]

Nature thus possesses within itself an inexhaustible source of motion, connected somehow with periodic central-force systems. Schelling later invoked an explicit analogy between universal gravitation and the opposing forces of the so-called chemical process, both of which reveal the general tendency in nature toward equilibrium:

> Since there is a universal striving [*Bestreben*] toward equilibrium in nature, every excited principle arouses necessarily and in accordance with a universal law the *opposing* principle with which it stands in equilibrium. One not unjustly regards this law as a modification of the universal law of gravitation; it is, along with the law of universal gravity, at least dependent upon a *common higher* law.
>
> One must assume that in every chemical process there predominates such a dualism of opposing, *mutually excited* forces. *For* in every chemical process there *arise* qualities which were not there previously, and which owe their origin merely to the striving of opposing forces to come into equilibrium with each other. It has long since been the ambition of philosophers and physicists to investigate the connection in which the chemical attraction of bodies stands to the universal attraction. One must affirm that both attractions stand under the same original law, namely that matter in general reveals its existence in space by means of a continual striving toward equilibrium, without which all material substances would be exposed to dissipation into infinity.[183]

Schelling further connected gravitational systems and the opposing forces making up matter with the production of light, regarded as streaming out from the center toward the periphery. The details are hazy, but the general tendency of his thinking recalls some of Mayer's early planetary speculations:

> The *positive* cause of all motion is the force that *fills* space. If motion is to be sustained, then this force must be excited. The phenomenon of every force is therefore a *material substance* [*eine Materie*]. The first phenomenon of the universal force of nature by means of which motion is kindled and sustained is *light*. That which streams toward us from the sun (since it maintains the motion) appears to us as the *positive* [factor], that which our earth (as merely *reacting*) opposes to that force appears to us as *negative*.
>
> . . .
>
> The matter that in every system streams from the center toward the periphery—i.e., *light*—is moved with such force and speed that some have even doubted its materiality because it lacks the universal characteristic of matter, inertia.
>
> . . .
>
> That therefore light disperses itself in all directions in rays must be explained from the fact that it finds itself to be in *continuous development* and in the *original* dispersion. That even light attains relative rest can already be concluded from the fact that the light of an infinite number of stars does not propagate its motion as far as us.[184]

Yet despite the similarities between these passages from Schelling and some of Mayer's early speculations, in the same book Schelling in fact explicitly considered the production of light in the sun as due to a decomposition of its atmosphere.[185] And the *Abhandlung* added to the second and third editions, whose title might seem to announce another source of cognate ideas—"On the Relationship between the Real and the Ideal in Nature, or Development of the first Principles [*Grundsätze*] of the Philosophy of Nature from the Principles [*Principien*] of Gravity and Light"—was for the most part a striking example of the kind of gibberish *Naturphilosophie* is famous for, and it provides no promising connections.[186]

Schelling's two major works of *Naturphilosophie* of 1799 elaborated on the above themes and connected them with the central concept of indifference, in particular creation via the annihilation of difference. In analyzing the nature of gravity (*Schwere*), Schelling noted that a body alone in empty space is not heavy (*schwer*), but becomes so only in the presence of another body outside itself: every body has the degree of its weight (*Schwere*) in itself, but the cause of its weight outside itself. There thus must be "an *original difference*."[187] (Recall Mayer's interpretation of gravitation in terms of the spatial difference between two bodies.) In imagining the original development of the universe in terms of an alternation between expansion and contraction, Schelling added that this alternation was nature's attempt to return to a state of indifference from the difference that is contrary to its original unity.[188] Though he did not there connect these ideas with gravitation, the general context was clear. The explicit connection with gravitation came in his second work of that year, in discussing the production of a sequence of antitheses out of the original productive identity of nature, antitheses that are in turn partially neutralized around particular points of indifference. Some of his language recalls that of the first paragraph of Mayer's first (unpublished) essay of 1841:

> The organization so determined is none other than the organization of the universe in the gravitational system.—The *force of gravity is simple*, but its *condition* is duplicity.—Indifference proceeds only out of difference.—The neutralized duality is matter, insofar as it is only *mass*.
>
> The *absolute* point of indifference exists nowhere, but is (as it were) distributed among several *individual* points.—The universe, which forms itself from the center toward the periphery, *seeks* the point where even the most extreme antitheses of nature neutralize each other; the impossibility of this neutralizing [*Aufheben*] assures the endlessness of the universe.
>
> Of every product A the unneutralized antithesis is transferred to a new product B; the former thereby becomes the cause of the duality and the gravitation for B. . . . Thus for example the sun, because it is only *relative* indifference, sustains, as far as its sphere of action reaches, the antithesis that is the condition of gravity on the subordinate heavenly bodies.
>
> The indifference is in every moment neutralized, and in every moment reestablished. . . . The universal reestablishment of duality and the renewed neutralization in every moment can only appear as a striving [*nisus*] toward a third thing [*ein Drittes*]; this third thing is abstracted from the tendency, [is] nothing, therefore merely *ideal* (only designating the direction)—a *point*.[1]
>
> . . .
>
> [1] It is precisely nought [*die Null*] into which nature strives continuously to return, and into which it would return if the antithesis were ever neutralized. Let us imagine the original state of nature as = 0 (lack of reality). Now the nought can of course be imagined as separating itself into 1 − 1 (for this = 0).[189]

Schelling argued that the production of light in the sun must be necessarily connected with its position at the center of the solar system, though again his solution to that problem did not here (in the *First Sketch*) involve gravitation.[190] That connection came later, in a passage reminiscent of Mayer's probing for connections:

> That the greatest and foremost portion of the heavenly bodies is intended for light processes does not indicate something accidental, but rather a *universal*, higher, and farther-reaching *law of nature*. The action of light must stand in secret connection with the action of gravity exerted by the central bodies.[2] The former will give the things of the world the dynamic, as the latter the static tendency.
>
> . . .
>
> [2] That is, through the action of gravity the indifference is continually neutralized—the condition of gravity reestablished. But we perceive in light nothing but this reestablishment of the antithesis, therefore it is already clear here in what connection the chemical action may well stand with the action of gravity.[191]

He expanded upon these connections—now also including chemical reactions—later that year in the *Introduction* in words likewise suggestive of the conceptions Mayer brought together. At issue was the dynamical process by which antitheses give rise to each other via the alternation of difference and indifference, of which process Schelling offered the following further illustration:

> [T]hus when the bond of gravity is dissolved in the chemical process, the phenomenon of light, which accompanies the chemical process in its greatest perfection (as the process of combustion), is a strange phenomenon that, when further pursued, confirms what is said on page 146 of the *Sketch*: "The action of light must stand in secret connection with the action of gravity exerted by the central bodies." For is not that indifference of gravity dissolved at every moment, since indeed gravity, as always active, presupposes a continuous neutralizing of indifference?—Thus does the sun bring about a universal dissociation of matter into the original antithesis (and thereby gravity) by means of the separation [*Vertheilung*] exerted on the earth. That universal *neutralizing of indifference* is what appears to us (animate beings) as *light*; *where*, therefore, that indifference is dissolved (in the chemical process), there light *must* appear to us.—According to the foregoing, it is one antithesis that, from magnetism on through electricity, finally loses itself in the chemical phenomena.[1]
>
> . . .
>
> [1] . . . What happens then in the chemical process? Two entire products gravitate toward each other. The *indifference* of *each one by itself as absolute* is neutralized. This absolute neutralizing of indifference puts the entire body into a state of light, just as the partial neutralizing in the electrical process puts it into a partial state of light. Thus the light that appears to stream to us from the sun will also in all probability be nothing but a phenomenon of the indifference neutralized at every moment. For since gravity never ceases to act, its condition—the antithesis—must also be regarded as arising again at every moment. We would thus have in the case of light a continuous visible phenomenon of the force of gravity, and it would be explained why precisely those bodies of the world system that are the principal seat of gravity are also the principal source of light, explained, then, in *what* connection the action of light stands with that of gravity.[192]

If there is, to be sure, no Mayer-like application of the parallelogram of forces to central-force motions in Schelling's works—the confusions surrounding which seem to have led Mayer to first entertain the possibility of the creation of force out of nothing—and even if Mayer never spoke of Schelling-like antitheses (much less of their successive production), nevertheless the nexus of issues and much of the language is quite similar to that of Mayer's earliest work.

One of the criticisms Schelling leveled against mechanical and atomistic systems of natural philosophy (such as that of the Genevan natural philosopher, Georges-Louis LeSage) was that they cannot explain the first origin of motion in the universe, only its continuance. According to his dynamical philosophy, however, motion originally arose out of a state of rest. Although he did not relate this creation of motion to central-force systems, nor speak in this context in terms of equal and oppositely directed motions arising out of zero à la Mayer, much of his language is otherwise strongly reminiscent of Mayer's:

> Since heterogeneity is the source of activity and motion, so would the cause of universal magnetism also be the ultimate cause of all activity in nature, thus the *original* magnetism would be for nature in general what sensibility is for organic nature—a *dynamic* source of activity; for in the realm of mechanism one sees motion arise out of motion. But what is then the first *source* of *all* motion? It cannot *again* be motion. It must be the opposite of motion. Motion must originate from rest. As in the chemical process, where the moved body does not move the resting or moved one, but the resting body moves the resting one. Likewise in the organism, where no motion immediately again produces motion, but where every motion is mediated by *rest* (by sensibility).
>
> . . .
>
> For since the first problem of this science, to investigate the *absolute* cause of motion (without which nature is nothing in itself whole and bounded), is simply not to be solved mechanically—because mechanically motion only arises out of motion *ad infinitum*—there thus remains open only one route for the actual construction of a speculative physics, the dynamical, with the presupposition that motion arises not only out of motion, but even out of rest as well, that there is therefore also motion in nature's rest, and that all mechanical motion is merely the secondary and derivative of the sole primitive and original, which originates already from the first factors of the construction of any kind of nature (from the fundamental forces).[193]

Once again we find certain similarities with Mayer—such as the creation of motion out of rest—in close connection with other themes quite absent from his work, such as the (in part implied) tripartite sequence of magnetism, electricity, and the chemical process, plus the further analogies with sensibility and so forth.

Nor was Schelling the only *Naturphilosoph* to deal with these issues. Consider the connection between light, gravity, and heat as expounded in Eschenmayer's *Outline of Nature Philosophy*:

> Light stands in proportion with the greatest degree of velocity and force, gravity with the greatest degree of inertia and weight [*Last*], but heat with motion of an intermediate degree. . . .
>
> Gravity seeks to reduce everything to a point—i.e., the center of gravity—and thereby to impede all motion. For that reason it is the negative force of nature, or rather it belongs to the negative order, where inertia has its domain.
>
> Light seeks to neutralize all centers of gravity and to potentiate every motion. For that reason it is the positive force of nature, or rather it belongs to the positive order, where velocity has its domain.
>
> Heat is the indifference of nature and mediates gravity and light, it stands as unity, as it were as the zeroth power between the negative and positive exponents of gravity and light.
>
> Light is *the integrating*, gravity *the differentiating*, and heat *the indifferentiating* principle of nature.

There are only three powers: light, heat, and gravity, out of whose reciprocal interaction, integration, indifferentiation, and differentiation all material substances receive their quantitative and qualitative relationships.[194]

If some of the specifics are different from the ideas Mayer toyed with, nevertheless there are suggestive generic similarities.

Before assessing the evidence for Mayer's having read and been influenced by one or more works of *Naturphilosophie*, let us look at one more work, Naumann's *Pathogeny*, published during the second half of 1840 while Mayer was on board the *Java*. Naumann's work illustrates both some of the affinities between aspects of Mayer's work and aspects of *Naturphilosophie* as well as some of the fundamental differences between them. As was not unusual in medical and physiological works of the day, the author developed at some length the ontological and epistemological views that underlay his thinking. Our knowledge of matter is thus only in terms of certain basic antitheses, and the "highest law of nature" is a vaguely enunciated "*law of antithesis* or *polarity*"; these antitheses and "the original polarity" have their basis "in the original difference [*Differenz*] of matter."[195] There was the standard *naturphilosophisch* talk of the "neutralization of all antitheses."[196] Naumann distinguished—idiosyncratically—two fundamental polarities, a *Direktionspolarität*, which referred somehow to the relationship between the density and extension of matter and corresponded roughly to "mechanism," and an *Affinitätspolarität*, which referred somehow to the particular properties of matter and corresponded roughly to "chemism."[197] Transitions between these two manifestations of polarity, accompanied by changes in the form and composition of matter, produce the phenomena of electricity, magnetism, etc. It was in discussing these two *Fundamentalpolaritäten* that Naumann treated the relationship betwen the sun and the earth in a way that recalls at least one of the motifs of Mayer's understanding of it:

That they [i.e., the two fundamental polarities] can be derived from one and the same cause—namely from the original difference of matter (§11)—can easily be proven from the relationship of the earth to the sun. That relationship clearly corresponds to the directional polarity insofar as it designates the activity of a force by means of which the distance of the planet from the central body and its motion and orbit are determined. But at the same time this relationship manifests itself as affinitative polarity in that it affords the opportunity for manifold changes in the matter of the terrestrial body. The polarity between sun and earth is, however, in essence of a dynamical nature; for it is continuously excited and just as continuously again annihilated [*aufgehoben*], so that the earth retains its relative autonomy, while nevertheless the antithesis between the mass and the elements [*Stoffe*] is continually called forth anew in its matter by means of the powerful external impulse. But the polar relationship of the earth to the sun is only determined by the higher antithesis as whose result one must regard the gradually occurring change in location within the universe of the sun with its whole planetary system, and which itself presupposes the continuous activity of a material force that, even if otherwise unknown, is nevertheless directed toward the sun.[198]

Thus amidst all the decidedly un-Mayer-like language we find Naumann talking about the continuous interruption and reestablishment of the polarity between the earth and the sun. This is still not really Mayer's language—polarities and antitheses were not two of his usual concepts—but the passage nevertheless shows that the general issue of relating sun-centered planetary motion to some kind of dynamical process of creation and annihilation was a common topos of *Naturphilosophie*. In terms of understanding Mayer's possible sources, however, it does not get us much to assume he read Naumann, nor does it seem very likely that he would have done so during the undoubtedly hectic four months between his return to Heilbronn in February 1841 and the completion of his first essay in June.[199]

Eschenmayer's *Outline of Nature Philosophy* was published in Tübingen in 1832, the year Mayer entered the University of Tübingen to study medicine. Thus timing and location make this a reasonable candidate for a work of *Naturphilosophie* Mayer might well have known. Unfortunately—as the case may be—it does not provide either very many or very strong leads for understanding the peculiarities of Mayer's work. Like Naumann's *Pathogeny*, Eschenmayer's book has served here mainly to illustrate the characteristically *naturphilosophisch* aspects of a peculiar way of relating light and gravity and to highlight the problem of identifying Mayer's potential sources.

As far as the works considered here go, that leaves Schelling, in particular his two works of *Naturphilosophie* from 1799. In this case the thematic similarities are indeed very strong: a connection between light and gravitation via the annihilation of differences; the all-important concept of indifference; and the creation of opposites out of some kind of zero or rest state. These points of contact remain striking even considering the more numerous ways in which Schelling's and Mayer's work and thought were vastly dissimilar.

But did Mayer actually read Schelling? Not only is there no direct evidence that he did—and Shelling's name never once appears in the entire corpus of Mayer's surviving writings—but other evidence, already cited, argues that he had little exposure to or appreciation of philosophical works in general. That Schelling's especially relevant works of 1799 were not reprinted until 1858 further lessens the likelihood that they would have come to Mayer's attention. In assessing Mayer's receptivity to *naturphilosophisch* ideas one may cite, too, his response to an essay entitled "Toward a Criticism of Contemporary Science" published in the *Deutsche Jahrbücher für Wissenschaft und Kunst* by an otherwise unidentified "Dr. N. Löwenthal" in October 1842—too late to have influenced him, but one of the only *naturphilosophisch* works Mayer specifically commented upon in his surviving writings. Löwenthal paid particular attention to the issue of the proper meaning and common misuse of the terms *Schwere* and *Schwerkraft*, the issue that caused Griesinger to call the article to Mayer's attention in December 1842.[200] Löwenthal wrote:

> The individual heavenly body's simultaneous dependence on and relative independence of the law of the unity of the entire material world reveals itself namely in a three-part process whose first moment is that the particular heavenly body posits itself

> [*setzt sich*] as autonomous and closed within itself. . . . The second moment is that the heavenly body that posits itself in its individual autonomy by means of the axial rotation proceeds to negate the codetermining influence of the universal center. It strives to tear itself loose—in the direction of the tangent perpendicular to the line connecting its center with the center of the world—from the connection with the universal ground out of which its particular existence has been formed. But because the universe never ceases to be the bearer of all particular configurations, this negating tendency of the individual is opposed by the universal principle of gravitation, which strives by means of a central force to maintain the individual heavenly body in its dependence on the universal center. Thus there finally results (thirdly) from the antithesis and the struggle of forces acting at the same time in tangential and centripetal directions on the individual heavenly body the revolution of the planet around the sun simultaneously with the rotation about its own axis.[201]

Not that Löwenthal's essay contained as suggestively similar language as Schelling's or even Eschenmayer's, but its overall tone was comparable and it touched upon certain conceptions, such as the combination of tangential and centripetal forces in the production of planetary motions, to which Mayer might have been responsive. In Mayer's reply to Griesinger, he said he had read Löwenthal's article but not paid much attention to it "because he begins more with philosophical reflections unintelligible to a layperson in this school than he provides a clear scientific viewpoint [*Anschauung*]."[202] As far as it goes, this suggests that Mayer had not had much exposure to, nor have much sympathy for, writings in a *naturphilosophisch* mode.

Yet, limited as they were to a few topics, the above-identified similarities between Mayer and Schelling remain strong and should not be brushed aside too easily. Given that the specific topics of relevance were reasonably well represented in the *Naturphilosophie* literature, it is possible that Mayer encountered them in some as yet unidentified and more 'scientific' work published closer in time to the years of his education and creative intellectual life. I personally regard this as the likeliest possibility. The only other explanation would seem to be either an admittedly possible coincidence of convergent considerations or an exceedingly problematic active *Zeitgeist* à la Rupert Sheldrake's morphic fields.

Even granting the first alternative, however, one must be careful to specify just what is being explained, and just what is being claimed. What is being explained are certain aspects of Mayer's early confusion surrounding central-force motions and the idea that force might be created in "organisms" like the solar system. Even at that, *Naturphilosophie* does not seem to provide sources for Mayer's all-important misunderstanding of the parallelogram of forces, derived from the standard scientific works of the day, the application of which to central-force motion apparently first led him to entertain the possibility that force may be created. *Naturphilosophisch* might then have been (merely?) the exposition and terminological packaging of those ideas. Even the ostensibly telltale concepts of indifference and the neutralization of differences had close analogs in contemporary physics and chemistry. What a Mayer-*Naturphilosophie* connection does *not* explain is Mayer's conception of force—including its uncreatability and indestructibility—let alone his notion of the quantitative equivalence between heat and motion—that is, the central features of his theory. Nor does it help in understanding the physiological ideas that provided the initial context for the development of his theory. For the most part Mayer's work can be situated quite adequately within the context provided by mainstream work in physiology, physics, and chemistry; such metaphysical issues as may have been important did not in any event belong to *Naturphilosophie*. Nor, in the main, does Mayer's writing *sound* at all *naturphilosophisch*. I would even venture to say that, insofar as *Naturphilosophie* exerted any influence on him, it was an influence that he had to get past on his way to clarity, not an influence that helped guide him to energy conservation. Remove all the suspected traces of *Naturphilosophie*, and one can still (so to speak) get Mayer to his mature conclusions.

Quite aside from that judgment, there remains the more general issue of what one is in fact claiming by (for example) identifying as *naturphilosophisch* certain aspects of Mayer's work. Even if that identification is valid, the fact remains that the relevant topics—indifference, the generation of light somehow in connection with gravitating bodies, etc.—make up a very small part of that larger exuberant philosophical whole known as *Naturphilosophie*, almost all of which is entirely irrelevant to an understanding of Mayer's ideas. Again, almost all of the most characteristic features of *Naturphilosophie*—*Steigerung*, antitheses, tripartite hierarchies, and the rest—are quite without traces in Mayer's work. It would thus seem to me problematic in any account to speak simply of the influence of *Naturphilosophie* on Mayer. The vexing issue of identity broached here will be addressed in chapter 8. For the present, I would only note that the traditional way of asking and answering such questions of influence operates at too great a level of abstraction. As this entire book has sought to illustrate, it is only by avoiding general identity labels and paying attention to specific problems within specific and historically reasonable contexts that the historian can adequately render the past.

On the other side of the claims equation—that is, with respect to what one claims to have shown about *Mayer*—even if the *naturphilosophisch* pedigree of the aspects of his work under consideration were solid and unobjectionable, it would still not be accurate to state simply that Mayer was influenced by *Naturphilosophie* without specifying *with regard to what* and *to what extent*. After all, the only reason anyone pays any attention to Mayer is because of his role in the Europe-wide formulation of the principle of the conservation of energy, and, if my judgment is correct, the role *Naturphilosophie* played in the development of his ideas was not only minor at best, but also largely negative. In an important regard Mayer's work was not a whole, but a mosaic of pieces drawn from a variety of sources, pieces whose shape and meaning changed as the more-or-less final picture emerged. Pieces labeled *Naturphilosophie* did not appear in that final picture, nor did their earlier presence, such as it may have been, facilitate its completion.

• CHAPTER EIGHT •

# Assessment and Conclusions

MY GOAL in this study has been to make Mayer historically intelligible, that is, to reconstruct his probable *context* and the likely *specific issues* around which his reflections crystallized in such a way as to illuminate both the what and the why of his (conventionally so-termed) discovery of the conservation of energy. It is in the location of Mayer's thinking within a rich context of contemporary problems and themes that, in my opinion, the book's strength lies. As it turns out, Mayer's work fell during a crucial period of accelerated change in the assumptions and identity of the life sciences in Germany, and the particulars of Mayer's story highlight important aspects of the larger state of affairs.

For example, during the few years extending from the late 1830s to the mid-1840s, the vital force went from being generally granted as essential to the explanation of organic processes to being widely criticized as empty and unscientific. At the same time, Liebig defended a conception of the vital force that more forcefully than ever brought into focus the differences between its developmentally directive and its physiologically operative—that is, work-producing—functions. Its directive functions receded in importance as physiology proper came to exclude *Entwicklungsgeschichte*, just as the vital force lost the ontological validation it had enjoyed from analogies with mind and soul as physiology came more and more to exclude such entities from its purview. And its implicitly work-producing functions ran afoul of the increasingly taken-for-granted assumption that all vital processes are underlain by processes of material exchange. Issues having to do with the scope and definition of science and the legitimacy of scientific concepts were in turn situated against the rise of a self-assertive scientific materialism. In such a context Mayer sought to occupy a precariously narrow position: to accept a physiology based on *Stoffwechsel* and the laws of physics and chemistry while rejecting a despiritualizing scientific materialism and (later) an antiteleological and antitheological Darwinism.

The course and meaning of Mayer's theory of force can be understood only against the backdrop of these issues. At first the vital force was not a critical problem for him; indeed, his acceptance of the creation of force in divinely ordered "organisms" like the solar system, his early restriction of his ideas to inanimate nature, and his peculiar reluctance (until around 1844/45) to criticize Liebig's conception of the vital force all strongly suggest that he at least implicitly accepted the legitimacy of a vital force for several years after 1840. After all, belief in a conventional vital force lent analogical legitimation to his interpretation of cosmic systems, giving him a vested interest in its acceptability. And his conviction that his theory of force refuted materialism—in vindicating the autonomous existence of a nonmaterial substance—and that it solved important problems of metaphysics suggests that he was responsive to the widespread discussion during the 1830s of the relationship between the vital force, the soul, and the imponderables. What caused him to change his mind about the creatability of force is not clear from the surviving evidence, though part of that change must have had to do with the resolution of his confusions concerning central-force motions. At the same time, increasing criticism of the vital force as unscientific just as Liebig was making unprecedentedly strong (if not always self-consistent) claims for its energetic efficacy may also have prompted Mayer to rethink the matter. As I have also argued, problems relating to the creation and destruction of the vital force, the soul, and the imponderables would have provided Mayer with a rich specific context within which to reflect on those crucial issues of conservation, issues not commonly broached in the physical and chemical literature of the day.

The importance of relating context to specific problems is highlighted by comparing Mayer to Liebig and Lotze. Mayer and Liebig shared much of the same scientific context, and Liebig even delivered himself of many an utterance that, taken by itself, might seem to make him a contender for the codiscoverer title; yet Liebig never made a sustained attempt to formulate a new conception of force—his toying with the vague notions of *Kraftmoment* and *Bewegungsmoment* was a feeble flash in the pan—and he was bafflingly slow to recognize it even after several others had. (Recall that both Mohr and Mayer published their relevant work in Liebig's *Annalen*.) In part that was (I suspect) because of his denial that forces can exist independently of matter, in part because Liebig was not consistently a deeply consequential thinker. Similarly, his appreciation of issues of causality as they related to his (sometime) conviction that all organic functions must be underlain by processes of material exchange did not translate into a clear general appreciation of energetic considerations. For the rest, Liebig's 'failure' underscores the crucial historical importance of the *particular* issues that occupied Mayer's thinking—including, too, his childhood failure to construct a perpetuum mobile. In somewhat the same way, it underscores the importance of historical contingency over what might appear to be internal logic to note that Lotze, a self-professed mechanist and antivitalist, could still accept the effective creation and destruction of force, if to be sure for his own contingent philosophical reasons.

If at times this study of Mayer has seemed like a study of Liebig, too, then that is due to the further fact that Mayer's life and work intersected repeatedly and intimately with Liebig's. Mayer reported that he was prompted to compose his "Remarks on the Forces of Inanimate Nature"—and to tender it to Liebig's *Annalen*—as a result of reading Liebig's essay on "The Vital Process in the Animal, and the Atmosphere"; and his book of 1845 is unintelligible except as a response, in part, to the questionable legitimacy of Liebig's handling of the vital force in his *Animal Chemistry*.

It has been my goal to reconstruct, as far as the evidence and reasonable speculation will allow, the path of Mayer's thinking from his initial surprise at the color of the blood he let on arriving in the Dutch East Indies, to the mature statement of his ideas in 1845, down to his late conception of *Auslösung*, the last representing the conclusion of a long struggle to come to terms with the proper scope and meaning of concepts like force and cause. Mayer's 'discovery' had a complex internal structure extending over many years, during which the scope and meaning of his theory underwent profound change. Notable is the extreme contraction of his effective sphere of interest, at least as it pertained to the ideas he was willing to make public: from a grand view of physics and metaphysics—recall his early conversations with Rümelin—to the mechanical equivalent of heat.[1] One of the reasons Mayer has been so hard to understand is that the context of his earliest reflections left few traces in his published writings. Just as Mayer's use of mathematical symbols was frequently symbolic and not literal—recall his use of zero as a pseudocoefficient in expressions like 0 2A*c*—so, too, was force for him more than a physical concept susceptible to quantitative determination: at another level, it symbolized an entire antimaterialistic worldview, and was a point of crystallization about which a new physics and a new physiology could organize themselves. If my reading of Mayer is correct, in an attempt to accommodate himself to contemporary norms in the natural sciences he kept the metaphysical meanings of force out of his more self-consciously scientific writings. It is ironic, I think, that by playing the game of hypothesis-free objective science, by refusing (for the most part) publicly to engage ontological questions, and by emphasizing logic and methodological considerations, Mayer essentially assured that an appropriately recast concept of force would be easily assimilable by a minimally readjusted scientific materialism.

In some ways Mayer's theory was composed of two strangely separate parts: the equivalence of heat and motion—his earliest and always cardinal insight—and the subsequent elaboration of an overarching theory of force. Such a distinction may explain an apparent inconsistency with respect to Mayer's conception of the ontology of force. On the one hand, I have argued the centrality of his reflections on the nature of the imponderables and the soul, and on the similarities and differences, ontological and otherwise, between force and matter. Yet he elsewhere denied attachment to any particular ontological conception of heat, and he otherwise grounded his theory of force on abstract general principles (*causa aequat effectum* and the like). Perhaps his abiding nonontological understanding of heat reflected the circumstances of his original deduction, soon after his bloodletting observations, of the necessity of a constant numerical relationship between heat and motion, a deduction wholly independent of any representation of the nature of heat. And perhaps it was precisely in his subsequent creation of a general theory of *force* that ontological considerations came—for the first time, but crucially—to guide his thinking. His appeal to general philosophical principles would appear to reflect both his process of discovery—What are causes? Can force ever be created?—and his later attempt to justify his theory by grounding it on ostensibly universally valid principles. The equivalence of heat and motion was thus both logically and autobiographically independent of any ontological conception of force or its universal principled uncreatability or indestructibility.

In his more mature writings Mayer never insisted on the importance of his having identified a new conceptual entity, force, even as physics *became* in a sense the science of force—that is, energy. He never adopted a phrase equivalent to the "conservation of energy"—in part to avoid suggesting that his theory was merely an extension of the principle of the conservation of vis viva—and he called his collected works *The Mechanics of Heat*. In some ways Mayer's status as codiscoverer of the conservation of energy represents a further crystallization of meaning with respect to the content of his work when assessed from a perspective broader than that of his work alone—that is, from the perspective of a new field of meaning.

Mayer's conception of physics as the science of force and chemistry as the science of matter captured more than one important insight. He forged his concept of force and the principle of its uncreatability and indestructibility in close conjunction with his contemplation of the nature of the imponderables and of the essential properties of matter—his guiding analogy par excellence. As I argued in chapter 4, there was no explicitly formulated and generally invoked principle of the conservation of matter in the physics and chemistry literature of Mayer's day: what he did was in essence to create, for himself, *both* principles simultaneously. (How the conservation of matter became the conservation of mass is another story.) Similarly, Mayer's understanding of force represented both a reconceptualization of the erstwhile imponderables—recall his terminological wavering in 1842—and a struggle to specify the proper meaning of cause in physics. To put it figuratively, on the one side he had to take the material aspect away from the imponderables, on the other side he had to make force more than a mere property of matter. Although Mayer personally was not an important enough figure in the world of science to have contributed much to its transformation, his insights nevertheless foreshadowed the growing importance of self-consciously enunciated fundamental principles as the conceptual foundations of scientific knowledge à la Rudolph Clausius. As Ernst Cassirer put it, "In its general structure nineteenth-century physics might be characterized as a physics not of images and models but of principles."[2] Johannes Müller's attachment to empiricism and reluctance to take a more principled (and energetic) approach to the question of animal heat symbolizes, in its own small way, an earlier conception of science.

The reader may have noticed a disharmony betweeen the historiographical style of the chapter on Mayer and *Naturphilosophie* and that of the rest of the book. For the most part I have striven to write an historical account in terms of particular people, particular works, and particular ideas without feeling any need to identify them as belonging to this or that tradition.[3] Indeed, the desire to assign Mayer to one or another named philosophical tradition, without due consideration of the historical particulars on both sides, has been the bane of much of the scholarship on Mayer. Thus few have noticed that the alleged "unity of forces" was not a pronounced motif of either Mayer's or of much *Naturphilosophie*. However, in devoting a chapter to *Naturphilosophie* I could not wholly escape the implicit dangers of 'identificationist' historiography short of writing a second and much longer book. The interlocking questions of who the representatives of *Naturphilosophie* were and what its principal characteristics are cannot be adequately answered short of a broad and deep analysis of a large mass of roughly like-spirited sources. In so doing, one would need to distinguish a Kantian strain of natural philosophy that shared certain conceptions with *Naturphilosophie* (such as dynamism) as it differed from it profoundly in others (such as with respect to efficient versus formal causality). One would also need to distinguish an architectonically luxuriant *Naturphilosophie* from antisystematic Romantics like Goethe. Nor should one expect clean divisions, since many individuals—such as Eschenmayer, Burdach, Oersted, and mineralogist Christian Samuel Weiss—drew inspiration from Kant, Schelling, and a wealth of other sometimes overlapping sources.[4] In any event, ideas can be classified into abstract conceptual groupings more easily than historically multifaceted people. The ever-present danger is that generalized descriptive categories, the use of which is all but indispensable in practice, very quickly become hypostasized into oversimplified *explanatory* classifications.

Even after that kind of classificatory analysis might have been accomplished, it would remain problematic to identify an individual *as* a representative of a particular tradition, as if that identity somehow covered and explained all aspects of that person's work, and as if any influence he or she might have on someone else could be tagged as due to that tradition. Sometimes one will in fact wish to make such claims, but other times the casual use of insufficiently considered identity labels results in a style of historiography dangerously abstracted from the complexities of historical reality and liable to impose ill-fitting categories on unruly data.[5]

Just as Mayer must be explained in terms of his context without being reduced to that context, so too must his work be understood at least in part in terms of his biography and his personality. Of crucial importance were, for example, his failed childhood attempt to construct a perpetuum mobile, his medical training, and the accident of fate that confronted him with unexpectedly red venous blood in a hot climate. Likewise his religiosity, in particular his belief in the immortality of the soul and his oppostion to materialism, contributed in important ways to the shaping of his theory of force as an autonomous entity on an ontological par with matter. His character, too, was not without influence on the kind of science he did. His strong will; his passionate attachment to slogans and quotations; his prodigious memory; the dogged consequentiality of his mind, which delighted in both the logical and the paradoxical; his preoccupation with principles, laws, and rules; his interest in causality; his quirkiness and originality—such traits all left their traces on his scientific work. Insofar as this study can be taken as typical, it illustrates not only the inseparability of the individual scientist from his or her social, intellectual, and scientific communities, but also the inappropriateness of trying to understand the dynamics of scientific change without due attention to biographical particulars.

Should historians venture to judge the quality of a scientist's work, or mind? Such a question is not usually relevant to the kinds of things we wish to say, though in fact Mayer has enjoyed a rather poor reputation as a thinker ever since anyone began to take notice of him. And not surprisingly, since his earliest literary endeavor was confused and obscure, since his forays into public speaking late in life were easy to put down as the embarrassingly Christian effusions of an intellectual lightweight, since he never did anything of substance other than his work on force, and since he did in fact suffer from fits of insanity. Jolly's oft-retold anecdote alone painfully reveals the low esteem in which he was held by many scientists during the second half of the century, despite his growing stature among a broader public as a hero of German science and the pride of his birthplace. Nor did a later historiographical association with *Naturphilosophie* lend much polish to Mayer's scientific reputation.

My own estimation of the quality of Mayer's mind has fluctuated over the years, only to gel into an abiding ambivalence. On the one hand his dogged lack of appreciation for some of the simplest facts of mathematical physics and the amazing consideration of various systems of coinage in the context of an address in 1870 on "the meaning of invariable magnitudes"—among other possible examples—do not bespeak much scientific sophistication.[6] On the other hand, many of his early confusions with respect to central-force motions and the measure of the quantity of motion (as $mv$) reflected the consensus of the best contemporary physics texts, and his calculation of the mechanical equivalent of heat in the only way then possible with existing data was in fact an impressive accomplishment, all the more considering his deficiencies in physics and mathematics. Then, too, it was Mayer and not, for example, Liebig who appreciated the need for a new concept of force, and who worked his way through the considerable terminological and conceptual confusion of his sources to forge just such a concept. Judged against contemporary norms, his achievements were brilliant. Insofar as Mayer refused to go the materialistic route of interpreting forces as properties of matter, but instead insisted on the ontological independence of force (read here energy), he was on surer ground than most of his preenergeticist contemporaries—allowing for the fact that neither he nor they anticipated mass-energy equivalence. The possible validity of Mayer's analogical linkage between force and soul—as I have reconstructed it, following Autenrieth's lead and Hirn's example—is precisely the kind of question that no longer had much respectable scientific currency in Germany after the 1840s. Ironically, the triumph of Helmholtzian conservation of energy contributed mightily to the securing of the new scientific dispensation.

# • APPENDIXES •

• APPENDIX ONE •

# Timeline of Robert Mayer's Life and Work[1]

1804 November 5: marriage of Robert Mayer's parents
1805 December 18: birth of his brother, Friedrich Ferdinand ("Fritz"), who took over their father's apothecary ca. 1832/33
1810 May 3: his father (since 1803 apothecary in Lorch) purchased the apothecary "Zur Rose" in Heilbronn
1810 August 22: birth of his brother, Carl Gustav, who became an apothecary in Messkirch and Sinsheim
1814 November 25: birth of Julius Robert Mayer in Heilbronn
1829 Spring: left home to attend the *evangelisch-theologisches Vorbereitungsseminar* in Schöntal, following his friend Gustav Rümelin, who had gone there the previous fall
1831 September: returned to Heilbronn; assisted in his father's apothecary till May 1832 and audited the highest class of the *Gymnasium* there
1831 September 12–13: took the *Studienkandidatenprüfung* in Stuttgart, the equivalent of the later *Abiturientenexamen*[2]
1832 May 17: entered the Eberhard-Karls-Universität in Tübingen to study medicine
1836 August 12: Mayer and Wilhelm Griesinger among the cofounders of the student corps "Guestphalia"
1837 March 13: expelled from the university for one year due to membership in a forbidden organization
1837 Summer–Fall: traveled to Switzerland; visited clinics in Munich and Vienna
1837 October 19–23: in a letter to Paul Friedrich Lang, a friend from Schöntal, Mayer discussed his planned trip to Java
1838 January 24: pardoned by King Wilhelm I of Württemberg to be able to take his medical exams
1838 March 2–3: took the first state medical exam (*erste medizinische Staatsprüfung*) in Tübingen; the oral portion of the exam followed on March 5
1838 March 6: took (oral) doctoral exam
1838 July 25: promotion to doctor of medicine and surgery in Tübingen
1838 August 15–20: took the second state medical exam in Stuttgart (medicine on August 15–16, surgery on August 17–18, and obstetrics on August 20)
1838 August 23: licensed to practice medicine
1838 November–December: attended medical lectures and demonstrations in Munich
1839 March 4: began practicing medicine in Heilbronn
1839 June 15: took exam in Amsterdam to become a ship's doctor in Dutch service
1839 September 25: admitted to Dutch service as ship's doctor
1839 Fall–1840 February 14: in Paris to study medicine; attended primarily clinical and surgical demonstrations; became friends with Carl Baur
1840 February 23: left Rotterdam aboard the ship *Java* to sail to the Dutch East Indies
1840 March 9: reunited with his chest of books
1840 June 8: ship reached the Sunda straits
1840 June 11–23: at the roads off Batavia

1840 June 20–24: let blood seven times (twice on the 20th, once each on the 21st and 22nd, and three times on the 24th)[3]
1840 July 4–August 30: ship anchored off Surabaja
1840 ca. December 25: *Java* arrived back in Rotterdam
1841 February: returned to Heilbronn; resumed medical practice
1841 May 28: named *Oberamtswundarzt* in Heilbronn (with primarily forensic medical responsibilities; resigned 1845)[4]
1841 June 16: sent paper "Über die quantitative und qualitative Bestimmung der Kräfte" to Poggendorff for publication in the *Annalen der Physik*; got no reply and the manuscript was not returned; sent follow-up letters on July 3 and 31
1841 July 24: began intensive correspondence with Baur
1841 between September 6 and 12: meeting with Johann Gottlieb Nörrenberg, professor of physics at Tübingen[5]
1841 September: composed an untitled manuscript headed "sapere aude"
1841 September 16 and 30: publication (anonymously) of the second and fourth of Liebig's "Chemische Briefe" in the Augsburger *Allgemeine Zeitung*
1841 ca. November: meeting with Philipp Gustav Jolly, professor of physics at Heidelberg
1841 December 15: publication of Liebig's "Der Lebensproceß im Thiere, und die Atmosphäre"
1842 March 31: sent the paper "Bemerkungen über die Kräfte der unbelebten Natur" to Liebig for publication in the *Annalen der Chemie und Pharmacie*
1842 April: date of the preface of Liebig's *Die organische Chemie in ihrer Anwendung auf Physiologie und Pathologie*, which appeared by July 16
1842 May 31: married by Rümelin to Wilhelmine Regine Caroline Closs (born in Winnenden on July 31, 1816) in a joint ceremony with Closs's sister and Rümelin's brother; on the same date was published the May *Heft* of Liebig's *Annalen* containing his first published paper
1842 October–1843 Spring: Baur was in Heilbronn to teach mathematics and physics at the *Gymnasium*, and he tutored Mayer in mathematics and mechanics
1842 November 30: began intensive correspondence with Griesinger
1843 May 18: birth of daughter Wilhelmine Elise (died 1919)
1844 January 6: death of his mother
1844 June 11: sent Griesinger an early version of what eventually became *Die organische Bewegung* (1845); requested its return on July 16
1844 July 20: sent Baur a new draft of this work
1844 July (?): composed an essay "Ueber medicinische Physiologie," possibly for submission to Wunderlich and Roser's *Archiv für physiologische Heilkunde*
1844 August 14: birth of second child, Christian Robert
1844 August 21 and 30: sent Baur updated drafts and revisions
late 1844/early 1845: completed a manuscript entitled "Theorie der physikalischen Principien als Grundlage physiologischer Lehrsätze"
1845 January 3: sent Liebig the manuscript of what eventually became *Die organische Bewegung*; returned by August Wilhelm Hofmann, Liebig's assistant, on January 6
1845 May 13: death of second child, Christian Robert; around this time he was revising the manuscript of *Die organische Bewegung*
1845 August: stay at spa in Wildbad
1846 April 10: Johannes Müller returned to him a manuscript he had submitted for publication in the *Archiv für Anatomie, Physiologie und wissenschaftliche Medicin* on the connection between oxidation and mechanical effect in the organism

1846 June 12: birth of third child, Anna
1846 July 27: the Paris Academy received his paper "Sur la production de la lumière et de la chaleur du soleil"
1847 June 25: birth of fourth child, Julie Wilhelmine
1847 July 22: named *Stadtarzt* in Heilbronn (charged with the care of the city's poor and the lower civil servants); resigned September 1873
1847 August 23: Joule's communication on the mechanical equivalent of heat read at the Paris Academy
1848 August 19: death of third child, Anna, from whooping cough
1848 August 25: death of fourth child, Julie Wilhelmine
1848 October 16: Mayer's priority claim vis-à-vis Joule read at the Paris Academy
1849 January 22: Joule's reply to Mayer read at the Paris Academy
1849 September 19: birth of fifth child, Paul Theodor (died 1909)
1849 November 12: Mayer's reply to Joule read at the Paris Academy
1850 May 28: severly injured himself after jumping out of a third-floor window and falling nine meters to the ground
1850 September 8: death of his father
1850 October 13: birth of sixth child, Emma Johanna (died 1894)
1851 Summer: stay at spa in Wildbad
1852 April: short stay at sanatorium in Kennenburg (near Esslingen)
1852 May–July 31: stay at Christophsbad asylum (near Göppingen)
1852 August 1–1853 September 1: forced stay at Winnenthal asylum (near Winnenden)
1854 January 12: birth of seventh child, Richard Friedrich
1856 April 27–May 30: stay at sanatorium in Kennenburg
1858 September: attended the meeting of Gesellschaft deutscher Naturforscher und Ärzte in Karlsruhe
1858 November 10: named corresponding member of the Naturforschende Gesellschaft in Basel (at the instigation of Christian Friedrich Schönbein)
1859 July 4: awarded an honorary doctorate by the philosophical faculty of the University of Tübingen
1859 November 28: named corresponding member of the Bavarian Academy of Sciences in Munich (subsequent honorary memberships and awards not listed here)
1860 January 1: death of seventh child, Richard Friedrich
1862 June 6: John Tyndall recognized Mayer's priority in a lecture at the Royal Institution in London
1863 September: attended the meeting of the Gesellschaft deutscher Naturforscher und Ärzte in Stettin
1864 August 22–24: attended the meeting of the Schweizerische Naturforschergesellschaft in Zürich, where he met Tyndall
1865 October 3–November 1: second stay at sanatorium in Kennenburg
1867 July: publication of *Die Mechanik der Wärme*
1867 November 5: awarded personal nobility with bestowal of the Württemberg *Ritterkreuz*
1869 September 18: delivered a talk "Über nothwendige Consequenzen und Inconsequenzen der Wärmemechanik" at the meeting of the Gesellschaft deutscher Naturforscher und Ärzte in Innsbruck
1871 August 8–November 15: last stay at sanatorium in Kennenburg
1878 March 20: death at age 63 from tuberculosis
1899 June 7: death of Mayer's wife, Wilhelmine Closs
1944 December 4: Heilbronn destroyed by Allied air attack

• APPENDIX TWO •

# Courses Mayer Took at the University of Tübingen, 1832–37

THE FOLLOWING information is based in the first instance on the *Belegbogen* (a sort of preregistration form) prepared by Mayer listing the courses he anticipated taking during the ensuing semester. However, not all the courses so listed (*belegt*) were actually offered or taken, and the entries below have been edited in accordance with the *Hörerlisten* (class rolls) of students attending the courses given and with the printed *Vorlesungsverzeichnis* (course list) for that semester.[1] The letters H and V in brackets after a course entry indicate that Mayer's name appeared on the appropriate *Hörerliste* and that the course was listed in the *Vorlesungsverzeichnis* for that semester. Sometimes there are variations in the name of the course as given in these three sources; that of the *Hörerliste* is given priority as representing most accurately the focus of the course as actually given. Variant course names from the *Belegbogen* are identified by the letter B.[2] Mayer's *Belegbogen* are in the Universitätsarchiv Tübingen ("UAT"), Personalakten, 40/141, Nr. 88. The *Hörerlisten* are UAT 51/n, where n is the particular number for each individual as given in the following list of the instructors with whom Mayer listed courses for the designated semesters:

Autenrieth, Hermann Friedrich (n=12): WS 1835/36, SS 1836
Autenrieth, Johann Heinrich Ferdinand ("Kanzler v. Autenrieth"; n=11): SS 1834, WS 1834/35, SS 1835
Baur, Christian Jakob (n=20): SS 1832, WS 1832/33, SS 1836
Elsässer, Carl Ludwig (n=121): WS 1832/33, SS 1833
Frank, Christian Friedrich (n=155): WS 1836/37
Gmelin, Christian Gottlob (n=187): SS 1832, WS 1832/33
Gmelin, Ferdinand Gottlob (n=186): WS 1833/34, SS 1834, WS 1834/35, SS 1835
Märklin, Gustav Friedrich (n=404): SS 1832
Ofterdinger, Ludwig Felix (n=464): SS 1832
Rapp, Wilhelm Ludwig (n=518): WS 1832/33, SS 1833, WS 1833/34, WS 1834/35
Riecke, Leopold Sokrates von (n=530): SS 1834, WS 1834/35, SS 1835, WS 1835/36, SS 1836, WS 1836/37
Schill, Albert Friedrich (n=571): WS 1835/36, SS 1836, WS 1836/37
Schübler, Gustav (n=606): SS 1834
(Note that Mayer seems not to have had courses with Frank, Schill, or Schübler.)

### SUMMER SEMESTER 1832

"Anatomie" with Baur [H, V]
["Osteologie" with Baur; no H found; not listed in V for that semester]
"Pharmaceutische Chemie" with C. G. Gmelin [H, V]

"Specielle Botanik" with Märklin [H, V]
"Physik (Theoretische und Experimental Physik)" with Ofterdinger [H; on B as "Physik"; V lists two courses, "Theorie der Kegelschnitte" and "Theorie der Gränzen," for neither of which an H is extant]

### Winter Semester 1832–33

"Allgemeine Chemie" with C. G. Gmelin [H, V]
"Vergleichende Anatomie" with Rapp [H, V]
"Das Präpariren am Leichnam" with Baur [H, V; on B as "Repetitionen der Anatomie"]
"Menschliche Physiologie" with Elsässer [H, V]

### Summer Semester 1833

"Allgemeine Pathologie" with Elsässer [H, V]
["Zoologie" with Rapp; Mayer's name absent from H; in V as "Naturgeschichte der Thiere"]

### Winter Semester 1833–34

"Der zweite Theil der speciellen Krankheitslehre" with F. G. Gmelin [H, V; on B and in V as "Specielle Therapie"]
"Pathologische Anatomie" with Rapp [H, V]
"Demonstrationen der Anatomie des Menschen" with Rapp [H, V]

### Summer Semester 1834

"Materia medica" with F. G. Gmelin [H, V]
"Allgemeine Chirurgie" with Riecke [H, V]
"Der erste Theil der Nosologie" with J.H.F. Autenrieth [H, V]
["Botanik" with Schübler; no H found—Schübler died on 8 September 1834]

### Winter Semester 1834–35

"Formulare" with F. G. Gmelin [H, V]
"Der zweite Theil der Nosologie" with J.H.F. Autenrieth [H, V]
"Geburtshülfliche Diagnostik," Erste Abtheilung, with Riecke [H; not in V]
"Chirurgisch-geburtshülfliche Clinik" with Riecke [H, V]
["Clinik der innerlichen Krankheiten" with F. G. Gmelin; Mayer's name absent from H; same course name in V; on B as "Medicinische Clinik"]
"Cursus operationum chirurgicarum" with Riecke [H, V]
"Demonstrationen der Anatomie des Menschen" with Rapp [H, V]
"Specielle Chirurgie, I. Theil" with Riecke [H, V]

### Summer Semester 1835

"Specielle Krankheitslehre" with F. G. Gmelin [H; in V as "Der erste Theil der speciellen Nosologie und Therapie"; on B as "Specielle Therapie"]

"Specielle Chirurgie, II. Theil" with Riecke [H; in V as "Der zweite Theil der speciellen Chirurgie sammt der Augenheilkunde"; on B as "Specielle Chirurgie" without an instructor's name]
"Das geburtshülflich-chirurgische Clinicum" with Riecke [H, V]
"Augenoperationen. Collegium Privatissimum" with Riecke [H, V]
["Medicina forensis" with "Prof. Autenrieth"—i.e., *not* "Prof. Kanzler v. Autenrieth; V lists "Gerichtliche Medicin" with Kanzler v. Autenrieth, for whom no H found—J.H.F. Autenrieth died on 3 May 1835; H. F. Autenrieth, his son, took over the course; Mayer's name absent from latter's H]

WINTER SEMESTER 1835–36

"Formulare" with H. F. Autenrieth [H, V]
["Semiotik und Diagnostik der Brustkrankheiten" with Schill; Mayer's name absent from H; on B as "Semiotik"; not in V]
"Geburtshülfe" with Riecke [H, V]
["Krankheiten der weiblichen Genitalien" with Riecke; no H found; not in V]
"Chirurgisch-geburtshülfliche Clinik" with Riecke [H; not in V; on B as "Gesammte Clinic" without an instructor's name]

SUMMER SEMESTER 1836

"Anatomie" with Baur [H, V]
"Gerichtliche Medicin" with H. F. Autenrieth [H, V; on B as "Medicina forensis"]
["Formulare" with Schill; no H found; not in V; Mayer's name is absent for the H of the course that Schill did teach that semester, "Pathologische Anatomie"]
"Chirurgisch-geburtshülfliches Clinicum" with Riecke [H, V; on B as "Gesammte Clinik" without an instructor's name]

WINTER SEMESTER 1836–37

"Cursus operationum chirurgicarum" with Riecke [H, V]
"Chirurgisch-geburtshülfliche Clinik" with Riecke [H, V; on B listed as "Gesammte Clinik"]
["Geburtshülfe" with Frank; Mayer's name absent from H; on B as "Phantom Uebungen"; in V as "Geburtshülfe . . . und Uebungen am geburtshülflichen Fantom"]
["Examinatorium" with Schill; no H found; not in V; Mayer's name absent from the H of the course that Schill did teach that semester, "Allgemeine Pathologie"]

• A P P E N D I X   T H R E E •

# The German Text of the Longer Passages Quoted from Manuscript

*Baur 1846: Letter to Mayer of 30 November [1846]*

Daß ich nun einmal derjenige nicht bin, der dich von deinen Paradoxen, denn daß du damit behaftet bist, war von jeher meine Ueberzeugung—Parallelogramm der Kräfte, Proportionalität der Kräfte und Geschwindigkeiten—und werde es stets bleiben, daß ich, sage ich nicht der bin, der dich von deinen Paradoxen kurirt, und auch nicht der bin, der dir sonderlich behilflich seyn kann, um deine neue Wahrheiten ans Licht der Welt zu fördern, das hättest du seit lang, anfänglich an der Art meiner Theilnahme an deinen Bestrebungen und hienach an meiner Nichttheilnahme bemerken können. ([1])

*Mayer 1838b: "Zweite medicinische Staatsprüfung"*

Die Wärme expandirt die Luft, sie wird dadurch dem volumen nach ärmer an Sauerstoff (so wie auch an Stickstoff); zugleich wird die Tention der Flüssigkeit vermehrt. Daher ist in der Wärme, durch dem [*sic*] verminderten Sauerstoffgehalt, der Athmungsprocess weniger energisch, und die Luft wird beim Athmen viel weniger zersetzt, das Blut wird kohlenstoffreicher, dunkler, der Unterschied zwischen rothem und schwarzem Blut ist geringer. Dafür muß die Leber in vicarirende Thätigkeit treten, sie sondert mehr Galle ab, welche ein sehr kohlenstoffreicher Stoff ist. Durch die mit der Wärme vermehrte Tention der Flüssigkeiten, wird die unmerkliche, gasförmige Haut- und Lungenausdünstung vermehrt, wodurch Wärme latent und der Körper also wieder abgekühlt sein. Wenn die Luft aber mit Wassergas nahezu gesättigt ist, so ist die gasförmige Ausdünstung sehr beschränkt, die Naturhilfe fällt weg und es entsteht das / drückende Gefühl, der schwülen Hitze; der Darmkanal tritt für die Haut- und Lungenausdünstung vicarirend auf. In der feuchten Wärme entsteht also ein Säftezufluß zur Leber und dem Darmkanal. Die Menschen werden bei der mangelhaften Blutoxydation träge, bleich; die Verdauung ist geschwächt, der Darmkanal zu Diarrhöe geneigt. Die Leber wird sehr zu Entzündungen geneigt, die bald acut, bald äußerst chronisch sind; erstere befällt häufig den linken Lappen. Strukturveränderungen der Leber sind die häufige Folge, nametlich Suppuration. Der Darmkanal ist zu Ruhren geneigt, die oft sehr bösartig und allgemein epidemisch werden, und bei der durch die Wärme entstandenen Sensibilität der Haut, oft bei der leisesten Erkältung entstehen. Bei der verminderten Oxydation des Blutes ist der Lebensprocess viel weniger Energisch, die Säfte Masse zur Sepsis geneigt, alle Krankheiten nehmen leicht diesen Character an. ($28^r$–$28^v$) [Instead of "durch dem [*sic*]" Mayer had originally written "bei dem."]

*Mayer 1843a: Undated and unsigned draft of a letter to a "Lieber Freund"—i.e., Carl Wilhelm Baur*

Bei gleichförm:[ig] beschl:[eunigten] Geschw:[indigkeiten] bilden die die Zeiten vorstellenden Ordinaten eine Parabel, deren parameter mit der Größe der sog:[enannten] beschleunigenden Kraft, mit der Größe von g umgekehrt wächst und fällt./ [Cf. figure 6.3] Unter andern Umständen bi[l]den die Ordinaten der Zeit natürlich andere Curven während die Axe der x unter allen Umständen fest bleibt. Bei Pendelschwingungen hätte die Curve etwa folgende Form [figure 6.4] was vielleicht eine Cycloide ist, u.s.f.—Was diese Darstellung die ich nicht weiter fortführen will, für einen Werth hat? den großen, daß der Begriff "Kraft" mit Consequenz durchgeführt ist; daß dann in der Mechanik nichts anderes unter Kraft verstanden wird, als in der übrigen Naturlehre, "Ursache der Bewegung"; daß man überhaupt weiß, was Kraft ist[;] das Product der Masse in ihre Höhe, das Product der Masse in das Quadr:[at] ihrer Geschw:[indigkeit,] die freie Wärme, das sind Objecte, keine Abstractionen. (3–4)

*Mayer 1860: Untitled, undated manuscript fragment*

Dem Chemiker ist es gestattet, seine Untersuchungsobjecte unmittelbar durch die Wage zu bestimmen. Die Chemie wurde aber erst vor hundert Jahren dadurch zu einer Wissenschaft, als es Lavoisier gelang, die zwischen den verschiedenen Stoffen bestehenden unveränderlichen Größenbestimmungen aufzufinden, welche man Mischungsgewichte, und deren Kenntniß man die Stöchiometrie nennt. Lavoisier gelangte hiezu bekanntlich dadurch, daß er die Zusammensetzung des Wassers entdeckte, welches aus einem Gewichtstheile Wasserstoff und 8 Gewichtstheilen Sauerstoff, oder 1 Volum Theil Wasserstoff und 16 Volumtheilen Sauerstoff besteht. Derartige unveränderliche Größenbestimmungen wie hier beim Wasser, also beispielsweise hier das Verhältniß 1:8, besteht aber bei allen übrigen, einfachen sowohl als zusammengesetzten, Stoffen, und erst die Kenntniß dieser unveränderlichen Verhältnißzahlen hat die Chemie zu dem Range einer Wissenschaft erhoben.

*Mayer 1865: Untitled autobiographical sketch sent to John Tyndall*

Es ist möglich, dass das Ausbleiben jedweder Anerkennung, auf die ich vorschnell gerechnet, das Seinige dazu beigetragen, meinen Eifer für die Wissenschaft zeitweise etwas abzukühlen;—gewiss ist, dass zu jener Zeit das Interesse für transcendentale, religiöse Wahrheiten mit unwiderstehlicher Gewalt bei mir hervorzutreten anfing . . . . Mit der leidenschaftlichen Hast, mit der Exclusivität, die ich als Temperaments-Fehler bei mir zu beklagen habe, warf ich mich sofort auch auf dieses Gebiet. Nun bin ich älter geworden und lasse mich gerne wieder zu den Jüngern der Wissenschaft zählen, aber der Eifer für die Wahrheiten der christlichen Religion ist dennoch bei mir keineswegs im Erkalten.

. . .

Was ich mir aber damals zu denken still verbot, will ich jetzt ohne Rückhalt bekennen. Es lebte in mir ein Verlangen nach Anerkennung, und so sehr ich auch ein solches Gefühl als sündigen Hochmuth niederzukämpfen bemüht sein mochte, so gieng es doch über meine Kräfte, mein wissenschaftliches Bewusstseyn in mir zu unterdrücken, . . . ([3], [4])

*Mayer ca. 1866(?): Untitled and undated manuscript fragment*

. . . Licht über die Vorgänge im Mikrokosmus auszubreiten; denn wenn sich nachweisen ließe, daß irgendwo im Makrokosmus, durch das Zusammenwirken seiner, mit so großer Weisheit geordneten Theile, eine mechanische Arbeit vollbracht, Licht oder Wärme entwickelt würde, so wäre dadurch die Möglichkeit, daß ein Gleiches auch in den lebenden Geschöpfen der Erde vor sich gehen könnte, unzweifelhaft dargethan; im andern Falle aber, wenn sich das für die anorganische Welt giltige Gesetz: keine Leistung ohne Verbrauch, auch im Planetensysteme selbst überall bewährt findet, so werden wir durch die bündigste Induction [berechtigt,] das nemliche Gesetz auch für die Muskeln und den Thierorganismus überhaupt in Anspruch zu nehmen, ein Schluß der fortan nur durch gesicherte Thatsachen, nicht aber durch naturwissenschaftlich-vitalistische Meditationen aufzuheben wäre.

Solcher makrokosmischen Erscheinungen sind es aber im Ganzen dreierlei: 1) die Bewegungen der Himmelskörper in ihren Bahnen um die Sonne, 2) die Licht- und Wärmeentwicklung im Planetensysteme, und 3) die Ebbe und Fluth. (57) [I have transcribed Mayer's use of the standard astronomical symbols as "Erde" and "Sonne."]

Was nun 2) die Licht- und Wärmeentwicklung auf der Sonne betrifft, so wäre es ganz im Geschmack der Vitalisten, wenn man diese Action als die Wirkung einer, dem Planetensysteme inhärirenden Lebenskraft hervorgebracht durch Polarität u.dgl. sich denken würde; da aber die exacten Wissenschaften sich bekanntlich mit solchen Phrasen nicht begnügen wollen, so habe ich im Sinne der letzteren versucht, diese Naturerscheinung zu erklären, oder dieselbe als ein gesetzmäßiger Proceß darzustellen./

Die Ebbe und Fluth endlich, derer hier noch einmal Erwähnung gethan werden muß, ist für die Biologie deßhalb eine interessante und wichtige Erscheinung, weil sich mit mathematischer Schärfe bestimmen läßt, ob die bei diesem makroskopischen Processe zu Tage geförderte Arbeit nur als das Resultat der Combination gewisser Organ-Theile des Planetensystems zu betrachten sey, oder ob auch hier wieder Leistung und Verbrauch Hand in Hand gehen? Die Frage ist nemlich die: kann durch die Gravitation an und für sich allein eine Arbeit vollbracht werden? Fällt diese Frage in Beziehung auf die Ebbe und Fluth bejahend aus, so darf nach diesem Vorgange die Möglichkeit, daß auch durch die Muskelirritabilität, ohne allen Verbrauch, Bewegung erzeugt werden könne, nicht ferner geleugnet werden. Soferne ich aber im Vorhergehenden nachgewiesen habe, daß die Arbeit der Ebbe und Fluth nur auf Kosten der Umdrehungsbewegung der Erde vollbracht werden kann, so bleibt keine Berechtigung mehr übrig, der organischen Gliederung eines Systems, oder dem Lebensprocesse die Fähigkeit zu vindiciren, an und für sich allein Actionen vollbringen zu können.

Hievon die Anwendung auf die geistigen Actionen gemacht, so kann nach physikalischen Gesetzen nicht ferner die Rede davon seyn, diese Actionen als Resultat der Combination der Gehirnfaser, oder der Lebenskraft dieses Organs ersehen zu wollen, es würde aber noch zwei Annahmen übrig bleiben. Bildet man sich nemlich eine Analogie zwischen Gehirn und Muskel, so kann man zu der Vorstellung gelangen, daß wie der Muskel mit Hilfe des chemischen Prozesses, und unter dem Verbrauche der durch denselben gelieferten Kraft seine mechanische Leistung vollbringt, so auch das Gehirn mit Hilfe desselben Prozesses seine geistige Thätigkeit produciren könne.

Auf der ersten Anblick könnte eine solche Ansicht etwas bestechendes haben, bei näherer Prüfung stellt es sich dieselbe als völlig unhaltbar heraus. Die Actionen der Secretions- und Bewegungs-Organe beruhen nemlich wesentlich darauf, daß die Leistung die sie hervorbringen, der Sekretionsprodukt oder die mechanische Arbeit, äquivalent ist dem Verbrauche, der in diesen Organen vor sich geht, eine Aequivalenz die durch die Erfahrungsgesetze nachzuweisen ist, und nachgewiesen werden kann. Die Kohlensäure, welche von dem Thiere ausgehaucht wird, ist äquivalent dem Kohlenstoff und Sauerstoff, den das Thier consumirt, seine mechanische Leistung äquivalent dem chemischen Verbindungsprozesse der in seinen Bewegungsorganen vor sich geht; wie aber sollte die von dem Gehirn producirte geistige Arbeit ebenfalls als Aequivalent mit dem chemischen Proceß, der Wärme, Electrizität etc. gesetzt werden dürfen? und man wollte es unternehmen hievon einen Beweis beizubringen?

Da nun somit die geistigen Thätigkeiten weder als das Resultat der Lebenskraft des Thiers, noch als das Resultat eines chemischen Prozesses in demselben angesehen werden können und dürfen, so werden wir nothwendig zu dem Satze zurückgeführt, den die gesunde Vernunft der ganzen Menschheit zu allen Zeiten und allen Orten als axiomatische Wahrheit anerkannt hat und [wofür sich der Ausdruck in allen Sprachen befindet], und wir können nicht anstehen, den Ausspruch zu thun, daß die Annahme einer metaphysischen [?] Existenz, die Annahme eines fühlenden, wollenden und denkenden Wesens, einer Seele, für den empirischen Naturforscher sich als unabweisliches Postulat geltend macht. (57-58) [The phrase in brackets was a marginal interpolation, possibly in another hand.]

Da es nun keinen Theil des Gehirns, und noch weniger des übrigen Körpers gibt, der durch seine Structurveränderung, nothwendig eine gewisse Seelenstörung bedingt, so folgt daraus wiederum, daß alle körperliche Organe nur per consensum Störungen in den Seelenthätigkeiten zu erregen vermögen, oder mit andern Worten, daß das Organ der Seelenthätigkeit keine körperliche ist.

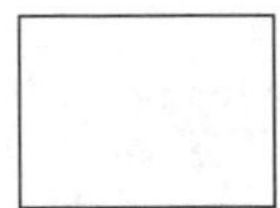

Soll ich nun noch über die Fortdauer der Seele nach dem Absterben ihrer materiellen Organe über die persönliche Unsterblichkeit des Menschen aussprechen, so kann ich das sehr oft gesagte, und in jeder Brust geschriebene wiederholen.

Der vollgültige Beweis für diese Fortdauer liegt in der göttlicher Vollkommenheit der ganzen Natureinrichtung. Jeder pflanzliche, und jeder Thier-Organismus ist auf solche Weise construirt, daß er den Zweck, zu dem er erschaffen, zu erfüllen, und die letzte Stufe der Vollkommenheit zu erreichen vermag, [und es wäre eine lächerliche Vermessenheit, an einem einzigen Organe [Organismus?] in der langen Reihe der Pflanzen- und Thier-Species eine Verbesserung auch nur für möglich zu halten.] Und doch klagt der Thor wie der Weise mit Recht über die Unvollkommenheit dieser Welt; das einzig unvollkommene auf derselben ist aber der menschliche Geist; ihm ist auf dieser Welt allein nicht gestattet, seinen Zweck, den der göttlichen Vollendung, zu erreichen, denn was ist die Weisheit des Menschen, und die Tugend des Tugendhaftesten dieser Erde[?] Der menschliche Geist ist der in den Eihüllen eingeschlossenen Luft zu vergleichen. Nehmen wir dem kindlichen Organismus mit der Trennung von der Nabelschnur, oder dem geistigen Organismus mit der Trennung vom Körper, seine Zukunft, so bleibt uns in beiden Fällen nichts als eine Fehlgeburt, unfähig die in sie gelegten Keime zur Reife zu bringen./

Es sind also nur zwei Annahmen möglich: entweder mußte die Natur es unterlassen haben in dieser sonst überall mit göttlicher Vollkommenheit construirten Welt, der Krone ihrer Schöpfung, dem Menschen, das Siegel der Vollendung aufzudrücken, oder es muß die geistige Individualität des Menschen auch nach der Trennung vom Körper eine Zukunft haben, eine Zukunft, in welcher ihr die Muße [?] wird, ihren Zweck, die göttliche Vollkommenheit zu erreichen.

Die Mittel, welche die Natur zur Erreichung ihrer Absichte wählt, sind uns freilich in sehr vielen Fällen gänzlich verborgen, und insbesondere sind die wichtigsten Epochen des menschlichen Daseyns, Zeugung, Geburt und Tod in undurchdringliche Geheimnisse [*sic*] gehüllt. Da wir aber wissen, daß die fernsten Fixsterne auf unerforschte Weise sich Licht gegenseitig zusenden, und daß ein geheimnißvolles Band der Anziehung alle Massen des Weltraumes untereinander verbindet, wie sollten wir in der moralischen Welt an der Möglichkeit eines Zusammenhanges ferner Räume und ferner Zeiten verzweifeln wollen? (60–61) [It is not clear whether the words in brackets were intended to be crossed out.]

Man achte eine Sternschnuppe für zu gering, um ihr [nicht] einen wichtigen Zweck in der Ökonomie des Planetensystemes zuzutrauen. Welch' kleines Ding ist es doch um ein Blutkörperchen, wovon über 100 Millionen in einem einzigen Blutstropfen schwimmen!

Und doch kann der Organismus leichter ganzer Gliedmassen, als dieser Blutkörperchen entbehren, von denen seine Lebensäußerungen, seine Wärme, seine mechanische Actionen abhängen. Im Mikrokosmus wie im Makrokosmus ist es die unermeßliche [?] Zahl, welche die winzigsten Objecte die höchste Bedeutung zu verleihen vermag. (66) [The word in brackets was crossed out, then reinstated (in what appears to have been Mayer's usual fashion) by underlining with dots. I have left it untranslated since it goes against the clear sense of the sentence.]

# • NOTES •

## INTRODUCTION

1. Kuhn 1959.
2. Cf. Holmes 1986.
3. Kuhn 1959, 322; Heimann 1976, 278–279.
4. Kuhn 1959, 323, 345; Heimann 1976, 295–296; Buck, in Mayer 1979a, 2–3, and 1979b, i–ii, 156.
5. Kremer 1984, 192, 221–223 (quote on 223).
6. See esp. Haas 1909, 26, 31, 47, and Heimann 1976, 287.
7. Pfister 1914, 495–496; Exner 1914, 1532; Jentsch 1916, 91; Hell 1925, 15–17; Ober 1968, 2453–2454.
8. Heimann 1976, 280; Breger 1982, 172–173; Kremer 1984, 216–220 (quotes on 216 and 218).
9. Kremer 1984, 230–231.
10. Cf. Elkana 1970, 278: "Mayer was no mere 'hunch' philosopher, though it is difficult to find the logical path from the change in the color of venous blood in the tropics to a full annunciation of the principle of conservation of energy." He repeated this view in Elkana 1974, 121. For "concepts in flux," such as Mayer's concept of force, "we cannot follow logically what happened between the stage when those vaguely formulated principles were conceived and the point when their final, general form was realized" (Elkana 1970, 280; cf. 1974, 122). I believe such resignation is premature.
11. Coley and Hall 1980, 66.
12. Caneva 1974.

## CHAPTER ONE
## MAYER THE PERSON

1. See the timeline of the major events relative to Mayer's life in Appendix 1 and the map on page 2.
2. Krusemarck 1942, 32; Titot 1865, 42–44; Weller and Weller 1972, 130, 246; population figures from Drees 1988, 98.
3. Rümelin 1878, 1761 ≈ 1881, 352–354.
4. Gerlach 1942, 2; 1963, 15, 19.
5. Mayer 1877b, 389–390. Weyrauch supplied the words in brackets from a similar passage in another of Mayer's autobiographical sketches (1872a, 392); cf. 1873/74a, 386.
6. Rümelin 1878, 1762 = 1881, 358.
7. Mayer 1876, 107 = 1893a, 445–446.
8. During the period 1835–40, 21.0% of the medical students at Tübingen had fathers who were apothecaries, dentists, or veterinarians; 17.0% were higher civil servants or foresters, 16.0% were physicians, and 12.0% were pastors; none were from the nobility. The corresponding percentages for all students were 3.8, 16.1, 4.0, 14.3, and 2.2 (Drees 1988, 127–131). Mayer's background was thus entirely unexceptional.
9. See Appendix 2 for a list of the courses he took.
10. Grüsser 1987, 2; Heubner 1878, 292–293; Sticker 1939–40, 32 (1939), 229 (quoting Wilhelm Roser), 230–236; Wunderlich 1869, 115–116; the opposing voice is Snelders 1973, 208 ("geheel bepaalde").
11. Riecke 1835, 535 = 1836, 37; Rothschuh 1941, 621; 1953, 95, 103; Wunderlich 1843, 292; 1859, 287–288. A more ambivalent view is that of Anonymous 1841, 207, 211.
12. Anonymous 1801, 853; 1802, 35; 1803a, 783; 1803b, 356–358.
13. F. G. Gmelin 1821, 8–9.
14. Sticker 1939–40, 32 (1939), 235; Anonymous 1841, 208.

15. Roser 1878, 323 (including the quote).

16. Heubner 1878, 294; Roser 1878, 323; Sticker 1939–40, 32 (1939), 235.

17. The attempt in Brunn 1965 to connect Mayer's thoughts on energy conservation with certain of Schill's teachings is based on a reading of Schill 1838 which I cannot confirm, and strikes me as otherwise unlikely.

18. Roser 1878, 321; Sticker 1939–40, 32 (1939), 218.

19. Heubner 1878, 293; Wunderlich 1869, 115–116.

20. Bauer 1972, 54–60.

21. Wunderlich 1869; Roser 1878; Heubner 1878.

22. Mayer ca. 1863/67, 380–381; cf. Rümelin 1878, 1779 = 1881, 385–386. Mayer had begun hearing nosology under Autenrieth in the summer semester of 1834.

23. Mayer 1869, in Mayer 1893a, 347 ≈ 1978, 341.

24. Bauer 1972, 56–57.

25. Rümelin 1878, 1763 = 1881, 371; Diepgen 1942, 62–63.

26. Letter to Paul Friedrich Lang of 19–23 October 1837, in Mayer 1893b, 10–12; Rümelin 1878, 1762 = 1881, 357.

27. Weyrauch, in Mayer 1893b, 108.

28. Weyrauch, in Mayer 1893a, 9.

29. Mayer 1840–41, 51, 70.

30. Letters of 24 July 1841 to Baur, 5/6 December 1842 to Griesinger, 6 August 1842 to Baur, and 14/16 June 1844 to Griesinger, in Mayer 1893b, 110, 194, 140, and 212, respectively.

31. Mayer 1845, 84 = 1893a, 105.

32. Mayer 1851a, 15 = 1893a, 243; Mayer 1872a, in 1893b, 391.

33. Drawing presumably from notes supplied him by Mayer in the late 1860s, Saint-Robert (1870, 455) reported that, during the Java voyage, Mayer "studied with zeal especially Müller's handbook of physiology." Callisen (1830–45, 30 [1842], 473) says twice that the third part of volume 2 appeared in 1839; cf. a letter to Müller from Alexander von Humboldt, probably written in February 1840, which comments in detail on the contents of this part (Haberling 1924, 229).

34. Letter of 25 March 1840 to his parents, in Mayer 1893b, 92.

35. Mayer 1840–41, 51, 52, 53, 59, 63, 66; Herschel 1838 is cited in Mayer 1848a, 2, 30, 33–35 = 1893a, 152, 180, 182–184.

36. A perusal of the *GVDS* under "Strauss" failed to net any other candidates. Unlike Strauss 1839a (with xxxiii+132 pages), Strauss 1839b (with x+459+[1] pages) is too big to have been called a *Schriftchen*. The preface to Strauss 1839a is dated 15 March 1839, and it had come out by 11 April 1839 (letter to the wife of Ludwig Georgii, in Maier 1912, 28).

37. Mayer 1840–41, 50.

38. Mülberger 1878–79, 164. Arthur Mülberger, a physician from Crailsheim, had gotten to know Mayer during the latter's last stay at the sanatorium in Kennenburg in 1871.

39. Mayer 1848, 31 = 1893a, 180.

40. Letters of 8/12 June, 22 June, and 25 July 1840 to his parents, in Mayer 1893b, 95, 96, 97.

41. Letters of 19 March 1844 to Lang, 25 August 1848 to his mother-in-law, and 26 June 1855 to his father-in-law, in Mayer 1893b, 20, 466, 476.

42. Mayer 1865, [8]: "schwärmerische-pietistische Sentimentalität"; cf. Mayer 1893a, 307, 363.

43. Rümelin 1878, 1795 = 1881, 394.

44. Letter of 2 December 1851 to Lang, in Mayer 1893b, 338.

45. Mayer 1865, [3], [4]; cf. Mayer 1893a, 303, 305, 363. See Appendix 3 for the German text of the passages quoted.

46. Mayer 1865, [8]–[9]: "ich betrachte / es als eine besondere Fügung Gottes, dass ich, um an dem Gedeihen meiner geistigen Kinder noch Freude zu erleben, Leiden überstehen konnte, denen leicht viel stärkere erlegen wären."

47. For his belief in the immortalilty of the soul, see also his letter of 3 August 1869 to Friedrich Mohr, in Mayer 1893b, 423, and an anecdote related to Hermann Hettich by Theobald Kerner in a letter of 4 March 1886, in Hettich 1886, 1, and retold by Weyrauch in Mayer 1893a, 431.

48. Weyrauch, in Mayer 1893a, 362.

49. Rümelin 1878, 1795 = 1881, 399.

50. Breger 1982, 178–179. With respect to Mayer's belief in the immortality of the soul, see the long passage from Mayer ca. 1866(?), 60–61, quoted in chapter 6, section 2.

51. Letter of 13 December 1867 to Moleschott, in Mayer 1893b, 361–362.

52. Schmolz and Weckbach 1964, 160.

53. Letter of [31 December] 1851 to Lang, in Mayer 1893b, 339–340.

54. Dürr 1914, 148.

55.* Mayer 1869, in 1978, 348 = 1893a, 357.

56. Mayer 1851a, 20–21 = 1893a, 248.

57. Schmolz and Weckbach 1964, [2]. Mayer included the motto in the continuation of a passage from his *Organic Motion* cited at the end of chapter 2, section 2.1.4: "aber wir wissen auch, dass die Natur in ihrer einfachen Wahrheit grösser und herrlicher ist, als jedes Gebild von Menschenhand und als alle Illusionen des erschaffenen Geistes" (Mayer 1845, 36 = 1893a, 74).

58. Schmolz and Weckbach 1964, 103, reproducing an undated manuscript sheet.

59. Letter of 22 December 1874 to Rudolf Schmid, in Mayer 1893b, 460; cf. R. Schmid 1878, 690.

60. Letter from Hüfner of 5 August 1873 to Friedrich Zöllner, in Zöllner 1881, 689 = Mayer 1893a, 416.

61. Letter of 22 December 1874 to Schmid, in Mayer 1893b, 460–461; R. Schmid 1878, 690–691, gives an appreciably edited version. On the passage from Haller, see chapter 5, section 1. The context of the biblical passage Mayer quoted is significant; Romans 1:20–22 reads, in the King James Version: "For the invisible things of Him from the creation of the world are clearly seen, being understood by the things that are made, even his eternal power and Godhead; so that they are without excuse: Because that, when they knew God, they glorified Him not as God, neither were thankful; but became vain in their imaginations, and their foolish heart was darkened. Professing themselves to be wise, they became fools." If the last part spoke to Mayer's feeling of sinfulness at being proud, the first part evokes the rationale of natural theology.

62. Strauss 1839a, ix.

63. Strauss 1839a, x–xi. Ruge (1839, 996) quoted the passage cited here in his review of Strauss's book after noting with approval that "Strauss correctly rejects the materialistic way of thinking of many of our contemporaries." Mayer would seem to have thought so, too.

64. Strauss 1839a, 82–83 = 1838c, 18–19.

65. Strauss 1839a, 88 = 1838c, 22.

66. Strauss 1839a, xx.

67. Strauss 1839a, xxiii–xxv.

68. Strauss 1872, 205, 206.

69. Strauss 1872, 206, 207, 210–216.

70. Hettich 1878a, 1225, col. 3.

71. Mayer 1872(?)b: "Herr Strauß erklärt in dieser seiner neuesten Schrift das Christenthum als 'Wunderglauben' für einen von dem Darwinismus überwundenen Standpunkt. 'Die Auferstehung sey die größte Lüge in der Weltgeschichte u.s.w.'—Ob die von Herrn Strauß u. Cons. [i.e., Consorten] geschriebenen Bücher im Stande sein werden, die heil. Schrift der Menschheit zu ersetzen, wird die Zeit lehren."

72. Cf. the remembrances of participants at the 1869 *Naturforscherversammlung* at Innsbruck that Weyrauch assembled in Mayer 1893b, 453–454.

73. Cf. Mülberger 1879, Nr. 23, 3 = 1896, 16.

74. Schmoller 1907, 599–600.

75. Weyrauch, in Mayer 1893b, 1–2.

76. Gerok 1892, 234; Rümelin 1878, 1763 = 1881, 367 (quote).

77. Martini 1982, 329.

78. Zeller 1908, 69, 84.

79. Weyrauch, in Mayer 1893b, 286–287; Cranz 1894, 486.

80. Gerok 1892, 234; Rümelin 1878, 1762 = 1881, 361; Weyrauch, citing a recollection of Baur's, in Mayer 1893a, 9–10.

81. Zeller 1908, 86–87.

82. Letters from Strauss of 20 November 1837 and 23 January 1841 to Friedrich Fischer, in Zeller 1895, 44, 97–98. Reuschle spoke at Strauss's funeral, and was one of his oldest and closest friends (Zeller 1874, 125).

83. Strauss 1838a, 7 = 1839a, 4. On the long-standing friendship between Kerner and Strauss, see Grüsser 1987, 200, 305–312; during the spring of 1841 they frequently discussed immortality (309).

84. K. Hermann 1966, 185; cf. Mayer's letter of 19 March 1844 to Lang, in Mayer 1893b, 20–21.

85. See Kerner's *Personalakten*, Universitätsarchiv Tübingen, 40/109, Nr. 91, and the *Hörerliste* for Christian Jakob Baur's anatomy lectures for summer semester 1836, UAT 51/20; T. Kerner 1866.

86. Letter from Justinus Kerner of 18 January [1854], in Schmolz and Weckbach 1964, 104. Grüsser (1987, 260, 363) implied that Rümelin was a frequent guest of Kerner's in Weinsberg, and cited unpublished letters to Rümelin and his wife from the years 1836 to 1862.

87. Ofterdinger 1879.

88. Rümelin 1878, 1762, 1761, 1763, and 1762 = 1881, 361, 356, 367, and 364, respectively.

89. Quoted by Weyrauch in Mayer 1893a, 9–10.

90. Gerok 1892, 234.

91. Mülberger 1879, Nr. 23, 2; Dürr 1914, 148.

92. Letter of 11 November 1851 to Lang, in Mayer 1893b, 337.

93. Timerding 1925, 44; cf. 63.

## CHAPTER TWO
## MAYER'S WORK

1. Mayer 1841a, 100–101 = 1978, 23. I have translated *Stoffe* as "material substances" since "matter" has no English plural and "substance" alone does not necessarily convey the qualification "material"—as shown by Mayer's insistence that force, too, is a *Substanz*.

2. Mayer 1841a, 102 = 1978, 24.

3. Mayer 1841a, 102 = 1978, 24.

4. Mayer 1841a, 102 = 1978, 24.

5. Mayer 1841a, 102 = 1978, 24–25.

6. Mayer 1841a, 103 = 1978, 25. In rendering Mayer's algebraic expressions I have consulted the facsimile of the manuscript in Zöllner 1881, which indicates that Mayer always put a space between the "qualitative symbol" 0 and the terms that followed. Note that the word *Bestimmung*, here (as in the title of the paper) translated as "determination," also has the sense of "definition."

7. Mayer 1841a, 104 = 1978, 26.

8. Mayer 1841a, 105 = 1978, 27.

9. Mayer 1841a, 105 = 1978, 27. Note that *Null* means both "zero" and "nothing," hence my usual translation as "nought."

10. Mayer 1841a, 105–107 = 1978, 27–29. The next-to-the-last sentence is hard to translate: "Anders ausgedrückt heisst dies: in gleichem Masse wie die peripherischen Teile als gegen das Centrum fallend sich verhalten, fällt letzteres gegen die Peripherie."

11. Letter of 16 August 1841 to Baur, in Mayer 1893b, 127.

12. Mayer 1842, 233 = 1893a, 23. In translating passages from this paper I consulted the translation in Lindsay 1973, 68–74, which should be used with caution.

13. Mayer 1842, 234 = 1893a, 24.

14. Mayer 1842, 234 = 1893a, 31. Mayer omitted the two paragraphs on matter and chemistry from the reprinting of his paper in the *Mechanik der Wärme* (passages that Weyrauch reinstated in a note).

15. Mayer 1842, 235 = 1893a, 24.

16. Mayer 1842, 235–236 = 1893a, 25.

17. Mayer 1842, 236 = 1893a, 25.

18. Mayer 1842, 238 = 1893a, 27.

19. Mayer 1842, 238–239 = 1893a, 28. The word "original" was added, in parentheses, to the *Mechanik der Wärme* reprinting.

20. Mayer 1842, 239 = 1893a, 28.

21. Mayer 1842, 240 = 1893a, 29.

22. Mayer 1842, 240 = 1893a, 29 (which supplied a missing *von*).

23. Heimann 1976, 278.

24. Letter of 1 August 1841 to Baur, in Mayer 1893b, 114.

25. Mayer 1844/45, 9: "Schwefel und Quecksilber stehen mit Zinnober in einem solchen genetischen Verhältnisse; Wasser verwandelt sich in Dampf oder Eis etc. etc.; die chemischen Vorgänge samt und sonders bestehen in Formveränderungen gegebener Objecte, wobei ein Aufgewendetes stets mit einem Neugebildeten im engsten, genetischen Zusammenhang steht."

26. Mayer 1844/45, 9–10: "Es ist kein Grund vorhanden, dieses / Axiom auf das Ponderable zu beschränken." Cf. Mayer 1845, 5 = 1893a, 47.

27. Heimann 1976, 284–286.

28. Letter of 24 July 1841 to Baur, in Mayer 1893b, 110–111; note that the symbol for zero, when read in German as *Null*, conveys also the sense of "nothing." Cf. letters to Baur of 1 and 16 August 1841 and to Griesinger of 30 November 1842 and 20 July 1844—especially that of 30 November 1842, Mayer's first letter to Griesinger—in Mayer 1893b, 115, 120, 176, 223.

29. Heimann 1976, 277, 286. Later, "the axis of all my thought and action" was the problem of relating velocity to the change in distance between two gravitating bodies (letter of 30 November 1844 to Baur, in Mayer 1893b, 170).

30. Letter of 16 August 1841 to Baur, in Mayer 1893b, 120, 121–122. The words Mayer quoted in the first passage were taken from Baur's letter of 11 August 1841, in Mayer 1893b, 118 ≈ Schmolz and Weckbach 1964, 75. The sentiment was that expressed at length in Mayer's first letter of 24 July 1841.

31. Heimann 1976, 285–286.

32. The quote is from Mayer's letter of 17 July 1842 to Baur, in Mayer 1893b, 135; cf. Mayer 1851a, 38 = 1893a, 262 and his letters of 16 August 1841 and 30 November 1842 to Baur and Griesinger, in Mayer 1893b, 123–124, 180.

33. Cf. Mayer's letter to Baur of 24 July 1841, in Mayer 1893b, 112.

34. Mayer 1845, 6–36 = 1893a, 49–74. Cf. Rümelin's observation on Mayer's working out of his insights during the years after 1841: "The relationship between heat and motion was for him *now* only an example of a general law" (Rümelin 1878, 1779 = 1881, 386–387; emphasis added).

35. Breger 1982, 118; cf. Kuhn 1959, 329, and Lindsay 1973, 30.

36. Mayer 1848b, 385 = 1893b, 275.

37. Mayer 1873/74a, in 1893b, 386; cf. Mayer 1873/74b.

38. Cf. Mayer 1869 (1893b, 347 = 1978, 341); 1873, 194; 1878, 236; and the discussion in chapter 6. Dürr's *Heilbronner Chronik* of 1895 identified Mayer as "the discoverer of the mechanical theory of heat" (Dürr 1895, 320).

39. Cf. Mittasch 1940b, 23.

40. See Griesinger's letter of 4 December 1842, Mayer's rejoinder of 5/6 December, and Griesinger's response of 14 December, in Mayer 1893b, 184, 190–192, 195–196. Evidence for contemporary definitions of force will be given in chapter 4.

41. Letters of 30 November and 5/6 December 1842 to Griesinger, in Mayer 1893b, 180–181, 187; cf. 188. Note that the words "was auf Rechnung der Kräfte der unbelebten Natur kommen kann und muss" leave open the possibility that some organic phenomena might require the assumption of forces other than those of inanimate nature.

42. Mayer 1845, 10 = 1893a, 51–52; 1851a, 37 = 1893a, 261–262.

43. Mayer 1851a, 44–45 = 1893a, 267; cf. 1845, 88 = 1893a, 108.

44. Letter of 16 August 1841 to Baur, in Mayer 1893b, 123–125.

45. Mayer 1845, 33 = 1893a, 71.

46. Mayer 1845, 33–34 = 1893a, 71–72.

47. Gillispie (1960, 380) identified Mayer's advance over classical physics in his breaking with the tradition of imponderable fluids and "in assigning fundamental ontological status to force." Stichweh (1984, 153–154) stressed the genetic connection between Mayer's concept of force and his reinterpretation of the imponderables.

48. Letter of 22 June 1844 to Griesinger, in Mayer 1893b, 217. Note, however, that neither in this letter nor in most of Mayer's correspondence with Griesinger that year did Mayer use the term *Kraft*.

Griesinger had objected to Mayer's inclusion of motion in the same class with the imponderables because he was attached to the notion of the latter as weightless fluids and could not imagine motion as the same kind of entity; see his letter of 14 December 1842, in Mayer 1893b, 195.

49. Mayer 1842, 234 = 1893a, 24 (quoted above in section 1.2); cf. 1841b, 2, col. 2 and 3, col. 1.

50. Letter of 16 December 1842 to Griesinger, in Mayer 1893b, 199.

51. Draft of Mayer's letter of 31 March 1842 to Liebig, in Schmolz and Weckbach 1964, 56–57; letters of 6 August 1842 to Baur and of 30 November, 5/6 December, and 16 December 1842 to Griesinger, in Mayer 1893b, 140, 177, 190–192, 198–201.

52. Letter of 5/6 December 1842 to Griesinger, in Mayer 1893b, 191–192.

53. Cf. Griesinger's letter of 14 December 1842, in Mayer 1893b, 194–197.

54. Letter of 16 December 1842 to Griesinger, in Mayer 1893b, 200–201.

55. Mayer 1851a, [3] = 1893a, 235; 1852, 443.

56. Mayer 1862, 385 = 1893a, 324.

57. E.g., Mayer 1842, 234 = 1893a, 24; letter of 16 December 1842 to Griesinger, in Mayer 1893b, 201. Cf. Heimann 1976, 287–288: "The central feature of his 1842 paper was his attempt to explicate a concept of force as a substantial, yet nonmaterial, entity."

58. Letter of 30 November 1842 to Griesinger, in Mayer 1893b, 177–178. For the long phrase in brackets, see p. 176; cf. Mayer 1842, 239 = 1893a, 29. Here, as usual, I render *Materie* as "matter" but *eine Materie* as "a material substance"; and here, as often, Mayer's calculation is off.

59. Mayer 1851a, 20, 23 = 1893a, 248, 250; for other instances see also pp. 34, 41 *bzw*. 259, 264–265.

60. Riehl (1900, 167) noted the significance of Mayer's "*substantial* conception of force": "If force is a substance, then the law of the persistence [*Beharrlichkeit*] of substance must apply to it, too." Heimann/Harman also stressed the importance of Mayer's conception of the substantiality (but immateriality) of force as providing a basis for belief in its conservation, and noted the "clear disharmony between his claim that he refused to speculate on or had knowledge of the essence of force and was only concerned to define the relations between forces and his statement that force was a nonmaterial, yet substantial, entity" (Heimann 1976, 289; cf. Harman 1982, 2–3). Harman rightly criticized Cassirer's stress on the purely functional or relational character of Mayer's conception of force (cf. Cassirer 1957, 461–462).

61. Mayer 1844/45, 18: "so ist eine Kraft gegeben, welche, gleich der Bewegung, Substanzialität, oder die Eigenschaft zu existiren hat, oder, welche nimmermehr zu Null werden kann." Schelling (1797/1803, 228) once glossed *Substantialität* parenthetically as *Selbstdaseyn*. The word would seem to have been generally so understood. Cf. Burdach's characterization of the immaterial soul "as a substance, as an autonomous entity [*als Substanz, als Selbstständiges*]" (Burdach 1837, 308).

62. Mayer 1873, 194 = 1878, 236; cf. 1872a (quoted in chapter 2, section 2.7).

63. Mayer 1845, 36 = 1893a, 73–74. In rendering Mayer's "Es gibt keine immateriellen Materien!" I was forced to translate the plural *Materien* as the singular "matter."

64. Letters of 22 June 1844 to Griesinger and 31 July 1844 to Baur, in Mayer 1893b, 217 and 155, respectively.

65. Mayer 1845, 22 = 1893a, 62. I have regularly (if not always) translated *aufheben* as "annihilate."

66. Mayer 1842, 234, 236 = 1893a, 31, 27; letters of 24 July and 16 August 1841 to Baur and of 5/6 December 1842 to Griesinger, in Mayer 1893b, 110–111, 120–121, 187.

67. Mayer 1842, 240 = 1893a, 30; cf. his letter of 30 November 1842 to Griesinger, in Mayer 1893b, 180.

68. Mayer 1863, [2]: "die ganze Lehre von der Erhaltung der Kraft, oder wie ich mich auch ausdrücken möchte, von der physikalischen Stöchiometrie" (cf. Mayer 1893b, 369, and Schmolz and Weckbach 1964, 113); Mayer 1873, 194 = 1878, 236.

69. Letter of 17 July 1842 to Baur, in Mayer 1893b, 137.

70. Letter of 17 July 1842 to Baur, in Mayer 1893b, 137–138.

71. Letter of 31 July 1844 to Baur, in Mayer 1893b, 155. Mayer's reference to page 45 of his manuscript does not correspond to any I have seen.

72. Mayer 1860; cf. 1870, 367, and 1871b, 57 = 1893a, 399. The German text of the passage quoted is given in Appendix 3.

73. Mayer 1841a, 100–101 = 1978, 23; letter to Griesinger of 30 November 1842, in Mayer 1893b,

177; 1845, 39, 46 = 1893a, 75, 81. With respect to my interpretation of Mayer's somewhat idiosyncratic wording, note that Fries spoke regularly of the "logischer Satz des Grundes," which he rendered in 1804 as "Die Aussage jedes Satzes muß einen hinreichenden Grund haben" (Fries 1804, 119; 1827, 41–42; 1837, 139).

74. In addition to the references in the previous note, cf. F. Fischer 1838b, 35–39. Fischer distinguished what he called "das Gesetz des zureichenden Grundes" (35) or simply "das Gesetz des Grundes" (36) from the law of causality: the former pertains to logic, the latter to reality.

75. Letter of 1 August 1841 to Baur, in Mayer 1893b, 115.

76. Letter of 30 November 1842 to Griesinger, in Mayer 1893b, 177.

77. Rümelin 1878, 1779 = 1881, 380, who cited the variant "ex nihilo nihil fit"; Mayer 1845, 5 = 1893a, 47.

78. Mayer 1845, 19, 38–39 = 1893a, 59, 75 (my emphasis); cf. Mayer 1851a, 20–21 = 1893a, 248.

79. Mayer 1851a, 20 = 1893b, 247–248. Mayer's 1842 paper had not, in fact, invoked that principle.

80. Letter to Griesinger of 20 July 1844, in 1893b, 226; cf. his letter to Griesinger of 16 December 1842, in Mayer 1893b, 204.

81. Mayer 1845, 15 = 1893a, 56; cf. Mayer ca. 1863/67, 381–382. The correct reduction of his numbers yields 366 meters.

82. Mayer 1851a, 5–14 = 1893a, 236–243 (many passages).

83. Letter from Baur of 7 September 1844, in Mayer 1893b, 159. Baur's frustration with Mayer's "paradoxes" was considerable; his letter to Mayer of 30 November 1846 began (Baur 1846, [1]):

> That I am not the one to cure you of the paradoxes afflicting you—then that you are afflicted by them has long since been my conviction (parallelogram of forces, proportionality of forces and speeds) and will always remain so—that I, I say, am not the one to cure you of your paradoxes, and am also not the one who can be particularly helpful to you in getting your new truths to see the light of day, you could have noticed that long ago, in the beginning from the quality of my participation in your endeavors and thereafter from my nonparticipation.

What set him off again was Mayer's new claim to have discovered a cause for the progressive retardation of the earth's rotation as a result of tidal action. The quoted passage is freely translated from the German text given in Appendix 3.

84. Mayer 1844/45, 18, 47–48.

85. Mayer 1845, 8, 19 = 1893a, 50, 60; Mayer 1851a, 24, 30, 39 = 1893a, 251, 256, 263.

86. Letter to Baur of 17 July 1842, in Mayer 1893b, 136. Mayer repeated his conviction that taking the measure of moving force as mass times velocity "leads to absurdities, to a perpetuum mobile" in his letter to Baur of 6 August 1842, in Mayer 1893b, 143. (He used the term *mobile perpetuum* in both passages.) These are his earliest recorded references to the impossibility of perpetual motion.

87. Letter to Griesinger of 5/6 December 1842, in Mayer 1893b, 193.

88. Letter to Baur of 24 July 1841, in Mayer 1893b, 112. The last phrase is difficult to translate: "es verhält sich der Zentralkörper gegen die Peripherie fallend, wie die peripherischen Körper sich als gegen das Zentrum fallend verhalten." Note that although Mayer's language (*die Neutralisierte*) did not require him to specify what was being neutralized, its feminine form (*die*) betrays its original reference to motion (*die Bewegung*); the otherwise regular form for substantivized adjectives would have produced the neuter *das Neutralisierte*.

89. Letter to Baur of 16 August 1841, in Mayer 1893b, 126 (my emphasis).

90. Letter to Baur of 17 July 1842, in Mayer 1893b, 138–139 (my emphasis in the two longer phrases).

91. Letters to Baur of 24 July 1841 and 6 August 1842 and to Griesinger of 16 December 1842, in Mayer 1893b, 112–113, 140, 204.

92. Letter to Baur of 18 July 1842, in Mayer 1893b, 139. Weyrauch called attention to the significance of this passage in Mayer 1893a, 35–36.

93. Rümelin 1878, 1779 = 1881, 380–384 (quoted at length in chapter 6, section 1).

94. Mayer 1869, in 1978, 346–347 = 1893a, 355–356, except that the latter has *können* instead of *dürfen* for the word translated as "must" just after the ellipsis. Mayer variously used *Princip* and *Satz* for what I have rendered simply as "principle."

95. Letter of 24 July 1841 to Baur, in Mayer 1893b, 110.

96. Mayer 1851a, 38 = 1893a, 262.

97. Although he did not provide much evidence or analysis, Wise (1981, 272) is one of the few to call attention to the importance of this aspect of Mayer's thinking: "In company with many German scientists of the nineteenth century, Mayer concerned himself not merely with the physical coherence of material nature but also with the relation between matter and mind, or better, between *Natur* and *Geist*, where *Geist* implies both mind and soul."

98. Mülberger 1878–79, 164, quoting Mayer 1842, 234 = 1893a, 24; cf. Mülberger 1879, Nr. 23, 2–3 ≈ 1896, 16–17.

99. Volkmann 1837, 76.

100. Mayer 1867, cited from Mayer 1893a, iv–v.

101. Mayer 1869, cited from Mayer 1978, 347 = 1893a, 356. Wise (1981, 272) quoted part of this passage as evidence that in mid-nineteenth century Germany "force [commonly occupied] . . . a middle position between inert matter and *Geist*." Hirn was clear in assigning credit for the mechanical theory of heat to Mayer; see Hirn 1864, 4, 11, 12, 15 = 1868, 18, 25, 26, 29. For those views of Hirn's which would have appealed to Mayer, see Hirn 1864, 19–20, 37–39, 103, 111–117 = 1868, 34, 54–56, 124–125, 133–141.

102. Hirn 1868, 11; cf. 364–371.

103. Mayer 1872a, end of fascicle 19, beginning of fascicle 9: "Selbstverständlich kann ein Mann, der als der erste den Grundsatz ausgesprochen und durchgeführt hat, daß nicht nur der Materie, sondern auch der lebendigen Kraft in ihren verschiedenen Formen, wie der Bewegung, der Wärme, dem Lichte und der Electricität die Eigenschaft der substanziellen / quantitativen Unzerstörbarkeit zukommen [*sic*], einem Materialismus, der diese Eigenschaft nur auf ponderablen Objecten [*sic*] beschränkt wissen will, nicht huldigen, und es mögen in so ferne diese Vorträge nicht nach dem Geschmack exclusiver Materialisten ausgefallen seyn." In place of "auf ponderabeln Objecten beschränkt wissen" Mayer had originally written "ponderabeln Objecten vindiciren"; he neglected to change the case of "ponderabeln Objecten" when he changed the verb and added "auf." The passage quoted from the manuscript sheets picks up immediately where Weyrauch's transcription in Mayer 1893b, 394, leaves off. Cf. Mayer 1871a, 358 = 1893a, 376, and 1873, 194 = 1878, 236. The addresses referred to were Mayer 1869, 1870, 1871a, and 1871b.

104. Letter of 15 June 1871 to Rudolf Schmid, in R. Schmid 1878, 691 = Mayer 1893b, 460. I cite from the *New English Bible*.

105. Mayer ca. 1866(?), 58, 60–61 (quoted in chapter 6, section 2).

106. Rümelin 1878, 1779 = 1881, 382.

## CHAPTER THREE
## PHYSIOLOGY AND MEDICINE

1. To cite only a few of the more important works on the blood from a medical standpoint, see Thackrah 1834 (widely cited in German works), H. Nasse 1836, and Magendie 1838, 1839a, 1839b. As a guide to more of the primary literature, see Rothschuh 1941. For contemporary statements of the importance of the scientific study of the blood, see Roser 1838, 3; Magendie 1839a, 15, 18; Hünefeld 1840, 21.

2. F. Hoppe 1835, 1.

3. On Rapp, see chapter 1, section 1, and the quotation there from Anonymous 1841, 208. See also Autenrieth 1792 and Schüz 1799. Autenrieth is reported to have been the effective author of most of the dissertations done under him (Callisen 1830–45, *1* [1830], 307; Riecke 1835, 535 = 1836, 36; Stübler 1948, 84; Wunderlich 1859, 289).

4. Autenrieth 1801–2, *1* (1801), 138–139, 311–314, 355.

5. According to Diepgen (1942, 60), who was apparently referring to the secondary literature on Mayer. For a contemporary reference, see H. Nasse 1836, 10. Mayer mentioned Autenrieth's observation in Mayer 1845, 86 = 1893a, 107.

6. Autenrieth 1801–2, *1* (1801), 313, 314. Autenrieth subscribed to the oxygen theory, but retained much of the old terminology.

7. Autenrieth 1801–2, *1* (1801), 349; cf. 3 (1802), 61.

8. Autenrieth 1801–2, *1* (1801), 354–355.

9. Autenrieth 1836 (1825), 336–337.

10. Krimer 1823, 286–288. Krimer, like many, cited Adair Crawford's experiments and observations (published between 1779 and 1788) on the effect of temperature on blood color; see also H. Nasse 1836, 176, and Brauss 1841, 13 (the latter confirming Crawford's original experiments on dogs). On Crawford, see Mendelsohn 1964a, 154. Krimer had begun his book as a translation of the 1819 first edition of Thackrah 1834.

11. Thrackrah 1834, 123; cf. 115. Würzburg professor of anatomy and physiology Carl Friedrich Heusinger (1836, 269) summarized these findings in his review of Thackrah's book; he noted that Thackrah's writings "were received with well-earned acclaim already at their first appearance, as was the first edition of the present work" (266). H. Nasse (1836, 364; cf. 176) reported Thackrah's bath observation and explanation of its cause; Nasse's *Nachtrag* (355–372) is essentially a critical précis of Thackrah 1834. Mayer (1845, 86 = 1893a, 106) wrote in his *Organic Motion*: "We recall here . . . the bright redness of the blood that Thackrah observed as he bled a patient in a warm bath."

12. Thackrah 1834, 181.

13. Thomas G. Wright, in Thackrah 1834, 116.

14. H. Nasse 1836, 171. Nasse also cited the usual examples of change in the color of the blood in baths, warm or cold weather, etc. (H. Nasse 1836, 10, 176, 296; cf. 1842, 169). Note that Nasse never discussed the connection between respiration and animal heat. (A *Privatdozent* in Berlin since 1831, in 1837 Nasse became professor and director of the Physiological Institute in Berlin.) London physician William Prout (1834, 523–524) also explained the color of arterial blood in terms of both oxygen and "saline matters."

15. F. G. Gmelin 1821, 64–65. This work was in Mayer's library.

16. Autenrieth 1834, *1*, 419, 422. This work was in Mayer's library. On the danger of hepatitis in the tropics and its treatment by bloodletting, see also Schönlein 1832, *1*, 405–406, 409; Schönlein said nothing about the color of the blood.

17. Mani 1956, 204–205, citing Autenrieth 1801–2, 2 (1802), 90–93, and Tiedemann and Gmelin 1826–27, 2, 49–65.

18. Autenrieth 1801–2, 2 (1802), 89, 90.

19. Hasper 1831, *1*, 7–8, 27; cf. *1*, xxviii, and Liebig 1841d, 209 = 1841e, 2866 = 1842b, 24 = 1844b, 240: "Among us, diseases of the liver (carbon diseases) predominate during the summer, diseases of the lungs (oxygen diseases) during the winter" (also paraphrased in Rose 1843, 270). For Liebig's somewhat confused picture of the role of bile in respiration, see 1842a, 259 ≈ 1842b, 62. Among those Hasper named were Adair Crawford and Armand Séguin; I have been unable to identify Copland and Pierson.

20. Hasper 1831, *1*, 322–323, 323; *Naturkraft* would seem to be elliptical for the commonly invoked *Naturheilkraft*. Tübingen *Privatdozent* Albert Schill explained the effect of elevated ambient temperature on the blood in similar terms, and stressed the blood's increased venosity (without, however, specifically noting a change in color) and the enhanced functioning of the liver (Schill 1840, 348, 355; cf. 283–284, where he discussed further possible causes of "excessive venosity" of the blood). Concomitantly, "the blood now needs to generate less heat. The organism must decrease the activities which produce animal heat, but those are the activities of the blood and those immediately connected therewith" (349). He did not pursue the implications of this line of reasoning. The first German physiological work I know of to point out the crippling deficiencies of the standard explanation in terms of the allegedly greater oxygen content of colder air, etc., was Lotze (1842, 424–425). It was, however, still being invoked two years later by Valentin (1844, *1*, 154) and as late as 1850 by Vogt (1850, 673).

21. Hasper 1831, *1*, 67; cf. 63. Hasper recommended ample prophylactic bloodletting as well (2, 344).

22. Müller 1835, 151 = 1833, 150 = 1837b, 162 = 1841, 132. I cite first from the edition in Mayer's library.

23. Müller 1835, 152 = 1833, 151 = 1837b, 163 ≈ 1841, 133 (which begins the sentence instead with "It is believed that"). Mayer's medical exam asked him to write on "the physiological aspects of the secretion of bile and the influence of the latter on the digestion process" ("Das Physiologische über die Absonderung der Galle, und den Einfluß dieser letztern auf den Verdauungsprocess"); in his response, Mayer noted that "particular emotions increase the secretion of bile, but Johannes Müller gives here no specific connection" ("einzelne Leidenschaften vermehren die Galle absonderung, Johannes Müller giebt aber hiebei keine specifische Beziehung zu") (Mayer 1838b, 23$^r$). Essentially the same physiology of the liver was given by Rudolphi 1821–28, 2, pt. 2 (1828), 164.

24. Mayer 1838b, 28$^r$–28$^v$: "Welcher Einfluß übt eine anhaltend feucht warme Witterung auf den Gesundheitszustand des Menschen aus[?]" See Appendix 3 for the German text of the longer quote. Diepgen discussed Mayer's exam answer (without quoting from it) in relation to the same texts discussed here (i.e., Autenrieth 1834–36 and Hasper 1831), and noted that Mayer explained the slight difference between arterial and venous blood in the tropics in terms of the excess of carbon resulting from the diminished oxygenation (Diepgen 1942, 59–61). He does not seem to have appreciated that Mayer expected the venous blood to be *darker* than normal. In basing his account on Diepgen's, Kremer (1984, 218) misinterpreted the latter to have Mayer asserting (rather paradoxically) "that the venous blood will lighten in color because decreased oxidation will leave excess carbon in it," and was thus wrong in his conclusion that "Mayer could not have been surprised by the bright venous blood he observed in Java."

25. Letter of 22 February 1839 to Lang, in Mayer 1893b, 16.

26. Müller 1835, 23, 46–47 = 1833, 23, 46 = 1837b, 23, 47 = 1841, 21, 44–45.

27. Müller 1835, 38 = 1833, 37 = 1837b, 38 = 1841, 35–36.

28. For accounts of the theories, experiments, and controversies relating to the question of animal heat during the late eighteenth and early nineteenth centuries, see Culotta 1972, Goodfield 1960, Holmes 1964, Kremer 1984, and Mendelsohn 1964a.

29. Magnus 1837; cf. Mohr 1837f and Mendelsohn 1964a, 176–177. For a good summary of the disputed question of whether arterial and venous blood contain (respectively) oxygen and carbon dioxide, see T.L.W. Bischoff 1837 and its one-page summary in Müller 1838a, clviii–clix; Bischoff said yes to both, and explained thereby the change in the color of the blood.

30. Brodie 1811 and 1812. Cf. Goodfield 1960, 93–99, and Kremer 1984, 106–110; both authors point out that Brodie himself never advocated regarding the action of the nervous system as a *source* of animal heat. The dependence of animal heat on the nervous system was the subject of somewhat inconclusive experiments by Friedrich Nasse (F. Nasse 1839).

31. Dulong 1841 and Despretz 1824. Dulong's paper was read in 1822; Despretz received the Academy of Science's physics prize in 1823.

32. Holmes 1964, xxxix.

33. Müller 1835, 80, 318 = 1833, 79, 318 = 1837b, 85, 327 = 1841, 79 (first reference only).

34. Müller 1835, 81 = 1833, 80 = 1837b, 86. In the last edition of his text he concluded instead: "However probable this theory is, one must nevertheless not forget that the production of heat does not depend on the chemical process alone, but is also subject to the influence of living parts" (Müller 1841, 81). Under the latter he included muscular motion, citing the experiments of Becquerel and Breschet. Kremer (1984, 184–186) stressed Müller's acceptance of a wide range of sources of animal heat. Even in 1843, when he had been influenced by Liebig's understanding of respiration, he still tempered the identification of respiration as a combustion process by stressing that the latter requires a high temperature (Müller 1843, 268). English physician C.J.B. Williams (1835, 871–872; 1836, 39) likewise ascribed the principal source of animal heat to the combustion of oxygen and carbon in the blood while doubting its sufficiency (à la Dulong and Brodie).

35. Müller 1835, 318 = 1833, 318 = 1837b, 327. One can imagine Liebig's pique at Müller's dismissal of chemists; he would later return the disfavor.

36. Kremer (1984, 164–189) made the reconceptualization of animal heat between 1823 and 1850 from a problem of respiration to one of the exchange of matter (*Stoffwechsel*) a major theme of his dissertation, though I believe he overemphasized the opposition between these two alternatives: respiration is, after all, a kind of *Stoffwechsel*.

37. Tiedemann 1830, 477–478. Tiedemann's text continued for another half page, running through

classes of animals from birds and mammals to mollusks and worms (478–479). Tiedemann, like many of his contemporaries, looked at the problem of animal heat as much if not more in terms of the comparative anatomy and physiology of different species, as in terms of the physiological processes of *one* organism (see also 400–477, esp. 453). Kremer (1984, 182) would have Tiedemann treat animal heat within the context of *Stoffwechsel* rather than respiration, though even in the section he cited for support it seems clear that for Tiedemann (1830, 374–375) the combustion theory of animal heat was simply an instance of *Stoffwechsel.* The latter was not an alternative to respiration, but embraced it. For the rest, Tiedemann's principal treatment of animal heat was not, as Kremer maintained, in the section on *Stoffwechsel*, but in a separate chapter, "On the Disengagement of Heat by Living Bodies," in the section headed "Development of Imponderables."

38. Berthold 1837, *1*, 121.

39. Volkmann 1837, 44–46; Hünefeld 1841, 147–150; Lotze 1842, 461–462. Wistinghausen (1837, 14–17) and Prout (1834, 519, 522–526, 535–536) also believed that respiration is the chief but not the only cause of animal heat; on Prout, cf. Holmes 1964, lii–liii. Alfred Wilhelm Volkmann moved from Leipzig to accept a professorship of physiology, pathology, and semiotics at Dorpat in 1837; Friedrich Ludwig Hünefeld was a professor of chemistry and mineralogy at Greifswald; Hermann Lotze was then an instructor in medicine and philosophy at Leipzig.

40. Volkmann 1837, 245; cf. 69–72.

41. Rudolphi 1821–28, *1* (1821), 188–191, and 2, pt. 2 (1828), 383. Mayer owned the 1830 edition of this work.

42. Simon 1840–42, 2 (1842), 53–58 (quote on 58).

43. Simon 1844, 73. Since 1842 a *Privatdozent* in Berlin and a chemist at the Charité hospital, Simon died in 1843.

44. Berzelius 1821–32, *4*, pt. 1 (1832), 112–123 ≈ 1825–31, *4*, pt. 1 (1831), 109–120 ≈ 1837–41, 9 (1840), 144–155. I cite first from the edition in Mayer's library.

45. Berzelius 1821–32, *4*, pt. 1 (1832), 120–121 = 1825–31, *4*, pt. 1 (1831), 117–118. The second paragraph is essentially the same in Berzelius 1837–41, 9 (1840), 152. Unlike Berzelius and Magendie, German authors rarely included friction among the causes of animal heat. Note that the term *Athmen*, here translated "respiration," means simply "breathing," and does not refer to the complex *physiological* process also termed "respiration."

46. Berzelius 1837–41, 9 (1840), 151.

47. In discussing Berzelius's views on animal heat, Kremer (1984, 181) noted a connection with his dualistic electrochemical theory. Yet Berzelius did not emphasize that connection, and it would hardly seem to explain his aversion to the combustion theory, which fit easily into his electrochemical theory.

48. Berzelius, letter of 12 April 1842 to Liebig, in Carrière 1893, 238.

49. Berzelius 1844, 579–580.

50. Berzelius 1844, 580–581.

51. Berzelius 1837–41, 9 (1840), 152.

52. Becquerel and Breschet 1835a, 257–258 ≈ 1835b, 114 ≈ 1839, 384.

53. Muncke 1829, 446–447. August Gottfried Ferdinand Emmert investigated the effect on respiration of cutting or impeding the operation of one or both vagus nerves of the sympathetic nervous system (Emmert 1809 and 1812). His conclusion was "that thereby the chemical process which takes place in the lungs during respiration—insofar as it can be judged from the color and coagulation of the blood and the condition of the exhaled air—undergoes no other change than can be explained from the diminished entry of air into the lungs and from the disturbed circulation" (1812, 130). Emmert had been a student of Autenrieth's (Kremer 1984, 124–125).

54. Muncke 1841, 396.

55. Kämtz 1839, 455.

56. Baumgartner 1836, 487–488; cf. 1832b, 474.

57. H. Davy 1830, 197–198 = 1840, 332 ≈ 1833, 211–212.

58. H. Davy 1830, 198–199, 199–200 = 1840, 332–333, 333–334 ≈ 1833, 212–213, 214–215. The bracketed gloss, "Mr. Brodie," was contained in a footnote to the 1830 edition absent from the 1840 reprinting. Davy was especially critical of the attempt to use electricity to explain organic functions.

59. Mayer 1845, 49 = 1893a, 84; the question mark is Mayer's interpolation. Along with the blood

itself, Reich (1843, 19) regarded "the *constant animal heat proper* to the human organism in general *likewise as inherited property*." Similarly obsolete by that time was Reich's conception of the vital force: "It [i.e., faithful observation of nature] has taught me that there is only *a single and indivisible vital force* (Blumenbach's *formative force*) which, like the vital heat and life itself, is inborn in the organism and which allows of no further explanation" (xviii).

60. Emile du Bois-Reymond's correspondence with Eduard Hallmann provides an inside look at some of these changes in attitude during the early 1840s; see in particular a letter of 9 August 1841 in which du Bois noted the conflicting views of Henle, Schwann, and Benedict Stilling on one side versus those of his esteemed teacher, the anatomist and embryologist Carl Bogislaus Reichert, on the other with respect to the sufficiency of chemistry and physics to explain organic phenomena, and a famous letter from around May 1842 in which du Bois announced that he and Ernst Wilhelm Brücke had sworn to defend the "truth" that no other forces are active in organisms than attractive and repulsive forces inherent in matter after the fashion of the common physical and chemical forces (E. du Bois-Reymond 1918, 98–99, 108; cf. Temkin 1946, 326). In his doctoral dissertation of 1843, du Bois declared publicly that "in neither inorganic nor organic nature do there exist forces whose ultimate components are not either attractive or repulsive" (quoted in a letter of March 1843, ibid., 113). Helmholtz would soon also adopt such a worldview. From Galaty's account, Reichert advanced the usual objections to a fully reductionist accounting of vital phenomena (Galaty 1974, 298). Galaty gave an insightful analysis of the "philosophical basis of mid-nineteenth century German reductionism," in particular that of du Bois-Reymond, Brücke, and Helmholtz.

61. Magendie 1836, 2, 509–519 = 1833, 2, 509–519 ≈ 1834–36a, 2 (1836), 343–350 ≈ 1834–36b, 2 (1836), 432–440.

62. Magendie 1836, 2, 519–520 = 1833, 2, 519–520 ≈ 1834–36a, 2 (1836), 350–351 ≈ 1834–36b, 2 (1836) 440. As Kremer (1984, 193) noted, by around 1844 Magendie had (privately) rejected the combustion theory of respiration, ridiculing its defense by Dumas and Liebig.

63. Dumas 1841a, 21 August, [3] = 1841b, 59 = 1842a, 43. For an excellent discussion of Dumas's work, his debt to Jean-Baptiste Boussingault, and his relationship with Liebig, see Holmes 1974, 15–47.

64. Dumas 1841a, 21 August, [3] = 1841b, 58–59 = 1842a, 42–43; cf. 1842a, 85 = 1842b, 126 and Holmes 1964, xliii–xliv.

65. Kremer 1984, 201–202. I am less certain than Kremer that this demonstrates "Liebig's complete ignorance of the research and discussions on respiration since Lavoisier." Gerhardt rendered *Verwesung* as "érémacausie" or "la pourriture sèche" (Liebig 1840a, xvi; cf. xxx); I have translated it simply as "decay."

66. Liebig 1840d, 281–282.

67. Liebig 1840d, 282.

68. Liebig 1840d, 301–302, 337, and Kremer 1984, 201 (citing Liebig 1841a, 52).

69. Liebig 1841b, 156; cf. Kremer 1984, 202, and Holmes 1974, 31–32.

70. Liebig 1841b, 160.

71. Liebig 1841d, 197 = 1842b, 10.

72. Liebig 1841d, 204, 204–205 = 1841e, 2836, 2865 = 1842b, 18, 19. Otto Kohlrausch (1844, 34–37) criticized as unproven Liebig's assumption that the same amount of heat is produced by similar chemical changes under all circumstances. Liebig's explanation of animal heat as a function of respiration was attacked by Schultz 1843.

73. Liebig 1841d, 218–219 = 1842b, 36–39.

74. The first paper contained only the barest allusion to the respiratory function of the skin: Liebig 1841d, 212 = 1841e, 2873 = 1842b, 29.

75. Liebig 1842a, 268 = 1842b, 78.

76. Liebig 1841d, 204 = 1841e, 2836 = 1842b, 18; cf. the context of the following remark (Liebig 1840d, 323 = 1841a, 329 ≈ 1840a, clxxxiv): "The greatest miracle [*Wunder*] in the living organism is precisely that an unfathomable wisdom has placed in the cause of an incessant destruction, in the maintenance of the respiration process, the source of the organism's renewal, the means to resist all other atmospheric influences, a change in temperature, and moisture." Such language expressing Liebig's belief in the divinely ordered harmony of the natural world was not uncommon; cf. Liebig

1842a, 242, 261 = 1842b, 40, 69: "a marvelous [*wunderbar*] wisdom" and "an admirable [*bewunderungswürdig*] wisdom."

77. Liebig 1842a, 273–276 = 1842b, 85–88.

78. Liebig 1842a, 277–278 = 1842b, 90–91. Liebig's casual prose style often violates the normal rules of syntax; my translations have kept some of that flavor. Kremer (1984, 208–209) called attention to Liebig's speculation that the formation of fat might be another source of bodily heat.

79. Liebig 1842a, 279 = 1842b, 92.

80. Liebig 1842a, 280, 281 = 1842b, 95, 96. Liebig never explicitly discussed the contribution to the body's oxygen supply of absorption through the skin. Typical was his passing reference to this phenomenon in his comment that the oxygen released via the production of fat does not leave the body as oxygen gas; rather, "it leaves in the same form as the oxygen absorbed out of the air by lungs and skin" (Liebig 1842a, 276 = 1842b, 89).

81. Liebig 1842a, 276 = 1842b, 89.

82. Valentin 1842, 63; Simon 1844, 71–73; Wunderlich 1845, 125. Holmes (1974, 38–39) discussed Liebig's theory that the body obtains oxygen from the production of fat (which had been deposited in the first place because of insufficient inspired oxygen), but did not consider it with respect to the production of heat.

83. Valentin 1844, *1*, 151. He remarked that it was especially Liebig who had recently laid stress on the *Verbrennungshypothese* (154, citing Liebig 1842b, 18ff). He also noted problems for that theory from the small proper heat of plants and cold-blooded animals (155).

84. Valentin 1846, 196–197 (quotes on 197).

85. Vierordt, 1845b, 911. Vogt was of the same opinion in 1850, though in 1845 he was less certain and more wide-ranging in his search for an explanation; compare Vogt 1850, 673, to Vogt 1845–47, pt. 1 (1845), 132–141.

86. Vierordt 1845a, 228. He acknowledged, nevertheless, that we still cannot determine or measure all of the possible sources of animal heat (230–233).

87. Vierordt 1845a, 226; cf. iii.

88. Vogt 1845–47, pt. 1 (1845), 142–143 (quotes on 143); Liebig 1845, 63, 63–64. Having lived and worked in Paris from roughly 1844 to 1847, in the latter year Vogt became professor of zoology in Giessen.

89. Kremer 1984, 254; cf. 237–238: "Physiological discussions of animal heat during the 1830–40s by no means represented a hotbed of vitalism, as the young (or old) Helmholtz might have thought." Kremer's conclusion—"that Helmholtz's views on *Lebenskraft* probably did not arise from an actual problem of physiological research" (254)—is probably correct.

90. Valentin 1847, 127–174. He cited Mayer 1845, 79 (= 1893a, 101) for Mayer's claim that oxidation takes place in the blood (160).

91. Valentin 1846a, 170.

92. Valentin 1847, 155.

93. Mayer 1845, 97 = 1893a, 116, quoting from Valentin 1844, *1*, 576.

94. Valentin 1847, 221, citing Mayer 1845, 96 (= 1893a, 115) in a footnote.

95. For a discussion of this issue and references to the contemporary literature, see Rothschuh 1941, 622–623; for the views of C. H. Schultz, see Liedmann 1966, 253.

96. Rudolphi 1821–28, 2, pt. 2 (1828), 328; he attributed such a notion to Carl Friedrich Kielmeyer, Gottfried Reinhold Treviranus, Ignaz Döllinger, Friedrich Ludwig Kreysig, and Carl Gustav Carus.

97. Volkmann 1837, 239.

98. Roser 1838, 3. Recall that Roser had ridiculed his and Mayer's professor, Wilhelm Rapp, for teaching the *Propulsionskraft des Blutes*.

99. Schleiden 1842–43, *1* (1842), 51.

100. Müller 1841, 176–178; cf. Müller 1833, 211–212; 1835, 211–212; 1837b, 222–223. The discussion was much expanded in the fourth edition.

101. Burdach 1826–40, *4* (1832), 408.

102. Cf. Mayer 1845, 93 = 1893a, 113: "while the venous blood of other organs has to struggle against a centrifugal pressure, the blood of the heart is drawn away in a centripetal direction."

103. Autenrieth 1801–2, *1* (1801), 48, 49.

104. F. G. Gmelin 1821, 216–217.

105. F. G. Gmelin 1821, 218.

106. Tiedemann 1830, 538; cf. Autenrieth 1801–2, *1* (1801), 57, and (especially) C.C.E. Schmid 1798–1801, 2 (1799), 296:

> Furthermore: no machine is a perpetuum mobile in the sense that the principle of its motion is determined and maintained in continuous action by the mechanism itself. In order to work purposefully it needs purposefully determined influences whose existence and purposefulness the machine itself does not determine. No first, original motion is explainable purely mechanically, but only the determinate continuation, the direction of the motion.—*The organic body is, during its entire organic existence, a true perpetuum mobile.*

Autenrieth singled out Schmid's book for special commendation (ibid., [vii]). In 1840 theologian David Friedrich Strauss (1840–41, *1*, 714) rejected Schelling's likening of an organism to a perpetuum mobile by pointing out that Schelling had failed to consider the friction, detrition, and destruction that attend an organism's relationship with the external world. Strauss there paraphrased and cited the following passage from Schelling 1809, 456 (= 1834, 67–68): "That all organic beings face dissolution cannot in any way appear to be an original necessity; the bond of forces constituting life could, as far as its nature is concerned, just as well be indissoluble, and a creature which restores what has become defective in it by means of its own forces appears destined—if anything is—to be a perpetuum mobile."

107. Vetter 1837, 89–90. The hypothesis of *fluidisirtes Nervenmark* was proposed the year before in Naumann 1835.

108. Vetter 1837, 79.

109. Vetter 1837, 92.

110. On Tiedemann's importance, see Kremer 1984, 181–182.

111. Tiedemann 1830, 26–27; cf. 45.

112. Tiedemann 1830, 34; cf. 129.

113. Tiedemann 1830, 370; he here spoke of the *Wechsel des Stoffs*.

114. Tiedemann 1830, 371–372. Like many of his contemporaries, he was more inclined to illustrate such dependencies by comparing different classes of animals than by investigating the physiological responses of one animal under different circumstances.

115. Tiedemann 1830, 374–375.

116. Tiedemann 1830, 662; cf. 665–667. *Thätigkeits-* and *Kraft-Aeusserungen* might also be translated "expressions" of activity and force. "The dependence of life on external agencies" (662) is a major theme of 662–680.

117. Tiedemann 1830, 668–669.

118. Tiedemann 1830, 686. The last sentence recalls Tiedemann's inclination to seek physiological enlightenment from comparative anatomy.

119. Müller 1835, 1 = 1833, 1 = 1837b, 1 = 1841, 1.

120. Müller 1835, 29 = 1833, 29 = 1837b, 29–30 = 1841, 27; Müller's use of the expressions *Stoffveränderungen, materielle Veränderungen*, and *Austausch ponderabeler und imponderabeler Materien* suggests the extent to which *Stoffwechsel* had not yet become the standard term.

121. Müller 1835, 30 = 1833, 29–30 = 1837b, 30 = 1841, 28. I have not been able to trace the comparison attributed to Richerand in his popular *New Elements of Physiology* (Richerand 1833).

122. Müller 1835, 46–47 = 1833, 46 = 1837b, 47 = 1841, 44–45.

123. Cf. the section "On the Causes of Animal Motion" in Müller 1837a, 46–62. The same judgment applies to Rudolphi's treatment of *Muskelthätigkeit* (Rudolphi 1821–28, 2, pt. 1 [1823], 289–336). Hall (1980, 31) has claimed, with some textual support, that Müller's "central idea was that the activities of nerves always entail expenditure of force or power which has to be replenished continually from external sources, especially by nutrition; nutrition was therefore seen as a dynamical process (i.e., as supplying the organism with *Energie*) as well as a material process (i.e., as supplying foodstuffs)." It should be noted that when Müller and his German contemporaries used the word *Energie* they meant by it not "energy" but (in the *OED*'s words) "vigour or intensity of action."

124. Cf. Holmes's characterization: "Although Müller treated thoroughly the numerous contemporary analyses of the chemical composition of tissues and body fluids, as well as investigations of the compounds present at different stages of digestion and absorption, he had no overall view of the meaning of the chemical reactions underlying the continuous exchanges of materials in the tissues" (Holmes 1964, liv).

125. Burdach 1837, 4.

126. Burdach 1837, 93–94.

127. Valentin 1837, 2–3.

128. Valentin 1844, *1*, 10–13; 1846b, 1–3; 1847, 221.

129. Valentin 1844, *1*, 11; cf. 1846b, 1–2.

130. Valentin 1844, *1*, 12; cf. 1847, 9–11.

131. Valentin 1844, *1*, 13.

132. Valentin 1844, *1*, 13–14, 16–17, 192–193; 1846b, 1–3.

133. Valentin 1847, 10–11.

134. Valentin 1844, *1*, 192; cf. 1846b, 1–2.

135. Valentin 1844, *1*, 192, 194–195; 1847, 220–221. He eliminated the more religiously flavored language from the second edition.

136. Schwann 1839, 229.

137. Schwann 1839, 237, 238. Another self-consciously programmatic author to stress the significance of *Stoffwechsel* was Lotze 1842, 401.

138. Dumas 1841a, 20 August, [2] = 1841b, 38 = 1842a, 9–10; cf. 1841a, 21 August, [2] = 1841b, 53 ≈ 1842a, 34. Holmes (1974, 24) called attention to Dumas's inclusion of force within the larger sphere of the economy of living things. Prout (1834, 538) had earlier described "the mutual relations and dependence of plants and animals" in terms of the balanced exchange of oxygen and carbon dioxide—a balance which evidenced "the general scheme of Providence"; cf. Holmes 1964, lii.

139. Holmes (1974, 37–38) well analyzed the significance of this paper. Müller (1843, 262–264) gave a full and favorable précis of it and Liebig 1842a.

140. Liebig 1841d, 198–199, 213, 214–215 = 1842b, 11–12, 30, 32–33 (except that the latter has "one and the same effect" instead of "a certain effect" in the fifth paragraph). Mayer told Griesinger in a letter of 5/6 December 1842 that it was a passage on p. 32 of Liebig's book—corresponding roughly to the third, fourth, and fifth paragraphs quoted here—which prompted him to compose his 1842 paper (Mayer 1893b, 190). Mayer had of course first read them in Liebig's *Annalen* paper (i.e., Liebig 1841d).

141. Liebig 1841d, 215–217 = 1842b, 33–35. Mayer quoted the third, fourth, and fifth paragraphs with approval in Mayer 1845, 57–58 = 1893a, 132–133. Cf. Liebig 1841d, 197 = 1842b, 10—but also *contrast* the following: "A plant contains no nerves. Heat and light are the remoter causes of motion in plants; in animals we recognize in the nervous system a source of force which is capable of renewing itself again at every moment of its life" (Liebig 1841d, 192 = 1842b, 3, except that the latter has "remotest" in place of the former's "remoter"). Müller (1843, 264–265) revised the section on respiration in the fourth edition of his *Handbook* in accordance with Liebig's views: the effect of oxygen on the organic substances in the organism is to put them into a state of *Kraftäusserung*—i.e., of motion—whereby the heat and electricity produced can be regarded as states of motion of matter.

142. Holmes 1964, xxxii; cf. xlvii–xlviii and Kremer 1984, 338–340. In 1856, after materialism had become a threat, Liebig (1856a, 386) could liken the *Stoffwechsel* that produces forces in the steam engine and the voltaic pile to the *Stoffwechsel* that produces mechanical force in an organism while pulling back from any physicochemically reductionist conclusion: "But the connection itself [between exchange of matter and mechanical force] is still entirely unknown to us. What we know is that the force in the organism is not produced as in the steam engine, that it is not explicable in terms of the known electrical laws." Nor, he added, does it make sense to try to attribute the activity of the brain to an underlying *Stoffwechsel*.

143. T.L.W. Bischoff 1842, xci–xcii. Bischoff's misreading of Liebig is all the more amazing since his remarks were directed to that part of Liebig's *Animal Chemistry*—the third, to be discussed in section 2.2 of chapter 3—which assigned the vital force causal priority in the production of organic motions.

144. Simon 1844, 73; Rose 1843; Kohlrausch 1844, 36–37.

145. In German, *Geist* means both "mind" and "spirit," while *Seele* is often closer to "psyche" than to the theologically tinged word "soul."

146. There are echoes here of the classical distinction between the vital anima and the *geistig* animus; cf. French 1981, 114, 129.

147. Cf. Kremer 1984, 237–238, esp. 237: "Nowhere is the axiom that 'the victors write history' better illustrated than in the historical interpretation of Helmholtz's early physiological researches. In an autobiographical speech delivered in 1891, Helmholtz initiated the view that his early work was motivated by the desire to destroy vitalism in physiology." Kremer was correct to insist that Helmholtz was guilty of misrepresenting his problem context, in that no one was then explaining *animal heat* in terms of a vital force. The fact that vitalism was not a term generally used in discussions of the vital force during this period is a further indication of the low-key nature of the issue, since it tended to be used principally by its self-conscious opponents.

148. Autenrieth 1801–2, *1* (1801), 49.

149. Autenrieth 1801–2, *1* (1801), 58; cf. 53–55. The functions Autenrieth assigned to the soul and to the vital force correspond closely to the traditional distinction between sensibility and irritability, although Autenrieth did not use these terms.

150. Autenrieth 1801–2, *1* (1801), 56.

151. Autenrieth 1801–2, *1* (1801), 55: "blos Maschineneinrichtung durch Lebenskraft in Thätigkeit gesetzt"; cf. 56, 57. Note that Autenrieth did not regularly use the definite article before "Lebenskraft."

152. Autenrieth 1801–2, *1* (1801), 57–58.

153. Autenrieth 1801–2, *1* (1801), 126–127 (quote on 126).

154. Valentin (1837, 39) cited Autenrieth's book under the brief entry "Lebenskräfte" in his report on the progress in physiology during 1836. Of the work itself he said: "Many ingenious [*geistvoll*] views, the particular elaboration of which often falls short of the present state of science" (18). Carus (1837, 953) mentioned this essentially dismissive citation as indicative of the prejudice against speculation in physiology that works like Autenrieth's had elicited.

155. Carus 1837, 953.

156. Autenrieth 1836, 348; cf. 395–396. Views very similar to Autenrieth's were expressed by Carl Traugott Kretschmar (1834, 2–5; 1835, 3–4), who was especially concerned to establish the role of God as *Geist* and Creator. Unlike Autenrieth, however, Kretschmar believed that "every manifestation of force is connected with exchange of matter" (1835, 3). Schill (1840, 329–331) also regarded the vital force as the link between the soul and the body while criticizing the analogy between the soul, the vital force, and galvanism.

157. Autenrieth 1836, 359–362; cf. 7 for a more tentative judgment. Then again, in still a third of the ten essays comprising the collection, he allowed that animal magnetism might be a possible exception, though "under unnatural circumstances," to the rule that organisms can only affect the external world via physicochemical means, the effects of the vital force being limited (normally) to the substance of the organism itself; the context of *this* qualification was the identification of an important difference between the vital force and the imponderables (273): "The relationship of heat, electricity, and magnetism to their corporeal conductors stands in opposition to this self-enclosure [*Insichgeschlossenseyn*] of the vital activity within the boundaries of the organic body. That is, these imponderables radiate out from the ponderable substances with which they entered into combination without any recognizable boundary to their activity."

158. Autenrieth 1836, 209–210, 35, and 282, respectively.

159. Erman 1812; Autenrieth 1836, 207; cf. 200.

160. Autenrieth 1836, 50; cf. 51.

161. Autenrieth 1836, 54–56 (quote on 55); cf. 103 for the significance of the balanced number of births of girls and boys. Recall Süssmilch's use of the same example to illustrate the workings of divine providence (chapter 1, section 3). The influential clinician and journal editor, Christoph Wilhelm Hufeland, published a short work on the subject (C. W. Hufeland 1820).

162. Autenrieth 1836, 102; cf. 40–42, 393–396, and 1836 (1825), 342–343. Most of the examples cited in this paragraph were repeated on 271–272. In general, virtually every point Autenrieth made in this book was reiterated ad nauseam in its 552 pages.

163. Autenrieth 1836, 393; cf. 343. Thus Autenrieth maintained that phenomena such as the spiral growth of plants and an animal's instincts cannot be an "immediate consequence of a mechanism" (203).

164. Autenrieth 1836, 267 (my emphasis).

165. Autenrieth 1836, 264–266.

166. Autenrieth 1836, 400 = 1816, 294. My phrase, a "merely mechanical arrangement of material parts," is intended to capture the sense of Autenrieth's *Maschineneinrichtung* (399–400 *bzw.* 293, among many places).

167. Autenrieth 1836, 444 = 1816, 368.

168. Carus 1837, 955. Autenrieth's ideas reveal certain affinities to speculations regarding the functions of the aether (cf. Cantor and Hodge 1981b, 27–29).

169. Tiedemann 1830, 26; cf. 35 for his views on the usefulness and limitations of chemistry to physiology.

170. Tiedemann 1830, 29–30.

171. Tiedemann 1830, 94–95; cf. 127–128. Mani (1956, 214) gave the following assessment of Tiedemann and Gmelin's *Digestion According to Experiments* (1826–27): "One finds no trace of that arrogant pride of many late nineteenth-century scientists who believed one might reduce all of life to physicochemical processes. Tiedemann and Gmelin granted full legitimacy to the 'vital force.' They precisely staked out the domain of methods and restricted themselves to investigating the chemical phenomena of life with chemical methods."

172. Tiedemann 1830, 98.

173. Tiedemann 1830, 390. The passage comes near the end of a long section headed "Bildungskraft, Ernährungskraft" (376–393).

174. Tiedemann 1830, 647 (both quotes).

175. Tiedemann 1830, 650–651.

176. Tiedemann 1830, 657, 658. Schultz (1831, 53) quoted the last sentence approvingly in his review of Tiedemann's book. Tiedemann named P. J. Barthez, J. F. Blumenbach, J. D. Brandis, and C. W. Hufeland; "Springel" and "Rosse" are probably Kurt Polycarp Joachim Sprengel and Theodor Georg August Roose; the otherwise mysterious "Fryer" may have been Friedrich Ludwig Kreysig.

177. Volkmann 1837, 70–72.

178. Volkmann 1837, 159, from the section "Von dem Seelenleben" ("On Psychic Life" or "On the Life of the Soul").

179. Volkmann 1837, 160–161.

180. Rudolphi 1821–28, *1* (1821), 245–246 (quotes on 246); quoted in full in Schlüter 1985, 147. Mayer owned the 1830 edition of Rudolphi's *Outlines of Physiology*.

181. F. G. Gmelin 1821, 19–20, 20 (reading "Lebensturgor," as elsewhere, instead of "Lebenturzor"). In 1835, however, he wrote of the vital force in words much like Müller had used two years before and Liebig would use five years later (F. G. Gmelin 1835, 44): "In a healthy state it is the vital force alone which maintains the human body in the struggle with the external world, and which so harmoniously orders the interplay of the complicated activities whose totality constitutes life that the unity of life in its psychic [*geistig*] and corporeal sphere is thereby maintained."

182. Schill 1840, 7–8.

183. Magendie 1836, *1*, v–vi = 1833, *1*, v–vi ≈ 1834–36a, *1* (1834), vi ≈ 1834–36b, *1* (1834), vii–viii. Magendie was elsewhere sometimes reluctant to invoke physical analogies in the explanation of vital phenomena [Magendie 1836–37, *1* (1834), 5–6 ≈ 1837, *1* (1834), 1].

184. Magendie 1836, *1*, 32–33 = 1833, *1*, 32–33 ≈ 1834–36a, *1*, 23–24 ≈ 1834–36b, *1*, 23–24. The passage quoted begins the section headed "Causes of the Vital Phenomena." Abraham Kaau Boerhaave published a large work in Leiden in 1745 entitled *Impetum faciens dictum Hippocrati per corpus consentiens philologice et physiologice illustratum observationibus et experimentis passim firmatum*. Magendie's stance was criticized by an anonymous reviewer (Lotze?) of Lehmann's *Textbook of Physiological Chemistry*, who argued that one cannot separate mechanical from teleological (or vital) explanations, since in all cases one has to do with both general laws and particular given circumstances (Anonymous 1842b, 417): "One cannot therefore agree with the division of tasks as employed by famous physiologists such as Magendie; one cannot attribute some phenomena of life to mechanical,

others to vital forces, as if the latter were different from the former in the manner of their action, whereas they only differ in the form of their effects because they are a specific ensemble [*bestimmte Zusammenfassung*] of mechanical processes not existing outside the living body." The reviewer criticized Magendie for pushing mechanical explanations as far as possible, then invoking "purposefully acting forces" not subject to mechanical laws (418).

185. Magendie 1836, *1*, 37 = 1833, *1*, 37 ≈ 1834–36a, *1* (1834), 26–27 ≈ 1834–36b, *1* (1834), 26.

186. Magendie 1839a, 16. The book appeared in a German translation of 1839 that I have not seen.

187. Magendie 1839a, 20, 21; cf. his condemnation of explanations in terms of "vitality" (41–42). I do not find that Magendie treated questions of animal heat and respiration in these lectures on the blood.

188. Elsässer, in Magendie 1834–36a, *1* (1834), 317–318, 319. The intervening text, dealing with the nature of matter, force, and the imponderables, is quoted in chapter 3, section 3.1.

189. I cite an illustrative selection of usages: "Lebenskraft": 1835, 18 = 1833, 18 = 1837b, 19 = 1841, 17; 1840, 553; "organische Kraft": 1835, 18, 23 = 1833, 18, 22 = 1837b, 18, 23 = 1841, 17, 23; "organisirende Kraft": 1835, 24–25 = 1833, 24 = 1837b, 25 = 1841, 22–23; 1840, 509; "Thätigkeit": 1835, 19 = 1833, 18 = 1837b, 19 = 1841, 17; "vernünftige Schöpfungskraft": 1835, 23 = 1833, 23 = 1837b, 23 = 1841, 21; "Seele": 1840, 510; "*primum movens*": 1835, 24, 48 = 1833, 24, 47 = 1837b, 24, 48 = 1841, 22, 47; "bewusstlos wirkende zweckmässige Thätigkeit": 1835, 25 = 1833, 24 = 1837b, 25 = 1841, 23; "*vis essentialis*": 1835, 48 = 1833, 47 = 1837b, 48 = 1841, 47; "Idee": 1840, 505; "bewegende Idee": 1840, 510, 511; "Lebensprincip": 1840, 506, 508.

190. Müller 1835, 18 = 1833, 18 (which has *Ausgabe* for *Aufgabe* in the third sentence) = 1837b, 18–19 = 1841, 17. One of Müller's contemporary reviewers objected to his thus limiting the scope of science, insisting that it is indeed the business of the physiologist to answer the question he posed (Anonymous 1835, 199). Cf. 1840, 522, for Müller's important conception of "reflective experience [*eine denkende Erfahrung*], which distinguishes the essential from the contingent in experiences and thereby discovers principles from which many experiences can be derived."

191. Müller elsewhere assigned to the vital force "the development of the germ [*Keim*]" and "the continuous preservation of the whole and its regeneration" (Müller 1835, 47 = 1833, 47 = 1837b, 48 = 1841, 46).

192. Müller 1840, 513; see chapter 3, section 3.1, for a further discussion of this issue.

193. Müller 1835, 19, 20–21 = 1833, 18, 20 = 1837b, 19, 21 = 1841, 17–18, 19. He invoked Kant's name explicitly.

194. Müller 1835, 23 = 1833, 22–23 = 1837b, 23 = 1841, 21–22. Müller variously used the words *Glieder* and *Theile* for what I have translated simply as "parts." He began the paragraph by again associating himself with Kant's view of the organism as "composed of dissimilar organs that have the ground of their existence in the whole, as Kant put it." The sentence beginning "The external impulse" was quoted in German by Kremer (1984, 203), who noted its affinity with the passage from Liebig 1841d, 212–213, quoted in chapter 3, section 3.2.

195. Although his conceptualization and terminology are different from mine, a similar point was one of the main theses of Lenoir 1982.

196. Müller 1835, 24 = 1833, 24 = 1837b, 24–25 = 1841, 22.

197. Müller 1835, 25 = 1833, 24 = 1837b, 25 = 1841, 23.

198. Müller 1835, 25 = 1833, 25 = 1837b, 25 = 1841, 23.

199. Müller 1835, 26 = 1833, 26 = 1837b, 26 = 1841, 24.

200. Prout had a similar conception of the directive role of what he usually called the "organic agent," which was, he insisted, "*not* the *result* of organization, but the *cause* of organization" (Prout 1834, 418, 429–430; 438). Prout's "organic agency" was a force opposing the general laws of inorganic matter; when its "efforts" are exhausted, the organism decays (529). For Lenoir, Müller was a "teleomechanist" for whom "organization is the primary starting point for biological research for which no mechanical explanation can be advanced" (Lenoir 1982, 154, *sic*, though I suspect that the sense requires a comma after "research"; cf. 159). According to him, Müller's vital force, "like the force of gravitation, does not exist independently of the material constituents" (154). Müller, it seems to me, *denied* the sufficiency of organization in the explanation of all vital phenomena and spoke consistently of the vital force as an ideal something superadded to matter; I am not aware that he ever invoked the

analogy with gravitation in this regard. Lenoir could have made his analysis more cogent if he had distinguished between the developmental and the more narrowly physiological aspects of Müller's complex vital force: in the latter case alone was Müller's research program plausibly "mechanical." See also Caneva 1990.

201. Müller 1835, 27 = 1833, 27 = 1837b, 27 = 1841, 25.

202. Müller 1835, 27 = 1833, 27 = 1837b, 28 = 1841, 25.

203. Müller 1835, 28 = 1833, 27–28 = 1837b, 28 = 1841, 26.

204. Müller 1835, 39 = 1833, 38–39 = 1837b, 39–40 = 1841, 37–38. Hall (1980, 32) quoted this passage (beginning with the words "the source of the increase in the organic force" in my translation) in support of his interpretation that "whatever its nature he [Müller] was certain that force in the living organism was never *sui generis* but always required replenishment from external sources of force. Between inorganic and organic forces he saw a fundamental correlation and interchangeability; to suppose the contrary, namely that organic force might be created by life itself, struck him as absurd." As I will argue, Hall mistook a specific and limited analogy for a general identity. Nor can I agree with Hall that Müller formulated "a version of the conservation of force or *Energie*"; at most Müller might have formulated a principle of conservation of *vital* force—though in fact he did not—but that force was not interconvertible with the other forces of nature. In seeking to uncover antecedents to the conservation of energy, many commentators have looked too exclusively at the *conservation* side, not fully appreciating that one of Mayer's triumphs was to have worked out a *concept of force* that was amenable to a general conservation principle.

205. Müller 1835, 52 = 1833, 51 = 1837b, 52 = 1841, 50.

206. Müller 1835, 54 = 1833, 53 = 1837b, 54–55 = 1841, 52. Müller used both *gebunden* and *latent* to describe the same phenomenon. The continuation of this passage (after a brief ellipsis) invoked a property of the vital force which Liebig would later make much of: the organism's resistance to external sources of disturbance.

207. The sixth book was headed "Vom Seelenleben" (Müller 1840, 505–588); the two major subsections of interest here, the first and the third, were headed "On the Nature of the Soul in General" (505–525) and "On the Interaction between the Soul and the Organism" (553–588).

208. Müller 1840, 505 (my emphasis).

209. Müller 1840, 506ff; he spoke of "die empfindende und vorstellende Seele der Thiere" (506).

210. Müller 1840, 509.

211. Müller 1840, 510.

212. Müller 1840, 511.

213. Müller 1840, 510.

214. Müller 1840, 510–511.

215. Müller 1840, 511. In the continuation of this passage Müller reiterated where the strength of this position lay (511): "On the other hand, according to this idea [*Vorstellung*] of the relationship of the motive ideas to matter, as emanations of the divinity, it is easy to see how the diverse organisms—the classes, orders, families, genera, and species—with all their independence from each other nevertheless express so completely one idea active over them in the beginning."

216. Müller 1840, 511–512.

217. Müller 1840, 513 (my emphasis).

218. Schwann 1839, 220–257. Schwann discussed at length the limits of legitimacy of his guiding analogy between cell formation and crystallization (239–256). Liebig (1846, 180–182, 190–193) attacked Henle's use of that analogy in his own consideration of the (in)appropriateness of analogies in scientific explanation.

219. Schwann 1839, 221–223; cf. 1879, 50. Schwann is perhaps Lenoir's best example of a "teleomechanist," one who accepted teleolgical explanations within a mechanistic ontology (Lenoir 1982, 126–127); but see Caneva 1990, 295–296.

220. Schwann 1839, 224.

221. Schwann 1839, 224, 226. Burdach (1826–40, 6 [1840], 527–528) was quick to attack what he took to be Schwann's materialism.

222. Letter from Griesinger of 14 December 1842, in Mayer 1893b, 197. Mayer replied on 16 December 1842 (205): "Through his microscopical and mechanical experiments Schwann has, to be sure,

earned himself lasting great credit and has swept aside many a foolish hypothesis about animal motion."

223. Schwann 1879, 50, 54.

224. Schwann 1879, 54.

225. Henle 1841, 216–217. Callisen (1830–45, 28 [1840], 474–475) says this volume began appearing in 1839. At least the first 112 pages and two plates of Schwann's book were published in 1838; its preface was dated March 1839. Despite the fact that Henle's programmatic plea for a "rational medicine" based on the empirical search for causal connections identified its "ultimate goal" as the reduction of physiological and pathological phenomena to physical and chemical processes, he nevertheless distanced himself from the "anathema" that many contemporaries, "blinded by the light given off by recent chemical discoveries," had pronounced against the vital force, an hypothesis he judged "just as good or weak as an explanation" as those of gravitation or elective affinity (Henle 1842, 31). Blameworthy was only its premature invocation, which cut short further investigation.

226. Henle 1841, 217–218, 218–219.

227. F. Fischer 1839, 107.

228. Muncke 1829, *1*, 359–360; cf. 361 and 1836b, 1450–1453. In his review of Fries's *Textbook of Physics*, Muncke (1827, 84) distinguished the vital force from the imponderables while noting that we don't know whether the latter are bound to a material substrate or exist independently.

229. Muncke 1831, 111.

230. Muncke 1831, 117.

231. Baumgartner 1837, 16–17.

232. Berzelius 1813, 4–5 ≈ 1814, 290–291 = 1815, 2–3. This work represents an address delivered in 1810 and first published in Swedish in 1812.

233. Berzelius 1813, 6 ≈ 1814, 292 = 1815, 4. Maintaining that "the human mind [*Geist*] cannot for now penetrate beyond a certain limit," Heinrich Rose (1843, 260) quoted this passage with approval in his review of Liebig's *Animal Chemistry*.

234. Berzelius 1813, 8 ≈ 1814, 294 = 1815, 6; this passage also quoted with approval in Rose 1843, 261.

235. Berzelius 1821–32, 3, pt. 1 (1828), 145 = 1825–31, 3, pt. 1 (1827), 135 ≈ 1837–41, 6 (1837), 3.

236. Berzelius 1821–32, 3, pt. 1 (1828), 146–147 = 1825–31, 3, pt. 1 (1827), 136–137 ≈ 1837–41, 6 (1837), 4–5.

237. Berzelius 1821–32, 3, pt. 1 (1828), 155 = 1825–31, 3, pt. 1 (1827), 145 = 1837–41, 6 (1837), 25.

238. Berzelius 1839b, 290–291 ≈ 1839c, 2–3; both are translations of Berzelius 1839a, 78–79. I translate "eine Hypothese . . . , deren Wirklichkeit noch unbewiesen ist und deren Annahme" from 1839c, 3, which more closely follows the original's "en hypotes, hvars tillvaro ännu är obevisad, och dess antagande" (1839a, 78–79) than does "eine hypothetische, ihrem Wesen nach noch unerwiesene ist, deren Voraussetzung" from 1839b, 290. Berzelius thus provided Lenoir with a good example of someone who conceived of the vital force "as the expression of a complex interrelation of material parts incapable of further analysis but inseparable form the order and arrangement of matter" (Lenoir 1982, 160).

239. A somewhat different view of Berzelius's changing attitude toward the vital force has been advanced by Alan Rocke (1992). As Rocke has demonstrated, Berzelius had come close to embracing an unqualified materialistic reductionism of organic processes to the usual physical and chemical forces already in his *Föreläsningar i djurkemien* (2 vols., 1806–8), though as Rocke himself pointed out, Berzelius had had at this time little first hand experience in animal chemistry. Perhaps Berzelius's apparent acceptance of a vital force in 1812 and 1827 reflected a chastened experimentalist's caution, while his assertion in 1839 that organisms' distinctiveness consists only in the complexity of their organization reflected chemists' increasing success in synthesizing organic compounds. In my opinion, Jørgensen (1965a, 266, 276, 278) imputed too "konsekvent" and unchanging a position to Berzelius on the subject of vital force over a period of four decades. He did, however, quote from a letter of Berzelius's to botanist Carl Adolf Agardh from 1831 which lends some support to such an interpretation (177): "To suppose that the elements in the organic world [*Natur*] are endowed with other fundamental forces than in the inorganic is an absurdity. The dissimilarity of products depends on the

dissimilar circumstances under which the forces act, and these can vary indefinitely; but if we are not capable of correctly comprehending the circumstances which exist in the organic world, this gives us no justified occasion to guess at other forces."

240. C. G. Gmelin 1835, 431.

241. C. G. Gmelin 1835, 447, 448. cf. 443: "The vital force appears rather as merely a condition favoring the coming together of certain elements into more complicated combinations [or compounds (*Verbindungen*)], more or less as the nascent state permits combinations between elements which without it would not take place." Gmelin's reference to light, heat, and electricity as "forces" is worth noting, since they were more commonly referred to as "imponderables"—though he also included the *status nascens* among those forces, that is, the condition of an element at the moment of liberation from a compound marked by greater-than-normal chemical activity. Gmelin's likening the vital force to the generally accepted nascent state had the effect of assimilating it to other bona fide chemical "forces."

242. Geiger 1833, 186, 655, 657; 1827, 226, 666, 668.

243. Lehmann 1842, xii.

244. Lehmann 1842, 40.

245. Lehmann 1842, 41. Lehmann's reluctance categorically to reject the reality of the "dynamic" vital force while rigorously excluding it from a science which deals only with physical laws and their material substrate found renewed expression in the course of his discussion of the assimilation of nutrients into the substance of different organs (99).

246. The views expressed—expecially the sythesis of teleological design and lawbound operation—are entirely in harmony with Lotze's. Lotze and Lehmann were close colleagues, and Lotze had published his 1839 review of Stark in the same journal (under its earlier title) as this review of Lehmann.

247. Anonymous 1842a, 417–418. See the next-to-the-last note for the gist of Lehmann's p. 99. If this review was by Lotze, it would have prepared him for the composition of his famous *Lebenskraft* article of 1843 discussed later in this section.

248. Anonymous 1842a, 418.

249. Simon 1844, 88. Simon devoted several pages of this review to a précis of the third part of Liebig's *Animal Chemistry*; there he reported without criticism Liebig's understanding of the physiological function of the vital force (Simon 1844, 79–80). Perhaps Liebig had convinced Simon that his vital force was close enough to the other physicochemical forces to be scientifically legitimate; compare here Gmelin's likening of the vital force to the nascent state and the analogy Liebig drew between vital and catalytic forces.

250. Rose 1843, 260 (italics added).

251. Berzelius 1813, 6, 8 (etc.), quoted earlier in this section.

252. Schleiden 1842–43, *1* (1842), 49; cf. 2 (1843), 436–441.

253. Kohlrausch 1844, 4.

254. Valentin 1844, *1*, 13, 17, 194; 1846b, 2–3; 1847, 13, 220.

255. Vogt 1845–47, pt. 1 (1845), 3, 143.

256. Vogt 1845–47, pt. 3 (1847), 460.

257. For a rich discussion of Liebig's lifelong attachment to a vital force and his opposition to materialism, see T.L.W. Bischoff 1874, 47–58, 88–100. Bischoff was in intimate contact with Liebig from 1843 on. A good overview of Liebig's understanding of vital force is Lipman 1967. Lenoir (1982, 14–15, 134–135, 162, 188–189) has seriously distorted Liebig's work by identifying him as an *opponent* of vital forces, though at other places he has him reinterpreting the vital force in terms of the order and arrangement (or motion) of matter (160, 163–164, 166), a claim I find unsupported by most of the evidence he cited; see, however, Liebig 1842b, 213, quoted by Lenoir on 166 and by me later in this section, which does appear to support his interpretation, and as such necessitates a revision of the criticism made in Caneva 1990, 296.

258. Liebig 1840b, 122 = 1840c, 32. Liebig elaborated these points in the context of a somewhat ambivalent discussion of the differences and similarites between the vital force and chemical forces in the first of his collected *Chemical Letters* (Liebig 1844a, 722 = 1844b, 22–23).

259. Liebig 1840a, ix.

260. Liebig 1840d, 34–35 = 1841a, 34.

261. Liebig 1840d, 35 = 1841a, 34–35, except the latter has *Von* instead of *Vor* as the first word of the second paragraph and *unantastbar* instead of *unanschaubar.*

262. Liebig 1840d, 49 = 1841a, 49 ≈ 1840a, xc; cf. Liebig 1840d, 53–55 = 1841a, 53–55 ≈ 1840a, xcii–xciii.

263. Liebig 1844a, 721 = 1844b, 18. Liebig went on immediately to use the term himself, making a show of distinguishing his scientifically defensible use of it from others' illegitimate use but without critically addressing the hard issues.

264. Hall 1980, 27. The first function, the regulation of chemical transformations, would have been better illustrated by a lower-level change than "reproduction"—for example, the body's synthesis of complex compounds.

265. Liebig 1840d, 322 = 1841a, 328 ≈ 1840a, clxxxiii. Hall (1980, 44) maintained that one of the "laws" Liebig believed held for both living organisms and the inorganic world was that "there is a 'metamorphosis' or interdependence among all forces." I have not found that Liebig used the word *Metamorphose* with respect to forces before 1858.

266. Liebig 1840d, 36 = 1841a, 36 ≈ 1840a, lxxxii.

267. Liebig 1840d, 39 = 1841a, 38 ≈ 1840a, lxxxiv.

268. Liebig 1840d, 320–321 = 1841a, 326–327 ≈ 1840a, clxxxii; cf. 1840d, 344 = 1841a, 350, and the first of his collected *Chemical Letters*, in which he was concerned to vindicate both the importance of chemistry to the elucidation of physiological questions and the necessity of assuming the existence of a vital force alongside the forces of chemistry and physics (Liebig 1844a, 723 = 1844b, 24–25).

269. Liebig 1840a, xvi; cf. 1840d, 43–44. Such views were widespread: Mayer (1838b, 25$^{r}$ and 27$^{v}$) included in his answer to the third question asked him on his medical exam—"What are the principal means which promote and impede animal putrefaction?" ("Welches sind die hauptsächlichste[n], die thierische Fäulniß befördende[n], und welches die dieselbe vorzüglich hemmende[n] Mittel?")—the observation that "the vital and putrefactive processes are the most extreme opposites in the realm of organic beings. The vital process is the most powerful limitation on putrefaction." ("Der Lebens- und der Fäulnißprocess sind die extremsten Gegensätze im Reiche der organischen Wesen. Der Lebensprocess ist die gewaltigste Beschränkung der Fäulniß.")

270. Liebig 1840d, 323 = 1841a, 329 ≈ 1840a, clxxxiii–iv.

271. Liebig 1840d, 344 = 1841a, 350.

272. Liebig 1840d, 323–325 = 1841a, 329–331 ≈ 1840a, clxxxiv–v; cf. 1842a, 284 = 1842b, 101, and 1846, 204.

273. Liebig 1846, 178–182.

274. Liebig 1856a, 370.

275. Liebig 1841d, 190 = 1842b, 1, except that the latter has "animal egg" (*Thierei*) instead of "animal" (the dative singular, *Thiere*) and "state of static equilibrium" instead of "static moment." He repeated the association of straight and curved lines with inorganic and organic processes in a lecture many years later (Liebig 1856a, 370, quoted in chapter 3, section 3.3).

276. Liebig 1841d, 196 = 1842b, 8 (both quotes).

277. Liebig 1841d, 193–194 = 1842b, 5.

278. Liebig 1841b, 195 = 1842b, 7.

279. Liebig 1841d, 194 = 1842b, 6.

280. Liebig 1842b, 203–204. For the ability of the vital force to impart motion to resting particles, see also Liebig 1844b, 148; Liebig pictured the vital force as overcoming the resistance of other forces such as cohesion, heat, and electricity in the production of organic compounds (149) and interpreted fermentation, putrefaction, and decay as processes of decomposition brought about by the slightest external influences after the organic substance has been removed from the protection of the vital force (149–150). Essentially the same views were repeated in some of the same words in Liebig 1846, 210, in the section headed "Opposition [*Gegensatz*] between Putrefaction and the Vital Process."

281. With respect to content and timing, see Liebig's letter of 24 May 1842 to Wöhler (in Kahlbaum 1904, 72), written in his usual run-on style: "My book is finished it will consist of 21–22 sheets I expect the finished copies any day; in it I've given a theory of the phenomena of [animal] motion as well as of fever and of disease (in general) but without knowledge of mechanics all this is

completely lost to the physician, these people don't know what a force is, they have no conception of effect and cause!" The possibility that Liebig's concern for mechanical concepts might have been stimulated in part by Mayer's 1842 paper was broached by Kremer (1984, 344; cf. 204). Liebig solicited Jolly's reaction to the third section of his book, in particular as to whether it violated any principles of mechanics; Jolly declined to give a detailed criticism but assured Liebig that he found no offense against said principles (Jolly 1842c, [1]).

282. Liebig 1842b, 199–237, 238–260.

283. Liebig 1842b, 199.

284. Müller (1843, 342) interpreted Liebig as saying that accelerated *Stoffwechsel*—in particular, increased oxidation—is the *result* of increased physical activity.

285. Liebig 1842b, 199; cf. 200.

286. Liebig 1842b, 204.

287. Liebig 1842b, 207, 208; cf. 1844a, 1153 = 1844b, 136: "Light, heat, vital force, the electric and magnetic force, the force of gravity [*Schwerkraft*] manifest themselves as forces of motion and of resistance, and as such change the direction and strength of the chemical force; they are capable of increasing, diminishing, or annihilating it." Citing the first sentence of the passage quoted in the text, Lenoir (1982, 165) maintained that "Liebig was struggling to formulate a conception of the *Lebenskraft* in terms similar to the framework that Helmholtz would later provide with the conservation of energy." Such a judgment ascribes too much clarity and prescience to Liebig's soon-to-be-abandoned foray into physics, and Lenoir goes too far in interpreting Liebig's vital force as a "form of motion" wholly explicable in the same terms as other mechanical forces (164). Lenoir wrongly implied that Bischoff adopted Liebig's concept of *Bewegungsmoment* (167, citing T.L.W. Bischoff 1847, 431).

288. Liebig 1842b, 208–209.

289. Liebig 1842b, 220, 224. Liebig noted the proportionality between the chemical activity of acid and zinc in a voltaic pile and the amount of force developed (222). Although we cannot know the underlying nature of the processes involved, we can be sure "[that the] ultimate cause is the chemical force itself" (223).

290. Liebig 1842b, 209–210. Hall's statement that "Liebig usually expressed this latter idea [i.e., the conservation of forces within the organism] as conservation of the momentum of force, (*Kraftmoment*), meaning the work that a moving force could do" extracts too much after-the-fact clarity from Liebig's inconclusive and contradictory musings, and assigns false precision to Liebig's pseudo-technical terminology (Hall 1980, 40).

291. Liebig 1842b, 209, 214; 1846, 192–193; cf. 1844a, 722 = 1844b, 22: "Indeed, there are no causes in nature which produce motion or transformations, no forces which are more closely allied than the chemical force and the vital force." Liebig (1842b, 282) later referred to the vital force as "a property of certain substances."

292. Liebig 1842b, 212.

293. Liebig 1842b, 212–213. Kremer's assertion that "Liebig allowed the conservation principle to be violated by chemical [e.g., catalytic] forces" suggests that Liebig otherwise *had* a conservation principle that he let be violated, as opposed to his simply having had unclear and inconsistent ideas (Kremer 1984, 347).

294. Liebig 1842b, 213, 214.

295. Liebig 1842b, 216. Liebig used that otherwise general term in a special sense in this work; cf. 227: "*Stoffwechsel*—that is, excretion [*Austreten*] in the form of an inanimate compound—only takes place by means of the chemical action of the oxygen." In accordance with these physiological views, Liebig later stated that the mechanical effect produced is proportional to the nitrogen content of the urine (251).

296. Liebig 1842b, 225–226. For Mayer's citation of it, see Mayer 1845, 58–59 = 1893a, 133, and the discussion in chapter 6, section 2.

297. Liebig 1842b, 227 (emphasis supplied); cf. 228, 231–232.

298. Liebig, 1842b, 234.

299. Liebig 1842b, 216–217; cf. 253–254.

300. Liebig, 1842b, 236.

301. Liebig 1842b, 237. For Liebig's support for the mode-of-motion theory of heat, see chapter 3, section 2.1. In this case, as often with Liebig, he appears more as the opportunistic advocate of a fragmentary cause than as someone who holds a coherent and well thought-out set of beliefs.

302. Liebig 1842b, 242.

303. Liebig, 1842b, 239, 244.

304. Liebig 1842b, 248, 248–249; for Mayer's criticisms, see Mayer 1845, 68–69 = 1893a, 135–136, and the discussion in chapter 6, section 2.

305. Wunderlich 1845, 38.

306. Kremer 1984, 339–340.

307. Lotze 1843, xxvi and xxv, respectively. For the dating of this article, see Lotze's letters of 14 January and 7 May 1843 to Wagner (Woodward and Rainer 1975, 365–366).

308. Lotze 1842, 8.

309. Lotze 1843, xxi–xxii. Although I find it odd to call Lotze a "functional morphologist," he provides Lenoir (1982, 14, 168–172) with a good example of a teleomechanist.

310. Lotze 1843, xliv–xlv (reading "eine selbst nach mechanischen Principien [er]folgende zufällige Entstehung"); the first sentence repeated an identical passage from Lotze 1842, 7–8.

311. Lotze 1843, il.

312. Lotze 1842, 19–20; cf. 1843, xlvii.

313. Letter from Griesinger of 17 May 1843, in Mayer 1893b, 206–207 (reading "des Keims" with Preyer 1889, 63). In print, Griesinger (1843, 287) praised Lotze's attack on "vitalism" and his defense of a physicalist approach to physiology; Lotze's approach was, he said, that already characteristic of the "younger generations."

314. Lotze 1843, xvii–xix.

315. Lotze 1843, xix, xxii–xxiii: "Bildungstrieb," "Selbsterhaltungstrieb" (both on xxii). Note that although *Trieb* is commonly translated "force," it more properly means "drive" in the sense of an organism's impetus or active motive power.

316. Lotze 1843, xx.

317. Cf. Wunderlich 1859, 285: "He [Treviranus] assumes a particular vital matter which, in that it becomes the vehicle of that fundamental force, contains the vital force. He parallelizes this vital matter completely with electricity and magnetism."

318. Lotze 1843, xxxvii, citing Henle 1841, 218 (i.e., 217).

319. Lotze 1843, xxxvii.

320. Lotze 1843, xxxviii.

321. Lotze 1842, 13; cf. 1843, xl–xlii. Griesinger (1843, 288) quoted with approval the first sentence cited here and the one immediately preceding.

322. Lotze 1842, 57–58 (quote on 58).

323. Lotze 1842, 58–59 ("Occasionalismus" on 58); he referred readers to book II, chapter 3, of his *Metaphysics* (i.e., Lotze 1841).

324. Lotze 1842, 59–60.

325. Lotze 1842, 60.

326. Lotze 1842, 62, 61, and 61, respectively.

327. Lotze 1843, xlviii and xxxix, respectively (the former italicized in the original).

328. Lotze 1843, xl–xli.

329. Lotze 1842, 325.

330. Wunderlich 1859, 285.

331. Autenrieth 1801–2, *1* (1801), 61.

332. Autenrieth 1801–2, *1* (1801), 67–68. Contagions provided another example of an autonomous agency capable of indefinite self-multiplication (136).

333. Autenrieth 1801–2, *1* (1801), 68–69; the paragraphs quoted comprise §§119–120.

334. Autenrieth 1801–2, *1* (1801), 69–70; cf. 102, where he reinvoked "the capability for multiplication which belongs to the vital force as to every imponderable substance" and "the observed disproportionality of the effect to the cause (§120) with these substances."

335. Autenrieth 1801–2, *1* (1801), 70.

336. Autenrieth 1801–2, *1* (1801), 93; cf. 84–85.
337. Autenrieth 1801–2, *1* (1801), 127–128.
338. Matthes 1811, 3.
339. Matthes 1811, 7 and 5–6, respectively.
340. Matthes 1811, 8–9.
341. Matthes 1811, 4; he used the word *distributio* in the sense of the German *Vertheilung*.
342. Matthes 1811, 9–10.
343. Matthes 1811, 25.
344. Wunderlich 1859, 286.
345. Burdach 1837, 95 and 94, respectively.
346. Müller 1835, 54 = 1833, 53 = 1837b, 54–55 = 1841, 52.
347. Müller 1840, 513.
348. Müller 1835, 28 = 1833, 27 = 1837b, 28 = 1841, 25. That Müller's vital principle was analogous to but distinct from the other physical and chemical forces was noted by Benton (1974, 29). Basically ignoring magnetism as irrelevant to organic phenomena, Müller called electricity, heat, and light *imponderable Materien* and usually spoke of their *Wirkungen*; that is, he followed common usage and did not usually refer to them as *Kräfte* (Müller 1835, 64–90 = 1833, 63–89 = 1837b, 64–98 = 1841, 62–91).
349. Müller 1840, 553.
350. Elsässer, in Magendie 1834–36a, *1* (1834), 318–319.
351. Autenrieth 1836, 3–4; cf. 24–25.
352. Autenrieth 1836, 4. Autenrieth did not regularly refer to heat, light, electricity, and magnetism as forces, but rather as imponderables or weightless fluids (6–7), although in one section he referred to both the imponderables and motion as "forces" (261–264).
353. Autenrieth 1836, 4–5.
354. Autenrieth 1836, 6. He also referred to that "außersinnliche Welt" as "jenes unsichtbare Jenseits" (65).
355. Autenrieth 1836, 9–10.
356. Autenrieth 1836, 110–111; cf. 270–271 and esp. 480.
357. Autenrieth 1836 (1825), 338.
358. Autenrieth 1836, 444 = 1816, 368–369.
359. Autenrieth 1836, 135–136.
360. Autenrieth 1836, 19.
361. Autenrieth 1836, 33–34; cf. 483–484 and 1836 (1825), 334–335. Autenrieth rarely (if ever) mentioned the generation of heat through friction.
362. Autenrieth 1836, 282.
363. Autenrieth 1836, 36–37. Autenrieth wrote that the vital force was "something essentially different from every other imponderable force or fluid" (37); the term *unwägbare Kraft* was not at all common in the German of the day, whereas *unwägbare Flüssigkeit* (or some equivalent) was in quite general use.
364. Autenrieth 1836, 383–384.
365. Autenrieth 1836, 295; cf. 1836 (1825), 339.
366. Autenrieth 1836, 296; cf. 1836 (1825), 340–341.
367. Autenrieth 1836, 391; cf. 14 for some of the similarities and differences between the polarities of the vital force and the imponderables.
368. Autenrieth 1836, 296; cf. 372–373 and (by the Erlangen professor of chemistry and physics) Kastner 1832–33, 2 (1833), 9: "In and of itself *motive* (absolutely active) but not movable is only the *spiritual* [*das Geistige*]; in and of themselves *motive* and simultaneously *movable* are *living organisms*, as is the *whole natural world*, in that the latter continuously moves, and allows to move themselves, all beings belonging to itself according to eternal laws, and continuously renews the requisite motive forces out of an inexhaustible source; on the other hand, only *movable* and not by themselves motive are all inanimate beings, enclosed in space and time."
369. Autenrieth 1836, 377–378.

370. Autenrieth 1836, 108.

371. Autenrieth 1836, 104–105. The property of light alluded to at the beginning of the paragraph was the noninterference of light rays passing through the same space (cf. 490–492).

372. Autenrieth 1836, 480, 483.

373. Autenrieth 1836, 482, 483, 495, 518, 519, 520.

374. Cf. Autenrieth 1836, 369: "Ponderable matter forms the beginning of the series; as far as our knowledge of creation goes, the human subject ends the series. The imponderables, and after them the organic vital force, come between the former and the latter."

375. Autenrieth 1836, 117; cf. 264, 389–390.

376. Autenrieth 1836, 261; cf. 392–394.

377. Autenrieth 1836, 264–266.

378. Autenrieth 1836, 486–487.

379. Autenrieth 1836, 487; cf. 515. Autenrieth used the word *Indifferenz* instead of the more usual *Indifferenzpunkt* for the point of "magnetic indifference" at the center of the magnet.

380. Autenrieth 1836, 487.

381. Autenrieth 1836, 493.

382. Autenrieth 1836, 357; cf. 357–358 for another full description of this series.

383. Autenrieth 1836 (1825), 333.

384. Autenrieth 1836, 370; cf. 478–479.

385. Autenrieth 1836 (1825), 333; cf. 1836, 484.

386. Autenrieth 1836, 495 and 494–495, respectively.

387. Autenrieth 1836 (1825), 334.

388. Autenrieth 1836, 380–381.

389. Autenrieth 1836 (1825), 335.

390. Autenrieth 1836 (1825), 335 and 1836, 387, 479, respectively.

391. Autenrieth 1836, 92; 1836 (1825), 335–337.

392. Autenrieth 1836, 93.

393. Autenrieth 1836, 375–376; cf. 382: "One kind of imponderable fluid can to be sure also arouse another, . . . but the arousing kind continues to manifest its presence in its particularity alongside the newly aroused kind . . . . Nowhere is there manifest an intermediate form between two kinds of imponderables, nor therefore any transformation of one kind of them into another."

394. Autenrieth 1836, 263.

395. Lotze 1843, xx.

396. Griesinger 1845, 6.

397. Burdach 1837, 109.

398. Tiedemann 1830, 378; cf. 34, 42. Tiedemann cited Berzelius 1825–31, 3, pt. 1 (1827), 138 = 1821–32, 3, pt. 1 (1828), 148 = 1837–41, 6 (1837), 6; cf. 135, 145, and 3, respectively, and Berzelius 1825–31, 4, pt. 1 (1831), 88 = 1821–32, 4, pt. 1 (1832), 90 = 1837–41, 9 (1840), 117. Prout (1834, 414) quoted the same passage from Berzelius.

399. Autenrieth 1801–2, *1* (1801), 55 (quote), 57–58. For other uses of *Maschine*, see 49, 54, 70; for *Maschineneinrichtung*, see also 49 and 1816, 293 (twice) = 1836, 399–400.

400. Müller 1835, 34 = 1833, 34 = 1837b, 34 = 1841, 32.

401. Brougham 1835a, 26–30 = 1835b, 24–29 (the latter being the German translation of the English original).

402. Prout 1834, 85. A German translation appeared in 1836.

403. Prout 1834, 89; see 539–540 for seven other examples of Prout's use of the language of "machinery," "mechanical arrangements," and the like.

404. Liebig 1841d, 212–213 = 1841e, 2873 = 1842b, 29–30. Kremer (1984, 203) quoted the first part of the first sentence.

405. Valentin 1842, 63.

406. Valentin 1844, *1*, 10–11.

407. Valentin 1844, *1*, 11; cf. 1846b, 1–2.

408. Valentin 1844, *1*, 11; cf. 1847, 10.

409. Liebig 1842b, 269.

410. Liebig 1846, 183.

411. Dumas 1841a, 20 August, [2] = 1841b, 38 = 1842a, 10: cf. 1841a, 21 August, [2] = 1841b, 54 = 1842a, 35. I, like Kremer (1984, 18), could find no earlier use of this metaphor in the physiological literature than Dumas's in 1841.

412. Dumas 1841a, 21 August, [3], etc. (as quoted more extensively in chapter 3, section 1). For Hirn's likening of animals to motors, see Hirn 1864, 19, 37, 111 = 1868, 34, 54, 133; for him, as for Mayer, the analogy broke down over the fact that animals possess not only matter and force, but also an active power of spontaneous choice.

413. The title of Breger 1982 is *Die Natur als arbeitende Maschine*. Cf. Kremer 1984, 18: "It is no coincidence that during the decades of the 1830–40s railroads began to cross Europe, the principle of energy conservation was formulated, and physiologists began to write of the human as steam engine."

414. Breger 1982, 153.

415. Breger 1982, 142, 154.

416. Anonymous 1842b, 2587.

417. Schlüter 1985, 59–60, quoting from Ludwig 1870 as cited in Schröer 1967, 282; I quote directly from Ludwig 1870, 359.

418. The first German railroad began operation between Nuremberg and Fürth on 7 December 1835; the first stretch of the Munich–Augsburg line opened on 1 September 1839; the Mannheim–Heidelberg line (the first in Baden) opened on 12 September 1840; Württemberg got its first railroad in 1845 (Anonymous 1936, 504–506). Steamboats began operating out of Heilbronn on 7 December 1841; the first train reached there on 25 July 1848 (Titot 1865, 107–108). I have not been able to determine when Mayer might have had his first direct experience with steam engines or railroads; unlike his early and important experiences with water-powered machines, it seems that his appreciation of steam-powered engines was predominantly secondhand.

419. Autenrieth 1836, 296.

420. Passavant 1837, 13.

421. Lotze 1843, xxxvi.

422. Lotze 1843, xxxix (*Seele* on xl–xli).

423. Burdach 1837, 125.

424. Burdach 1837, 126.

425. Volkmann 1837, 74–75 (quote on 75).

426. Schwann 1839, 222.

427. Brougham 1835a, 33–35 = 1835b, 30–32.

428. Schleiden 1842–43, *1* (1842), 24–25. The second problem was to be addressed after the model of crystallization, as Schwann had already indicated.

429. Fries 1826, 12.

430. Lotze 1839, 1549–1550 (quote on 1550).

431. Lotze 1843, xxvii; cf. li, where he developed an analogy between the mechanically acting solar system and the nervous system, and 1842, 347–349.

432. Lehmann 1842, 44. For yet another example of the metaphor of solar system as organism, see the anonymous review—possibly by Lotze—of Lehmann's book (Anonymous 1842a, 421).

433. Valentin 1844, *1*, 194–195; 1846b, 3–4. For his similar appeal to God as first cause and creator, see Valentin 1844, *1*, 195, and the evidence cited in chapter 3, sections 2.1 and 3.2.

434. Kastner 1832–33, 2 (1833), 66.

435. Liebig 1856a, 370; cf. 1841d, 190 (quoted in chapter 3, section 2.2).

436. Volkmann 1837, 76; cf. 163–167 for a further discussion of materialism vs. spiritualism.

437. Volkmann 1837, 295.

438. Volkmann 1838, 6–7. Volkmann's booklet was cited by Valentin 1840, 30.

439. Volkmann 1838, 13; cf. 21.

440. Volkmann 1838, 21.

441. Wunderlich 1859, 279. He was of the opinion that homeopaths had contributed greatly to the loss of physicians' credibility and their lowered public esteem in Germany during the 1830s; part of the problem, he thought, was with the notion that the lay public is capable of judging the effectiveness of medical treatment (297), as the populist and anti-establishment homeopaths maintained. The medi-

cal reforms sought by Wunderlich et al. would thus seem to be in part an attempt to reestablish the professional status of medicine vis-à-vis its public.

442. As Eschenmayer (1853, 414) wrote to Justinus Kerner on 4 August 1838: "You're right, [animal] magnetism and homeopathic dynamism are very close to each other and recognize ultimately one principle." Eschenmayer (1834, 107) had earlier made an explicit connection between homeopathy and somnambulism.

443. Liebig 1844a, 721 = 1844b, 17.

444. Eschenmayer 1834, 103–104.

445. Eschenmayer 1834, 105 (in boldface italics). F. G. Gmelin (1835, 183) quoted and rejected this assertion.

446. Letter from Eschenmayer of 4 August 1838 to Kerner, in Eschenmayer 1853, 414; the parenthetic "(Tübingen)" is probably Kerner's interpolation.

447. Anonymnous 1834, *1*, 30; cf. 29–33 and Eschenmayer 1834, 100–101.

448. Anonymous 1834, *1*, 35.

449. As quoted in Anonymous 1834, *1*, 37.

450. Anonymous 1834, *1*, 38–39; cf. Eschenmayer 1834, 105.

451. See chapter 4, section 1.2.

452. F. G. Gmelin 1835, 204–205, 217–219; on 220, however, he *excluded* contagions and frog semen from this conclusion.

453. F. G. Gmelin 1835, 220. This passage notwithstanding, Gmelin criticized Eschenmayer for claiming that, through continued dilution, matter ceases to have weight (206–207). I do not find Eschenmayer actually to have made such a claim.

454. I doubt that teleological issues per se were of much significance to Helmholtz, as Lenoir (1982, 15–16, 197) has urged. Here again I think Lenoir has conflated the developmental/teleological and the physiological/energetic aspects of the vital force; Helmholtz was concerned with the latter. See also Caneva 1990, 297.

## CHAPTER FOUR
## PHYSICS AND CHEMISTRY

1. Muncke 1829, 27; cf. 1842, 4. The first three editions of Muncke 1842 appeared in 1825, 1829, and 1833, and hence were available to Mayer before 1840. Although I have not been able to consult them or the early editions of one or two other texts, in virtually every case where I *have* looked at the several editions of a given work I have found little or no difference with respect to the issues of concern.

2. Muncke 1836b, 1394; cf. 1833b, 494.

3. E. G. Fischer 1827, 32–33 = Fischer and August 1837, 33 (reading *trennbare* with the fourth edition in place of *trennbar* with the third).

4. Eisenlohr 1841, 1, 2, 3; I was not able to consult the first two editions of this work, published in 1836 and 1839.

5. In addition to the references otherwise cited, see also Anonymous 1839b, 210–211; Fries 1822, 450–451; Lehmann 1842, 45; and Muncke 1825, 921, 1830b, 957.

6. Letter of 6 August 1842 to Baur, in Mayer 1893b, 142. Mayer enclosed the words "dunkle Ursache der Bewegung" within quotation marks; I have not been able to identify their possible source. The following three citations are from this same page of the letter.

7. Baumgartner 1836, 7 (as cited by Mayer) ≈ 1832b, 6–7; cf. 1842, 6, 1836, 63 = 1832, 62, and 1837, 4.

8. Lamé 1836–37, *1* (1836), 18 (as cited by Mayer) = 1840, *1*, 18.

9. Mayer 1893b, 142, citing Biot 1828–29, *1* (1828), 28.

10. Whewell 1836, 1; cf. Mayer 1843a, 2, citing Whewell 1841, 1. It was probably Baur who called Mayer's attention to the 1841 German edition of Whewell. A remark in Mayer's letter to Baur of 10/11 July 1843 suggests that Mayer 1843a was written before 18 May 1843 (Mayer 1893b, 147–148). Since Baur was in Heilbronn until the spring of 1843, it probably dates from April or May.

11. Whewell 1836, 138; cf. Mayer 1843a, 1.

12. Herschel 1833, 10.

13. Herschel 1833, 232 = 1838, 268. Both Whewell and Herschel connected the concept of force with the human will and our internal sense of exertion, though that was not a common feature of German works.

14. Geiger 1833, 13 = 1827, 14 ≈ Liebig 1837–43, 7–8.

15. Geiger 1833, 16 = 1827, 19–20 = Liebig 1837–43, 17.

16. E. G. Fischer 1827, 30–31 = Fischer and August 1837, 31–32.

17. Biot 1828–29, *1* (1828), 28.

18. Biot 1828–29, *1* (1828), 22–24.

19. Kämtz 1839, 4.

20. Fries 1822, 451–452.

21. Muncke 1829, 36–37; 1830b, 959–960; 1842, 5–6.

22. Muncke 1830b, 958.

23. Muncke 1830b, 961.

24. Tiedemann 1830, 23–25.

25. In his famous attack on the vital force, Emil du Bois-Reymond (1848, xliii) rejected the concept "of a force as an autonomous thing that maintains an independent existence vis-à-vis matter." He cited Helmholtz's "Erhaltung der Kraft" with approval.

26. L. Gmelin 1817, 1 = 1827, 1–2. Poppe (1837, 35) considered forces of cohesion, adhesion, attraction, and elasticity as the "forces that act in and on the bodies themselves and produce all kinds of phenomena."

27. Baumgartner 1836, 7 = 1832b, 7; Baumgartner and Ettingshausen (1842, 7) adds "of a vital force and the like" after "force of adhesion" and "of the life of organic beings and the like" after "adhesion" in the last sentence.

28. Baumgartner 1836, 10 ≈ 1832b, 10; cf. Eisenlohr 1841, 5.

29. Fries 1826, 70–71, 218.

30. Fries 1826, 37.

31. Geiger 1833, 73–74 = 1827, 83 = Liebig 1837–43, 84–85.

32. My conclusion applies to textbooks of *physics*, and not necessarily also to those of *mechanics*, which I did not systematically review. Physics texts seldom mentioned the conservation of vis viva.

33. Lamé 1836–37, *1* (1836), 22–23 = 1840, *1*, 22–23. Mayer (1893b, 143) quoted the last group of italicized words in his letter of 6 August 1842 to Baur. My translation of *vitesse* (and *Geschwindigkeit*) in this and other passages as "velocity" should not be taken as implying that the authors distinguished that concept from "speed."

34. Lamé 1836–37, *1* (1836), 23 = 1840, *1*, 23, quoted without a specific citation in Mayer's letter of 6 August 1842 to Baur (Mayer 1893b, 142). A similar definition is found in Pouillet 1832, 48.

35. Biot 1828–29, *1*, 87; Whewell 1836, 141. So did Poppe 1811–21, *1* (1811), 209 (motive force); Fries 1822, 501, and 1826, 81 (quantity of motion); Eisenlohr 1841, 50–52 (force); and Baumgartner and Ettingshausen 1842, 20 (quantity of motion). The earlier editions of Baumgartner's text presumably used the same measure. Whewell, however, wrote two pages later that "*with uniform accelerating forces, the velocity generated in any time is equal to the product of the force and the time*"; by "force" he here meant what we term "$g$," writing $v = tf$.

36. E. G. Fischer 1827, 37 = Fischer and August, 1837, 38. He noted that the theory of motion has nothing to say about "the internal essence of the motive force" (39 in both editions).

37. E. G. Fischer 1827, 34–35 ≈ Fischer and August 1837, 35–36.

38. E. G. Fischer 1827, 43 = Fischer and August 1837, 42.

39. Muncke 1830b, 963–965. In another article he defined *quantitas motus* as $mv$ (Muncke 1825, 930).

40. Anonymous 1839b, 212.

41. Schubert 1842, as quoted in Jolly 1842b, 853.

42. Jolly 1842b, 854.

43. Berzelius 1821–32, *3*, pt. 1 (1828), 147–148 = 1825–31, *3*, pt. 1 (1827), 137-138 = 1837–41, *6* (1837), 5–6, citing first from the edition in Mayer's library.

44. H. Davy 1830, 259–260 = 1840, 369–370 ≈ 1833, 280–282.

45. Mayer 1841a, 101 = 1978, 23.

46. My use here of the imprecise term "motion" reflects contemporary usage, which did not regularly employ a technical term comparable to "momentum."

47. Geiger 1833, 18 = 1827, 20 = Liebig 1837–43, 17–18, except that the 1827 edition lacked the word *Tangentialkraft* in parentheses.

48. Poppe 1811–21, *1* (1811), 153–154 and pl. IV, fig. 2.

49. Poppe 1815, 60–61 and pl. I, fig. 24; 1834, 89 and pl. III, fig. 2 = 1837, 81–82 and pl. III, [fig. 2].

50. Baumgartner 1836, 199–200 = 1832b, 196–197 ≈ 1842, 237–238 and the accompanying pl. III, fig. 108 (1836 and 1832b) = fig. 125 (1842); E. G. Fischer 1827, 101–105 and pl. I, fig. 17 ≈ Fischer and August 1837, 141–143 and pl. III, fig. 73; Siber 1837, 31 and pl. I, fig. 15. Some authors who gave similar accounts and diagrams opposed centripetal and tangential forces without calling the latter a centrifugal force: Kämtz 1839, 110 and pl. I, fig. 58; Eisenlohr 1841, 77–80 and pl. II, fig. 53. Cf. Brandes 1826, 62; Fries 1833, 193–194 and pl. III, fig. 39; Kastner 1832–33, 2 (1833), 61–64 and pl. II, fig. 25; Brougham 1835a, 33-34 ≈ 1835b, 30–31.

51. Lamé's analysis of central-force motion was very different: he clearly saw it as the result of a deflecting force on an original *motion* (velocity), and drew a much simpler diagram that did not employ a parallelogram of forces; see Lamé 1836–37, *1* (1836), 27–29 and pl. I, fig. 5 = 1840, *1*, 27–29 and pl. I, fig. 5. He did, however, speak of the centrifugal force as constantly destroyed by the opposing centripetal force. Biot said that the simultaneous action of at least two forces (*sic*) is necessary to produce curved motion; thus his account of central-force motion applied the word force to the sideways impulse imparted to the body, but in his diagram the sides of the parallelogram were clearly and explicitly *distances* the body would traverse in a given instant; see Biot 1828–29, *1* (1828), 90–91 and pl. II, fig. 50 ≈ Biot 1824, *1*, 86–87 and pl. II, fig. 50. Neither Pouillet 1832 nor Whewell 1836 treated circular motion.

52. Muncke 1829, 57–59 and pl. I, fig. 3.

53. Fries 1826, 117 and pl. I, fig. 21.

54. Fries 1826, pl. I, fig. 22.

55. Fries 1826, 119.

56. Fries 1826, 124.

57. Fries 1826, 125. This passage and the first paragraph of the next were quoted (with the inclusion of two sentences omitted here) in Muncke 1827, 88.

58. Fries 1826, 125–126; cf. 1822, 582–586.

59. Fries 1826, 83.

60. Fries 1826, 94–96.

61. Fries 1826, 95–96; cf. 81.

62. Muncke 1827, 78–93, esp. 87–88.

63. Muncke 1833a, 410.

64. Muncke 1833a, 410.

65. Muncke 1833a, 412.

66. Muncke 1833a, 413.

67. Mittasch 1940a, 38.

68. Mayer 1876, 104 = 1893a, 440. Mittasch's method of citation—"M.I.440"—did not specify either the specific source or its date, and obscured this fact.

69. Letter of 17 July 1842 to Baur, in Mayer 1893b, 134.

70. Mitscherlich 1833–34, *1*, pt. 1 (1833), 105 = 1834, 281; cf. 1833–34, *1*, pt. 2 (1834), 415–416, for further use of the term *Contactsubstanz*.

71. Berzelius 1836a, 242–243 ≈ 1836b, 94–95; cf. 1837–41, *6* (1837), 19–25. I have translated "in anderen Verhältnissen" in the first sentence as if it were in the accusative. Mittasch (1935, 174–178, 191) argued convincingly that Berzelius based his concept of catalytic force in part on Mitscherlich's idea of contact reactions, and he suggested that for Berzelius it functioned, in one of the roles usually assigned to the vital force—i.e., the production of complex substances—as a kind of conceptual link between certain inorganic and many organic reactions.

72. Liebig 1837–43, 84.

73. Letters from Wöhler of 30 May 1837 to Liebig and from Liebig of 2 June 1837 to Wöhler, in Hofmann 1888, *1*, 102, 104 ≈ Carrière 1893, 132, 133. (The two editors gave remarkably different readings of the same letter, though the sense is the same.) Wöhler was likely the author of the implicitly favorable reference to Berzelius's catalytic force in their joint paper on the oil of bitter almonds (Wöhler and Liebig 1837, 22).

74. Letter from Liebig of 3 June 1839 to Wöhler, in Hofmann 1888, *1*, 150; cited in Volhard 1909, 2, 74.

75. Liebig 1841c, 2070. Liebig mailed the first five "Letters" to Georg von Cotta, the editor, on 9 August 1841 (letter in Kleinert 1979, 4).

76. T.L.W. Bischoff 1847, 422, 424, 426–427, 432, 435, 436. I owe reference to Bischoff's paper to Lenoir 1982, 167; the words he there quoted showing Bischoff's approval of Liebig's use of catalytic forces to provide "the internal level for the transfer of motion" within the organism would appear to be a combined misprint and mistranslation of the original's "dem gemeinschaftlichen inneren Hebel in der Mittheilung einer inneren Bewegung" (427). Note that Bischoff, like Liebig, also invoked a vital force to explain the original creation of fibers and cells and the restoration of their exhausted state (432–433, 434, 435).

77. Liebig 1839a, 250.

78. Liebig 1839a, 253. Liebig used the word *Atom* where we would say molecule. He would later say that the elements of organic molecules capable of undergoing metamorphosis maintain their state solely as a result of their inertia (Liebig 1839b, 366–367).

79. Liebig 1839a, 254.

80. Liebig 1839a, 262.

81. Liebig 1839a, 262.

82. Liebig 1839a, 279.

83. Liebig 1839b, 364.

84. Liebig 1839b, 365–366. His examples now are the sudden crystallization of a supercooled solution and the throwing of a precipitate by a solution when given a small shake.

85. Liebig 1839b, 366.

86. Letter from Liebig of 3/4 June 1839 to Wöhler, in Hofmann 1888, *1*, 148–151. This letter would thus seem to come between Liebig 1839a and 1839b.

87. Letter from Liebig of 3 June 1839 to Wöhler, in Hofmann 1888, *1*, 148.

88. Letter from Liebig of 3 June 1839 to Wöhler, in Hofmann 1888, *1*, 149.

89. Letter from Liebig of 4 June 1839 to Wöhler, in Hofmann 1888, *1*, 150: "Was heißt denn das?" The interchange was discussed in Volhard 1909, 2, 74. Liebig quoted from Berzelius 1837–41, 6 (1837), 24.

90. Letter from Liebig of 4 June 1839 to Wöhler, in Hofmann 1888, *1*, 150. In a later work, Liebig (1840d, 220–224, 233, 252) was more inclined to assimilate complex processes of fermentation to relatively much simpler processes of inorganic decomposition.

91. Letter from Liebig of 4 June 1839 to Wöhler, in Hofmann 1888, *1*, 151.

92. In Kremer's account, Liebig assigned the cause of fermentation to "transmitted vibrations"—an idea he suggests Liebig derived from his correspondence with Mohr in 1837 on the nature of heat—which then "led him to think about the conservation of 'forces' in chemical and other phenomena" (Kremer 1984, 198–200; quotes on 199). But Liebig never spoke in this context in terms of vibrations, and I find no evidence that these reflections moved him closer to an appreciation of the conservation of motion, force, or anything; for the rest, his letter to Wöhler seems to me to isolate the progress of his thinking without invoking a problematic reference to his interchange with Mohr two years before.

93. Liebig 1840d, 206; relevant here is the entire section headed "The Chemical Process of Fermentation, Putrefaction, and Decay" (197–345). I was unable to verify the corresponding pages in Liebig 1841a, the copy I used having become inaccessible to me.

94. Liebig 1840d, 211; cf. 234 and 1842b, 212–213 (quoted in part in chapter 3, section 2.2).

95. Liebig 1840d, 247–248 = 1841a, 253–254. He had already broached this subject in Liebig 1839a and 1839b. Liebig gave considerable attention to general considerations of causality (e.g., in 1840d, 167 = 1841a, 167).

96. Liebig 1840d, 249 = 1841a, 255. I have profited from Kremer's discussion of this aspect of

Liebig's work (Kremer 1984, 201). In a later section of the book Liebig sought to explain the action of poisons, contagions, and miasmas in a like fashion in terms of molecular motions (Liebig 1840d, 311):

> This cause can be expressed most simply in terms of the following principle laid down long ago by Laplace and Berthollet, but which has only recently been proven for chemical phenomena: "A molecule (*molécule*) *set in motion by any force can communicate its own motion to another molecule in contact with it*."
>
> This is a law of dynamics, provable for all cases where the resistance (the *force, affinity*, or *cohesion*) which opposes the motion does not suffice to neutralize it.

Contrary to Hall's contention, this passage, whose sense Liebig (1844b, 186) restated essentially unchanged four years later, does not imply the conservation of motion because there is no suggestion that the molecule—Liebig said *Atom*—that produces motion in another substance thereby loses any of its own motion (Hall 1980, 26–27); see, however, Liebig 1840d, 344–345 (quoted in chapter 3, section 2.2). Six years later, Liebig (1846, 200–201) was still interpreting fermentation and putrefaction as a propagation of motion in a way not easy to harmonize with an intuitive notion of conservation of energy.

97. Liebig 1840d, 234; cf. 252.

98. Anonymous 1842a, 419–420. The reviewer did not explicity apply his remarks on the catalytic force to his anti-*Lebenskraft* argument a page or two earlier, but they seem to be composed in the same spirit.

99. Anonymous 1842a, 420. I neglected to note Lehmann's treatment of catalytic force when I had access to the book.

100. This subject is in need of a major historical study. For an in-depth introduction, see Ostwald 1896, 426–492. I have dealt with it briefly in Caneva 1973, 36; 1974, 89–91; 1978, 84–85.

101. See the passage from §113 of Roget's treatise on *Galvanism* quoted in Faraday 1840, 126; Roget's 32-page work was published in 1829 and reprinted in his *Treatises on Electricity, Galvanism, Magnetism, and Electro-Magnetism* (1832) as well as in the second volume of the collected treatises (by various authors) on "Natural Philosophy" in the *Library of Useful Knowledge* published by the Society for the Diffusion of Useful Knowledge. Roget was a physician and longtime secretary of the Royal Society.

102. Mohr 1837b, 117.

103. Mohr 1837d, 157. Johann Müller (1836, 22–25) also regarded Volta's contact theory as by then wholly refuted. In his preface to Müller's book Liebig also noted the insufficiency of Volta's theory (iii); but cf. Liebig 1840e, where he seconded Fechner's experimental defense of the contact theory. Once again it proves difficult to determine what Liebig 'really' believed.

104. Valentin 1837, 3.

105. Faraday 1840, 126, 126–127 ≈ 1841, 568–569, 570. His specific mention of the two electric fish was probably intended to address and exclude the possibility that living organisms might be capable of producing force.

106. See also the important work published between 1834 and 1839 by Moritz Hermann Jacobi and Heinrich Friedrich Lenz on the efficiency of electrochemically powered motors, discussed in Caneva 1974, 315–323.

107. Liebig 1841c, 2177 = 1844b, 115–116, with the "[not]" supplied from the latter.

108. Liebig 1841c, 2177 = 1844b, 117.

109. Liebig's favorable reporting of Fechner's alleged *experimentum crucis* in support of the contact theory was innocent of energetic considerations and reveals his weak appreciation of their *principled* significance (Liebig 1840e). It is striking the consistency with which Liebig coupled considerations of mechanical or electrochemical efficiency with blatently *economic* considerations. In his *Chemical Letters* of 1844 he wrote (Liebig 1844b, 270): "Civilization is the economy of force; science teaches us the simplest means of achieving the greatest effect with the least expenditure of force, and with given means to produce a maximum of force. Each unnecessary manifestation of force, each waste of force in agriculture, industry, and science, as well as in the state, characterizes the rudeness or lack of civilization [of a society]." In his first public discussion of the "transformation of forces" he argued that

we cannot destroy or create force out of nothing, otherwise we could build a perpetuum mobile and with it, "without spending any money," earn an unlimited amount (Liebig 1858b, 585 = 1859, *1*, 207).

110. Liebig 1841c, 2177 = 1844b, 118.

111. Liebig 1844a, 722 ≈ 1844b, 18–19. The passage quoted is very similar to one Liebig had published four years earlier; cf. Liebig 1840b, 119 = 1840c, 28–29. In that earlier version he called electricty both an "activity" and a "state of matter"; it is not clear whether his use of the word "force" was there intended to apply to electricity, to the vital force of the next paragraph, or to both.

112. The increasing prominence of discussions of heat as mode-of-motion in the late 1830s, prompted in particular by Macedonio Melloni's work on radiant heat after 1831, will be discussed in chapter 4, section 2.

113. Pfaff 1845, 106–115 = Mayer 1893b, 232–239.

114. As quoted in Pfaff 1845, 104 from L. Gmelin 1843, 455, where the passage forms an independent paragraph followed by the word "Faraday." The previous edition of Gmelin's *Handbook of Chemistry* had not discussed theories of the pile (L. Gmelin 1827, 166–176).

115. Pfaff 1845, 105–106 = Mayer 1893b, 231–232; "If we nevertheless . . . accelerate existing motions" quoted *after* "If it is completely true . . . rekindles its activity" in Mayer 1851, 32–33 = 1893a, 257–258. Pfaff's thinking recalls that of Lotze 1843, xxxvi (quoted in chapter 3, section 3.3).

116. Mayer 1851, 35 = 1893a, 260.

117. Autenrieth (1836, 264) once called heat, light, and electricity *selbstständige Kräfte* in a passage in which his purpose was to urge the autonomous existence of an organic vital force. Using a word somewhat more common than the (in this context) seldom-encountered *Kraft*, Liebig (1841b, 193–194 = 1842b, 5) once referred to heat, light, electricity, and magnetism as *immaterielle Potenzen* in a context whose purpose was likewise to establish the existence of a vital force.

118. L. Gmelin 1817, 57 = 1827, 67 = 1843, 156; C. G. Gmelin 1835, 9–10. L. Gmelin (1827, 9–10 = 1843, 7–8) divided substances (*Stoffe*) into four classes: *unwägbare Materien*, *wägbare elastische Flüssigkeiten*, *tropfbare Flüssigkeiten*, and *feste Körper*. Cf. Rudolphi 1821–28, *1* (1821), 167: "So hat man unsern [körperlichen] Bestandtheilen eine eigene Klasse von *Imponderabilien* oder unwägbaren Stoffen beigesellt, namentlich den *Wärmestoff* (Thermogenium, Caloricum), den *Lichtstoff* (Photogenium), die *electrische Materie* (Electrogenium)." For a rich discussion of the place of the imponderables in the chemistry and physics of the period, see Stichweh 1984, 94–172.

119. Poppe 1811–21, *2* (1813), 169; *3* (1814), 76; *4* (1815), 103 and 57, respectively.

120. Muncke 1829, 369–371.

121. Muncke 1842, 48.

122. Muncke 1830a, 765–769. In one of his texts he wrote that "the absolute weightlessness of the so-called imponderables [is] as yet in no way proven" (Muncke 1829, 36).

123. Fischer and August 1837, 8.

124. Fries 1826, 75. He mentioned in this context only light, heat, and electricity. On 370 he repeated his view that light is a kind of "higher form of aggregation" of matter; he did not refer to it as a force. Cf. Fries 1822, 556.

125. Berzelius 1821–32, *1*, pt. 1 (1821), 6–7 ≈ 1825–31, *1*, pt. 1 (1825), 7–8 ≈ 1835–36, *1* (1835), 11.

126. Biot 1828–29, *4* (1829), 2 ≈ 1824, *1*, 2, to cite first from the edition known to Mayer; the original French had *principe* and *calorique* for Fechner's *Grundursache* and *Wärme*, and the words translated here as "nature" (first occurrence) and "qualities" were *Beschaffenheit* and *Beschaffenheiten* in German, *nature* and *qualités* in French.

127. Lamé 1836–37, *1* (1836), 291, 297–299 = 1840, *1*, 293, 299–301.

128. Eisenlohr 1841, 221–222, 361, 457, 495.

129. Siber 1837, 83ff.

130. Siber 1837, 83–84.

131. E. G. Fischer 1827, 156; cf. Fischer and August 1837, 413.

132. Fischer and August 1837, 550–551.

133. Lamé 1836–37, *1* (1836), 491–493 = 1840, *1*, 481–483.

134. Baumgartner 1836, 500–502 = 1832b, 486–488.

135. Baumgartner 1836, 502 = 1832b, 488. Prout, who regarded heat and light as "imponderable

matter" (or "imponderable forms of matter"), likewise noted the difficulty of explaining the "extrication" of heat by friction: "but whence the heat is derived does not appear to be capable of satisfactory explanation, unless we suppose a perpetual decomposition and recomposition to take place, which is not improbable" (Prout 1834, 80–81; cf. 130, 151, 554).

136. Baumgartner (1837, 132) insisted on this last reason in the popular abridgement of his text: "that heat, like light, most probably consists of vibrations of the aether, perhaps also of the smallest particles of bodies themselves." He argued at length that the rules of analogy require one to conclude that the nature of heat, like that of sound and light, consists of a vibratory motion (Baumgartner 1836, 502–503 = 1832b, 488–489), "but whether the vibrations of the aether or those of the particles of bodies or both together constitute the cause [*Grund*] of the phenomena of heat is something even the defenders of this viewpoint are not in agreement on" (1836, 503).

137. Baumgartner 1836, 549–550; cf. 1832b, 531, where Ampère's theory was not mentioned.

138. Baumgartner 1836, 507 = 1832b, 491. The views expressed on this point in the 1836 edition of his text would appear to be an uncorrected holdover from earlier editions, since in other writings he had already clearly expressed his attachment to a vibratory-motions explanation of magnetism (Baumgartner 1832a, 87; 1835, 73).

139. Baumgartner and Ettingshausen 1842, 339, 506.

140. Baumgartner 1836, 486 = 1832b, 473. Morosi's data (from a paper read in 1816) supported Baumgartner's reading, though he himself did not draw that explicit conclusion from them; see Morosi 1824, 147.

141. Baumgartner 1836, 438 = 1832b, 427; cf. Baumgartner and Ettingshausen 1842, 661.

142. Geiger 1833, 73–110; Geiger 1827 and Liebig 1837–43 use the same phrase.

143. Geiger 1833, 74 = 1827, 83–84 = Liebig 1837–43, 85.

144. Geiger 1833, 91 = 1827, 106 = Liebig 1837–43, 103 (both quotes).

145. Letter of 16 August 1841 to Baur, in Mayer 1893b, 124. In discussing heat, Geiger (1833, 86 = 1827, 99) asserted that "nothing is explained by the term *property*." Mayer, too, denied that heat is a property of matter.

146. Kämtz 1839, 460; cf. 458–459.

147. Kämtz 1839, 461–462.

148. Mohr 1837b, 116-117. I quote the passage cited from Baumgartner 1837 in an earlier note in this section. Citing Mohr's work, Stephen Brush (1970, 159–161) noted the role played by the wave theory of heat in the discovery of energy conservation. During most of the 1830s Melloni lived and worked in Paris without any institutional affiliation.

149. Mohr 1837c, 141.

150. Mohr 1837a, 420.

151. Mohr 1837c, 142–143. Morosi (1824, 137) grouped together heat, magnetism, light, electricity, sound, *and motion* as entities (he called heat an *essere*) whose nature is unknown and which can be known only through their effects—and which, he speculated, might show a common origin.

152. Mohr 1837a, 421; cf. 422: "The expansion of rigid, fluid, and gaseous bodies by heat; these are phenomena of force of the most stupendous magnitude, and [they are] occasioned by heat, *but that which produces a force must itself be a force.*"

153. Mohr 1837a, 423; 1837c, 143.

154. Mohr 1837a, 424.

155. Mohr 1837a, 441–442.

156. Letter from Liebig of 25 January 1837 to Mohr, in Kahlbaum 1904, 22; cf. Liebig's letter of 31 January 1837, where he ran through these considerations again (23–24). On the inconceivability of immaterial entities, see also Liebig 1841e, 195 = 1842b, 7 (quoted in chapter 5, section 1).

157. Muncke 1841, 74, 83.

158. Muncke 1841, 90.

159. Mohr 1837e.

160. Fries 1826, 438; cf. 445. Fries treated light and heat as *Stoffe* of different degrees of aggregation (446).

161. Fries 1826, 445.

162. Letter of 24 July 1841 to Baur, in Mayer 1893b, 111.

163. Kuhn 1959, 335.

164. Letters of 5/6 December 1842 to Griesinger and of 31 July 1844 to Baur, in Mayer 1893b, 189 and 154–155, respectively. Note also Mayer's use of the French term *quantité de mouvement* in his letter of 12 September 1841 to Baur (Mayer 1893b, 129).

165. Lamé 1836–37, *1* (1836), 419 = 1840, *1*, 401.

166. Lamé 1836–37, *1* (1836), 428 = 1840, *1*, 409–410, except that the latter version omitted the word "completely" from the next-to-the-last sentence.

167. Lamé 1836–37, *1* (1836), 496, 498–500 = 1840, *1*, 486, 488–490. For the original experiment see Gay-Lussac 1807, 184–203. Gay-Lussac and Welter (1822, 436) later reported that the air escaping under pressure from a container did not show a change in temperature despite the fact that it underwent expansion; they explained this result by supposing that the air on streaming out was heated by an amount exactly equal to the cooling produced by the expansion.

168. Lamé 1836–37, *1* (1836), 503 (as cited by Mayer) = 1840, *1*, 493–494. Mayer (1893b, 189) abbreviated the beginning of his citation as "M. Dulong . . . a trouvé . . . cette loi [*etc.*]" in a letter of 5/6 December 1842 to Griesinger; he again quoted Lamé in a letter of 31 July 1844 to Baur (Mayer 1893b, 154–155). For the original report (read in 1828), see Dulong 1831.

169. Lamé 1836–37, 2, pt. 1 (1836), 88 = 1840, 2, 88–89.

170. Baumgartner 1836, 485 = 1832b, 472.

171. Kämtz 1839, 453–454.

172. Although there is no evidence Mayer read Prout's Bridgewater Treatise on *Chemistry Meteorology and the Function of Digestion*—published in a German edition in 1836—its discussion of the thermal properties of gases called explicit attention to the fact that "under the same temperature and pressure they [gaseous bodies] all undergo equal expansion by an equal increase of heat" and that "under the same pressure the same volume of all gases have the same capacity for heat" (Prout 1834, 62, 64).

173. Riehl 1900, 165. Mayer (1845, 5 = 1893a, 48) wrote that "die *quantitative* Unveränderlichkeit des Gegebenen ist ein oberstes Naturgesetz, das sich auf gleiche Wiese über Kraft und Materie erstreckt."

174. It is significant that Lavoisier's oft-quoted statement of the conservation of matter, though principled and explicit, was tucked away in a section of his *Traité* dealing with vinous fermentation. Note that Lavoisier had also included light and caloric in his "Table of Simple Substances."

175. Such works include Baumgartner 1832b, 1836, 1837; Berzelius 1821–32; Biot 1828–29; Eisenlohr 1841; Geiger 1833; L. Gmelin 1817, 1827, 1843; Kämtz 1839; Kastner 1832; Mitscherlich 1833–34; Muncke 1842; Pfaff 1824–25; Pouillet 1832, 1839; Siber 1837; Wöhler 1840a, 1840b. With the qualifications given in the text, the following works also warrant inclusion in this list: C. G. Gmelin 1835; Fischer and August 1837; Lamé 1836–37, 1840; Muncke 1829. The exclusion from this list of a work cited elsewhere does not imply that it does not necessarily belong, but rather (more likely) that I neglected to check it or make the proper notation when I had the chance to consult it. It is, of course, hard to be sure that a work absolutely does *not* say something, but I am confident my findings cannot be far off. I have not failed to mention any positive counterinstances known to me.

176. Brandes 1836.

177. Muncke 1836b, 1420, 1464.

178. Muncke 1829, 358.

179. C. G. Gmelin 1835, 14.

180. Fischer and August 1837, 16–17; cf. Autenrieth 1836, 375, for an even more general and incidental statement of the *implicit* conservation of matter.

181. Muncke 1833b, 497–498; cf. Burdach 1826–40, 2 (1828), 700 = 1835–38, 2 (1837), 785: "We experience everywhere that neither new matter arises out of nothing nor can already existing matter be annihilated [*vernichtet*] and driven out of existence."

182. Fries 1822, 501 (both quotes).

183. Fries 1826, 22.

184. Fries 1826, 22. *Wesen* has a multitude of meanings and is a notoriously difficult word to translate; it can mean essence, being(s), existence, nature, substance, etc.

185. Fries 1826, 65; cf. 1822, 502. Fries (1826, 80) stated as the first of his six laws of motion that

"mass and force are invariable in matter; all change is change of motions," but he did not pursue this passing reference to the invariability of mass.

186. Kant 1786, 541.

187. Kant 1786, 541.

188. C.C.E. Schmid 1798–1801, 2 (1799), 51; Eschenmayer 1796, 20.

189. Kant 1786, 496–535.

190. Cf. Fries 1822, 508: "Schelling made the mistake of leaving out of the Kantian construction the material substance, the mass as a fundamental concept, and of wishing to complete the construction with only opposing forces. But this attracting and repelling without anything that is attracted or repelled affords no determinate concept and is a mathematically useless idea." Friedrich Hufeland (1834, 336), professor of medicine in Berlin and a well-known author on mesmerism, noted that Schelling follower C.H.E. Bischoff presupposed (in his book on chemical medicaments) "that force and matter [*Materie*] are inextricably combined with each other, that matter [*Stoff*] is only the material representation of force."

191. C.C.E. Schmid 1798–1801, 2 (1799), 26 and passim.

192. Autenrieth 1801–2, *1* (1801), [vii]; he quoted from C.C.E. Schmid 1798–1801, *1* (1798), 16.

193. Autenrieth 1801–2, 2 (1802), 189, 189–190.

194. Fries 1826, 65–67.

195. Mayer 1845, 6 = 1893a, 48; cf. 1845, 24, 34, 41, 82, 100 = 1893a, 65, 72, 77, 103, 118.

196. Mayer 1845, 88 = 1893a, 108.

197. Mayer 1845, 79 = 1893a, 101.

198. Berzelius 1837–41, 9 (1840), 6 and passim; Liebig 1840a, lxxxiv; 1840d, 39, 48ff, 201 = 1841a, 38–39, 48ff, 207; Wöhler 1840b, 118; Lehmann 1842, ix, 6.

199. Pietsch 1942, 282. Thus for L. Gmelin (1843, 92; cf. 1817, 26) the neutralization of acids and bases produces "chemical indifference."

200. Mayer 1841b, 10; Baumgartner 1836, 43, 50 ≈ 1832, 50, 57; 1837, 17; Liebig 1840d, 25, 58, 339, 341 = 1841a, 25, 58, 345, 347. The word was of entirely general usage.

201. Baumgartner 1837, 17; Liebig 1839a, 253; Berzelius 1821–32, 3, pt. 1 (1828), 76 = 1825–31, 3, pt. 1 (1827), 70 = 1835–36, 5 (1836), 65.

202. Geiger 1833, 15–16 = 1827, 17.

203. Prout 1834, 155–156. A German edition was published in Stuttgart in 1836.

204. Prout 1834, 169.

205. Prout 1834, 161.

206. Cf. Mayer's talk of the "magnetizing [*Magnetischwerden*] of the previously indifferent steel rod" (Mayer 1845, 26 = 1893a, 66); for "Indifferenzpunkt," see, for example, Poppe 1811–21, *4* (1815), 251; Muncke 1836a, 799; and Autenrieth 1836, 38–39.

207. Autenrieth 1836, 515–516.

208. Autenrieth 1836, 516.

209. Autenrieth 1836, 517.

210. Autenrieth 1836, 518, 518–519.

211. Autenrieth 1836, 519, 520, and 519, respectively.

212. The notion of "clear conceptual space" is a close analog of Kuhn's notion of the empty perceptual space around objects seen to belong to the same class (Kuhn 1974, 474, 480).

## CHAPTER FIVE
## SCIENCE CIRCUMSCRIBED

1. Mayer 1841a, 100 = 1978, 23.

2. Mayer 1844(?), 431; the sentence begins the second paragraph.

3. Mayer 1845, 10 = 1893a, 51–52; 1851a, 37 = 1893a, 261–262.

4. Muncke 1833b, 496; cf. 507.

5. Muncke 1833b, 497.

6. Muncke 1833b, 498; cf. Baumgartner 1836, 8 = 1832b, 8; Baumgartner and Ettingshausen 1842,

8; and esp. Baumgartner 1837, 20: "Our endeavor to explain the internal course of the phenomena of nature always leads in the most successful case only to forces, and we think we have achieved our goal when we have so clearly recognized the law according to which a force acts that we can represent it in a mathematical expression."

7. Baumgartner 1836, 7 ≈ 1832b, 6–7 (quoted in chapter 4, section 1); Berzelius 1813, 6, 8 ≈ 1814, 292, 294 = 1815, 4, 6 (quoted in chapter 3, section 2.2); 1821–32, *1*, pt. 1 (1821), 22 = 1825–31, *1*, pt. 1 (1825), 24 = 1835–36, *1* (1835), 25–26; Geiger 1833, 15 = 1827, 17; Muncke 1833b, 497; 1836b, 1394, 1426.

8. Fries 1822, 505. Cf. the motto of Mayer's quoted in chapter 1, section 3. Timerding (1925, 72) suggested that Mayer's "erschaffener Geist" was an allusion to this verse of Haller's, which continued: "Zu glücklich, wann sie noch die äußre Schale weis't"—"Too fortunate, if she even shows her outer covering." The lines are from Haller's poem, "Die Falschheit menschlicher Tugenden," written in 1730 and published in 1732; quoted from Haller 1762, 100. Duttenhofer (1842) used the first line of Haller's verse as a motto. Invoking the same sentiment, Mayer (1845, 88 = 1893a, 108) appealed to the limitations of human knowledge to cover *his* ignorance as to how oxidation *in the blood* brings about the performance of mechanical work *by the muscle*. Epistemology often serves defensive ends.

9. Liebig 1841e, 195 = 1842b, 7. For "letzte Ursachen" see 195 *bzw.* 7–8. Cf. Liebig's comment to Mohr in 1837 that "force without matter is inconceivable" (quoted in chapter 4, section 2).

10. Tiedemann 1830, 29.

11. Tiedemann 1830, 391.

12. Tiedemann 1830, 393, citing "*Biblia naturae* T. 2. p. 867." Jan Swammerdam's two-volume work in Latin and Dutch was published in 1737–38.

13. Autenrieth 1836, 197.

14. Autenrieth 1836, 198.

15. Autenrieth 1836 (1825), 540–541.

16. Autenrieth 1836 (1825), 541; cf. 544–545.

17. Mayer 1869, in 1978, 348 = 1893a, 357.

18. Müller 1840, 522.

19. Müller 1840, 522. Rudolphi (1821–28, 2, pt. 1 [1823], 242) also insisted that we know the soul (and matter) only through their effects, their ultimate nature being completely unknown to us.

20. Müller 1840, 522.

21. Fries 1826, 23–26.

22. The historical factors prompting this transformation, such as the possible influence of French models of scientific medicine, are not relevant to this study, whose only need is to mark its occurrence.

23. Temkin (1946, 324) dated the rise of a materialistic movement in German physiology to around 1838, and assigned special importance to Schwann's critique of the vital force in 1839.

24. Schleiden 1842, 5. Schleiden was here following what he elsewhere termed "the spirit of the genuine German Friesian natural philosophy" (34). Among the members of the younger generation, Helmholtz's subscription to such a worldview was an essential component of *his* route to energy conservation.

25. Schleiden 1842, 6; cf. 5: "That which is of the spirit and alone free [*das allein freie Geistige*], arising as it does out of a completely different source of knowledge, also remains eternally foreign to the *scientific* (theoretical) knowledge of nature, and incompatible with it." As scientists became more and more the self-appointed arbiters of the true and the real, such epistemological/methodological exclusion of *das Geistige* from science was tantamount to its ontological exclusion from the natural world.

26. Vogt 1845–47, pt. 2 (1846), 206, as cited by Gregory 1977, 64. Gregory cited Pierre-Jean Georges Cabanis as an earlier source for such views; see in particular his *Rapports du physique et du moral de l'homme*, first published in 1802.

27. Vogt 1845–47, pt. 3 (1847), 450–465, esp. 456.

28. Vogt 1845–47, pt. 3 (1847), 460.

29. Vogt 1845–47, pt. 3 (1847), 458; he spoke of "Form und Mischung" (457), "Form und Zusammenstellung" (459), and "Struktur und Mischung" (460).

30. Vogt 1845–47, pt. 3 (1847), 458–459.

31. Vogt 1850, 685.

32. Vogt 1845–47, pt. 3 (1847), 459. According to Gregory (1977, 159), "Vogt said little or nothing on the subject [of the conservation of energy]."

33. Vetter 1837, 65–66 (quote on 66). Vetter, too, stressed the limitations to our knowledge of the world.

34. Valentin 1839, 9; cf. Link 1840, 481.

35. Anonymous 1842a, 413.

36. Sticker 1939–40.

37. Roser and Wunderlich 1843, 5. Roser (1838, 29) had looked forward in his doctoral dissertation to the fruitful application of organic chemistry to medical questions.

38. Roser and Wunderlich 1841, 5 and 4 = 1842, iii and ii, respectively. Roser and Wunderlich's program was criticized by a Stuttgart physician who rejected what he saw as the French emphasis on the material localization of disease to the neglect of a more holistic (and *naturphilosophisch*) understanding of the disease process. He made a plea for the research-worthiness of presently inexplicable phenomena like animal magnetism, and he accused his opponents of subscribing to the principle "quod scalpellum anatomicum non tangit, non est!" (Duttenhofer 1842, 194). In his response, Wunderlich (1842, 216) denied that "physiological medicine" is the same as what Duttenhofer called the "French School." A physician in Esslingen criticized Roser and Wunderlich for treating life as a collection of vital phenomena "without a higher ruling principle" and for treating disease as a collection of phenomena "without the intellectual [*geistig*] unity of the concept of disease" (Späth 1842, 230).

39. Roser and Wunderlich 1841, 9, 13–14 = 1842, vii, xi–xii.

40. Closely paraphrasing Sticker 1939–40, 32 (1939), 237, 245. Julien Offray de La Mettrie published his infamous *L'Homme machine* in Holland in 1747.

41. Marx 1970, 355–358; Griesinger quoted (with no source given) on 357.

42. Letter from Griesinger of 14 December 1842, in Mayer 1893b, 197. According to Ackerknecht (1968, 63), "Griesinger was responsible for the triumph of mechanism [in German psychology]."

43. Griesinger 1844, 900; cf. 908.

44. Griesinger 1844, 902; cf. 901–907. It is a goal of modern medicine to understand fever "in the actual mechanism of its course [*im eigentlichen Mechanismus seiner Hergänge*]," not to explain its function teleologically (907).

45. Griesinger 1844, 908–910.

46. Griesinger 1845, iii, v–vi.

47. Griesinger 1845, 1–9, esp. 5–6.

48. Lotze 1852, 30–45. Wagner (1854, 18–21) also protested against what he saw as the increasing tendency of recent physiologists to subscribe to the "materialistic viewpoint" and thus to deny the existence and legitimacy of an immortal "substantial soul." He quoted several contemporaries (anonymously) in support of his contention; according to Gregory (1977, 35–38; cf. 72–75), Wagner's target was Vogt.

49. Virchow 1907, 7–8.

50. Rümelin 1878, 1795 = 1881, 399.

51. Letter from Griesinger of 4 December 1842, in Mayer 1893b, 183. I thus do not agree that "Griesinger had from the beginning seen in Mayer's new physics a weapon against vitalism" (Kremer 1984, 225). On the contrary, Griesinger thought that Mayer's concept of an autonomous force entity smacked of what was then called "ontology."

52. Letter of 5/6 December 1842 to Griesinger, in Mayer 1893b, 188; cf. the passages from Mayer's letters of 30 November and 5/6 December 1842 quoted in chapter 2, section 2.1.3.

53. Letters of 14 and 22 June 1844 to Griesinger, in Mayer 1893b, 208-209, 217–218 (quotes on 209).

54. Mayer, after 1851(?), 427 = 1893b, 250, except that instead of *Subtilitäten* the latter transcribed *Substitutionen*. It is not obvious how the body of his essay prepared for this conclusion. For the dating of this essay I follow Weyrauch (Mayer 1893b, 243–244); but cf. Johannes Müller's letter of 10 April 1846 regarding a paper Mayer had submitted to Müller's *Archiv* (Mayer 1893a, 143).

55. Cf. Wise 1981, 295. Helmholtz's understanding of force (i.e., energy), which interpreted its conservation on the basis of a mathematical analysis of a system of mass points endowed with attrac-

tive and repulsive forces, also worked against the acceptance of a Mayer-like force as an ontologically autonomous entity.

56. Popper 1876; Mach 1960, 602–608.

57. Hall 1980, 56.

58. Hall 1980, 57.

59. Liebig 1844a, 777 = 1844b, 29–30. I have left unedited Liebig's typically run-on sentence, but have changed the singular *Mensch/er* to the plural "human beings"/"they."

60. Liebig 1846, 238. Hall (1980, 51) quoted this passage from an English translation of 1846 but did not identify the original place of publication; its translation of the first paragraph seriously distorted Liebig's meaning, who clearly saw himself as that perceptive "chemist."

61. Liebig 1856a, 370.

62. Liebig 1856a, 370. According to Gregory (1977, 93–94), Liebig's 1856 address was directed against the polemical materialist Jacob Moleschott, who had attacked Liebig's defense of the vital force in his 1852 book, *Der Kreislauf des Lebens. Physiologische Antworten auf Liebig's Chemische Briefe*. Gregory gives a valuable account of the materialist debates of the 1850s, especially as they related to physiological issues.

63. Liebig 1856a, 386.

64. Liebig 1858b, 585–586 = 1859, *1*, 207. In what may be a reporter's addition to the account of Liebig's talk published a few days later in the *Allgemeine Zeitung*, one reads: "Most likely as a result of the ideas and suggestions that Liebig himself gave in the *Chemical Letters*, fourteen [*sic*] years ago Dr. Meyer [*sic*], a young physician in Heilbronn, sent him a paper for publication in the *Annalen der Chemie*" (Liebig 1858a, 1481).

65. Lenoir 1982, 4, critiqued in Caneva 1990, 298–299.

66. Tiedemann 1830, 80–81.

67. Liebig 1844a, 723 = 1844b, 26, the latter of which added the word "multicolored"; cf. the sentiment of Mayer's "Nature in its simple truth is grander and more magnificent than any creation of human hands and than all illusions of the created mind" (Schmolz and Weckbach 1964, [2]).

68. Liebig 1844a, 777 = 1844b, 27.

69. Liebig 1856a, 370.

70. Hermann Friedrich Autenrieth, in Autenrieth 1836, iii.

71. Autenrieth 1801–2, *1* (1801), 57; 1836 (1825), 343.

72. Berzelius 1813, 8 (etc.), and 1821–32, 3, pt. 1 (1828), 146–147 (etc.), as quoted at length in chapter 3, section 2.2.

73. Berzelius 1821–32, 3, pt. 1 (1828), 147 = 1825–31, 3, pt. 1 (1827), 137 ≈ 1837–41, 6 (1837), 5. I rendered Berzelius's *Mensch* as "human beings" in order to employ a gender-neutral pronoun.

74. Cf. Brougham 1835a, 164, and Prout 1834, 4–5.

75. Brougham 1835a, 191–192 = 1835b, 173–175 (quote on 192 *bzw.* 174–175).

76. Brougham 1835a, 89, 91–92 = 1835b, 79, 81–82.

77. Brougham 1835a, 226, 228 = 1835b, 211, 213.

78. Brougham 1835a, 229 = 1835b, 213. From Lucretius's *De rerum natura*, Brougham quoted i, 151, 217, 249 (= i, 149–150, 215–216, 248 in the Loeb edition). According to Hennemann (1959, 84), "It is reported that he [Mayer] read Lucretius during his long and monotonous sea voyage," but he cited no source for this report and I have been unable to trace it.

79. A German reviewer of Brougham's book called specific attention to this point (Göschel 1836, 274): "In the *seventh* note the doctrine of the ancients concerning *God* is criticized as incomplete only insofar as they taught the eternity of matter and had not progressed to the Christian concept of the Creation: *De nihilo nihil, in nihilum nil posse reverti*."

80. Brougham 1835a, 195 = 1835b, 178. Baron d'Holbach published his controversial, widely condemned, and often reprinted *Système de la nature; ou, des lois du monde physique et du monde moral* in London in 1770 under the name of [Jean-Baptiste de] Mirabaud; it appeared in German translation in 1783–91 and again in 1841.

81. Strauss 1840–41, *1* (1840), 627. Strauss sided rather with the philosophers than with traditional Christianity.

82. Schelling 1809, 451 = 1834, 61–62. He, too, put it in terms of "Schöpfung aus Nichts."
83. Süssmilch 1765, *1*, 52.
84. But see Muncke 1833b, 497–498 (quoted in chapter 4, section 3).
85. The second through fourth editions of Kerner's book appeared in 1832, 1838, and 1846; the second through fourth editions of Strauss's book appeared in 1837, 1838–39, and 1840. For contemporaries' testimony to the excitement and controversy that surrounded both men, see Rümelin 1862, 2814, 2829, 2845 = 1894, 352, 355, 362 (on Kerner); Mohl 1902, *1*, 210 (on Strauss); and Zeller 1874, 39–40 (on Strauss). Even Fries was prompted to review Kerner's book; he was contemptuous of the author's "magical cosmology" (Fries 1831, 336).
86. Hagen 1963, 150–160.
87. J. Kerner 1843; Mayer 1843b.
88. Dechent 1887, 203–205.
89. Anonymous 1838 and 1839a.
90. Alberti 1877, 349. In later years he lived in Kirchheim unter Teck.
91. Müller 1835, 28 = 1833, 27–28 = 1837b, 28 = 1841, 26; Autenrieth 1836, 359–362, 364.
92. Autenrieth 1836, 363–364; cf. 273.
93. Burdach 1837, 608.
94. Passavant 1837, 22–26; cf. 27–28. Strauss (1838b, 231–232 = 1839b, 341–343) gave an account of this strategy in his review of Passavant's book and noted Passavant's use of an analogy between the imponderables and the nervous force: both can act at a certain distance.
95. Passavant 1837, 63.
96. F. Fischer 1838a, 293–296, 305 (quote); cf. 1839, 107, 196.
97. F. Fischer 1838a, 317; cf. 1839, 104.
98. Strauss 1838a, 10ff = 1839a, 9ff.
99. Strauss 1830, 399 = 1839b, 390.
100. Strauss 1830, 404 = 1839b, 394, citing J. Kerner 1829, *1*, 263.
101. Strauss 1830, 416 = 1839b, 402–403.
102. Strauss 1838b, 249–251 = 1839b, 361–363.
103. J. Kerner 1836, vii*ff*, as quoted in Strauss 1836b, 899 = 1839b, 332–333.
104. Strauss 1836b, 894–895 = 1839b, 328–329; cf. 814–816 *bzw.* 303–306.
105. That was one of the points of his *Life of Jesus*; cf. the discussion of his *Two Irenic Essays* in chapter 1, section 3.
106. Eschenmayer 1835, viii.
107. Eschenmayer 1837, 186.
108. Eschenmayer 1837, 204.
109. Eschenmayer 1837, 208.
110. Eschenmayer 1837, 209.
111. Strauss 1838b, 254–255 = 1839b, 366–367.
112. Passavant 1837, 1; cf. v–vi.
113. Passavant 1837, 2.
114. Passavant 1837, v, 3.
115. Passavant 1837, 4.
116. Passavant 1837, 4–7, 9–10, 12. Passavant believed he was following the drift of recent physics, and cited Baumgartner's suppport for the explanation of electricity and magnetism in terms of opposing vibratory motions (citing Baumgartner 1832a, 87, and 1835, 73). In an explanation quite different from Mayer's, he theorized that the production of sunlight is due to the continual neutralization and reestablishment of an electrical tension created by "a polar tension between the sun and the peripheral heavenly bodies" (Passavant 1837, 7–9).
117. Passavant 1837, 10.
118. Passavant 1837, 13 (applies also to the next two sentences).
119. Passavant 1837, 15, 16; cf. Strauss 1838b, 231 = 1839b, 341–342, which emphasized this point. The analogies Passavant specified, exhibiting "the double relationship of center and periphery and of polar opposition" (15–16), are opaque and reminiscent of the analogies of *Naturphilosophie*.

## CHAPTER SIX
## A CONTEXTUAL RECONSTRUCTION

1. Cf. George Sarton's characterization of Mayer's "great discovery": "In the course of his service as a doctor aboard a Dutch ship he discovered the law of conservation of energy, by a sudden intuition" (Sarton 1929, 19).

2. Mayer 1845, 84–85 = 1893a, 105–106. Mayer's account is corroborated (as far as it goes) by his medical log (Mayer 1840).

3. Thackrah 1834, 118, citing Richerand and Magendie; J. Davy 1840, 9; South 1849, 123. There was some difference of opinion.

4. Schönlein 1832, *1*, 209–210, citing also Berzelius's opinion; Schill 1840, 441; J. Davy 1840, 67.

5. Schönlein 1832, *1*, 210; Thackrah 1834, 200; J. Davy 1840, 67; Schill 1840, 281–282, 441; Schreger 1823, 66–67; Anonymous 1890, 56. Emmert (1809, 419) recorded that the blood of horses suffering from *Entzündungsfieber* failed to show the usual buffy coat.

6. Schill 1840, 441; J. Davy 1840, 67–68. On the other hand, Davy noted a (less-than-universal) correlation between the appearance of a buffy coat and a contracted crassamentum (12, 27), and C.J.B. Williams (1835, 718) implied that the recession of the "coagulum" from the sides of the containing vessel is a normal phenomenon.

7. Mayer 1872a, in Mayer 1893b, 391; cf. 1845, 85–86 = 1893a, 106.

8. Mayer 1851a, 15–16 = 1893a, 243–244; cf. his letter of 14/16 June 1844 to Griesinger, in Mayer 1893b, 212–213.

9. Fries 1826, 440.

10. Mayer 1851a, 16–17 = 1893a, 244–245.

11. Letter of 14/16 June 1844 to Griesinger, in Mayer 1893b, 210.

12. Mayer 1848, 3 = 1893a, 153.

13. Letter of 5/6 December 1842 to Griesinger, in Mayer 1893b, 194.

14. Mayer 1872a, in Mayer 1893b, 391–392.

15. Letter of 30 November 1842 to Griesinger, in Mayer 1893b, 181–182. In another letter to Griesinger (of 14/16 June 1844, in Mayer 1893b, 210), after considering the work done and the heat generated by an animal through the combustion of a certain amount of carbon, he added parenthetically: "(Not unsuitable parallels with steam engines can be drawn on all sides.)" See also Mayer 1845, 105 = 1893a, 122–123, and Kremer 1984, 355. Mayer's use of the words *Lebensäther*, *Nervengeister* and *Muskelkraft* is curious, since none of those terms (with the partial exception of the last) was at all common in the physiological literature; perhaps he, like Liebig and a number of other contemporaries, felt a need to paint a distortedly *naturphilosophisch* picture of the immediately preceding state of physiology. Mayer clearly wished to align himself here with progressive thinkers like Griesinger.

16. Breger 1982, 182; cf. 144: "For Mayer the formulation of the energy principle stood in close connection with a mechanical conception of the living organism." Breger developed his argument on 182–184 with references to Mayer's works; his conclusion again identified "the core of Mayer's theory" with the idea that "the human being and the animal are equivalent to [*gelten als*] machines, which in the first instance are characterized by the capability to raise weights" (184).

17. Breger 1982, 127.

18. Breger 1982, 129–134 (phrase on 134); cf. 158.

19. Breger 1982, 183.

20. Mayer 1851a, 17–19 = 1893a, 245–246.

21. Mayer 1873/74b: "daß sich Wärme und Bewegung oder mechanische Arbeit nach einem unveränderlichen Verhältnisse in einander umsetzen müssen."

22. Mayer ca. 1863/67, 378 (italics in original).

23. Letter of 24 July 1841 to Baur, in Mayer 1893b, 110; cf. his letter to Baur of 6 August 1842 (Mayer 1893b, 140):

> With respect to your admonition, "ne sutor ultra crepidam" [i.e., "shoemaker, stick to your last"], I must remind you that the physiologist has the right to demand from the chemist, from

the physicist, clear information about what takes place in inanimate nature. The most important task for the last-named remains in this connection to clearly expound to the physiologist the theory of the agencies [*Agentien*], the imponderables, of lifeless nature. The physiologist must know what the "forces" of dead nature are, otherwise he remains from the start in hopeless darkness with respect to *those* forces of which he *must* speak. I must remind you further that I felt that desideratum more than ever through the unremitting study of a specialized area of physiology, but at the same time was led to physical principles which, as far as I am aware, have not heretofore been recognized but which nevertheless appear to be of unqualified importance for the study of living nature.

24. Letter of 14/16 June 1844 to Griesinger, in Mayer 1893b, 212–213.

25. Letter of 16 August 1841 to Baur, in Mayer 1893b, 121–122 (emphasis added).

26. A similar point was made by Riehl (1900, 174), who argued that "the substantial conception of cause," applied to both matter and force as "the two kinds of substantially conceived cause," yielded Mayer the concept of the indestructibility of force. Heimann (1976, 284–286) also stressed the importance to Mayer of the analogy between force/physics and matter/chemistry (Heimann 1976, 284–286): it was "not merely a heuristc metaphor, but expressed a fundamental feature of his conception of nature: nature constituted a duality of matter and force" (286). It is not clear whether he recognized the *creative* role this analogy played for Mayer.

27. The preface to the sixth edition was dated May 1839, hence it probably appeared no earlier than June of that year.

28. Mayer ca. 1863/67, 378. In letters to his friends he reported only that he had returned from his voyage with a new "system of physics" (letters of 24 July 1841 and 30 November 1842 to Baur and Griesinger, in Mayer 1893b, 110 and 175).

29. Letter of 16 August 1841 to Baur, in Mayer 1893b, 125–126: "das Leuchten der Sonne ist die den peripherischen Bewegungen entsprechende zentrale Bewegung im Planetensystem."

30. See his letter of 16 August 1841 to Baur, in Mayer 1893b, 122, and the references cited in chapter 2, section 2.2.

31. Muncke 1827, 81; Anonymous 1828, 337.

32. See chapter 4, section 1.1 (central-force motion and mechanical vs. organic systems), section 2 (light and heat), and section 3 (quantitative/qualitative).

33. Fries 1826, 11–12. Fries identified morphology as the third major division of natural science.

34. Fries 1826, 540.

35. Fries 1826, 542.

36. Letter of 30 November 1842 to Griesinger, in Mayer 1893b, 177–178.

37. Turner (1974, 237) noted the general interest during the 1840s in the efficiency of steam engines and in "the many new conversion processes which were being discovered in electricity, magnetism, and chemistry," and then added: "Mayer's early papers show little interest in these problems but instead suggest that philosophical and conceptual considerations largely guided Mayer's theorizing. One of these considerations was his constant identification of force and cause; another was his intuitive understanding of force as a substantial, quantitative entity. The source of these ideas of Mayer's and their relationship to the larger context of German science and philosophy remain unsolved historical problems."

38. Rümelin 1878, 1778–79 = 1881, 380–382; the words "at his request" in the third sentence were added to the latter version. Rümelin's account of his conversations with Mayer was written in indirect discourse, which my translation has not captured. Rümelin remarked apropos of the title of Mayer's first published paper: "My above-mentioned objections had perhaps some part in the addition of the word 'inanimate,' which he indeed later regarded as not entirely correct" (1779 *bzw*. 384).

39. Rümelin 1878, 1779 = 1881, 382–383.

40. Rümelin 1878, 1778 = 1881, 378. Aside from Rümelin's statement that Fritz advised Mayer about his experiments, no information survives as to the content of Mayer's interchanges with his brother and with Kehrer (on whom see Mayer 1893b, 110).

41. Letter of 6 August 1842 to Baur, in Mayer 1893b, 140–141.

42. Mayer 1841b is hard to date precisely. His reference to having performed water-shaking experiments suggests it dates from after his visit with Nörrenberg, which took place between 6 and 12

September 1841. On the other hand, it seems significantly less advanced than his letter to Baur of 12 September—in which he first spoke of trying to calculate the mechanical equivalent of heat on the basis of the compression and expansion of gases—and would thus seem to antedate that letter. But it is also hard to imagine his having written such a formal essay over a few days' time when his thinking was in considerable flux. On the other hand, the sentence in which he mentioned those experiments is stylistically something of a tag-on, and he may have added it to an already existing manuscript before copying it over neatly. The copy is clean and carefully written, as if intended for publication. The script is unlike Mayer's usual handwriting, but is identical to that of another 'formal' document, his letter of 7 April 1841 to the shipping company De Cock in Rotterdam (RM-Archiv, fasc. 28). In his letter to Baur of 12 September 1841 he said he had sent Christian Gottlob Gmelin "a mathematical development of my chief principle" but had not gotten a reply (Mayer 1893b, 133); perhaps that otherwise unidentified manuscript was a somewhat earlier version of Mayer 1841b. "Aude sapere" was the motto on the title page of Hahnemann 1819.

43. Mayer 1841b, 1, col. 1: "Eine Kette von Ursachen u. Wirkungen mag noch so weit fortlaufen, so liegt in dem gesagten, daß die ursprünglich gegebene Größe in derselben weder wachsen noch abnehmen könne."

44. Mayer 1841b, 2, col. 1: "ein logisches Verzeichniß der Resultate heißt Wissenschaft."

45. Mayer 1841b, 2, col. 1: "Der Schluß der antiphlogistischen Chemie Fe + O = $\dot{\text{F}}$e enthält die Emancipation der wegbaren Stoffe von den unwägbaren."

46. Mayer 1841b, 3, col. 1: "Imponderabilien oder Kräfte"; "räumliche Differenz"; "von Bewegungs-Ursache, von Kraft."

47. Mayer 1841b, 9, col. 2, to 10, col. 1: "Der Ausdruck dessen man sich hier bedient: Wärme dehnt die Körper aus, ist nichts / als eine Paraphrase des Satzes: Wärme ist Ursache entgegengesetzter Bewegungen, (positive u. negative Bewegungen, Bewegungen mit dem Zeichen 0) als nothwendigem Corollarium zu dem Satze: entgegengesetzte Bewegungen sind Ursache von Wärme." Cf. 1841a, 105 = 1978, 27, and letters to and from Baur of 24 July and 11 August 1841, in Mayer 1893b, 111 and 118–119. The wording of Mayer's letter is very close to that of 1841b.

48. Mayer 1841b, 6, col. 1: "durch einen präsumirten festen Punkt."

49. Mayer 1841b, 7, cols. 1–2: "Aggregations-Zustand."

50. Mayer 1841b, 8, col. 2: "[D]a aber dieselbe [die Volumens-Verminderung] bei den Erscheinungen, welche man auf diese Art zu erklären versucht hat, oft sehr bedeutend ist, so sehen wir hieraus, daß diese Erklärungsweise, welche noch die einzige trostreiche wird, u. die von einer richtigen Idee ausgeht, selbst da nicht viel taugt, wo sie noch am meisten hinpaßt."

51. Mayer 1841b, 10, col. 1: "Der activen Expansion elastischer Körper mittelst derer sie die auf sie applicirte Kraft wieder abgeben, steht demgemäß die Fähigkeit anderer Körper, die dieser Art von Expansion nicht fähig sind, die empfangene Kraft unter der Form von Wärme wiederzugeben, zur Seite."

52. Mayer 1841b, 9, col. 1: "Zum Schlusse sey hier noch erwähnt, daß nicht nur mittelst der Bewegung fester Körper, sondern auch tropfbare[r] Flüssigkeiten—durch Schütteln von Wasser—wie der Verfasser durch viele sorgfältige Versuche fand, sich Wärme entwickeln läßt, womit auch die Beobachtung aller Seefahrer, daß das bewegte Meerwasser wärmer als das ruhige, zusammenhängt. vergl. den Rumford. Versuch." The phrase "vergl. den Rumford. Versuch" is in appreciably smaller letters than the rest of the manuscript and appears to be a later addition.

53. Letter of 16 August 1841 to Baur, in Mayer 1893b, 124. In one of his autobiographical sketches, Mayer (ca. 1863/67, 382) wrote: "Already during my sea voyage I had heard from a well-travelled helmsman, *upon my questioning*, that storm-tossed waves are warmer than the calm sea" (emphasis added). Kremer's suggestion that Mayer came across the idea of the convertibility of heat and motion "from the sailors' chatter about storms warming sea water" is neither necessary nor tenable (Kremer 1984, 221–223; quote on 223). Mayer was already seeking confirmation of a hunch when he so inquired of the helmsman.

54. Letter of 30 November 1842 to Griesinger, in Mayer 1893b, 175.

55. Letter of 12 September 1841 to Baur, in Mayer 1893b, 128–129; Weyrauch called attention to the significance of this passage. Apparently drawing upon reminiscences supplied to him by Baur, Weyrauch added: "According to Baur, Nörrenberg made demands with respect to proofs via calcula-

tion and experiment that Mayer was then not yet up to" (Mayer 1893b, 129). The only other information bearing upon the session with Nörrenberg is Mayer's quoting him as having said "I regard motion not as a force, but only as the effect of a force" (letter of 30 November 1844 to Baur, in Mayer 1893b, 170).

56. Gerlach 1942, 26, and 1963, 24–25; A. Hermann 1972a, 229, and 1980, 18.

57. Letter of 12 September 1841 to Baur, in Mayer 1893b, 129–130. Weyrauch supplied the two sketches missing in the original.

58. Letter of 12 September 1841 to Baur, in Mayer 1893b, 131–132.

59. Kuhn 1958, 138. Following Weyrauch and the explicit references in several of Mayer's other letters, Kuhn noted that Mayer undoubtedly knew of Gay-Lussac's and Dulong's results from Lamé's well-known physics text (136).

60. Letter of 12 September 1841 to Baur, in Mayer 1893b, 132.

61. Letter of 12 September 1841 to Baur, in Mayer 1893b, 132–133.

62. Letter of 30 November 1842 to Griesinger, in Mayer 1893b, 175–176; cf. his letter of 6 August 1842 to Baur: "Professor Jolly told me . . . repeatedly that the matter was very interesting to him, and he encouraged me to 'carry on'" (Mayer 1893b, 141).

63. Mayer ca. 1863/67, 379.

64. Mach 1895, 171 = 1897, 274 ≈ 1896a, 289.

65. Mach 1896b, 247 and 246, respectively.

66. Lippmann 1897, 5 = 1906, 531. As if to enhance the legend, Lippmann added dual emphasis to his quotation of Mayer's words in the 1906 reprinting: "Es *ischt* aso!"

67. Voit 1885, 124.

68. Jolly 1841, 3; cf. 1842b.

69. Voit 1885, 121-122; T.L.W. Bischoff 1837.

70. Jolly 1842a (letter from Jolly of 29 May 1842 to Liebig).

71. Jolly 1842c, [2]–[5].

72. Jolly 1842d. Liebig's assertion to Eduard Vieweg that Jolly was reviewing his book for the *Allgemeine Zeitung* was apparently wishful thinking; see Liebig's letter of 6 January 1843 in Schneider and Schneider 1986, 130. (The letter's date of 1842 must be a mistake for 1843.) I was unable to find anything in the *Allgemeine Zeitung* by Jolly.

73. See Kremer 1984, 344, and chapter 3, section 2.2.

74. Holmes 1964, xv; 1974, 45. See the letters from Liebig of 26 April 1840 and 17 May 1841 to Berzelius and of 9 March 1841 to Wöhler, in Carrière 1893, 211, 230–231, and Hofmann 1888, 173; see also Holmes's reference to a letter from Liebig of 26 April 1840 to Théophile-Jules Pelouze (Holmes 1974, 45, 490).

75. Letter of 5/6 December 1842 to Griesinger, in Mayer 1893b, 190. For the text of Liebig's "p. 32"—i.e., Liebig 1842b, 32 = 1841d, 214—see chapter 3, section 2.1. The preface to Liebig's *Animal Chemistry* is dated April 1842; the book had appeared by 16 July 1842 (letter to Eduard Vieweg, in Schneider and Schneider 1986, 141). Holmes said Liebig "announced its publication" in a letter to Pelouze of 2 July 1842 (Holmes 1974, 490). Liebig expounded his physically imprecise concepts of *Kraftmoment* and *Bewegungsmoment* in Liebig 1842b, 206–207 (discussed in chapter 3, section 2.2). See also the letter from Griesinger of 2 December 1842 to Mayer, in Mayer 1893b, 183–184.

76. Mayer 1842, 237 = 1893a, 26–27.

77. Mayer 1842, 236 = 1893a, 25.

78. Mayer 1842, 238–239, 239 = 1893a, 28.

79. Mayer 1845, 14, 17 = 1893a, 55, 58; cf. 87 *bzw.* 108 for the becoming "latent" of that portion of the heat potentially available from the oxidation process in the blood vessels of a muscle which is used to do work—i.e., lift a weight.

80. Mayer 1843a, 2–3, 4 (letter of ca. April/May 1843 to Baur).

81. Mayer 1843a, 3: "Es handelt sich hier offenbar um die Veränderung des Coordinaten System's, oder vielmehr um die Einführung eines solchen, als fester Axe für sämtl:[iche] physikal:[ische] Wissenschaften."

82. Mayer 1843a, 3: "Die Kraft sey z.B. die Bewegung, oder vielmehr die lebendige Kraft der Bewegung."

83. Mayer 1843a, 3–4. See Appendix 3 for the German text.

84. Letters of 22 June 1844 to Griesinger and of 31 July and 21 August 1844 to Baur, in Mayer 1893b, 218, 150, and 156, respectively: *Polemik* in each.

85. See the letters cited in the previous note, and compare Mayer 1844/45, 32–38 ("force instantanée" on 36). In that work he tried introducing a distinction between "mathematical" and "physical" force: the former is (for example) the cause of acceleration, while the latter is the cause of motion (32).

86. Letter of 22 June 1844 to Griesinger, in Mayer 1893b, 216; cf. the title of Mayer 1844/45: "Theory of Physical Principles as a Basis for Physiological Propositions."

87. Letter of 22 June 1844 to Griesinger, in Mayer 1893b, 216.

88. Letters of 22 June 1844 to Griesinger and of 31 July 1844 to Baur, in Mayer 1893b, 217 and 156, respectively.

89. Mayer 1845, 5 = 1893a, 47.

90. Mayer 1845, 6 = 1893a, 48; cf. 38–39 *bzw*. 75. In 1846 he called his favored Latin slogans "these two axioms of logic" (Mayer 1846, 262; quoted by himself in 1848b, 385 = 1893b, 274).

91. Mayer 1845, 40 = 1893a, 76.

92. Mayer 1845, 6 = 1893a, 48.

93. Mayer 1844/45, 66: "nach dem Axiom der Unzerstörlichkeit der Kraft." The manuscript, which probably dates from late 1844 or early 1845, is clean and carefully written, with the individual sheets folded, cut, and sewn into a little booklet. It may be the version sent to Liebig on 3 January 1845 and returned to Mayer by A. W. Hofmann three days later; see Hofmann's letter of 6 January 1845 in Mayer 1893a, 139. The first eight pages of the manuscript correspond closely, albeit in reordered paragraphs, to the "Einleitung" of Mayer 1845. The first page of the manuscript is reproduced in facsimile in Mayer 1978, 455. Griesinger had received a copy of Mayer 1845 some time before his letter of 7 September 1845 (Mayer 1893b, 226–227).

94. Mayer 1844/45, 66: "Princip"; "positiven Bewegungen"; "Erhaltung der Kraft." With respect to his notion of "positive motions"—a term he soon dropped—cf. 49–50: "The theory presented here is based on the principle that the magnitude of the effect of a motive force is completely independent of the direction that the moved mass takes. One can call this principle the principle of positive motions." ("Die hier vorgetragene Theorie huldigt dem Grundsatze: daß die / Größe der Wirkung einer bewegenden Kraft durchaus unabhängig ist, von der Richtung, welche die bewegte Masse einschlägt. Diesen Grundsatz kann man Princip der positiven Bewegungen nennen.") As another example of Mayer's constant search for appropriate terminology, see his coinages "das Erzeugende" and "das Erzeugte" for the causes and effects in the quantitatively invariable chain of transformations of force (10–11). As he expressed his new (and short-lived) principle there (11), "A given object, an *Erzeugendes*, cannot cease to exist without having produced an equally great *Erzeugtes*. Nil fit ad nihilum." ("Ein gegebenes Object, ein Erzeugendes, kann zu Seyn nicht aufhören, ohne ein ihm gleichgroßes Erzeugtes hervorgebracht zu haben. Nil fit ad nihilum.") See also his letter of 20 July 1844 to Griesinger, in Mayer 1893b, 223.

95. Mayer 1862, 385 = 1893a, 324.

96. Mayer 1871a, 355; cf. 1866, 865 = 1893b, 253 = 1978, 333. Only one other time, and in passing, did Mayer speak of "das . . . Princip der Umwandlung und der Erhaltung der Kraft" (Mayer 1851a, 46 = 1893a, 268).

97. Mayer 1877a, 525 = 1893b, 440.

98. Mayer 1877a, 524 = 1893b, 439.

99. Mayer 1842, 236 = 1893a, 25.

100. Mayer 1845, 32 = 1893a, 71.

101. Mayer 1845, 33–34 = 1893a, 71–72; where I have "chemical force" Mayer wrote "Chemisches Getrenntseyn gewisser Materien. Chemisches Verbundenseyn gewisser anderer Materien," and he included under "chemische Kräfte" both these and electricity. In Mayer 1893a it is not clear, as it is in the original, that magnetism and electricty belong together under Roman numeral IV.

102. Mayer 1845, 23–26 = 1893a, 63–66.

103. Mayer 1848a, 5–6 = 1893a, 155–156.

104. Mayer 1851a, 14 = 1893a, 243.

105. Joule 1847b; Mayer 1848b; Joule 1849, title; Mayer 1849, 534 = 1893b, 281. In *his* first 'conser-

vation of energy' paper, Joule had spoken of the "mechanical value of heat" (Joule 1843, title and 435, 441 = 1884, 149, 156). He used the term "mechanical equivalent of heat" in a paper dated June 1844, published in May 1845, though not in one published in September of that year (Joule 1845a, 369 = 1884, 172; cf. Joule 1845b). The only one of his relevant papers to be translated into German was Joule 1847a.

106. Kremer (1984, 223) maintained that "Liebig's forays into animal chemistry . . . increasingly prod[d]ed Mayer to the physiological issues, and greatly influenced his treatment of them." As evidence of Liebig's influence, Kremer cited Mayer's first letter to Griesinger, adding that "Mayer first applied his theory to animals" in November 1842 (224; see the letter of 30 November 1842 in Mayer 1893b, 181–182). However, the issues Mayer put before his friend did not go beyond those basic questions of respiration, etc., that had occupied him at the very start of his theorizing; they surfaced then because he was bringing his friend up to date on the overall sweep of his work in what was first letter to him. In asserting that "refuting Liebig's use of *Lebenskraft* to explain muscle action would be the primary task of Mayer's 1845 essay" (225), Kremer exaggerated the role that that criticism played in Mayer's essay—important as it was—and missed the essential role Griesinger played in late 1844 in encouraging Mayer to add to his draft essay polemics against Liebig and Lotze (on which more presently). The use of the cited letter of Mayer's to demonstrate that it was Liebig he was responding to is in any event weakened by the fact that in it Mayer did not use Liebig's word *Lebenskraft*; see the passage from it quoted in chapter 6, section 1.

107. Letter from Griesinger of 17 May 1843, in Mayer 1893b, 206.

108. Letter of 11 June 1844 to Griesinger, in Mayer 1893b, 207. Perhaps the death of Mayer's mother on 6 January 1844 helps explain the tardiness of his reply, especially if she had been ailing for some time.

109. Letter of 11 June 1844 to Griesinger, in Mayer 1893b, 208.

110. Letter from Griesinger of 18 June 1844, in Mayer 1893b, 214.

111. Letter of 22 June 1844 to Griesinger, in Mayer 1893b, 216 and 218.

112. Letter from Griesinger of 15 July 1844, in Mayer 1893b, 220. For a while Mayer intended to write such a second and separate article; see his letter of 16 July 1844 to Griesinger, in Mayer 1893b, 221. The unfinished essay I have described as "Mayer 1844(?)" may be an aborted draft of that intention. For its dating, see the letters from and to Griesinger of 15 and 16 July 1844, in Mayer 1893b, 220–221. It was possibly intended for Wunderlich and Roser's *Archiv für physiologische Heilkunde*—cf. Mayer's use of the phrase "physiologische Heilkunde" within quotation marks on 431.

113. Letter of 31 July 1844 to Baur, in Mayer 1893b, 156.

114. Letter of 5/6 December 1842 to Griesinger, in Mayer 1893b, 190; Mayer cited Liebig 1842b, 206–207. Cf. Mayer 1845, 71–72 = 1893a, 137.

115. Letter of 22 June 1844 to Griesinger, in Mayer 1893b, 218, citing Liebig 1842b, 204.

116. Mayer 1845, 54–59 = 1893a, 88–90, 132–133; cf. Mayer's letter of 14/16 June 1844 to Griesinger, in Mayer 1893b, 210–211. Kremer's contention that "the chief purpose of Mayer's essay . . . was to attack Liebig's theory of muscle contraction as the decomposition of muscle tissue" seriously oversimplifies Mayer's intentions (Kremer 1984, 354).

117. Mayer 1845, 57 = 1893a, 132.

118. Mayer 1845, 57–58 = 1893, 132–133; the passage from Liebig 1842b, 34 (= 1841d, 216) that Mayer cited is quoted in chapter 3, section 2.1.

119. Mayer 1845, 58–59 = 1893a, 133; the passage from Liebig 1842b, 225–226, that Mayer cited is quoted in chapter 3, section 2.2.

120. Mayer 1845, 59, 68 = 1893a, 133, 135. John Hunter had been a surgeon and anatomist at St. George's Hospital in London.

121. Mayer 1845, 68–69 = 1893a, 135–136; the passage from Liebig 1842b, 248–249, that Mayer cited is quoted in chapter 3, section 2.2.

122. Mayer 1845, 70–71 = 1893a, 137. In calling attention to the importance of this passage, Kremer added (Kremer 1984, 357): "Mayer further tried to show that *Lebenskraft* was not required to protect living tissue from relentless oxidation, perceptively realizing that this problem, more than any, had driven Liebig to posit *Lebenskräfte*."

123. Mayer 1845, 98 = 1893a, 117.

124. Mayer 1845, 100 = 1893a, 118–119.
125. Mayer 1845, 101 = 1893a, 119.
126. Mayer 1845, 104–105 = 1893a, 122.
127. Mayer 1845, 106 = 1893a, 123.
128. Mayer 1845, 61 = 1893a, 90.
129. Mayer 1871a, 67 = 1893a, 406–407.
130. Mayer 1848a, 2 = 1893a, 152; cf. Herschel 1833, 212: "The great mystery, however, is to conceive how so enormous a conflagration (if such it be) can be kept up. Every discovery in chemical science leaves us completely at a loss, or rather, seems to remove farther the prospect of probable explanation." In Prout's opinion, the nature of the sun and the manner in which it produces heat and light "are quite unknown to us, and will probably always remain so" (Prout 1834, 93).
131. Mayer 1846, 264.
132. Mayer 1848a, 29 = 1893a, 178: "Das Strahlen der Sonne ist die einer centripetalen Bewegung äquivalente centrifugale Action."
133. Cf. Timerding 1925, 63.
134. Mayer 1848a, 2 = 1893a, 152. In close association with this theory Mayer developed the idea that the retardation of the earth's rotation due to the effect of the tides is exactly compensated by the acceleration due to the earth's thermal contraction (Mayer 1846, 270 = 1978, 166; 1848a, 38–65 = 1893a, 187–212; 1851b). In 1848 he invoked an analogy between the tides and a pendulum.
135. The manuscript is clearly a draft, rather sloppily written and full of additions and cancelations. Allusion to the impossibility of species change on p. 60 suggests it dates from after 1859, when Darwinism had become an issue. I have tentatively dated it to around 1866 from the similarity of some of its (for Mayer) peculiar language to that of Mayer 1866, in particular the microcosm-macrocosm terminology and the analogy between meteors and blood corpuscles; compare pp. 57 and 66 of the manuscript with Mayer 1866, 333, 335 = Mayer 1893b, 253, 256.
136. Mayer ca. 1866(?), 57.
137. Mayer ca. 1866(?), 57–58.
138. Mayer ca. 1866(?), 58; the phrase in brackets was a marginal interpolation, possibly in another hand. This passage is continuous with the preceding.
139. Mayer ca. 1866(?), 60–61; it is not clear whether the words in brackets were intended to be crossed out. Mayer's argument for the immortality of the soul is not unlike Prout's (Prout 1834, 411–412).
140. Mayer ca. 1866(?), 61–66: "kosmische Massen" on 62 and 63. On Parrot, professor of physics at Dorpat, cf. Mayer 1846, 266–267, 270 = 1978, 164, 166.
141. Mayer ca. 1866(?), 66. On his use of this analogy, cf. Mayer 1866, 335 = 1893b, 256: "As astronomers are beginning to see, the mechanical theory of heat affords simple and sufficient insight into these phenomena of light and heat in the macrocosm. The relatively tiny bodies, for example, that provide us with the show of meteorites and fireballs play the same important role in the organism of a solar system as do the blood corpuscles in the animal organism that, compensating for their smallness by their incalculable number [*unermeßliche Anzahl*], constitute an indispensable factor in the process of life." He repeated it in 1871a, 60–61 = 1893a, 401–402.
142. Such examples were then common; see Autenrieth 1801–2, *1*, (1801), 67–68, 68–69, 69–70; 1836, 135–136 (quoted or paraphrased in chapter 3, section 3.1), and Burdach 1837, 93–94. Cf. Tiedemann 1830, 668–669 (quoted in chapter 3, section 2.1).
143. Mayer 1851a, 35 = 1893a, 260.
144. Mayer 1876, 104 = 1893a, 440; cf. Mittasch 1940b, 239: "According to E. v. Lippmann (in a letter) there is no evidence that the word *Auslösung* was then common 'in the science of the day'; one would accordingly surmise that E. Mayer [*sic*] introduced the word himself—as a translation of 'catalysis'!" I have not encountered the word elsewhere, but am not familiar enough with the literature from that later period to know whether it was in more general use. Mittasch's hunch is plausible.
145. Letter of 20 July 1844 to Griesinger, in Mayer 1893b, 223–224. In a footnote to this passage Weyrauch remarked (224): "Here appears for the first time in Mayer the concept of *Auslösung* sharply separated from cause and effect as equivalents of work."

146. Mayer 1845, 53–54, 106–107 = 1893a, 87, 124; "notable" (*merklich*) was absent in the first edition.

147. Mayer 1871a, 67 = 1893a, 407, and passim.

148. Mayer 1845, 80 = 1893a, 101–102; in the latter the last sentence of the footnote ends cryptically with ". . . the law of causality, which is paralyzed by same [i.e., the force]"—apparently a pun on his etymologically motivated use of the word "paralytic." Mittasch (1839, 1114) identified this passage as signaling Mayer's protracted struggle to come to terms with causes that bear no quantitative relationship to their effects.

## CHAPTER SEVEN
## MAYER AND *NATURPHILOSOPHIE*

1. Friedlaender 1905, 68, 91; cf. 86 and 88 for his distancing Mayer from *Naturphilosophie*.

2. Friedlaender 1905, 89.

3. Mittasch 1940b, 37; Diepgen 1942, 54–55; Hennemann 1959, 83–85; cf. Hennemann 1975, 108.

4. Kuhn 1959, 336. The scientists Kuhn mentioned include the Dane, Ludvig August Colding, and the Frenchman, Marc Séguin.

5. Kuhn 1959, 337.

6. Kuhn 1959, 338. Note that for Mayer, the key was not an organic metaphor applicable to the whole universe, but a mechanical metaphor applicable to the organism.

7. Kuhn 1959, 339. Although Kuhn footnoted this sentence with a reference to Hell 1914—the entire article—I can find there no mention of Mayer's friends, let alone of their knowledge of *Naturphilosophie*. Hiebert (1959, 394) supported Kuhn's view: "Mayer cannot possibly conceal his Schelling influence."

8. Kuhn 1959, 339. The fifth German was Karl Holtzmann. Kuhn had cited Stauffer 1957 for Oersted's ties to *Naturphilosophie*. His citation for Hirn was neither informative nor very helpful: "Much biographical and bibliographical material for the study of Hirn's life and work can be found in the *Bulletin de la société d'histoire naturelle de Colmar*, I (1899), 183–335."

9. Kuhn 1959, 339. Following Stauffer and Kuhn, Mendelsohn (1964b, 41) accepted that "Mayer may be considered a disciple of romantic biology" while noting that "the relationship between his biological and medical researches and the new physical theory has not yet received adequate attention." Albarracín Teulón (1985, 211, 216–220) adopted Kuhn's assessment.

10. A. Hermann 1972b, 249.

11. A. Hermann 1972a, 230; 1976, 312, 314, 315, 318; 1977, 51.

12. A. Hermann 1977, 52.

13. Snelders 1973, 78, 208; Mende 1975, 481–482; Hartkopf 1979, 349; Krafft 1982, 82–84, 100; Breger 1982, 177–178; Hartkopf 1984, 84–85, 103; Wiederkehr 1986, 238. Mende, Hartkopf, Krafft, and Breger all make the unity-of-forces argument.

14. Breger 1982, 126–127.

15. Wise 1981, 274.

16. On Heimann 1976, see the Introduction.

17. Wise 1981, 275. Wise further noted the significance of Mayer's resurrection by late nineteenth-century energeticists.

18. This point was stressed by Heimann 1976, 293.

19. Gower 1973, 349.

20. Breger 1982, 101, 127.

21. Heimann 1976, 293–294.

22. Heimann 1976, 294.

23. Heimann 1976, 295.

24. Schelling 1797/1803, 70–71, quoted in Schlüter 1985, 91–92; translation adapted from Schelling 1988, 53. Note that in German the term *Naturphilosphie* applies not only to Schelling's philosophy of nature, but is otherwise the long-standing general term for "natural philosophy."

25. Rümelin 1878, 1778–79 = 1881, 380.

26. Letter of 30 November 1842 to Griesinger, in Mayer 1893b, 181; 1845, 10 = 1893a, 52; 1851a, 37 = 1893a, 261–262.

27. My selection is based on broad reading in the primary and secondary literature; see in particular B. Hoppe 1967, 380–382; Caneva 1974, 134–146; Holstein 1979, 6–7; Kamphausen and Schnelle 1982, 64–72.

28. Burdach 1837, 319.

29. Caneva 1974, 135, 140, 144–145, 153–154, 156.

30. Oken 1831, 158–159.

31. Caneva 1974, 144, citing Yelin 1819, passim.

32. Goethe 1823, 75 *bzw.* 295, quoted by Schlüter 1985, 95.

33. One thus suspects that Liebig's (for him atypical) description of "das Thierleben" as "[eine] unendlich höher gesteigerte Lebensthätigkeit" may be a trace of his youthful enthusiasm for Schelling's *Naturphilosophie* (Liebig 1841b, 155).

34. Walther 1807–8, *1* (1807), 35.

35. Schelling 1799a, 195; the original phrase was in italics.

36. Schelling 1799b, 321.

37. Schelling 1799a, 13.

38. Schelling 1799a, 15–17.

39. For typical examples of *Indifferenz, Differenzirung,* and the like, see Schelling 1799a, 161–162, 258, and many other examples to be cited later in this chapter; Walther 1807–8, *1* (1807), 70; 2 (1808), 9–10; Burdach 1826–40, 2 (1828), 702–706 = 1835–38, 2 (1837), 787–791; 1826–40, *4* (1832), 481.

40. Schelling 1799a, 90.

41. Burdach 1826–40, *5* (1835), 700, 721. Carus (1838–40, *1* [1838], 41–42) described the origination of the organism "as the result of a self-activating idea by means of differentiation from something indifferent [*aus einem Indifferenten*]"; he identified "the absolutely indifferent" with his concept of aether.

42. Walther 1807–8, *1* (1807), 11–12.

43. Autenrieth 1836, 42–44, 66, 495.

44. Autenrieth 1801–2, 3 (1802), 72; 1836, 38–39, 114–118, 125–127, 250, 283–293, 383–384.

45. Schelling 1799a, 9, 203; 1799b, 321–326.

46. Schelling 1799a, 203; cf. 85. F. C. Gmelin (1821, 20) discussed the "relationship of the various forces among themselves"—i.e., sensibility, irritability, and the formative force (*bildende Kraft*)—in similar terms.

47. Walther 1807–8, *1* (1807), 9.

48. Walther 1807–8, *1* (1807), 67–77 (all quotes from this and the next sentence on 68).

49. Eschenmayer 1832, 4–5; cf. his table on 28. A similar system was elaborated in Eschenmayer 1834, 1–3.

50. Eschenmayer 1832, 16–17.

51. Eschenmayer 1832, 30–31.

52. Schelling 1797/1803, 7 (from the preface to the second edition).

53. Schelling 1797/1803, 4–5 (from the preface to the first edition), 228 (the wording of which is slightly different in the second edition); cf. 231–237; translation adapted from Schelling 1988, 4, 182–183. Schelling cited Kant 1786 frequently.

54. Schelling (1799a, 101; 1799b, 281) criticized the limitations of Kantian dynamism, as (for example) unable to account for the specific differences of substances.

55. Schelling 1797/1803, 23. Schelling's *Naturphilosophie* did not escape the tendency of dynamical philosophies to effectively eliminate matter/mass as an ontologically independent entity; cf. Schelling 1799b, 313 and the Kant follower, C.C.E. Schmid (1798–1801, *2* [1799], 26), who so constructed matter out of the two *Grundkräfte* of primitive repulsion and attraction that it (here!) disappeared as an independent entity.

56. Schelling 1797/1803, 111 and 74, respectively; cf. 1798/1806/1809, 432, where he used the word *Conflict*.

57. Schelling 1797/1803, 85; translation adapted from Schelling 1988, 68.

58. Schelling 1797/1803, 86–87.

59. Schelling 1797/1803, 88, 95, 96.
60. Schelling 1797/1803, 98.
61. Schelling 1797/1803, 167–168.
62. Schelling 1797/1803, 168.
63. Schelling 1797/1803, 122; translation adapted from Schelling 1988, 96.
64. Schelling 1797/1803, 124, 128, 137, 138, 139, 143 (*elektrische Materie*); 139 (*elektrisches Fluidum*).
65. Schelling 1797/1803, 136, 139.
66. Schelling 1797/1803, 311-312, quoting Alexander Nicolaus von Scherer.
67. Schelling 1797/1803, 156–166: "eine (*wissenschaftliche*) *Fiktion*" (161).
68. Schelling 1797/1803, 157.
69. Schelling 1797/1803, 278; translation adapted from Schelling 1988, 223.
70. Schelling 1797/1803, 281–282.
71. Schelling 1797/1803, 282–283, 283–284; translation adapted from Schelling 1988, 226, 227. Elsewhere I have usually translated *Materie* as "substance" or "material substance."
72. Schelling 1797/1803, 289, 289, and 284, respectively.
73. Schelling 1797/1803, 300–301, 306.
74. Schelling 1797/1803, 301.
75. Schelling 1798/1806/1809, 384, 386.
76. Schelling 1798/1806/1809, 406–407.
77. Schelling 1798/1806/1809, 432.
78. Schelling 1798/1806/1809, 435.
79. Schelling 1798/1806/1809, 435.
80. Schelling 1798/1806/1809, 450.
81. Schelling 1798/1806/1809, 386.
82. Schelling 1798/1806/1809, 386.
83. Schelling 1798/1806/1809, 481; on 482 he spoke of "the material principle of magnetism."
84. Schelling 1798/1806/1809, 482.
85. Schelling 1798/1806/1809, 482.
86. Schelling 1798/1806/1809, 564–565; in his *Anmerkung* to this paragraph Schelling invoked support from Kielmeyer's seminal lecture of 1793, "Über die Verhältnisse der organischen Kräfte untereinander in der Reihe der verschiedenen Organisationen, die Gesetze und Folgen dieser Verhältnisse."
87. Schelling 1799a, 25 (all three quotes); for the second, Schelling wrote "keine besonderen Materien."
88. Schelling 1799a, 31.
89. Schelling 1799a, 33, 34.
90. Cantor and Hodge 1981a, passim.
91. Schelling 1799a, 35–36.
92. Schelling 1799a, 72 (five occurrences of *Wärmestoff*), 76, 83, 132.
93. Schelling 1799a, 207; the stars are as in the original, and do not represent an ellipsis; cf. 195–196 for another statement of the "one force" that appears in the three levels of the organic realm. In a superficially like-spirited passage, the force alluded to that holds sway through all of nature is nothing more than the force of gravitation (105).
94. Schelling 1799a, 218–219.
95. Schelling 1799a, 249–250, 250–251. Burdach's conception of magnetism and electricity owed much to Schelling (Burdach 1826–40, 6 [1840], 530-531, 535).
96. Schelling 1799a, 257.
97. Schelling 1799a, 258.
98. Schelling 1799b, 321; see 1799a, 15–16, for the concept of *Hemmung*.
99. Walther 1807–8, *1* (1807), 32–33.
100. Oken 1831, 35–36.
101. Oken 1831, 52.
102. Oken 1831, 53, 61.
103. Oken 1831, 61–62.

104. Oken 1831, 62.
105. Oken 1831, 71.
106. Oken 1831, 71.
107. Eschenmayer 1832, 22.
108. Carus 1837, 955–957.
109. Carus 1838–40, *1* (1838), xi.
110. Carus 1838–40, *1* (1838), 13, 14. Plato (*Cratylus*, 410b) and Aristotle (*De Caelo*, 270b22) had invoked the same etymological interpretation of "aether."
111. Carus 1838–40, *1* (1838), 17.
112. Carus 1838–40, *1* (1838), 17–18; my translation of "mit der wir zwar . . . zu [einem] bequemern, kürzern Ausdruck immerhim gebahren . . . mögen" takes some liberties with a difficult passage.
113. Link 1840, 483.
114. Naumann 1840, 16–17; cf. 13 for his somewhat qualified inclusion of magnetism.
115. Naumann 1840, 17.
116. Elsewhere he was more opaque—"Heat is only the expression of the intimacy with which the tendency toward change of the original form in substances reveals itself"—and less materialistic: "Light cannot, as matter, stream out of the sun and arrive at the earth, rather light is to be regarded only as the expression of the *Affinitätspolarität* existing between the two heavenly bodies" (Naumann 1840, 18 and 17, respectively).
117. On Pohl's *Naturphilosophie*, see Caneva 1974, 12, 74–75, 137–138, 145, 148–157.
118. Pohl 1829, iv–v, 8–9, 46, and passim.
119. Pohl 1829, 36.
120. Pohl 1837, 4.
121. Pohl 1837, 5; *efficientia* would appear to be his translation of *Wirksamkeit*.
122. Pohl 1837, 5: "ut . . . actionum unitas sive *efficientia primaria* firmiter animo et intuitu teneatur," where *intuitus* clearly stands for *Anschauung*.
123. Pohl 1837, 5–6; cf. 21.
124. Pohl 1837, 6.
125. Pohl 1837, 8–10.
126. Pohl 1837, 26–27.
127. Pohl 1837, 21.
128. Pohl 1837, 18, 19, and 19, respectively.
129. Pohl 1837, 19 (*effectus electricus*), 60 (*impulsio electrica, conatus, nisus*), 61 (*electrica incitatio*).
130. Pohl 1837, 65.
131. Such forays into Hegel's writings as I have dared have not netted anything that would lead me to suspect him as a possible source for any of Mayer's ideas. Although Hegel's treatment of light, heat, electricity, and magnetism is too complicated and idiosyncratic to summarize here, suffice it to say they were not embraced by a single concept, nor did *Kraft* play a central or unifying role in Hegel's physical thinking; see Hegel 1830 as represented by Hegel 1959. Recall, too, Mayer's utter incomprehension of the two works by Hegel Rümelin had lent him.
132. See also Vogt 1850, 647–648.
133. Schlüter 1985, 28–29, 92–93 (quote on 29); cf. Liedman 1966, 249, 250. For *Naturphilosophen*'s criticism of vital force, Schlüter cited Schelling 1798/1806/1809, 526ff, 619ff; Schelling 1799a, 80*f*, 151f; and Carus 1838–40, *1* (1838), 17ff (Schlüter 1985, 169, n. 128)—material I will discuss presently. According to Schlüter, the chief argument of *Naturphilosophie* against Newtonian physics lay in the *Nicht-Anschaulichkeit* of the latter's premises and conclusions (idem).
134. Schelling 1797/1803, 49.
135. Schelling 1798/1806/1809, 501; cf. 554, 566.
136. Schelling 1798/1806/1809, 496 and 502; cf. 567: "The cause of magnetism is thus present everywhere, and nevertheless acts on only a few bodies. The magnetic current finds the inconspicuous needle on the open sea as easily as in the closed room, and wherever it finds it it gives it the polar direction. Thus does the current of life encounter, from whence it comes, the organs receptive to it and it gives them, where it encounters them, the activity of life."

137. Schelling 1798/1806/1809, 526–527, 527; "suspend" and "eliminate" both render *aufheben*.
138. Schelling 1798/1806/1809, 527; cf. Engels 1980, 126, for Schelling's rejection of *Lebenskraft* and his conception of *Bildungstrieb*.
139. Schelling 1798/1806/1809, 528.
140. Schelling 1798/1806/1809, 529–530 (quote on 530).
141. Schelling 1798/1806/1809, 529.
142. Schelling 1798/1806/1809, 567–569 (quote on 569).
143. Schelling 1798/1806/1809, 568.
144. Schelling 1799a, 79–80.
145. Schelling 1799a, 81–82; cf. Pohl 1829, 79 (quoted in chapter 7, section 3).
146. Schelling 1799a, 90; cf. 82, 144–152.
147. Schelling 1799a, 152 (both quotes).
148. Schelling 1799a, 90.
149. In addition to the examples discussed here, see Schultz 1831, 52–53, in particular his favorable quotation of an anti-vital-force statement from Tiedemann 1830, 658. For Schultz's allegiance to Schelling, Oken, and Hegel, see Schultz 1833, 16, 20–23. Burdach's stance vis-à-vis the vital force is hard to pin down. Although he rejected the term *Lebenskraft* as providing no satisfactory explanation of life, he had a standard conception of a vital force performing the functions of growth, nutrition, etc. (Burdach 1840, 101–106; cf. 1826–40, 5 [1835], 720). Elsewhere he had a rough conception of the progressive *Potenzirung* of the forces of nature from the fundamental physical and chemical forces to a *Lebensprincip* to the soul (Burdach 1837, 307–308). He took matter and the organic *Form, Idee*, or *Gedankenbild* to be the two essential determinants of the organism.
150. Walther 1807–8, *1* (1807), 25.
151. Walther 1807–8, *1* (1807), 26.
152. Walther 1807–8, *1* (1807), 38.
153. Carus 1837, 956.
154. Carus 1838–40, *1* (1838), x–xi.
155. Carus 1838–40, *1* (1838), xii.
156. Carus 1838–40, *1* (1838), xiii; "material substances" and "matter" render *Stoffe* and *Stoff*.
157. Carus 1838–40, *1* (1838), xiii, xiv.
158. Carus 1838–40, *1* (1838), 36.
159. In addition to the works discussed here, see Reich 1828.
160. Schelling 1797/1803, 168, 169.
161. Schelling 1799a, 215–217.
162. Burdach 1826–40, *6* (1840), 451, citing Wilbrand 1827, 11ff, 22ff.
163. Pohl 1829, 79. We glimpse here again the possible Romantic roots of Liebig's view of the vital force as the agency which opposes the deleterious effects of atmospheric oxygen.
164. Walther 1807–8, *2* (1808), 13.
165. Walther 1807–8, *2* (1808), 228–229, 229–230, 231.
166. Oken 1831, 337.
167. Naumann 1835, 62. In another work, the "Gegensatz von Blut und Nervenmark" in animals paralleled the antithesis between root and crown in plants, while both were expressions of the fundamental antithesis between planetary and solar laws (Naumann 1840, 48–49). Recall the dismissive allusion to Naumann's hypothesis in Vetter 1837, 89 (quoted in chapter 3, section 2).
168. Naumann 1835, 62–63.
169. Carus 1838–40, *1* (1838), 269.
170. Carus 1838–40, *1* (1838), 270.
171. Carus 1838–40, *1* (1838), 273.
172. Carus 1838–40, *1* (1838), 273.
173. Burdach 1837, 234.
174. Burdach 1837, 96.
175. Burdach 1837, 23–24
176. Burdach 1826–40, *6* (1840), 540; see 540–555 and 579–581.
177. Burdach 1826–40, *6* (1840), 557.

178. Burdach 1826–40, 6 (1840), 565, citing Müller 1835, 84; cf. 566 and 581, where an *elektrische Wechselwirkung* between muscles and nerves is the source of animal heat.

179. Burdach 1826–40, 6 (1840), 567.

180. Mayer 1841a, 101 = 1978, 23: "Alle Erscheinungen können wir von einer Urkraft ableiten, welche dahin wirkt, die bestehenden Differenzen aufzuheben, alles Seiende zu einer homogenen Masse in einem mathematischen Punkte zu vereinigen."

181. Letter of 30 August 1844 to Baur, in Mayer 1893b, 157. As far as I know, Mayer never employed the term *Grundkräfte*.

182. Schelling 1798/1806/1809, 381.

183. Schelling 1798/1806/1809, 433.

184. Schelling 1798/1806/1809, 395, 382, and 384, respectively.

185. Schelling 1798/1806/1809, 387–394.

186. Schelling 1798/1806/1809, 357–378 (in the second and third editions). Schelling there usually spoke of *das Lichtwesen*; *Kraft* was not a leading concept.

187. Schelling 1799a, 106. Note that for Schelling—in *this* work!—it actually takes *three* bodies to form a gravitational system, the original mass and the two that arise from it via a kind of polar separation (117–118). Mayer posited only two.

188. Schelling 1799a, 116–117.

189. Schelling 1799b, 312–313 (omitting the editor's interpolations from Schelling's manuscript annotations).

190. Schelling 1799a, 132. Arguing that the sun is in a positive (chemical) state with respect to the earth, he concluded "*that light is a phenomenon of the chemical action of the sun on the earth*" (134; cf. 105–106).

191. Schelling 1799a, 136; from a later internal reference (cited in the next quote), p. 146 of the original edition.

192. Schelling 1799b, 318–319.

193. Schelling 1799a, 251–252; 1799b, 274–275.

194. Eschenmayer 1832, 17–18, 22. Wilbrand regarded light and gravity as the two antagonistic *Grundkräfte* (Probst 1966, 158).

195. Naumann 1840, 11.

196. Naumann 1840, 15.

197. Naumann 1840, 12–13.

198. Naumann 1840, 13–14; his §§10–11 covered the material cited above from pp. 11–13.

199. Mayer's formal complaint to the shipping company of the *Java* seems to have occupied much of his time and energy in early 1841; see Krusemarck 1942, 17–20, which prints the text of Mayer's letter to them of 7 April 1841. In a letter to Griesinger of 16 December 1842, Mayer ridiculed a "Naumann in seiner 'Allgemeinen Pathologie als Regulativ' etc., Heft 1" for talking nonsense about a "Lichtform der Materie" (Mayer 1893b, 199). Such a conception of light does not jibe with that of Naumann 1840, nor have I been able to identify any such work by Moritz Ernst Adolph Naumann (or anyone else); Naumann 1824, 1829–30, 1840 (including its *Fortsetzungen*), and 1851 do not confirm Mayer's statement.

200. Letter from Griesinger of 2 December 1842, in Mayer 1893b, 184.

201. Löwenthal 1842, 965–966.

202. Letter of 16 December 1842 to Griesinger, in Mayer 1893b, 205.

## CHAPTER EIGHT
## ASSESSMENT AND CONCLUSIONS

1. Mayer's struggle to create a coherent theory of force provides a good example of what Elkana (1970) termed "concepts in flux," though I disagree with his assessment of Helmholtz: Helmholtz may have retained the *term* "Kraft" as he formulated the notion of the conservation of energy, but I do not find him to have been confused *conceptually* about its several meanings.

2. Cassirer 1957, 460.

3. In rejecting Jørgensen's attempt to distinguish between *physiological* and *chemical* vitalism and to attribute only the former to Berzelius (Jørgensen 1965b, 395), Lipman (1965, 396) made the important observation that "we can often designate certain aspects of the thought of an individual as vitalistic, but to term such a person a vitalist would be to do him an injustice and accomplish no useful analytical function."

4. It was all but impossible for there to have been any generally accepted canonical version of *Naturphilosophie* because there was no objective way of deciding between differences of interpretation even among people who largely shared the same principles. (Compare the tripartite schematizations of Schelling, Eschenmayer, and Walther discussed in chapter 7, section 1.) Its looseness is evident in the case of Oersted's discovery of electromagnetism: he was inspired by *Naturphilosophie* to expect a connection between electricity and magnetism, but he received no guidance from it as to precisely where or how to look or as to precisely what he would find. *Naturphilosophie* was a program for understanding and interpreting, not for producing new knowledge on the basis of clear expectations.

5. I have explored some of these issues in a more personal way in Caneva 1991.

6. Mayer 1870, 367–368 = 1893a, 390–391.

## APPENDIX ONE
## TIMELINE

1. Much of the basic information in this appendix was drawn from the *Zeittafel* in Schmolz and Weckbach 1964, 161–176, and the documentation published elsewhere in that book. Most of the rest came from Mayer 1893a and 1893b.

2. Scheurlen 1948, 105; another source says Mayer took the *Reifeprüfung* in Stuttgart in the spring of 1832 (Schmolz and Weckbach 1964, 30).

3. Mayer 1840; the first two dates confirmed in Mayer 1840–41, 52 and 66–67. Previous venesections had been on 12 March and 30 April; the next were on 15 July and 10 August.

4. For a description of the positions of *Oberamtsarzt* and *Oberamtswundarzt*, see Drees 1988, 34–38. The position had been announced in the *Schwäbischer Merkur* on 28 April 1841.

5. A. Hermann 1972a, 228–229, and 1980, 17, give the date as 10 September, but without citing the source of information. Their meeting may have been closer to the sixth, since Mayer must have required time to perform the "many always successful careful experiments" that Nörrenberg demanded proving that water is heated by shaking (see the letter of 12 September 1841 to Baur, in Mayer 1893b, 129).

## APPENDIX TWO
## COURSES

1. *Verzeichniß der Vorlesungen, welche von den öffentlichen und Privat-Lehrern der königl. württembergischen Universität Tübingen in dem Sommer[Winter]-Halbjahr 1832 [1832/33] gehalten werden.* Tübingen: Ernst Traugott Eifert, 1832, etc.

2. Some of this information was published by Diepgen (1942, 57–62), who apparently followed the "Studien- und Sittenzeugniß" issued by the rector's office on 31 January 1838 and now among Mayer's *Personalakte* in the Staatsarchiv Ludwigsburg (Bestand E 162 II, Büschel 507, fol. 17ʳ–17ᵛ, 20ʳ–20ᵛ). It follows B and fails to take into consideration that Mayer did not hear all the courses there *belegt*. In principle, each *Belegbogen* has columns headed "Fleiß" and "Kenntnisse," each *Hörerliste* columns headed "Fleiß" and "Resultat der Prüfungen," in which the instructor marked the student's performance in each course. Sometimes such marks are missing from one or both lists, sometimes all students received the same (or no) mark, and sometimes it is unclear which one was intended for a given student. Such evidence must thus be used with caution in drawing realistic conclusions about an individual's performance, and I have judged its inclusion here not worth the elaborate apparatus necessary to make it even marginally meaningful. For none of the courses listed above in square brackets were any marks recorded on the relevant *Belegbogen*.

# • BIBLIOGRAPHY •


Ackerknecht, Erwin. 1968. *A Short History of Psychiatry*. 2d ed. Trans. Sula Wolff. New York and London: Hafner.

Albarracín Teulón, Agustín. 1985. "El tránsito de la *Naturphilosophie* a la *Naturwissenschaft*." *Asclepio; Archivo Iberoamericano de Historia de la Medicina y Antropología Médica*, 37, 209–220.

Alberti, Eduard. 1877. "Eschenmayer, Adam Karl August." In *ADB*, 6, 349–350.

Anonymous. 1801. Short notice of Autenrieth 1801–2, *1*. *Allgemeine medizinische Annalen des neunzehnten Jahrhunderts*, [St. 11], November, col. 853.

______. 1802. Review of Autenrieth 1801–2, *1 and 2*. *Journal der Erfindungen, Theorien und Widersprüche in der Natur- und Arzneiwissenschaft*, Bd. 9 (1800–1802), St. 35, Intelligenzblatt Nr. 31, 34–36.

______. 1803a. Short notice of Autenrieth 1801–2, 3. *Allgemeine medizinische Annalen des neunzehnten Jahrhunderts*, 1803, [St. 10], October, cols. 782–783.

______. 1803b. Review of Autenrieth 1801–2, 3. *Medicinisch-chirurgische Zeitung*, Bd. 1 [des Jahres], Nr. 21 (15 March), 355–366.

______. 1828. Review of Fries 1826. *Leipziger Literatur-Zeitung*, Bd. 1 [des Jahres], Nr. 43 (18 February), cols. 337–340.

______. 1834. *Die Homöopathik, der gesunden Vernunft, sowie dem Staats- und Privatrechte gegenüber*. 2 vols. Quedlinburg: L. Hanewald.

______. 1835. Review of Müller 1833. *Medicinisch-chirurgische Zeitung*, Bd. 2 [des Jahres], Nr. 39–40 (14, 18 May), 193–206, 209–218.

______. 1838. Review of Passavant 1837. *Jenaische allgemeine Literatur-Zeitung*, [Jg. 34], Bd. 2 [des Jahres], Nr. 115, cols. 433–438 (June). Signed "H."

______. 1839a. Review of F. Fischer 1839. *Jenaische allgemeine Literatur-Zeitung*, [Jg. 35], Bd. 2 [des Jahres], Nr. 97–98, cols. 289–296, 297–303 (May). Signed "R**."

______. 1839b. Review of Kämtz 1839. *Jenaische allgemeine Literatur-Zeitung*, [Jg. 35], Bd. 4 [des Jahres], Nr. 207, cols. 209–214 (November). Signed "S."

______. 1841. "Die Universität Tübingen." *Hallische Jahrbücher für deutsche Wissenschaft und Kunst*, Jg. 4, [1. Halbjahr], Nr. 111–119 (10–15, 17–19 May), 441–443, 445–447, 449–452, 453–456, 457–460, 461–463, 465–467, 469–471, 473–474; [2. Halbband], Nr. 52–57 (31 August; 1–4, 6 September), 205–208, 209–212, 213–216, 217–218, 221–223, 225–228.

______. 1842a. Review of Lehmann 1842. *Deutsche Jahrbücher für Wissenschaft und Kunst*, Jg. 5, Nr. 104–106 (3–5 May), 413–414, 417–420, 421–422. I suspect the author was Hermann Lotze.

______. 1842b. "Ueber die Gründung physiologischer Institute und das physiologische Institut zu Göttingen." *Allgemeine Zeitung* (Augsburg), Beilage, Nr. 324 (20 November), 2587–2589.

______. 1890. "Blut." In *Meyers Konversations-Lexikon*. 4th ed. (19 vols.; Leipzig: Verlag des Bibliographischen Instituts, 1885–92), 3, 54–58.

______. 1935. *Hundert Jahre deutsche Eisenbahnen*. [Leipzig]: Deutsche Reichsbahn.

Autenrieth, Johann Heinrich Ferdinand. 1792. *Dissertatio inauguralis medica exhibens experimenta et observata quaedam de sanguine praesertim venoso*. Stuttgart: Typis Academicis.

Autenrieth, Johann Heinrich Ferdinand. 1801–2. *Handbuch der empirischen menschlichen Physiologie.* 3 vols. Tübingen: Jacob Friedrich Heerbrandt. In Mayer's library.

———. 1816. "Gründe gegen den Materialismus." *Tübinger Blätter für Naturwissenschaften und Arzneykunde,* Bd. 2, Hft. 3, 289–383. Reprinted with minor changes in Autenrieth 1836, 397–453.

———. 1825. *Ueber den Menschen und seine Hoffnung einer Fortdauer vom Standpunkte des Naturforschers aus.* Tübingen: Heinrich Laupp. Essays "Natürliche Geschichte des Menschen" (1–27), "Wissenschaft des Menschen; seine angeborne Beschränktheit hierin" (28–49), and "Welche Erscheinung ist der Mensch in der Natur?" (50–77) reprinted in Autenrieth 1836 (314–328, 540–552, 329–344) and cited from there as Autenrieth 1836 (1825).

———. 1834–36. *Specielle Nosologie und Therapie. Nach dem Systeme eines berühmten deutschen Arztes und Professors.* Ed. Carl Ludwig Reinhard. 2 vols. Würzburg: C. C. Etlinger. In Mayer's library (saw vol. 1). Published without mention of Autenrieth's name.

———. 1836. *Ansichten über Natur- und Seelenleben.* Ed. Hermann Friedrich Autenrieth. Stuttgart and Augsburg: J. G. Cotta. Essays cited include "Die Verhältnisse des Lebens und der ihm zu Grund liegenden Kraft" (1–168), "Der Instinct und seine Begründung in dem Bildungstriebe der vegetativen Lebenskraft" (169–313), "Verbindung der Seele mit dem organischen Körper, Entwicklung des Charakters der Persönlichkeit in der Reihe der Wesen" (345–396), and "Die Raumwelt und die Unräumlichkeit der Seele" (478–520); cf. Autenrieth 1816, 1825.

Bauer, Erich. 1972. "Die Guestphalia I und II zu Tübingen. (I: 12.2.1831–7.7.1832. II: 12.8.1836–24.2.1852.)." *Einst und Jetzt. Jahrbuch des Vereins für corpstudentische Geschichtsforschung, 17,* 53–65.

Baumgartner, Andreas. 1832a. "Über neue magneto-electrische Erscheinungen." *Zeitschrift für Physik und verwandte Wissenschaften,* Bd. 1, Hft. 1, 74–87.

———. 1832b. *Die Naturlehre nach ihrem gegenwärtigen Zustande, mit Rücksicht auf mathematische Begründung.* 4th ed. Vienna: J. G. Heubner.

———. 1835. "Über den Einfluss der Gleichförmigkeit der Masse auf ihre Empfänglichkeit für Magnetismus." *Zeitschrift für Physik und verwandte Wissenschaften,* Bd. 3, Hft. 1, 66–73.

———. 1836. *Die Naturlehre nach ihrem gegenwärtigen Zustande mit Rücksicht auf mathematische Begründung.* 5th ed. Vienna: J. G. Heubner.

———. 1837. *Anfangsgründe der Naturlehre.* Vienna: J. G. Heubner.

Baumgartner, Andreas, and Ettingshausen, Andreas von. 1842. *Die Naturlehre nach ihrem gegenwärtigen Zustande mit Rücksicht auf mathematische Begründung.* 7th ed. Vienna: Carl Gerold.

Baur, Carl Wilhelm. 1846. Letter to Mayer of 30 November [1846]; from Ulm. In RM-Archiv, fasc. 11d. The partly illegible postmark appears to read "Ulm, 1 Dec. 1[84]6."

Becquerel, Antoine-César, and Breschet, Gilbert. 1835a. "Prémier Mémoire sur la chaleur animale." *Annales des sciences,* 2d ser., 3, [Partie de la] Zoologie, 257–273 (May).

———. 1835b. "Prémier Mémoire sur la Chaleur animale." *Annales de chimie et de physique,* 59, 113–136 (June).

———. 1839. "Recherches sur la chaleur animale, au moyen des appareils thermo-électriques. Première Partie." *Archives du Muséum d'histoire naturelle, 1,* 383–403.

Benton, E. 1974. "Vitalism in 19th-century scientific thought: A typology and reassessment." *Studies in History and Philosophy of Science,* 5 (1974–75), 17–48.

Berthold, Arnold Adolph. 1837. *Lehrbuch der Physiologie des Menschen und der Thiere.* 2d ed. 2 vols. Göttingen: Vandenhoeck & Ruprecht.

Berzelius, Jöns Jakob. 1813. *A View of the Progress and Present State of Animal Chemistry.* Trans. Gustavus Brunnmark. London: J. Skirven.

______. 1814. "Uebersicht der Fortschritte und des gegenwärtigen Zustandes der thierischen Chemie." Trans. G.C.L. Sigwart from the English ed. *Journal für Chemie und Physik,* Bd. 12, Hft. 3, 289–341; Hft. 4, 361–399.

______. 1815. *Uebersicht der Fortschritte und des gegenwärtigen Zustandes der thierischen Chemie.* Nuremberg: Johann Leonhard Schrag. A separate printing of Berzelius 1814.

______. 1821–32. *Lehrbuch der Chemie.* Trans. Karl August Blöde (*1*), Karl Palmstedt (2), and Friedrich Wöhler (3 and 4). 4 vols., each in 2 pts. (4, pt. 2 possibly never published). Reutlingen: J. J. Mäcken. In Mayer's library (saw).

______. 1825–31. *Lehrbuch der Chemie.* Trans. Friedrich Wöhler. 4 vols., each in 2 pts. Dresden: Arnold.

______. 1835–36. *Lehrbuch der Chemie.* Trans. F. Wöhler. 4th ed. Vols. 1–5. Dresden and Leipzig: Arnold.

______. 1836a. "Einige Ideen über eine bei der Bildung organischer Verbindungen in der lebenden Natur wirksame, aber bisher nicht bemerkte Kraft." *Jahres-Bericht über die Fortschritte der physischen Wissenschaften.* Trans. F. Wöhler. Jg. 15, 237–245.

______. 1836b. "Einige Ideen über eine bei Hervorbringung organischer Verbindungen in der lebenden Natur bisher nicht beachtete, mitwirkende Kraft." Trans. H. C. Schumacher. *Jahrbuch für 1836,* 88–97.

______. 1837–41. *Lehrbuch der Chemie.* Trans. F. Wöhler. 3d ed. Vols. 6–10. Dresden and Leipzig: Arnold.

______. 1839a. "Om några af dagens frågor i den organiska Kemien." *Kongl. Vetenskaps-Academiens Handlingar för År 1838* (publ. 1839), 77–111.

______. 1839b. "Ueber einige Fragen des Tages in der organischen Chemie." *Annalen der Physik und Chemie,* Bd. 47, St. 2, 289–322 (≈ June).

______. 1839c. "Ueber einige Fragen des Tages in der organischen Chemie." *Annalen der Pharmacie,* Bd. 31, Hft. 1, 1–35 (≈ July).

______. 1844. Review of Liebig 1842b. *Jahres-Bericht über die Fortschritte der Chemie und Mineralogie,* Jg. 23, 575–583.

Bianco, Bruno. 1980. *J. F. Fries. Rassegna storica degli studi (1803–1978).* Naples: Bibliopolis.

Biot, Jean-Baptiste. 1824. *Précis élémentaire de physique expérimentale.* 3d ed. 2 vols. Paris: Déterville.

______. 1828–29. *Lehrbuch der Experimental-Physik oder Erfahrungs-Naturlehre.* Trans. and ed. Gustav Theodor Fechner. 2d German ed. 5 vols. Leipzig: Leopold Voß. In Mayer's library (saw).

Bischoff, Christoph Heinrich Ernst. 1825–34. *Die Lehre von den chemischen Heilmitteln oder Handbuch der Arzneimittellehre.* 3 vols. plus *Supplementband.* Bonn: Eduard Weber.

Bischoff, Theodor Ludwig Wilhelm. 1837. *Commentatio de novis quibusdam experimentis chemico-physiologicis ad illustrandam doctrinam de respiratione institutis.* Heidelberg: J.C.B. Mohr.

______. 1842. "Bericht über die Fortschritte der Physiologie im Jahre 1841." *Archiv für Anatomie, Physiologie und wissenschaftliche Medicin,* lxi–cxxxiii.

______. 1847. "Theorie der Befruchtung und über die Rolle, welche die Spermatozoïden dabei spielen." *Archiv für Anatomie, Physiologie und wissenschaftliche Medicin,* 422–442.

Bischoff, Theodor Ludwig Wilhelm. 1874. *Ueber den Einfluss des Freiherrn Justus von Liebig auf die Entwicklung der Physiologie*. Munich: Verlag der K. B. Akademie.

Blüh, Otto. 1952. "The value of inspiration. A study on Julius Robert Mayer and Josef Popper-Lynkeus." *Isis*, *43*, 211–220.

Brandes, Heinrich Wilhelm. 1826. "Centralbewegung." In Gehler 1825–45, 2, 60–75.

———. 1836. "Masse." In Gehler 1825–45, 6, pt. 2, 1392–1393.

Brauss, Richard. 1841. *De caloris in organismum actione observationes et experimenta nonnulla*. Berlin: Typis Nietackianis.

Breger, Herbert. 1982. *Die Natur als arbeitende Maschine: Zur Entstehung des Energiebegriffs in der Physik 1840–1850*. Frankfurt am Main and New York: Campus Verlag.

Brodie, Benjamin Collins. 1811. "The Croonian Lecture, on some Physiological Researches, respecting the Influence of the Brain on the Action of the Heart, and on the Generation of animal Heat." *Philosophical Transactions of the Royal Society of London*, [*101*], pt. 1, 36–48.

———. 1812. "Further Experiments and Observations on the influence of the Brain on the generation of Animal Heat." *Philosophical Transactions of the Royal Society of London*, [*102*], pt. 2, 378–393.

Brougham, Henry Peter. 1835a. A *Discourse of Natural Theology, Showing the Nature of the Evidence and the Advantages of the Study*. London: Charles Knight; New York: William Jackson.

———. 1835b. *Gott und Unsterblichkeit aus dem Standpuncte der natürlichen Theologie und ihrer Beweiskraft*. Trans. Johann Sporschil. Leipzig: Otto Wigand.

Brunn, Walter L. von. 1965. "Albert Friedrich Schill und die allgemeine Pathologie vor Beginn der Aera der genetischen Zelltheorie." *Sudhoffs Archiv für die Geschichte der Medizin und der Naturwissenschaften*, *49*, 1–16.

Brush, Stephen G. 1970. "The wave theory of heat: A forgotten stage in the transition from the caloric theory to thermodynamics." *British Journal for the History of Science*, *5*, 145–167.

Burdach, Karl Friedrich. 1826–40. *Die Physiologie als Erfahrungswissenschaft*. 6 vols. Leipzig: Leopold Voss.

———. 1835–38. *Die Physiologie als Erfahrungswissenschaft*. 2d ed. Leipzig: Voss, 1835 (*1*), 1837 (*2*), 1838 (*3*).

———. 1837. *Anthropologie für das gebildete Publikum*. (= *Der Mensch nach den verschiedenen Seiten seiner Natur*.) Stuttgart: P. Balz. In Mayer's library (saw).

Callisen, Adolph Carl Peter. 1830–45. *Medicinisches Schriftsteller-Lexicon der jetzt lebenden Aerzte, Wundärzte, Geburtshelfer, Apotheker, und Naturforscher aller gebildeten Völker*. 33 vols. Copenhagen: Auf Kosten des Verfassers gedruckt im Königl. Taubstummen-Institute zu Schleswig.

Caneva, Kenneth L. 1973. "LaRive, Arthur-Auguste de." In *DSB*, *8*, 35–37.

———. 1974. "Conceptual and Generational Change in German Physics: The Case of Electricity, 1800–1846." Ph.D. diss., Princeton University.

———. 1978. "From galvanism to electrodynamics: The transformation of German physics and its social context." *Historical Studies in the Physical Sciences*, *9*, 63–159.

———. 1990. "Teleology with regrets." Review of the 1989 edition of Lenoir 1982. *Annals of Science*, *47*, 291–300.

———. 1991. "'Why are you a vegetarian?' On the historicity of becoming vs. the rationality of being and other practical matters." *Mad River: A Journal of Essays*, no. 2 (spring), 5–10. (MHller on l. 13 of p. 10 is a misprint for "Müller.")

Cantor, Geoffrey N., and Hodge, M.J.S., eds. 1981a. *Conceptions of Ether: Studies in the History of Ether Theories 1740–1900*. Cambridge, U.K.: Cambridge University Press.

______. 1981b. "Introduction: Major themes in the development of ether theories from the ancients to 1900." In Cantor and Hodge 1981a, 1–60.

Carrière, Justus, ed. 1893. *Berzelius und Liebig. Ihre Briefe von 1831–1845 mit erläuternden Einschaltungen aus gleichzeitigen Briefen von Liebig und Wöhler sowie wissenschaftlichen Nachweisen*. Munich and Leipzig: J. F. Lehmann.

Carus, Carl Gustav. 1837. Review of Autenrieth 1836. *Jahrbücher für wissenschaftliche Kritik*, Bd. 2 [des Jahres], Nr. 120, cols. 953–958 (December).

______. 1838–40. *System der Physiologie*. 3 vols. Dresden and Leipzig: Gerhard Fleischer.

Cassirer, Ernst. 1957. *The Philosophy of Symbolic Forms*. Trans. Ralph Manheim. Vol. 3, "The Phenomenology of Knowledge." New Haven: Yale University Press.

Clagett, Marshall, ed. 1959. *Critical Problems in the History of Science*. Madison: University of Wisconsin Press.

Coley, Noel G., and Hall, Vance M. D., eds. 1980. *Darwin to Einstein: Primary Sources on Science and Belief*. Harlow, Essex, U.K.: Longman, in association with the Open University Press.

Cranz, Carl. 1894. "Zum Andenken an C. W. Baur." *Neues Korrespondenz-Blatt für die Gelehrten- und Realschulen Württembergs*, *1*, 485–498.

Culotta, Charles A. 1972. *Respiration and the Lavoisier Tradition: Theory and Modification, 1777–1850*. ("Transactions of the American Philosophical Society," N.S., 62, pt. 3.) Philadelphia: American Philosophical Society.

______. 1974. "German biophysics, objective knowledge, and romanticism." *Historical Studies in the Physical Sciences*, *4*, 3–38.

Davy, Humphry. 1830. *Consolations in Travel, or the Last Days of a Philosopher*. London: John Murray. Reprinted in H. Davy 1839–40, 9 (1840), 207–388 (cited as H. Davy 1840).

______. 1833. *Tröstende Betrachtungen auf Reisen; oder die letzten Tage eines Naturforschers*. Trans. Carl Fr. Ph. von Martius. Nuremberg: J. L. Schrag. A second edition was published in 1839 (not seen).

______. 1839–40. *Collected Works*. Ed. John Davy. 9 vols. London: Smith, Elder & Co.

Davy, John. 1840. *Researches, Physiological and Anatomical*. Philadelphia: A. Waldie.

Dechent, Hermann. 1887. "Passavent, Johann Karl." In *ADB*, 25, 203–207.

Despretz, César-Mansuète. 1824. "Recherches expérimentales sur les Causes de la chaleur animale." *Annales de chimie et de physique*, *26*, 337–364 (August).

Diepgen, Paul. 1942. "Robert Mayer und die Medizin seiner Zeit." In Pietsch and Schimank 1942, 51–96.

Drees, Annette. 1988. *Die Ärzte auf dem Weg zu Prestige und Wohlstand: Sozialgeschichte der württembergischen Ärzte im 19. Jahrhundert*. Münster: F. Coppenrath.

du Bois-Reymond, Emil Heinrich. 1848. *Untersuchungen über thierische Elektricität*. Vol. 1. Berlin: G. Reimer.

du Bois-Reymond, Estelle, ed. 1918. *Jugendbriefe von Emil du Bois-Reymond an Eduard Hallmann*. Berlin: Dietrich Reimer (Ernst Vohsen).

Dürr, Friedrich. 1895. *Heilbronner Chronik*. Heilbronn: Eugen Salzer.

______. 1914. "Erinnerungen an Robert Mayer." *Heilbronner Unterhaltungsblatt*. Beilage zur Neckar-Zeitung, Nr. 97 (23 November), 385 (not seen). Reprinted (in part) in Schmolz and Weckbach 1964, 16–17, 147–148 (cited from there).

Dulong, Pierre-Louis. 1831. "Recherches sur la chaleur spécifique des fluides élastiques." *Mémoires de l'Académie royale des Sciences de l'Institut de France*, *10*, 147–191.

Dulong, Pierre-Louis. 1841. "Mémoire sur la chaleur animale." *Annales de chimie et de physique*, 3d ser., *1*, 440–455, pl. III (April).

Dumas, Jean-Baptiste. 1841a. "Sur la Chimie organique." *Journal des Débats Politiques et Littéraires*, 20 August, [1]–[3]; 21 August, [1]–[3].

———. 1841b. "Leçon sur la Statique chimique des êtres organisés, Professée par M. Dumas, Pour la clôture de son Cours à l'École de Médecine [le 20 Août 1841]." *Annales des sciences naturelles*, 2d ser., *16*, [Partie de la] Zoologie, 33–61 (July).

———. 1841c. *Leçon sur la statique chimique des êtres organisés*. Paris: Fortin, Masson et Cie. Not seen.

———. 1842a. *Essai de statique chimique des êtres organisés*. 2d ed. Paris: Fortin, Masson et Cie. The body of the text appears to be an exact reprinting of 1841c.

———. 1842b. "Additions à la leçon sur la statique chimique des êtres vivans, par M. Dumas (Extrait)." *Annales des sciences naturelles*, 2d ser., *17*, [Partie de la] Zoologie, 122–128 (February).

———. 1842c. "Essai de statique chimique des êtres organisés (1)." *Annales de chimie et de physique*, 3d ser., *4*, 115–126 (January). The footnote to the title says: "(1) Paris, chez *Fortin* et *Masson*, 2e édition, 1842." This piece is *not* the same as the other like-named entries above.

Duttenhofer, Friedrich Martin. 1842. "Ueber die sogenannte physiologische Schule der Medicin." *Medicinisches Correspondenz-Blatt des Württembergischen ärztlichen Vereins*, Bd. 12, Nr. 24–25 (2, 8 August), 185–191, 193–199.

Eisenlohr, Wilhelm Friedrich. 1841. *Lehrbuch der Physik*. 3d ed. Mannheim: Heinrich Hoff.

Eisert, Gisela. 1978. *Robert-Mayer Bibliographie*. Heilbronn: Stadtarchiv.

Elkana, Yehuda. 1970. "Helmholtz' 'Kraft': An illustration of concepts in flux." *Historical Studies in the Physical Sciences*, 2, 263–298.

———. 1974. *The Discovery of the Conservation of Energy*. Cambridge, Mass.: Harvard University Press; London: Hutchinson Educational.

Emmert, August Gottfried Ferdinand. 1809. "Ueber den Einfluß der herumschweifenden Nerven auf das Athmen." *Archiv für die Physiologie*, Bd. 9, Hft. 2, 380–420.

———. 1812. "Nachtrag zu den Beobachtungen über den Einfluß des Stimmnervens auf die Respiration, nebst einigen Bemerkungen über den sympathischen Nerven bey den Säugethieren und Vögeln." *Archiv für die Physiologie*, Bd. 11, Hft. 2, 117–130.

Engels, Eve-Marie. 1980. "Lebenskraft." In Joachim Ritter [and Karlfried Gründer], eds., *Historisches Wörterbuch der Philosophie*, vols. 1–6 (Basel and Stuttgart: Schwabe, 1971–1984), 5, 122–127.

Erman, Paul. 1812. "Einige Bemerkungen über Muskular-Contraction." *Annalen der Physik*, Bd. 40, St. 1, 1–30 (≈ January).

Eschenmayer, Carl August von. 1796. *Principia quaedam disciplinae naturali, inprimis chemiae ex metaphysica naturae substernenda*. Tübingen: Litteris Fuesianis.

———. 1830. *Mysterien des innern Lebens; erläutert aus der Geschichte der Seherin von Prevorst, mit Berücksichtigung der bisher erschienenen Kritiken*. Tübingen: Zu-Guttenberg.

———. 1832. *Grundriß der Natur-Philosophie*. Tübingen: Heinrich Laupp.

———. 1834. *Die Allöopathie und Homöopathie verglichen in ihren Principien*. Tübingen: Ludwig Friedrich Fues.

———. 1835. *Der Ischariothismus unserer Tage. Eine Zugabe zu dem jüngst erschienenen Werke: Das Leben Jesu von Strauß, I. Theil*. Tübingen: Ludw. Friedr. Fues.

———. 1837. *Conflict zwischen Himmel und Hölle, an dem Dämon eines besessenen Mädchens beobachtet. Nebst einem Wort an Dr. Strauß*. Tübingen and Leipzig: Zu-Guttenberg.

———. 1853. "Carl August v. Eschenmayer. (Von Justinus Kerner mitgetheilt.)" *Magikon. Archiv für Beobachtungen aus dem Gebiete der Geisterkunde und des magnetischen und magischen Lebens*, Bd. 5 (1851–53), Hft. 4, 383–422.

Exner, Sigmund. 1914. "Julius Robert v. Mayer." *Wiener klinische Wochenschrift*, 27, 1529–1533.

Faraday, Michael. 1840. "Experimental Researches in Electricity.—Sixteenth [*bzw.* Seventeenth] Series. §24. On the source of power in the voltaic pile." *Philosophical Transactions of the Royal Society of London*, [*130*], pt. 1, 61–91, 93–127.

———. 1841. "Sechszehnte [*bzw.* Siebzehnte] Reihe von Experimental-Untersuchungen über Elcktricität. §24. Ueber die Quelle der Kraft in der Volta'schen Säule." *Annalen der Physik und Chemie*, Bd. 52, St. 1, 149–177 (≈ January); St. 4, 547-573 (≈ April); Bd. 53, St. 2, 316–335 (≈ June); St. 3, 479–498 (≈ July); St. 4, 548–572 (≈ August).

Fischer, Ernst Gottfried. 1827. *Lehrbuch der mechanischen Naturlehre*. 3d ed. Pt. 1. Berlin and Leipzig: G. C. Nauck.

Fischer, Ernst Gottfried, and August, Ernst Ferdinand. 1837. *Ernst Gottfried Fischer's Lehrbuch der mechanischen Naturlehre*. Ed. E. F. August. 4th ed. Pt. 1. Berlin: Nauck.

Fischer, Friedrich. 1838a. "Ueber den Somnambulismus." *Deutsche Vierteljahrs Schrift*, Hft. 1, 293–322.

———. 1838b. *Lehrbuch der Logik für academische Vorlesungen und Gymnasialvorträge*. Stuttgart: J. B. Metzler.

———. 1839. *Der Somnambulismus*. Vol. 1. Basel: Schweighauser.

French, Roger K. 1981. "Ether and physiology." In Cantor and Hodge 1981a, 111–134.

Friedlaender, Salomo. 1905. *Julius Robert Mayer*. Leipzig: Theod. Thomas.

Fries, Jacob Friedrich. 1804. *System der Philosophie als evidente Wissenschaft*. Leipzig: Johann Conrad Hinrichs; reprinted photomechanically in Fries 1967–82, 3 (1968).

———. 1822. *Die mathematische Naturphilosophie nach philosophischer Methode bearbeitet*. Heidelberg: Christian Friedrich Winter; reprinted photomechanically in Fries 1967–82, *13* (1979).

———. 1826. *Lehrbuch der Naturlehre*. Pt. 1, "Experimentalphysik." Jena: Cröker; reprinted photomechanically in Fries 1967–82, *15* (1973).

———. 1827. *Grundriß der Logik*. 3d ed. Heidelberg: Chr. Friederich [*sic*] Winter; reprinted photomechanically in Fries 1967–82, 7 (1971).

———. 1831. Review of J. Kerner 1829. *Hermes, oder Kritisches Jahrbuch der Literatur*, Bd. 35, 333–344. Published anonymously; for authorship, see Bianco 1980, 80.

———. 1833. *Populäre Vorlesungen über die Sternkunde*. 2d ed. Heidelberg: Christian Friedrich Winter; reprinted photomechanically in Fries 1967–82, *16* (1973).

———. 1837. *System der Logik*. 3d ed. Heidelberg: Christian Friedrich Winter.

———. 1967–82. *Sämtliche Schriften*. Ed. Gert König and Lutz Geldsetzer. 25 vols. Aalen: Scientia Verlag.

Galaty, David. 1974. "The philosophical basis of mid-nineteenth century German reductionism." *Journal of the History of Medicine and Allied Sciences*, 29, 295–316.

Gay-Lussac, Joseph-Louis. 1807. "Premier essai Pour déterminer les variations de température qu'éprouvent les gaz en changeant de densité, et considérations sur leur capacité pour le calorique." *Mémoires de physique et de chimie de la Société d'Arcueil*, *1*, 180–203.

Gay-Lussac, Joseph-Louis, and Welter, Jean-Joseph. 1822. "Sur la Dilatation de l'air." *Annales de chimie et de physique*, *19*, 436–437 (April). An unsigned note reporting their work. Journal volume erroneously bears the date 1821.

Gehler, Johann Samuel Traugott. 1825–45. *Johann Samuel Traugott Gehler's Physikalisches*

*Wörterbuch*. Ed. H. W. Brandes, L. Gmelin, J. C. Horner, J. J. Littrow, G. W. Muncke, and C. H. Pfaff. 11 vols. in 20 pts. Leipzig: E. B. Schwickert.
Geiger, Philipp Lorenz. 1827. *Handbuch der Pharmacie*. Vol. 1. 2d ed. Heidelberg: August Osswald; Vienna: J. G. Heubner.
______. 1833. *Handbuch der Pharmacie*. Vol. 1. 4th ed. Heidelberg: C. F. Winter; Vienna: C. Gerold.
Gerlach, Walther. 1942. "Julius Robert Mayer, Leben und Werk." In Pietsch and Schimank 1942, 1–51.
______. 1963. "Julius Robert Mayer—Das Gesetz der Erhaltung der Energie. Entstehung, Geschichte, Bedeutung." *Historischer Verein Heilbronn, 24. Veröffentlichung*, 9–36.
Gerok, Karl Friedrich. 1892. *Jugenderinnerungen*. 5th ed. Bielefeld and Leipzig: Velhagen and Klasing.
Gillispie, Charles C. 1960. *The Edge of Objectivity*. Princeton, N.J.: Princeton University Press.
Gmelin, Christian Gottlob. 1835. *Einleitung in die Chemie*. Vol. 1, pt. 1. Tübingen: H. Laupp. In Mayer's library (saw).
Gmelin, Ferdinand Gottlob. 1821. *Allgemeine Pathologie des menschlichen Körpers*. 2d ed. Stuttgart and Tübingen: J. G. Cotta. In Mayer's library.
______. 1835. *Critik der Principien der Homöopathie*. Tübingen: C. F. Osiander.
Gmelin, Leopold. 1817. *Handbuch der theoretischen Chemie*. Vol. 1. Frankfurt am Main: In Commission bei Franz Varrentrapp.
______. 1827. *Handbuch der theoretischen Chemie*. 3d ed. Vol. 1, pt. 1. Frankfurt am Main: Franz Varrentrapp.
______. 1843. *Handbuch der Chemie*. 4th ed. Vol. 1. Heidelberg: Karl Winter.
Göschel, Carl Friedrich. 1836. Review of Brougham 1835a and 1835b. *Jahrbücher für wissenschaftliche Kritik*, Bd. 1 [des Jahres], Nr. 34–35, cols. 265–272, 273–280 (February).
Goethe, Johann Wolfgang von. 1823. "Problem und Erwiderung." *Zur Morphologie. Erfahrung, Betrachtung, Folgerung durch Lebensereignisse verbunden*, Bd. 2, Hft. 1, 28–45 (not seen). Reprinted in *Goethes Werke*, hrsg. im Auftrage der Großherzogin Sophie von Sachsen, Abth. 2, Bd. 7 (Weimar: Hermann Böhlau, 1892), 74–92, and in Goethe, *Die Schriften zur Naturwissenschaft*, Abt. 1, Bd. 9 (Weimar: Hermann Böhlaus Nachfolger, 1954), 295–306.
Goodfield, June. 1960. *The Growth of Scientific Physiology*. London: Hutchinson.
Gower, Barry. 1973. "Speculation in physics: The history and practice of Naturphilosophie." *Studies in History and Philosophy of Science*, 3 (1972–73), 301–356.
Gregory, Frederick. 1977. *Scientific Materialism in Nineteenth Century Germany*. Dordrecht and Boston: D. Reidel.
Griesinger, Wilhelm. 1843. "Bemerkungen zur neuesten Entwicklung der allgemeinen Pathologie." *Archiv für physiologische Heilkunde*, Jg. 2, Hft. 2, 278–289.
______. 1844. "Medicinische Briefe. Erster Brief." *Jahrbücher der Gegenwart*, Bd. 2, Hft. 10 (October), 898–910. Signed "W.G."; for authorship, see Griesinger's letter of 3 June 1843 to Albert Schwegler, the editor (Universitätsarchiv Tübingen, Md753/I 84).
______. 1845. *Die Pathologie und Therapie der psychischen Krankheiten*. Stuttgart: Adolph Krabbe.
Grüsser, Otto-Joachim. 1987. *Justinus Kerner 1786–1862: Arzt—Poet—Geisterseher*. Berlin: Springer.
Haas, Arthur Erich. 1909. *Die Entwicklungsgeschichte des Satzes von der Erhaltung der Kraft*. Vienna: Alfred Hölder.

Haberling, Wilhelm. 1924. *Johannes Müller. Das Leben des rheinischen Naturforschers.* Leipzig: Akademische Verlagsgesellschaft.

Hagen, Walter. 1963. "Justinus Kerner." *Lebensbilder aus Schwaben und Franken*, 9, 145–173.

Hahnemann, Samuel. 1810. *Organon der rationellen Heilkunde.* Dresden: Arnold.

———. 1819. *Organon der Heilkunst.* 2d ed. Dresden: Arnold.

Hall, Vance M. D. 1977. "Some Contributions of Medical Theory to the Discovery of the Conservation of Energy Principle during the Late 18th and Early 19th Centuries." Ph.D. diss., University of London (University College).

———. 1980. "The role of force or power in Liebig's physiological chemistry." *Medical History*, 24, 20–59.

Haller, Albrecht von. 1762. *Versuch Schweizerischer Gedichte.* 9th ed. Göttingen: Verlegts Abram Vandenhoeks sel. Witwe.

Harman, Peter M. 1982. *Metaphysics and Natural Philosophy: The Problem of Substance in Classical Physics.* Brighton, Sussex, U.K.: Harvester Press; Totowa, N.J.: Barnes & Noble.

Hartkopf, Werner. 1979. "Schellings Naturphilosophie." *Philosophia Naturalis, 17* (1978–79), 349–372.

———. 1984. "Denken und Naturentwicklung. Zur Aktualität der Philosophie des jungen Schelling." In Hans Jörg Sandkühler, ed., *Natur und geschichtlicher Prozeß: Studien zur Naturphilosophie F.W.J. Schellings* (Frankfurt am Main: Suhrkamp), 83–126.

Hartmann, Max. 1942. "Das Gesetz der Erhaltung der Energie in seinen Beziehungen zur Philosophie." In Pietsch and Schimank 1942, 303–328.

Hasper, Moritz. 1831. *Ueber die Natur und Behandlung der Krankheiten der Tropenländer durch die medizinische Topographie jener Länder erläutert nebst der in den Tropenländern zur Verhütung derselben zu beobachtenden Diätetik.* 2 vols. Leipzig: C.H.F. Hartmann. In Mayer's library (saw).

Hegel, Georg Wilhelm Friedrich. 1830. *Encyclopädie der philosophischen Wissenschaften im Grundrisse.* 3d ed. Heidelberg: Verwaltung des Oßwald'schen Verlags (C. F. Winter). Not seen.

———. 1959. *Enzyklopädie der philosophischen Wissenschaften im Grundrisse (1830).* Ed. Friedhelm Nicolin and Otto Pöggeler. Hamburg: Felix Meiner.

Heimann, Peter M. 1976. "Mayer's concept of 'force': The 'axis' of a new science of physics." *Historical Studies in the Physical Sciences*, 7, 277–296.

Hell, Bernhard. 1914. "Robert Mayer." *Kantstudien*, 19, 222–248.

———. 1925. *J. Robert Mayer und das Gesetz von der Erhaltung der Energie.* Stuttgart: Fr. Frommanns Verlag (H. Kurtz).

Helmholtz, Herman. 1847. *Über die Erhaltung der Kraft.* Berlin: G. Reimer.

———. 1877. *Das Denken in der Medicin.* Berlin: August Hirschwald. Not seen.

Henle, Jakob. 1841. *Allgemeine Anatomie. Lehre von den Mischungs- und Formbestandtheilen des menschlichen Körpers* (= *Samuel Thomas von Sömmering vom Baue des menschlichen Körpers.* Ed. W. Th. Bischoff, J. Henle, E. Huschke, F. W. Theile, G. Valentin, J. Vogel, and R. Wagner. Vol. 6). Leipzig: Leopold Voß, [1839]–1841.

———. 1842. "Medizinische Wissenschaft und Empirie." *Zeitschrift für rationelle Medizin*, Jg. 1, Hft. 1, 1–35.

Hennemann, Gerhard. 1959. *Naturphilosophie im 19. Jahrhundert.* Freiburg and Munich: Karl Albers.

———. 1975. *Grundzüge einer Geschichte der Naturphilosophie und ihrer Hauptprobleme.* Berlin: Duncker & Humblot.

Hermann, Armin, ed. 1971–72. *Geschichte der Physik*. 2 vols. Cologne: Aulis Verlag Deubner & Co.

______. 1972a. "Mayer, Julius Robert." In Hermann 1971–72, 2, 228–231.

______. 1972b. "Naturphilosophie." In Hermann 1971–72, 2, 247–249.

______. 1976. "Dynamismus und Atomismus—Die beiden Systeme der Physik in der 1. Hälfte des 19. Jahrhunderts." *Erkenntnis, 10*, 311–322.

______. 1977. "Schelling und die Naturwissenschaften." *Technikgeschichte, 44*, 47–53.

______. 1978. "Die Entdeckung des Energie-Prinzips: Wie der Arzt Julius Robert Mayer die Physiker belehrte." *Bild der Wissenschaft, 15*, 140–142, 144, 146–148.

______. 1980. "Physik an der Eberhard-Karls-Universität." In Wolf von Engelhardt, ed., *Physik, Physiologische Chemie und Pharmazie an der Universität Tübingen* (Tübingen: J.C.B. Mohr [Paul Siebeck]), 13–39.

Hermann, Karl. 1966. "David Friedrich Strauß in Sontheim und Heilbronn." *Historischer Verein Heilbronn, 25. Veröffentlichung*, 179–197.

Herschel, John Frederick William. 1833. *A Treatise on Astronomy*. London: Longman, Rees, Orme, Brown, Green & Longman, and John Taylor.

______. 1838. *Die Lehren der Astronomie für Gebildete faßlich dargestellt*. Ed. F.B.G. Nicolai. Heilbronn and Leipzig: J. D. Claß, [1835]–1838. In Mayer's library.

Hettich, Hermann Otto Friedrich. 1878a. "Plaudereien über Robert Mayer." *Schwäbische Kronik, des Schwäbischen Merkurs zweite Abtheilung* (Stuttgart), Nr. 144 (19 June), 1225.

______. 1878b. "Das Doctoriren und die Inaugural-Dissertationen nebst einer eingeflochtenen weiteren 'Plauderei' über Robert Mayer." *Medicinisches Correspondenz-Blatt des Württembergischen ärztlichen Vereins, 48*, 209–212.

______. 1886. "Erinnerungen an Dr. J. Robert v. Mayer, geb. in Heilbronn 5. Nov. 1814, †daselbst 20. März 1877 [*sic*]." *Schwäbische Chronik, des Schwäbischen Merkurs zweite Abteilung* (Stuttgart), Nr. 67 (21 March), 1. Published anonymously; identified in Eisert 1978, 12.

Heubner, Otto. 1878. "C. A. Wunderlich. Nekrolog." *Archiv der Heilkunde, 19*, 289–320.

Heusinger, Carl Friedrich. 1836. Review of Thackrah 1834. *Jahrbücher der in- und ausländischen gesammten Medicin*, Bd. 9, Nr. 2 (February), 266–270.

Hiebert, Erwin. 1959. "Commentary on the papers of Thomas S. Kuhn and I. Bernard Cohen." In Clagett 1959, 391–400.

Hirn, Gustav-Adolphe. 1864. "Esquisse élémentaire de la théorie mécanique de la chaleur et de ses conséquences philosophiques." *Bulletin de la Société d'histoire naturelle de Colmar*, 4e année, 1863 (publ. 1864), 3–126.

______. 1868. *Conséquences philosophiques et métaphysiques de la thermodynamique. Analyse élémentaire de l'univers* (= *Théorie mécanique de la chaleur*, [pt. 2]). Paris: Gauthier-Villars.

Hofmann, August Wilhelm, ed. 1888. *Aus Justus Liebig's und Friedrich Wöhler's Briefwechsel in den Jahren 1829–1873*. 2 vols. Braunschweig: Friedrich Vieweg & Sohn.

Holmberg, Arne, ed. 1933–53. *Bibliographie de J. J. Berzelius*. Pt. 1 (1933); supplement 2 (1953). Stockholm and Upsalla: Almqvist & Wiksells.

Holmes, Frederic L. 1964. "Introduction" to Justus Liebig, *Animal Chemistry or Organic Chemistry in its Application to Physiology and Pathology*. A Facsimile of the Cambridge Edition of 1842 (New York and London: Johnson Reprint Corp.), vii-cxvi.

______. 1974. *Claude Bernard and Animal Chemistry: The Emergence of a Scientist*. Cambridge, Mass.: Harvard University Press.

______. 1986. "Patterns of scientific creativity." *Bulletin of the History of Medicine, 60*, 19–35.

Holstein, Klaus. 1979. *Die Psychiatrie A.K.A. Eschenmayers (1768–1852): Ein Beitrag zur Entstehungsgeschichte der Psychiatrie in Deutschland*. Frankfurt am Main, Bern, Las Vegas: Peter Lang.

Holtzmann, Karl. 1864. "Nekrolog des Professors v. Nörrenberg in Stuttgart." *Jahreshefte des Vereins für vaterländische Naturkunde in Württemberg, 20*, 24–28.

Hoppe, Brigitte. 1967. "Polarität, Stufung und Metamorphose in der spekulativen Biologie der Romantik." *Naturwissenschaftliche Rundschau, 20*, 380–383.

Hoppe, F. 1835. *Die Eröffnung der Blutadern, eine vollständige Beschreibung des Aderlasses nebst den Indicationen, ein Leitfaden zum Gebrauch für Wundärzte und Chirurgen-Gehülfen*. Neisse and Leipzig: Theodor Hennings.

Hünefeld, Friedrich Ludwig. 1840. *Der Chemismus in der thierischen Organisation*. Leipzig: F. A. Brockhaus.

______. 1841. *Chemie und Medicin in ihrem engeren Zusammenwirken*. 2 vols. Berlin: Theod. Christ. Friedr. Enslin.

Hufeland, Christoph Wilhelm. 1820. *Ueber die Gleichzahl beider Geschlechter im Menschengeschlecht. Ein Beitrag zu der höhern Ordnung der Dinge in der Natur*. Berlin: G. Reimer.

Hufeland, Friedrich. 1834. Review of C.H.E. Bischoff 1825–34. *Jahrbücher für wissenschaftliche Kritik*, Bd. 2 [des Jahres], Nr. 41–43, cols. 329–336, 337–342, 345–348 (September).

Jentsch, Ernst. 1916. "Zur Geschichte der Entdeckung Julius Robert Mayers." *Die Naturwissenschaften, 4*, 90–93.

Jolly, Philipp Gustav. 1841. *Specimen primum ad doctrinam de machinarum effectu pertinens*. Heidelberg: Sumptibus Christiani Friderici Winter.

______. 1842a. Letter of 29 May 1842 to Liebig; from Heidelberg. 2 pp. In Bayerische Staatsbibliothek (Munich), Liebigiana II.B.Jolly, no. 1.

______. 1842b. Review of Schubert 1842. *Heidelberger Jahrbücher der Literatur*, Jg. 35, 2. Hälfte, 6. Doppelheft (November–December), Nr. 54, 851–855.

______. 1842c. Letter of 12 November 1842 to Liebig; from Heidelberg. 7 pp. In Bayerische Staatsbibliothek (Munich), Liebigiana II.B.Jolly, no. 2.

______. 1842d. Letter of 12 December 1842 to Liebig; from Heidelberg. 2 pp. In Staatsbibliothek Preußischer Kulturbesitz (Berlin), F1a 1850(2).

Jørgensen, Bent Søren. 1965a. "Berzelius und die Lebenskraft." *Centaurus, 10* (1964–65), 258–281.

______. 1965b. "More on Berzelius and the vital force." *Journal of Chemical Education, 42*, 394–396.

Joule, James Prescott. 1843. "On the Calorific Effects of Magneto-Electricity, and on the Mechanical Value of Heat." *The London, Edinburgh, and Dublin Philosophical Magazine and Journal of Science*, 3d ser., 23, nos. 152–154 (October–December), 263–276, 347–355, 435–443. Reprinted in Joule 1884, 123–159.

______. 1845a. "On the Changes of Temperature produced by the Rarefaction and Condensation of Air." *The London, Edinburgh, and Dublin Philosophical Magazine and Journal of Science*, 3d ser., 26, no. 174 (May), 369–383. Reprinted in Joule 1884, 172–189.

______. 1845b. "On the Existence of an Equivalent Relation between Heat and the ordinary Forms of Mechanical Power." *The London, Edinburgh, and Dublin Philosophical Magazine and Journal of Science*, 3d ser., 27, no. 179 (September), 205–207. Reprinted in Joule 1884, 202–205.

______. 1847a. "On the Mechanical Equivalent of Heat, as determined by the Heat evolved by the Friction of Fluids." *The London, Edinburgh, and Dublin Philosopical Magazine and Journal of Science*, 3d ser., 31, no. 207 (September), 173–176. Reprinted in Joule 1884, 277–281.

Joule, James Prescott. 1847b. "Expériences sur l'identité entre le calorique et la force mécanique. Détermination de l'équivalent par la chaleur dégagée pendent la friction du mercure." *Comptes rendus hebdomadaires des séances de l'Académie des sciences*, 25, pt. 2, 309–311. Reprinted in Joule 1884, 283–286 and Mayer 1893b, 271–273.

———. 1848. "Ueber das mechanische Aequivalent der Wärme, bestimmt durch die Wärme-Erregung bei Reibung von Flüssigkeiten." *Annalen der Physik und Chemie*, Bd. 73, Nr. 3, 479–484.

———. 1849. "Sur l'équivalent mécanique du calorique." *Comptes rendus hebdomadaires des séances de l'Acdémie des sciences*, 28, pt. 1, 132–135. Reprinted in Mayer 1893b, 276–280.

———. 1850. "On the Mechanical Equivalent of Heat." *Philosophical Transactions of the Royal Society of London*, [140], pt. 1, 61–82. Reprinted in Joule 1884, 298–328.

———. 1884. *The Scientific Papers of James Prescott Joule*. London: Taylor & Francis.

Kämtz, Ludwig Friedrich. 1839. *Lehrbuch der Experimentalphysik*. Halle: Gebauer.

Kahlbaum, Georg W. A., ed. 1904. *Justus von Liebig und Friedrich Mohr in ihren Briefen von 1834-1870*. Leipzig: Johann Ambrosius Barth.

Kamphausen, Georg, and Schnelle, Thomas. 1982. *Die Romantik als naturwissenschaftliche Bewegung. Zur Entwicklung eines neuen Wissenschaftsverständnisses*. Bielefeld: Forschungsschwerpunkt Wissenschaftsforschung, Universität Bielefeld, 1982 [B. Kleine Verlag, 1981].

Kant, Immanuel. 1786. *Metaphysische Anfangsgründe der Naturwissenschaft*. Riga: Johann Friedrich Hartknoch. Not seen. Reprinted in (and cited from) *Kant's gesammelte Schriften*, ed. Königlich Preußische Akademie der Wissenschaften, Abt. 1, Bd. 4 (Berlin: Georg Reimer, 1911), 465–565.

Kastner, Karl Wilhelm Gottlob. 1832–33. *Grundzüge der Physik und Chemie*. 2d ed. 2 vols. Nuremberg: Johann Adam Stein; Vienna: J. B. Wallishauser.

Kerner, Justinus. 1829. *Die Seherin von Prevorst. Eröffnungen über das innere Leben des Menschen und über das Hereinragen einer Geisterwelt in die unsere*. 2 vols. Stuttgart and Tübingen: J. G. Cotta.

———. 1834. *Geschichten besessener neuerer Zeit. Beobachtungen aus dem Gebiete kakodämonisch-magnetischer Erscheinungen. Nebst Reflexionen von C. A. Eschenmayer über Besessenseyn und Zauber*. Karlsruhe: G. Braun.

———. 1836. *Eine Erscheinung aus dem Nachtgebiete der Natur, durch eine Reihe von Zeugen gerichtlich bestätigt und den Naturforschern zum Bedenken mitgetheilt*. Stuttgart and Tübingen: J. G. Cotta.

———. 1843. "[Über die Heilung durch Sympathie]." Makes up most of Mayer 1843b, 109–116.

Kerner, Theobald. 1866. Letter of 3 April [1866] to Mayer, in RM-Archiv, RG III; the year supplied from a letter of 13 May 1866 to Mayer from Graf Alexander von Württemberg (RM-Archiv, fasc. 32e).

Kleinert, Andreas, ed. 1979. *Justus von Liebig, "Hochwohlgeborner Freyherr": die Briefe an Georg von Cotta und die anonymen Beiträge zur Augsburger Allgemeinen Zeitung*. Mannheim: Bionomica-Verlag; Heidelberg: Carl Winter Universitätsverlag; Vienna: Kommissionsverlag Wilhelm Braumüller.

Kohlrausch, Otto Ludwig Bernhard. 1844. *Physiologie und Chemie in ihrer gegenseitigen Stellung beleuchtet durch eine Kritik von Liebigs Thierchemie*. Göttingen: Dieterich.

Krafft, Fritz. 1982. "Bedingungen und Voraussetzungen für das Entstehen moderner

Physik in der ersten Hälfte des 19. Jahrhunderts." In Luboš Nový, ed., *Impact of Bolzano's Epoch on the Development of Science* (Prague: Czechoslovak Academy of Sciences, Institute of Czechoslovak and General History), 75–101.

Kremer, Richard L. 1984. "The Thermodynamics of Life and Experimental Physiology, 1770–1880." Ph.D. diss., Harvard University; reprinted photomechanically with minor corrections and a new preface, New York and London: Garland Publishing, 1990.

Kretschmar, Carl Traugott. 1834. *Streitfragen aus dem Gebiete der Homöopathie*. Leipzig: J. C. Hinrichs.

———. 1835. *Allöopathie und Homöopathie Hand in Hand*. Leipzig: Robert Friese.

Krimer, Johann Franz Wenzel. 1823. *Versuch einer Physiologie des Blutes*. Pt. 1. Leipzig: Carl Cnobloch.

Krusemarck, Götz. 1942. *Neues aus Robert Mayers Lebenskreis*. ("Veröffentlichungen des Archivs der Stadt Heilbronn a. N.," Hft. 2.) Heilbronn am Neckar: Eugen Salzer.

Kuhn, Thomas S. 1958. "The caloric theory of adiabatic compression." *Isis*, 49, 132–140.

———. 1959. "Energy conservation as an example of simultaneous discovery." In Clagett 1959, 321–356.

———. 1974. "Second Thoughts on Paradigms." In Frederick Suppe, ed., *The Structure of Scientific Theories* (Urbana, Chicago, London: University of Illinois Press), 459–482.

Lamé, Gabriel. 1836–37. *Cours de physique de l'École polytechnique*. 2 vols. (2 in 2 pts.). Paris: Bachelier.

———. 1840. *Cours de physique de l'École polytechnique*. 2d ed. 3 vols. Paris: Bachelier.

Lehmann, Carl Gotthelf. 1842. *Lehrbuch der Physiologischen Chemie*. Vol. 1. Leipzig: Wilh. Engelmann; Rotterdam: Adolph Baedeker, [1841]–1842.

Lenoir, Timothy. 1982. *The Strategy of Life: Teleology and Mechanics in Nineteenth Century German Biology*. Dordrecht, Boston, London: D. Reidel; reprinted photomechanically with minor corrections, Chicago and London: University of Chicago Press, 1989.

Liebig, Justus. 1837–43. *Handbuch der Chemie mit Rücksicht auf Pharmacie* (= Lorenz Geiger, *Handbuch der Pharmacie*. Vol. 1. 5th ed. Ed. Justus Liebig). 2 vols. Heidelberg: C. F. Winter; Vienna: C. Gerold, [1837]–1843.

———. 1839a. "Ueber die Erscheinungen der Gährung, Fäulniß und Verwesung und ihre Ursachen." *Annalen der Pharmacie*, Bd. 30, Hft. 3, 250–287 (≈ March).

———. 1839b. "Ueber Gährung, Verwesung und Fäulniß. Nachtrag zu der Abhandlung Seite 250 d. Bds." *Annalen der Pharmacie*, Bd. 30, Hft. 3, 363–368 (≈ March).

———. 1839c. "Ueber die Erscheinungen der Gährung, Fäulniß und Verwesung, und ihre Ursachen." *Annalen der Physik und Chemie*, Bd. 48, St. 1, 106–150 (≈ September). The main memoir on 106–144 and the *Nachtrag* on 144–150 reprint Liebig 1839a and 1839b.

———. 1840a. *Traité de chimie organique*. Vol. 1. Paris: Fortin, Masson et Cie., 1840–[1841].

———. 1840b. "Der Zustand der Chemie in Preußen." *Annalen der Chemie und Pharmacie*, Bd. 34, Hft. 1, 97–136 (≈ April).

———. 1840c. *Ueber das Studium der Naturwissenschaften und über den Zustand der Chemie in Preußen*. Braunschweig: Friedrich Vieweg & Sohn.

———. 1840d. *Die organische Chemie in ihrer Anwendung auf Agricultur und Physiologie*. Braunschweig: Friedrich Vieweg & Sohn.

———. 1840e. "Rechtfertigung der Contacttheorie." *Annalen der Chemie und Pharmacie*, Bd. 36, Hft. 2, 153–161 (≈ November).

———. 1841a. *Die organische Chemie in ihrer Anwendung auf Agricultur und Physiologie*. 2d ed. Braunschweig: Friedrich Vieweg & Sohn.

Liebig, Justus. 1841b. "Ueber die stickstoffhaltigen Nahrungsmittel des Pflanzenreichs." *Annalen der Chemie und Pharmacie*, Bd. 39, Hft. 2, 129–160 (≈ August).

———. 1841c. "Chemische Briefe. I [II, III, IV, V]." *Allgemeine Zeitung* (Augsburg), Beilage, Nr. 256, 259, 265, 273, 281 (13, 16, 22, 30 September; 8 October), 2044–2045, 2069–2070, 2115–2118, 2177–2179, 2241–2242. Published anonymously. Letters IV and V reprinted with minor changes as "Zehnter Brief" and "Eilfter [*sic*] Brief" in Liebig 1844b, 114–126 and 127–134.

———. 1841d. "Der Lebensproceß im Thiere, und die Atmosphäre." *Annalen der Chemie und Pharmacie*, Bd. 41, 1842 [*sic*], Hft. 2 (publ. 15 December 1841), 189–219. Largely the same as Liebig 1842b, 1–39 (§§I–VI).

———. 1841e. "Chemische Briefe. VI [VII, VII (Beschluß)]." *Allgemeine Zeitung* (Augsburg), Beilage, Nr. 355, 359, 360 (20, 24, 25 December), 2835–2836, 2865–2866, 2873. Published anonymously. Letters VI (beginning with line 24 on p. 2835), VII, and VII (Beschluß) are largely the same as Liebig 1842b (§II, p. 12, line 21–p. 18; §III, pp. 18–25; and §IV, pp. 25–30).

———. 1842a. "Die Ernährung, Blut- und Fettbildung im Thierkörper." *Annalen der Chemie und Pharmacie*, Bd. 41, Hft. 3 (publ. 31 March 1842), 241–285. Reprinted with some slight changes and a few appreciable additions in Liebig 1842b, 39–101 (§§VII–XX).

———. 1842b. *Die organische Chemie in ihrer Anwendung auf Physiologie und Pathologie*. Braunschweig: Friedrich Vieweg & Sohn.

———. 1843. *Die Thier-Chemie oder die organische Chemie in ihrer Anwendung auf Physiologie und Pathologie*. 2d ed. xvi,382,[1] pp. Braunschweig: Friedrich Vieweg & Sohn. There is also an otherwise identically described edition (not seen) with xviii,344,[1] pp.

———. 1844a. "Chemische Briefe. XVI [XVI (Beschluß), XVII, XVIII, XVIII (Beschluß), XIX, XX, XX (Beschluß), XXI]." *Allgemeine Zeitung* (Augsburg), Beilage, Nr. 90, 91, 98, 112, 113, 127, 135, 136, 145 (30, 31 March; 7, 21, 22 April; 6, 14, 15, 24 May), 713–715, 721–723, 777–778, 889–891, 897–899, 1009–1010, 1073–1074, 1081–1082, 1153–1155. Published anonymously but, as an editor's note on p. 713 makes clear, Liebig's anonymity was something of a fiction. XVI and XVI (Beschluß) reprinted with minor changes as "Erster Brief" in Liebig 1844b, 1–12 and 13–26; XVII reprinted as "Zweiter Brief" in Liebig 1844b, 27–32; XX bears the additional title "Isomorphismus. Atomismus"; XXI bears the additional title "Unterschied der organischen Körper von den Mineralsubstanzen," reprinted as "Zwölfter Brief" in Liebig 1844b, 135–147.

———. 1844b. *Chemische Briefe*. Heidelberg: Akademische Verlagshandlung C. F. Winter. In Mayer's library (saw).

———. 1845. "Ueber die thierische Wärme." *Annalen der Chemie und Pharmacie*, Bd. 53, Hft. 1 (publ. 8 February 1845), 63–77.

———. 1846. "Das Verhältniß der Physiologie und Pathologie zur Chemie und Physik, und die Methode der Forschung in diesen Wissenschaften." *Deutsche Vierteljahrs Schrift*, Hft. 3 [des Jahres], 169–243.

———. 1856a. "Ein Vortrag Liebigs über anorganische Natur und organisches Leben." *Allgemeine Zeitung* (Augsburg), Nr. 24–25 (24–25 January), 369–370, 385–386.

———. 1856b. *De onbewerktuigde natuur en het bewerktuigde leven. Eene redevoering tegen het materialisme dezer dagen*. Trans. H. Kloete Nortier. Rotterdam: Otto Petri.

———. 1858a. "Liebig über die Metamorphose der Kraft." *Allgemeine Zeitung* (Augsburg), Nr. 93 (3 April), Beilage, 1481.

———. 1858b. "Über die Verwandlung der Kräfte." In T.L.W. Bischoff et al., *Wissenschaftliche Vorträge gehalten zu München im Winter 1858* (Braunschweig: Friedrich

Vieweg & Sohn), 581-596. Reprinted with some changes as "Dreizehnter Brief" in Liebig 1859, *1*, 205–220.

______. 1859. *Chemische Briefe*. 4th ed. 2 vols. Leipzig and Heidelberg: C. F. Winter.

Liedman, Sven-Eric. 1966. *Det organiska livet i tysk debatt 1795–1845*. Lund: Berlinska Boktryckeriet. Contains a German summary, "Die deutsche Debatte ueber das organische Leben 1795–1845," on 242–257.

Lindsay, Robert Bruce. 1973. *Men of Physics: Julius Robert Mayer, Prophet of Energy*. Oxford: Pergamon Press.

Link, Heinrich Friedrich. 1840. Review of Carus 1838–40, *1*. and *2*. *Jahrbücher für wissenschaftliche Kritik*, Bd. 1 [des Jahres], Nr. 61–62, cols. 481–488, 489–491 (April).

Lipman, Timothy O. 1965. "More on Berzelius and the vital force." *Journal of Chemical Education*, *42*, 396–397.

______. 1967. "Vitalism and reductionism in Liebig's physiological thought." *Isis*, *58*, 167–185.

Lippmann, Edmund Oskar von. 1897. "Robert Mayer und das Gesetz von der Erhaltung der Kraft." *Zeitschrift für Naturwissenschaften*, *70* (1897–98), 1–36.

______. 1906. "Robert Mayer und das Gesetz von der 'Erhaltung der Kraft.'" In his *Abhandlungen und Vorträge zur Geschichte der Naturwissenschaften* (Leipzig: Veit), 527–566.

Löwenthal, N. 1842. "Zur Kritik der heutigen Naturwissenschaft." *Deutsche Jahrbücher für Wissenschaft und Kunst*, Jg. 5, Nr. 240–246 (8, 10–15 October), 959–960, 961–964, 965–966, 969–970, 973–975, 977–979, 981–983.

Lotze, Rudolf Hermann. 1839. Review of Stark 1838. *Hallische Jahrbücher für deutsche Wissenschaft und Kunst*, Jg. 2, Nr. 194–199 (14–17, 19–20 August), cols. 1545–1550, 1553–1557, 1561–1564, 1569–1573, 1577–1582, 1585–1592.

______. 1841. *Metaphysik*. Leipzig: Weidmann.

______. 1842. *Allgemeine Pathologie und Therapie als mechanische Naturwissenschaften*. Leipzig: Weidmann.

______. 1843. "Leben. Lebenskraft." In Wagner 1842–53, *1* (1842–[1843]), ix–lviii; reprinted photomechanically in William Coleman, ed., *Physiological Programmatics of the Nineteenth Century* (New York: Arno Press, 1981), separate pagination.

______. 1852. *Medicinische Psychologie oder Physiologie der Seele*. Leipzig: Weidmann.

Ludwig, Carl. 1870. "Leid und Freude in der Naturforschung." *Die Gartenlaube*, 340–344, 358–360. Reprinted in Schröer 1967, 272–286.

Mach, Ernst. 1895. "On the part played by accident in invention and discovery." Trans. Thomas J. McCormack. *The Monist*, *6* (1895–96), 161–175. Reprinted in Mach 1897, 259–281.

______. 1896a. "Über den Einfluß zufälliger Umstände auf die Entwickelung von Erfindungen und Entdeckungen." In his *Populär-wissenschaftliche Vorlesungen* (Leipzig: Johann Ambrosius Barth), 275-296.

______. 1896b. *Principien der Wärmelehre, historisch-kritisch entwickelt*. Leipzig: Johann Ambrosius Barth.

______. 1897. *Popular Scientific Lectures*. Trans. Thomas J. McCormack. 2d ed. Chicago: Open Court; London: Kegan Paul, Trench, Truebner & Co.

______. 1960. *The Science of Mechanics: A Critical and Historical Account of Its Development*. Trans. Thomas J. McCormack. 6th ed. LaSalle, Ill.: Open Court.

Magendie, François. 1833. *Précis élémentaire de physiologie*. 3d ed. 2 vols. Paris: Méquignon-Marvis.

______. 1834–36a. *Lehrbuch der Physiologie*. 3d ed. Trans. and ed. Carl Ludwig Elsässer. 2 vols. Tübingen: C. F. Osiander.

Magendie, François. 1834–36b. *Handbuch der Physiologie*. 3d ed. Trans. and ed. Carl Friedrich Heusinger. 2 vols. Eisenach: Johann Friedrich Bärecke; Vienna: Gerold.

______. 1836. *Précis élémentaire de physiologie*. 4th ed. 2 vols. Paris: Méquignon-Marvis.

______. 1836–37. *Leçons sur les phénomènes physiques de la vie, professées au Collège de France*. Ed. C. James. (Half-title: "Phénomènes physiques de la vie.") 3 vols. Paris: Ébrard et Cie. (*1*); J. Angé (*2–3*). Vols. 2 and 3 not seen, but cf. Magendie 1842.

______. 1837. *Vorlesungen über die physikalischen Erscheinungen des Lebens*. Trans. Baswitz. 2 vols. Cologne: M. DuMont-Schauberg.

______. 1838. *Leçons sur le sang, et les altérations de ce liquide dans les maladies graves*. Ed. G. Funel. (= *Leçons sur les phénomènes physiques de la vie*, vol. 4.) Paris: J. Angé et Cie. Not seen, but cf. Magendie 1842.

______. 1839a. *Leçons sur le sang, et les altérations de ce liquide dans les maladies graves*. Ed. G. Funel. Brussels: Société Encyclographique des Sciences médicales.

______. 1839b. *Magendie's Vorlesungen über das Blut*. Trans. Gustav Krupp. (= *Magendie's Vorlesungen über die physikalischen Erscheinungen des Lebens*, vol. 4.) Leipzig: Christian Ernst Kollmann. Not seen.

______. 1842. *Phénomènes physiques de la vie*. 4 vols. Paris: J.-B. Baillière. This edition appears to be identical to Magendie 1836–37 and 1838 with new title pages.

Magnus, Gustav. 1837. "Ueber die im Blute enthaltenen Gase, Sauerstoff, Stickstoff und Kohlensäure." *Annalen der Physik und Chemie*, Bd. 40, St. 4, 583–606 (≈ April).

Maier, Heinrich, ed. 1912. *Briefe von David Friedrich Strauss an L[udwig]. Georgii*. Tübingen: J.C.B. Mohr (Paul Siebeck).

Mani, Nikolaus. 1956. "Das Werk von Friedrich Tiedemann u. Leopold Gmelin: 'Die Verdauung nach Versuchen', und seine Bedeutung für die Entwicklung der Ernährungsslehre in der ersten Hälfte des 19. Jahrhunderts." *Gesnerus*, *13*, 190–214.

Martini, Fritz. 1982. "Kurz, Hermann." In *NDB*, *13*, 329–332.

Marx, Otto M. 1970. "Nineteenth-century medical psychology: Theoretical problems in the work of Griesinger, Meynert, and Wernicke." *Isis*, *61*, 355–370.

Matthes, Johann Christian. 1811. *Dissertatio inauguralis medica de differentia, quae naturam vis organicae et fluidorum imponderabilium indolem intercedit*. Tübingen: Ludov. Fried. Fues.

Mayer, Julius Robert. 1838a. *Ueber das Santonin*. Heilbronn: Maximilian Müller. Reprinted in (and cited from) Mayer 1893b, 23–44.

______. 1838b. "Zweite medicinische Staatsprüfung." Manuscript in Staatsarchiv Ludwigsburg, Bestand E 162 II, Büschel 507, fol. 21r–72v. 20.3 x 32 cm. Included here are only the pages dealing with medicine (i.e., excluding those on surgery and obstetrics), written on 15–16 August.

______. 1840. "Schip Java. Sieken Lijst." Manuscript in RM-Archiv, fasc. 9. 19 pp. 21 × 23 cm.

______. 1840–41. "Tagebuch der Reise nach Ostindien." In Mayer 1893b, 47–76.

______. 1841a. "Ueber [die] quantitative und qualitative Bestimmung der Kräfte." Manuscript published in facsimile in Zöllner 1878–81, *4* (1881), between pp. 688 and 689. Cited from Mayer 1893b, 100–107; also in Mayer 1978, 21–29 (in a new transcription).

______. 1841b. Untitled, undated, and unsigned manuscript bearing motto "sapere aude." In RM-Archiv, fasc. 16. 10 pp. 22 x 28 cm.

______. 1842. "Bemerkungen über die Kräfte der unbelebten Natur." *Annalen der Chemie und Pharmacie*, Bd. 42, Hft. 2, 233–240. Reprinted in Mayer 1893a, 23–30, and (in facsimile) Sarton 1929, 35–42, and Mayer 1978, 33–40.

______. 1843a. Undated and unsigned draft of a letter to a "Lieber Freund"—i.e., Carl Wilhelm Baur; never completed or sent. In RM-Archiv, fasc. 9. 4 pp.

______. 1843b. "[Bericht über die] Versammlung der Wundärzte aus den Oberamtsbezirken Besigheim, Brackenhem, Heilbronn, Neckarsulm und Weinsberg [am 30. Mai 1842]." *Medicinisches Correspondenz-Blatt des Württembergischen ärztlichen Vereins*, Bd. 13, Nr. 14–15 (29 May, 2 June), 109–112, 113–116.

______. 1844(?). "Ueber medicinische Physiologie." Manuscript published in (and cited from) Mayer 1978, 431–433. First published in Zaunick 1956.

______. 1844/45. "Theorie der physikalischen Principien als Grundlage physiologischer Lehrsätze." Undated manuscript. In RM-Archiv, fasc 27. 70 pp. 10.5 x 16.5 cm.

______. 1845. *Die organische Bewegung in ihrem Zusammenhange mit dem Stoffwechsel. Ein Beitrag zur Naturkunde*. Heilbronn: C. Drechsler. Reprinted in Mayer 1893a, 45–128 and (in facsimile) Mayer 1978, 43–155.

______. 1846. "Sur la production de la lumière et de la chaleur du soleil." First published in (and cited from) Mayer 1893b, 261–270; reprinted from there in Mayer 1978, 159–166.

______. 1848a. *Beiträge zur Dynamik des Himmels in populärer Darstellung*. Heilbronn: Johann Ulrich Landherr. Not seen. Reprinted in Mayer 1893a, 151–216, and (in facsimile) Mayer 1978, 169–236.

______. 1848b. "Sur la transformation de la force vive en chaleur, et réciproquement." *Comptes rendus hebdomadaires des séances de l'Académie des sciences*, 27, pt. 2, 385–387. Reprinted in Mayer 1893b, 274–276.

______. 1849. "Réclamation de priorité contre M. Joule, relativement à la loi de l'équivalence du calorique." *Comptes rendue hebdomadaires des séances de l'Académie des sciences*, 29, pt. 2, 534. Reprinted in Mayer 1893b, 280–281.

______. 1851a. *Bemerkungen über das mechanische Aequivalent der Wärme*. Heilbronn: Johann Ulrich Landherr. Reprinted in Mayer 1893a, 235–276, and (in facsimile) Mayer 1978, 255–308.

______. 1851b. "De l'influence des marées sur la rotation de la terre." First published in Mayer 1893b, 282–285.

______. after 1851(?). "Ueber die physiologische Bedeutung des mechanischen Aequivalents der Wärme." First published in Mayer 1893b, 247–250. Retranscribed from the manuscript in Mayer 1978, 425–427 (cited from this).

______. 1852. "Wahrung literarischer Eigenthumsrechte." First published in Mayer 1978, 443–445.

______. 1860. Untitled, undated manuscript fragment. In RM-Archiv, no fascicle number, in folder "Undatierte Konzeptfragmente u. mathemat. Übungen." 1 p.

______. 1862. "Ueber das Fieber. Ein iatromechanischer Versuch." *Archiv der Heilkunde*, Jg. 3, [Hft. 4], 385–394. Reprinted in Mayer 1893a, 324–336, and (in facsimile) Mayer 1978, 317–326.

______. 1863. Letter of 31 May 1863 to John Tyndall. In the library of the Royal Institution of Great Britain (London). 3 pp.

______. 1865. Untitled autobiographical sketch sent to John Tyndall with a one-page cover letter of 30 December 1865. In the library of the Royal Institution. The manuscript of 18 unnumbered pages is a transcription of the original, which was returned to Mayer's sister-in-law, Emma Closs, in 1897; see the letter from her of 21 March 1897 to Tyndall's widow, also in the Royal Institution.

______. ca. 1863/67. "Aufzeichnungen aus den sechziger Jahren," in Mayer 1893b, 378–385.

Mayer, Julius Robert. 1866. "Ueber temporäre Fixsterne." *Das Ausland. Ueberschau der neuesten Forschungen auf dem Gebiete der Natur-, Erd- und Völkerkunde*, Jg. 39, Nr. 37 (11 September), 865–866 (not seen). Reprinted in Mayer 1893b, 253–257, and (in repaged facsimile) Mayer 1978, 333–337 (cited from this).

______. ca. 1866(?). Untitled and undated manuscript fragment. In RM-Archiv, fasc. 9. 11 pp. (i.e., pp. 57–67 as marked). 22 x 27 cm.

______. 1867. *Die Mechanik der Wärme in gesammelten Schriften*. Stuttgart: J. G. Cotta.

______. 1869. "Ueber nothwendige Consequenzen und Inconsequenzen der Wärmemechanik." *Tageblatt der 43. Versammlung deutscher Naturforscher und Aerzte in Innsbruck vom 18. bis 24. September 1869*, 40–44. Reprinted (in slightly different versions) in Mayer 1893a, 347–357, and Mayer 1978, 341–348; title taken from the latter.

______. 1870. "Ueber die Bedeutung unveränderlicher Größen." *Heilbronner Unterhaltungsblatt* (Beilage zur Neckar-Zeitung), Nr. 131–132 (9, 11 November), 521–522, 525–527 (not seen). Reprinted in Mayer 1893a, 381–393, and (in repaged facsimile) Mayer 1978, 361–370 (cited from this).

______. 1871a. "Ueber Erdbeben." *Heilbronner Unterhaltungsblatt* (Beilage zur Neckar-Zeitung), Nr. 20–21 (15, 17 February), 77–78, 81–82 (not seen). Reprinted (with the addition of a new paragraph added to Mayer 1874) in Mayer 1893a, 367–376, and (in repaged facsimile) Mayer 1978, 351–358 (cited from this).

______. 1871b. "Ueber die Ernährung." First published in Mayer 1871c, 51–76. Reprinted in Mayer 1893a, 396–413, and (in facsimile from Mayer 1871c) Mayer 1978, 373–396.

______. 1871c. *Naturwissenschaftliche Vorträge*. Stuttgart: J. G. Cotta. Reprints Mayer 1869, 1870, 1871a, 1871b.

______. 1872a. Autobiographical sketch. Manuscript in two parts in RM-Archiv: fascicle 19 contains the autobiographical sketch published with a few omissions as "C" in Mayer 1893b, 391–394 (the published portions cited from this); fascicle 9 contains its continuation on one side of a loose sheet dated 8 April 1872.

______. 1872(?)b. A short review of Strauss 1872 (cf. Mayer 1893b, 424–440). In Deutsches Literatur-Archiv/Schiller Nationalmuseum (Marbach am Neckar), Handschriften-Abteilung, Sig. 27618. 1 p. 14 x 11 cm.

______. 1873. "Mayer (Jul. Rob.)." *Supplement zur elften Auflage des Conversations-Lexikon. Encyklopädische Darstellung der neuesten Zeit nebst Ergänzungen früherer Artikel* (2 vols.; Leipzig: F. A. Brockhaus, 1872–73), 2, 193–194. Published anonymously; for authorship see Hettich 1878a, 1225.

______. 1873/74a. Autobiographical sketch. First published in Mayer 1893b, 386-389, as "Aufzeichnungen aus den siebziger Jahren," "A. Mayer und die mechanische Wärmetheorie" (cited from this). Published in a new transcription from the manuscript in Schmolz and Weckbach 1964, 9–11.

______. 1873/74b. Autobiographical sketch. One-page manuscript in RM-Archiv, fascicle 25. Similar but not identical to Mayer 1873/74a.

______. 1874. *Die Mechanik der Wärme in gesammelten Schriften*. 2d ed. Stuttgart: J. G. Cotta.

______. 1876. "Ueber Auslösung." *Staatsanzeiger für Württemberg*, 22 March, Besondere Beilage Nr. 7, 104–107 (not seen, but original page numbers supplied from the 1978 facsimile). Reprinted in Mayer 1893a, 440–446, and (in repaged facsimile) Mayer 1978, 413–416.

______. 1877a. Review of Helmholtz 1877. *Memorabilien. Monatshefte für rationelle praktische Aerzte*, Jg. 22, Hft. 11, 524–525. Reprinted in Mayer 1893b, 438–440.

______. 1877b. Autobiographical sketch. First published as "B" in Mayer 1893b, 389–390.

______. 1878. "Mayer (Jul. Rob. von)." *Conversations-Lexikon. Allgemeine deutsche Real-Encyklopädie*. 12th ed. (15 vols.; Leipzig: F. A. Brockhaus, 1875–79), *10*, 236. Published anonymously, but almost identical to Mayer 1873.

______. 1893a. *Die Mechanik der Wärme in gesammelten Schriften*. 3d ed. Ed. Jacob J. Weyrauch. Stuttgart: J. G. Cotta.

______. 1893b. *Kleinere Schriften und Briefe von Robert Mayer, nebst Mittheilungen aus seinem Leben*. Ed. Jacob J. Weyrauch. Stuttgart: J. G. Cotta.

______. 1978. *Die Mechanik der Wärme. Sämtliche Schriften*. Ed. Hans Peter Münzenmayer. Heilbronn: Stadtarchiv Heilbronn.

______. 1979a. *Robert Mayer—Dokumente zur Begriffsbildung des Mechanischen Äquivalents der Wärme*. Ed. Peter Buck. Bad Salzdetfurth: Verlag Barbara Franzbecker, Didaktischer Dienst, 1979 (c. 1980).

______. 1979b. *Robert Mayer—Die universelle Anwendung des Prinzips vom Mechanischen Äquivalent der Wärme*. Ed. Peter Buck. Bad Salzdetfurth: Verlag Barbara Franzbecker, Didaktischer Dienst, 1979 (c. 1980).

Mende, Erich. 1975. "Der Einfluß von Schellings 'Princip' auf Biologie und Physik der Romantik." *Philosophia naturalis*, *15* (1974–75), 461–485.

Mendelsohn, Everett. 1964a. *Heat and Life: The Development of the Theory of Animal Heat*. Cambridge, Mass.: Harvard University Press.

______. 1964b. "The biological sciences in the nineteenth century: Some problems and sources." *History of Science*, *3*, 39–59.

Mitscherlich, Eilhard. 1833–34. *Lehrbuch der Chemie*. Vol. 1 in 2 pts. 2d ed. Berlin: Ernst Siegfried Mittler.

______. 1834. "Ueber die Aetherbildung." *Annalen der Physik und Chemie*, Bd. 31, Nr. 18, 273–282. Corresponds to 1833–34, *1*, pt. 2 (1834), 97–107.

Mittasch, Alwin. 1935. "Berzelius und die Katalyse. Zum Gedächtnis der Aufstellung des Katalyse-Begriffes 1835." *Chemische Novitäten*, *25*, 169–191.

______. 1939. "Auslösungskausalität: ein vergessenes Kapitel Robert Mayer?" *Die Umschau*, *43*, 1114–1116.

______. 1940a. "Robert Mayer und die Katalyse." *Chemiker-Zeitung*, *64*, 38–40.

______. 1940b. *Julius Robert Mayers Kausalbegriff: Seine geschichtliche Stellung, Auswirkung und Bedeutung*. Berlin: Julius Springer.

______. 1940c. "Julius Robert Mayers Stellung zur Chemie." *Angewandte Chemie*, *53*, 113–118.

______. 1940d. "Robert Mayers Begriff der Naturkausalität, mit Beziehung auf Arthur Schopenhauers Kausallehre." *Die Naturwissenschaften*, *28*, 193–196.

______. 1940e. "Robert Mayers Lehre über das Wirken in der Natur." *Forschungen und Fortschritte*, *16*, 178–180.

Mohl, Robert. 1902. *Lebenserinnerungen*. 2 vols. Stuttgart and Leipzig: Deutsche Verlags-Anstalt.

Mohr, Carl Friedrich. 1837a. "Ueber die Natur der Wärme." *Zeitschrift für Physik und verwandte Wissenschaften*, Bd. 5, Hft. 9, 419–432; Hft. 10, 433–445.

______. 1837b. "Jahresbericht über neue Entdeckungen und Erweiterungen im Gebiete der Pharmacie und der dahin einschlagenden Wissenschaften, für das Jahr 1837." *Annalen der Pharmacie*, Bd. 24, 113–240, 241–352 (all of *Hefte* 2 and 3; ≈ November and December).

______. 1837c. "Ansichten über die Natur der Wärme." *Annalen der Pharmacie*, Bd. 24, Hft. 2, 141–147. Part of 1837b.

Mohr, Carl Friedrich. 1837d. "Kampf der Contact-Theorie mit der chemischen, im Felde des Galvanismus." *Annalen der Pharmacie*, Bd. 24, Hft. 2, 157–163. Part of 1837b.

———. 1837e. "Santonin." *Annalen der Pharmacie*, Bd. 24, Hft. 3, 245–246. Part of 1837b.

———. 1837f. "Ueber die im Blute enthaltenen Gase." *Annalen der Pharmacie*, Bd. 24, Hft. 3, 293–301. Part of 1837b.

Morosi, Giuseppe. 1824. "Di alcuni sperimenti sull'eccitamento del calorico mediante la confricazione de' corpi." *Memorie dell'Imperiale Regio Istituto del Regno Lombardo-Veneto*, 3 (1816–17 [publ. 1824]), pt. 2, 137–147, and pl. III.

Mülberger, Arthur. 1878–79. "Robert Mayer." *Die Neue Welt*, 4 ([1878]–1879), 140–141, 153–155, 163–164.

———. 1879. "Zur Erinnerung an Robert Mayer." *Frankfurter Zeitung und Handelsblatt*, Morgenblatt, Nr. 21 (21 January), 1–2; Nr. 23 (23 January), 1–3.

———. 1896. "Robert Mayer. Ein Lebensbild." *XIII. u. XIV. Jahresbericht (1894 u. 95) des Württembergischen Vereins für Handelsgeographie und Förderung deutscher Interessen im Auslande* (publ. 1896), 3–20.

Müller, Johann. 1836. *Kurze Darstellung des Galvanismus*. Pref. by Justus Liebig. Darmstadt: Ludwig Pabst.

Müller, Johannes. 1833. *Handbuch der Physiologie des Menschen für Vorlesungen*. Vol. 1, pt. 1. Coblenz: J. Hölscher. viii,390 [i.e., 406] pp. Pt. 2 appeared in 1834 (xvi,[407]–852 pp.).

———. 1835. *Handbuch der Physiologie des Menschen für Vorlesungen*. Vol. 1. 2d ed. Coblenz: J. Hölscher. iv,856 [i.e., 850] pp. In Mayer's library (saw).

———. 1837a. *Handbuch der Physiologie des Menschen für Vorlesungen*. Vol. 2, pt. 1. Coblenz: J. Hölscher. [iii],247 pp. In Mayer's library (saw).

———. 1837b. *Handbuch der Physiologie des Menschen für Vorlesungen*. Vol. 1, pt. 1. 3d ed. Coblenz: J. Hölscher. 421 pp. Pt. 2 appeared in 1838 (vi,423–867 pp.).

———. 1838a. "Jahresbericht über die Fortschritte der anatomisch-physiologischen Wissenschaften im Jahre 1837." *Archiv für Anatomie, Physiologie und wissenschaftliche Medicin*, xci–cxcviii.

———. 1838b. *Handbuch der Physiologie des Menschen für Vorlesungen*. Vol. 2, pt. 2. Coblenz: J. Hölscher. [iv],[249]–502,[2] pp. In Mayer's library (saw).

———. 1840. *Handbuch der Physiologie des Menschen für Vorlesungen*. Vol. 2, pt. 3. Coblenz: J. Hölscher. [viii],[505]–780 pp.; I pl. Though dated 1840, this part appears to have been published in 1839. In Mayer's library (saw).

———. 1841. *Handbuch der Physiologie des Menschen für Vorlesungen*. Vol. 1, pt. 1. 4th ed. Coblenz: J. Hölscher. [iv],221 pp.

———. 1843. *Handbuch der Physiologie des Menschen für Vorlesungen*. Vol. 1, pt. 2. 4th ed. Coblenz: J. Hölscher. [iv],225–412 pp. Pt. 3 appeared in 1844 ([i],413-741 pp.).

Muncke, Georg Wilhelm. 1825. "Bewegung." In Gehler 1825–45, *1*, pt. 2, 914–972.

———. 1827. Review of Fries 1826 and three other physics texts. *Heidelberger Jahrbücher der Literatur*, Jg. 20, 1. Hälfte, Hft. 1 (January), Nr. 4–6, 50–64, 65–80, 81–93.

———. 1829. *Handbuch der Naturlehre*. Pt. 1, "Experimentalphysik." Heidelberg: C. F. Winter.

———. 1830a. "Inponderabilien [*sic*]." In Gehler 1825–45, 5, pt.2, 765–770.

———. 1830b. "Kraft." In Gehler 1825–45, 5, pt. 2, 956–1019.

———. 1831. "Lebenskraft." In Gehler 1825–45, 6, pt. 1, 111–123.

———. 1833a. "Perpetuum mobile." In Gehler 1825–45, 7, pt. 1, 408–423.

———. 1833b. "Physik." In Gehler 1825–45, 7, pt. 1, 493–573.

———. 1836a. "Magnetismus." In Gehler 1825–45, 6, pt. 2, 639–1162.

———. 1836b. "Materie." In Gehler 1825–45, 6, pt. 2, 1393–1472.

______. 1841. "Wärme." In Gehler 1825–45, *10*, pt. 1, 52–1179.

______. 1842. *Die ersten Elemente der gesammten Naturlehre*. 4th ed. Heidelberg: C. F. Winter.

Nasse, Friedrich. 1839. "Neue Versuche über die Abhängigkeit der thierischen Wärme vom Nervensysteme." In Friedrich Nasse and Hermann Nasse, *Untersuchungen zur Physiologie und Pathologie*, vol. 2, pt. 1 (Bonn: T. Habicht), 115–122.

Nasse, Hermann. 1836. *Das Blut in mehrfacher Beziehung physiologisch und pathologisch untersucht*. Bonn: T. Habicht.

______. 1842. "Blut." In Wagner 1842–53, *1* (1842–[1843]), 75–220.

Naumann, Moritz Ernst Adolph. 1824. *Skizzen aus der allgemeinen Pathologie*. Leipzig: Adolph Wienbrack.

______. 1829-39. *Handbuch der medicinischen Klinik*. 8 vols. in 11. Berlin: August Rücker (*1–6*); Rücker & Püchler (*7–8*).

______. 1830. *Versuch eines Beweises für die Unsterblichkeit der Seele aus dem physiologischen Standpunkte, zugleich als Einleitung in die Lehre von den sogenannten Geisteskrankheiten*. Bonn: Eduard Weber.

______. 1835. *Die Probleme der Physiologie, oder der Gegensatz von Nervenmark und Blut*. Bonn: Eduard Weber.

______. 1840. *Pathogenie*. Berlin: Rücker & Püchler.

______. 1851. *Allgemeine Pathologie und Therapie*. Pt. 1. Berlin: G. Reimer.

Ober, William B. 1968. "Robert Mayer, M.D. (1814–1878) and mechanical equivalent of heat." *New York State Journal of Medicine*, *68*, 2447–2454.

Ofterdinger, Ludwig Felix. 1879. Letter of 28 September 1879 to Theobald Kerner; from Ulm. In Deutsches Literaturarchiv/Schiller Nationalmuseum (Marbach am Neckar), Handschriften-Abteilung, Sig. 18773. 3 pp.

Oken, Lorenz. 1831. *Lehrbuch der Naturphilosophie*. 2d ed. Jena: Friedrich Frommann.

Ostwald, Wilhelm. 1896. *Elektrochemie; ihre Geschichte und Lehre*. Leipzig: Veit & Comp.

Paolini, Carlo. 1968. *Justus von Liebig. Eine Bibliographie sämtlicher Veröffentlichungen mit biographischen Anmerkungen*. Heidelberg: Carl Winter.

Passavant, Johann Carl. 1837. *Untersuchungen über den Lebensmagnetismus und das Hellsehen*. 2d ed. Frankfurt am Main: Heinrich Ludwig Brönner.

Pfaff, Christoph Heinrich. 1824–25. *Handbuch der analytischen Chemie für Chemiker, Staatsärzte, Apotheker, Oekonomen und Bergwerks Kundige*. 2d ed. 2 vols. Altona: J. F. Hammerrich.

______. 1845. *Parallele der chemischen Theorie und der Volta'schen Contacttheorie der galvanischen Kette, mit besonderer Rücksicht auf die neuesten Einwürfe Faraday's, Leop. Gmelin's und Schönbein's gegen letztere, nebst allgemeinen Betrachtungen über das Wesen einer physischen Kraft und ihrer Thätigkeit*. Kiel: Universitäts-Buchhandlung. §23, "Nähere Bestimmung des Wesens einer wahren physischen Kraft an der Schwere. Beleuchtung einer falschen Ansicht der Schwerkraft und der Ursache des freien Falls der Körper" (104–115), reprinted in Mayer 1893b, 230–239.

Pfister, Edwin. 1914. "Über den Schiffarzt Julius Robert Mayer." *Archiv für Schiffs- und Tropen-Hygiene*, *18*, 493–497.

Pietsch, Erich. 1942. "Julius Robert Mayer und der Erhaltungsgedanke in der Chemie." In Pietsch and Schimank 1942, 253–280.

Pietsch, Erich, and Schimank, Hans, eds. 1942. *Robert Mayer und das Energieprinzip 1842–1942*. Berlin: VDI-Verlag.

Pohl, Georg Friedrich. 1829. *Ansichten und Ergebnisse über Magnetismus, Elektricität und Chemismus*. Berlin: Ferdinand Dümmler.

Pohl, Georg Friedrich. 1837. *Principiorum tam in physice universa quam praesertim in eiusdem parte chemica adhuc desideratorum commentatio*. Breslau: Typis ex officina Richteriana.

Poppe, Johann Heinrich Moritz von. 1811–21. *Der physikalische Jugendfreund oder faßliche und unterhaltende Darstellung der Naturlehre*. 8 vols. Frankfurt am Main: Friedrich Wilmans (1–4); Gebrüder Wilmans (5–8). Vols. 1-6 in Mayer's library (saw).

______. 1815. *Lehrbuch der reinen und angewandten Mathematik*. Vol. 2, "Angewandte Mathematik." Frankfurt am Main: Johann Christian Hermann.

______. 1834. *Neue und ausführliche Volks-Naturlehre dem jetzigen Standpunkte der Physik gemäß*. 2d ed. Tübingen: C. F. Osiander; Vienna: Carl Gerold.

______. 1837. *Neue und ausführliche Volks-Naturlehre dem jetzigen Standpunkte der Physik gemäß*. Pt. 1, "Allgemeine und Experimental-Naturlehre." 3d ed. Tübingen: C. F. Osiander.

Popper, Josef. 1876. "Ueber J.R. Mayer's 'Mechanik der Wärme.' (Zweite Auflage.)" *Das Ausland. Ueberschau der neuesten Forschungen auf dem Gebiete der Natur-, Erd- und Völkerkunde*. Jg. 49, Nr. 35 (28 August), 681–685.

Pouillet, Claude-Servais-Mathias. 1832. *Élémens de physique expérimentale et de météorologie*. 2d ed. Vol. 1, pt. 1. Paris: Béchet jeune.

______. 1839–43. *Lehrbuch der Experimentalphysik und der Meteorologie*. Trans. and ed. C. H. Schnuse from the 3d French edition of 1837. 2 vols. Quedlinburg and Leipzig: Gottfr. Basse.

Preyer, William, ed. 1889. *Robert von Mayer über die Erhaltung der Energie. Briefe an Wilhelm Griesinger nebst dessen Antwortschreiben aus den Jahren 1842–1845*. Berlin: Gebrüder Paetel.

Probst, Christian. 1966. "Johann Bernhard Wilbrand (1779–1846) und die Physiologie der Romantik." *Sudhoffs Archiv. Vierteljahrsschrift für Geschichte der Medizin und der Naturwissenschaften, der Pharmazie und der Mathematik*, 50, 157–178.

Prout, William. 1834. *Chemistry Meteorology and the Function of Digestion Considered with Reference to Natural Theology*. ("The Bridgewater Treatises on the Power Wisdom and Goodness of God as Manifested in the Creation," Treatise VIII.) London: William Pickering.

______. 1836. *Chemie, Meteorologie und verwandte Gegenstände, als Zeugnisse für die Herrlichkeit des Schöpfers*. Trans. Gustav Plieninger. Stuttgart: P. Neff. Not seen.

Reich, Gottfried Christian. 1828. *Die Grundlage der Heilkunde*. Berlin: Duncker & Humblot.

______. 1843. *Lehrbuch der praktischen Heilkunde nach chemisch rationellen Grundsätzen*. Vol. 1, "Die Krankheiten des Respirationsapparats im gewöhnlichen Verhältnisse des Lebens." Berlin: Verlag von Oehmigke's Buchhandlung (Julius Bülow), [1842]–1843.

Reuschle, Carl Gustav. 1874. *Philosophie und Naturwissenschaft. Zur Erinnerung an David Friedrich Strauß*. Bonn: Emil Strauß.

Richerand, Anthelme-Balthasar. 1833. *Nouveaux élémens de physiologie*. 10th ed. 3 vols. Paris: Béchet jeune.

Riecke, Leopold Sokrates von. 1835. "Nekrolog Autenrieths." *Schwäbische Chronik. Zweite Abtheilung des Schwäbischen Merkurs* (Stuttgart), Nr. 132–134 (15–17 May), 531–532, 533, 535–536. Published anonymously; for authorship, see Riecke 1836, 30.

______. 1836. "[Nekrologische Erinnerungen an] Joh. Herrm. [*sic*] Ferd. v. Autenrieth." [Sachs'] *Medicinischer Almanach für das Jahr 1836*, Jg. 1, [3. Abtheilung], 30–47.

Riehl, Alois. 1900. "Robert Mayers Entdeckung und Beweis des Energieprincips." In Benno Erdmann et al., *Philosophische Abhandlungen. Christoph Sigwart zu seinem siebzigsten*

*Geburtstage 28. März 1900.* (Tübingen, Freiburg i. B., Leipzig: J.C.B. Mohr [Paul Siebeck]), 159–184.

Rocke, Alan J. 1992. "Berzelius's animal chemistry: From physiology to organic chemistry (1805–14)." In Evan M. Melhado and Tore Frängsmyr, eds., *Enlightenment Science in the Romantic Era: The Chemistry of Berzelius and Its Cultural Setting* (Cambridge, U.K.: Cambridge University Press, 107–131).

Rohlfs, Heinrich. 1879. "Julius Robert von Mayer, sein Leben und sein Wirken." *Deutsches Archiv für Geschichte der Medizin und medizinischer Geographie,* 2, 318–364, 405–462.

Rose, Heinrich. 1843. Review of Liebig 1842b. *Jahrbücher für wissenschaftliche Kritik,* Bd. 1 [des Jahres], Nr. 33–36, cols. 257–261, 262–272, 273–280, 281–287 (February).

Rosen, George. 1959. "The conservation of energy and the study of metabolism." In Chandler McC. Brooks and Paul F. Cranefield, eds., *The Historical Development of Physiological Thought* (New York: Hafner), 243–263.

Roser, Wilhelm. 1838. *Die Humoral-Aetiologie.* Stuttgart: Fr. Brodhag (Eduard Jung).

———. 1878. "Zur Erinnerung an C. A. Wunderlich." *Archiv der Heilkunde,* 19, 321–339.

Roser, Wilhelm, and Wunderlich, Carl August. 1841. *Ueber die Mängel der heutigen deutschen Medicin und über die Nothwendigkeit einer entschieden wissenschaftlichen Richtung in derselben. Programm einer medicinischen Vierteljahrschrift: Archiv der physiologischen Heilkunde.* Stuttgart: Ebner & Seubert.

———. 1842. "Ueber die Mängel der heutigen deutschen Medicin und über die Nothwendigkeit einer entschieden wissenschaftlichen Richtung in derselben." *Archiv für physiologische Heilkunde,* Jg. 1, Hft. 1, i–xxx.

———. 1843. "Ueber die jetzige Lage der physiologischen Medicin." *Archiv für physiologische Heilkunde,* Jg. 2, Hft. 1, 1–5.

Rothschuh, Karl E. 1941. "Zur Geschichte der Physiologie des Blutes in der ersten Hälfte des neunzehnten Jahrhunderts." *Klinische Wochenschrift,* 20, 621–624.

———. 1953. *Geschichte der Physiologie.* Berlin, Göttingen, Heidelberg: Springer-Verlag.

Rudolphi, Karl Asmund. 1821–28. *Grundriss der Physiologie.* 2 vols. in 3 pts. Berlin: Ferdinand Dümmler.

———. 1830. *Grundriss der Physiologie.* 2 vols. in 3 pts. Reutlingen: J. J. Mäckenschen. In Mayer's library (saw).

Rümelin, Gustav. 1862. "Justinus Kerner." *Allgemeine Zeitung* (Augsburg), Beilage, Nr. 163–166, 168–171 (12–15, 17–20 June), 2713–2714, 2729–2731, 2746–2747, 2761–2763, 2793–2794, 2813–2814, 2829–2830, 2845–2847. Published anonymously. Reprinted in Rümelin 1894, 303–374.

———. 1878. "Erinnerungen an Robert Mayer." *Allgemeine Zeitung* (Augsburg), Beilage, Nr. 120–122 (30 April; 1–2 May), 1761–1763, 1778–1780, 1795–1796. Reprinted in Rümelin 1881, 350–405.

———. 1881. *Reden und Aufsätze.* New ser. Freiburg i. B.: J.C.B. Mohr (Paul Siebeck).

———. 1894. *Reden und Aufsätze.* 3d ser. Freiburg i. B. and Leipzig: J.C.B. Mohr (Paul Siebeck).

Ruge, Arnold. 1839. Review of Strauss 1839a. *Hallesche Jahrbücher für deutsche Wissenschaft und Kunst,* Jg. 2, Nr. 124–126 (24, 25, 27 May), cols. 985–988, 993–1000, 1001–1004.

Saint-Robert, Paul de. 1870. *Principes de thérmodynamique.* 2d ed. Turin and Florence: Hermann Loescher.

Sarton, George. 1929. "The discovery of the law of conservation of energy." *Isis,* 13 (1929–30), 18–44.

Schelling, Friedrich Wilhelm Joseph von. 1797. *Ideen zu einer Philosophie der Natur*. Leipzig: Breitkopf & Härtel. A second edition appeared in 1803 (not seen). Reprinted in Schelling 1856–61, 2 (1857), 1–343, and Schelling 1927–59, *1* (1927), 653–723, *Ergänzungsband 1* (1956), 77–350; cited from the former as Schelling 1797/1803.

———. 1798. *Von der Weltseele; eine Hypothese der höhern Physik zur Erklärung des allgemeinen Organismus*. Hamburg: Friedrich Perthes. Not seen. 2d and 3d eds. appeared in 1806 and 1809 (not seen). Reprinted in Schelling 1856–61, 2 (1857), 345–583, and Schelling 1927–59, *1* (1927), 413–651; cited from the former as Schelling 1798/1806/1809.

———. 1799a. *Erster Entwurf eines Systems der Naturphilosophie*. Jena and Leipzig: Christian Ernst Gabler. Reprinted in Schelling 1856–61, 3 (1858), 1–268, and Schelling 1927–59, 2 (1927), 1–268; cited from the former.

———. 1799b. *Einleitung zu seinem Entwurf eines Systems der Naturphilosophie. Oder: Ueber den Begriff der speculativen Physik und die innere Organisation eines Systems dieser Wissenschaft*. Jena and Leipzig: Christian Ernst Gabler. Reprinted in Schelling 1856–61, 3 (1858), 269–326, and Schelling 1927–59, 2 (1927), 269–326; cited from the former.

———. 1809. "Philosophische Untersuchungen über das Wesen der menschlichen Freyheit und die damit zusammenhängenden Gegenstände." In *F.W.J. Schelling's philosophische Schriften*, vol. 1 (Landshut: Philipp Krüll), 397–511.

———. 1834. *Philosophische Untersuchungen über das Wesen der menschlichen Freiheit und die damit zusammenhängenden Gegenstände*. Reutlingen: J. N. Enßlin.

———. 1856–61. *Friedrich Wilhelm Joseph von Schellings sämmtliche Werke*. 14 vols. in 2 divisions. Stuttgart and Augsburg: J. G. Cotta.

———. 1927–59. *Schellings Werke*. Ed. Manfred Schröter. 12 vols. Munich: C. H. Beck & R. Oldenbourg.

———. 1988. *Ideas for a Philosophy of Nature as Introduction to the Study of This Science*. Trans. Errol E. Harris and Peter Heath, intro. by Robert Stern. Cambridge, U.K.: Cambridge University Press.

Scheurlen, Ernst von. 1948. "Julius Robert Mayer, praktischer Arzt, Oberamtswundarzt, dann Stadtarzt in Heilbronn. Der Entdecker des Gesetzes von der Erhaltung der 'Kraft' und des mechanischen Wärmeäquivalents (1814–1878)." In *Schwäbische Lebensbilder*, vol. 4 (Stuttgart: W. Kohlhammer), 101–133.

Schill, Albert Friedrich. 1838. *Ueber die Irritation*. Tübingen: H. Laupp.

———. 1840. *Dr. Albert Friedrich Schill's allgemeine Pathologie*. Ed. Victor Adolph Riecke. Tübingen: H. Laupp. The preface is a biography of Schill by Leopold Sokrates von Riecke (iii–xviii).

Schleiden, Matthias Jacob. 1842. *Herr Dr. Justus Liebig in Giessen und die Pflanzenphysiologie*. Leipzig: Wilh. Engelmann.

———. 1842–43. *Grundzüge der Wissenschaftlichen Botanik nebst einer Methodologischen Einleitung als Anleitung zum Studium der Pflanze*. 2 vols. Leipzig: Wilhelm Engelmann.

Schlüter, Hermann. 1985. *Die Wissenschaften vom Leben zwischen Physik und Metaphysik: Auf der Suche nach dem Newton der Biologie im 19. Jahrhundert*. Weinheim: Acta Humaniora, VCH.

Schmid, Carl Christian Erhard. 1798–1801. *Physiologie philosophisch bearbeitet*. 3 vols. Jena: Akademische Buchhandlung.

Schmid, Rudolf. 1878. "Robert Mayer, der große Förderer unserer heutigen wissenschaftlichen Welterkenntnis, seine wissenschaftliche Entdeckung und sein religiöser Standpunkt." *Theologische Studien und Kritiken*, Jg. 51, Bd. 2 [des Jahres], 677–692.

Schmoller, Gustav. 1907. "Rümelin, Gustav." In *ADB*, 53, 597–635.

Schmolz, Helmut, and Weckbach, Hubert, eds. 1964. *Robert Mayer. Sein Leben und Werk in Dokumenten*. Weißenhorn: Anton H. Konrad.

Schneider, Margarete, and Schneider, Wolfgang, eds. 1986. *Justus Liebig. Briefe an [Eduard] Vieweg*. Braunschweig and Wiesbaden: Friedr. Vieweg & Sohn.

Schönlein, Johann Lucas. 1832. *Allgemeine und specielle Pathologie und Therapie*. Nach J. L. Schönlein's . . . Vorlesungen niedergeschrieben und hrsg. von einem seiner Zuhörer [i.e., Carl Ludwig Reinhard?]. 2d ed. 4 vols. Würzburg: In Commission der C. Etlinger'schen Buchhandlung.

______. 1842. *Schönlein's klinische Vorträge in dem Charité-Krankenhause zu Berlin*. Ed. L. Güterbock. Berlin: Veit.

Schreger, Theophilus. 1823. "Blut." *Allgemeine Encyclopädie der Wissenschaften und Künste in alphabetischer Folge*. Ed. Johann Samuel Ersch and Johann Gottfried Gruber, [1st section, A–G], (99 vols.; Leipzig: Johann Friedrich Gleditsch [etc.], 1818–82), *11*, 58–71.

Schröer, Heinz. 1967. *Carl Ludwig, Begründer der messenden Experimentalphysiologie 1816–1895*. Stuttgart: Wissenschaftliche Verlagsgesellschaft.

Schubert, Johann Andreas. 1842. *Versuch einer neuen Begründung der Grundlehren der Mechanik*. Dresden: Arnold. Not seen.

Schüz, Gottlieb Friedrich. 1799. *Dissertatio inauguralis medica sistens experimenta circa calorem foetus et sanguinem ipsius instituta*. Tübingen: Typis Fuesianis.

Schultz, Carl Heinrich. 1831. Review of Tiedemann 1830. *Jahrbücher für wissenschaftliche Kritik*, Bd. 1 [des Jahres], Nr. 6–7, cols. 44–48, 49–55 (January).

______. 1833. *Grundriß der Physiologie*. Berlin: August Hirschwald.

______. 1843. Review of Liebig 1842b. *Jahrbücher für wissenschaftliche Kritik*, Bd. 2 [des Jahres], Nr. 105–108, cols. 838–840, 841–848, 849–856, 857–859 (December).

Schwann, Theodor. 1839. *Mikroskopische Untersuchungen über die Uebereinstimmung in der Struktur und dem Wachsthum der Thiere und Pflanzen*. Berlin: Sander'sche Buchhandlung (G. E. Reimer), [1838]–1839.

______. 1879. "[Réponse de M. Schwann]." In *Manifestation en l'honeur de M. le professeur Th. Schwann* (Düsseldorf: L. Schwann), 47–54.

Siber, Thaddaeus. 1837. *Grundlinien der Experimental-Physik*. Munich: Georg Franz.

Simon, Franz. 1840–42. *Handbuch der angewandten medizinischen Chemie*. 2 vols. Berlin: Albert Förstner.

______. 1844. "Bericht über die Leistungen im Gebiete der physiologischen u. pathologischen Chemie im Jahre 1842." *Jahresbericht über die Fortschritte der gesammten Medicin in allen Ländern*, Jg. 2, Bd. 3 des Jahrgangs, 67–144.

Snelders, H.A.M. 1973. "De invloed van Kant, de romantiek en de 'Naturphilosophie' op de anorganische natuurwetenschappen in Duitsland." Ph.D. diss., Rijksuniversiteit Utrecht.

South, John Flint. 1849. "Zoology." *Encyclopaedia Metropolitana; or, Universal Dictionary of Knowledge*. Ed. Edward Smedley, Hugh James Rose, and Henry John Rose (30 vols., London: John Joseph Griffin & Co.), 7, 109–392.

Späth. 1842. "Einige Bemerkungen über das Archiv für physiologische Medicin von Roser und Wunderlich." *Medicinisches Correspondenz-Blatt des Württembergischen ärztlichen Vereins*, Bd. 12, Nr. 29 (30 September), 228–232.

Stark, Karl Wilhelm. 1838. *Allgemeine Pathologie oder allgemeine Naturlehre der Krankheiten*. 2 vols. Leipzig: Breitkopf & Haertel.

Stauffer, Robert C. 1957. "Speculation and experiment in the background of Oersted's discovery of electromagnetism." *Isis*, *48*, 33–50.

Stichweh, Rudolf. 1984. *Zur Entstehung des modernen Systems wissenschaftlicher Disziplinen: Physik in Deutschland 1740–1890*. Frankfurt am Main: Suhrkamp.

Sticker, Georg. 1939–40. "Wunderlich, Roser, Griesinger, 'die drei Schwäbischen Reformatoren der Medizin.'" *Sudhoffs Archiv für Geschichte der Medizin und der Naturwissenschaften*, 32, 217–274; 33, 1–54.

Strauss, David Friedrich. 1830. "Kritik der verschiedenen Ansichten über die Geistererscheinungen der Seherin von Prevorst." *Hesperus. Encyclopädische Zeitschrift für gebildete Leser*, Nr. 100–104 (27–30 April, 1 May), 399–400, 403–404, 408, 411–412, 415–416. Reprinted in Strauss 1839b, 390–404.

______. 1835–36. *Das Leben Jesu, kritisch bearbeitet*. 2 vols. Tübingen: C. F. Osiander.

______. 1836a. Review of J. Kerner 1834. *Jahrbücher für wissenschaftliche Kritik*, Bd. 1 [des Jahres], Nr. 102–105, cols. 812–816, 817–821, 825–832, 833–840 (June). Reprinted with some changes in Strauss 1839b, 301–327.

______. 1836b. Review of Wirth 1836 and J. Kerner 1836. *Jahrbücher für wissenschaftliche Kritik*, Bd. 2 [des Jahres], Nr. 110–113, cols. 880, 881–888, 889–896, 897–904 (December). The review of Wirth ends and that of Kerner begins in col. 894; only the review of Kerner reprinted in Strauss 1839b, 328–338.

______. 1837. *Streitschriften zur Vertheidigung meiner Schrift über das Leben Jesu und zur Charakteristik der gegenwärtigen Theologie*. Zweites Heft: Die Herren Eschenmayer und Menzel. Tübingen: C. F. Osiander.

______. 1838a. "Justinus Kerner." *Hallische Jahrbücher für deutsche Wissenschaft und Kunst*, Jg. 1, Nr. 1–5, 7 (1–5, 8 January), cols. 6–8, 9–13, 22–24, 25–32, 33–40, 49–51. Reprinted in Strauss 1839a, 3–57.

______. 1838b. Review of Passavant 1837 and Eschenmayer 1837. *Jahrbücher für wissenschaftliche Kritik*, Bd. 1 [des Jahres], Nr. 29–33, cols. 228–232, 233–240, 241–248, 249–256, 257–264 (February). The review of Passavant ends and that of Eschenmayer begins in col. 243; reprinted in Strauss 1839b, 339–354 and 355–376, respectively.

______. 1838c. "Vergängliches und Bleibendes im Christenthum." *Der Freihafen*, [Jg. 1], Hft. 3, 1–48. Reprinted as "Ueber Vergängliches und Bleibendes im Christenthum" in Strauss 1839a, 61–132.

______. 1839a. *Zwei friedliche Blätter*. Altona: J. F. Hammerich.

______. 1839b. *Charakteristiken und Kritiken*. Leipzig: Otto Wigand.

______. 1840–41. *Die christliche Glaubenslehre in ihrer geschichtlichen Entwicklung und im Kampfe mit der modernen Wissenschaft*. 2 vols. Tübingen: C. F. Osiander; Stuttgart: F. H. Köhler.

______. 1872. *Der alte und der neue Glaube. Ein Bekenntniß*. Leipzig: S. Hirzel.

Stübler, Eberhard. 1948. *Johann Heinrich Ferdinand v. Autenrieth, 1772–1835, Professor der Medizin und Kanzler der Universität Tübingen*. Stuttgart: August Schröder.

Süssmilch, Johann Peter. 1765. *Die göttliche Ordnung in den Veränderungen des menschlichen Geschlechts, aus der Geburt, dem Tode und der Fortpflanzung desselben erwiesen*. 3d ed. 2 vols. Berlin: Im Verlag der Buchhandlung der Realschule. In Mayer's library (saw).

Temkin, Owsei. 1946. "Materialism in French and German physiology of the early nineteenth century." *Bulletin of the History of Medicine*, 20, 322–327.

Thackrah, Charles Turner. 1834. *An Inquiry into the Nature and Properties of the Blood, in Health and in Disease*. New ed. Ed. Thomas G. Wright. London: Longman [etc.].

Tiedemann, Friedrich. 1830. *Physiologie des Menschen*. Vol. 1, "Allgemeine Betrachtungen der organischen Körper." Darmstadt: Carl Wilhelm Leske.

Tiedemann, Friedrich, and Gmelin, Leopold. 1826–27. *Die Verdauung nach Versuchen*. 2 vols. Heidelberg and Leipzig: Karl Groos. Not seen.

______. 1831. *Die Verdauung nach Versuchen*. 2d ed. 2 vols. Heidelberg: Karl Groos.

Timerding, Heinrich. 1925. *Robert Mayer und die Entdeckung des Energiesatzes*. Leipzig and Vienna: Franz Deuticke.

Titot, Heinrich. 1865. *Beschreibung des Oberamts Heilbronn*. Ed. Königliches statistisch-topographisches Bureau. ("Beschreibung des Königreichs Württemberg," Hft. 45.) Stuttgart: H. Lindemann. For authorship, see the last (unnumbered) page.

Turner, R. Steven. 1974. "Julius Robert Mayer." In *DSB*, 9, 235–240.

Valentin, Gabriel Gustav. 1837. "Die Fortschritte der Physiologie im Jahre 1836." *Repertorium für Anatomie und Physiologie*, Bd. 2, 1–241.

______. 1839. "Die Fortschritte der Physiologie im Jahre 1838." *Repertorium für Anatomie und Physiologie*, Bd. 4, 1–358.

______. 1840. "Die Fortschritte der Physiologie im Jahre 1839." *Repertorium für Anatomie und Physiologie*, Bd. 5, 1–391.

______. 1842. "Die Fortschritte der Physiologie im Jahre 1841." *Repertorium für Anatomie und Physiologie*, Bd. 7, 1–448.

______. 1844. *Lehrbuch der Physiologie des Menschen*. 2 vols. Braunschweig: Friedrich Vieweg & Sohn. In Mayer's library (saw).

______. 1846a. "Bericht über die Leistungen in der Physiologie." *Jahresbericht über die Fortschritte der gesammten Medicin in allen Ländern im Jahre 1845*, [Jg. 5], Bd. 1 [des Jahres], 167–259.

______. 1846b. *Grundriß der Physiologie des Menschen*. Braunschweig: Friedrich Vieweg & Sohn.

______. 1847. *Lehrbuch der Physiologie des Menschen*. 2d ed. Vol. 1. Braunschweig: Friedrich Vieweg & Sohn.

Vetter, August. 1837. "Ueber das rein Physikalische und seine Grenzen im Organismus." *C. W. Hufeland's Journal der practischen Heilkunde*, Bd. 84 (= *C. W. Hufeland's Neues Journal der practischen Arzneikunde und Wundarzneikunst*, Bd. 1), St. 1 (January), 65–97.

Vierordt, Karl. 1845a. *Physiologie des Athmens, mit besonderer Rücksicht auf die Ausscheidung der Kohlensäure*. Karlsruhe: Christian Theodor Groos.

______. 1845b. "Respiration." In Wagner 1842–53, 2 (1844–[1845]), 828–916.

Virchow, Rudolf. 1907. "Über das Bedürfnis und die Richtigkeit einer Medizin vom mechanischen Standpunkt." *Virchows Archiv für pathologische Anatomie und Physiologie und für klinische Medizin*, *188*, 1–21.

Vogt, Carl. 1845–47. *Physiologische Briefe für gebildete aller Stände*. 3 pts. Stuttgart and Tübingen: J. G. Cotta. Letters 1–8, 9–16, and 17–29 (pp. 1–143, 145–279, and 281–492).

______. 1850. "Die Physiologie des Menschen auf dem Standpunkte der heutigen Wissenschaft." *Die Gegenwart. Eine encyklopädische Darstellung der neuesten Zeitgeschichte für alle Stände*, Bd. 4, 646–707. Published anonymously; for authorship, see Gregory 1977, 269.

Voit, Carl. 1885. "[Nekrolog auf Philipp Johann (*sic*) von Jolly]." *Sitzungsberichte der mathematisch-physikalischen Classe der k. b. Akademie der Wissenschaften zu München*, *15*, 119–136.

Volhard, Jakob. 1909. *Justus von Liebig*. 2 vols. Leipzig: Johann Ambrosius Barth.

Volkmann, Alfred Wilhelm. 1837. *Die Lehre von dem leiblichen Leben des Menschen*. Leipzig: Breitkopf & Härtel.

______. 1838. *Die Physiologie als Gegnerin der Lehre des Materialismus von der Identität des Leibes und der Seele*. Dorpat: Lindfors Erben.

Wagner, Rudolph, ed. 1842–53. *Handwörterbuch der Physiologie mit Rücksicht auf physiologische Pathologie*. 4 vols. in 5. Braunschweig: Friedr. Vieweg & Sohn.

Wagner, Rudolph. 1854. *Menschenschöpfung und Seelensubstanz.* Göttingen: Georg H. Wigand.

Walther, Philipp Franz. 1807–8. *Physiologie des Menschen mit durchgängiger Rücksicht auf die comparative Physiologie der Thiere.* 2 vols. Landshut: Philipp Krüll.

Weller, Arnold, and Weller, Karl. 1972. *Württembergische Geschichte im südwestdeutschen Raum.* 7th ed. Stuttgart and Aalen: Konrad Theiss.

Whewell, William. 1836. *An Elementary Treatise on Mechanics.* 5th ed. Cambridge, U.K.: J. & J. J. Deighton; London: Whittaker & Arnot.

______. 1841. *Elementar-Lehrbuch der Mechanik.* Trans. C. H. Schnuse from the 5th English ed. Braunschweig: G.C.E. Meyer sen. Not seen.

Wiederkehr, Karl Heinrich. 1986. "Mayer, Julius Robert." In Fritz Krafft, ed., *Große Naturwissenschaftler.* 2d ed. (Düsseldorf: VDI-Verlag), 238–239.

Wilbrand, Johann Bernhard. 1827. *Die Natur des Athmungs-Prozesses.* Frankfurt: Sauerländer. Not seen.

Williams, Charles James Blasius. 1835. "Observations on the Changes Produced in the Blood in the Course of its Circulation: with Experiments." *London Medical Gazette,* vol. 16, nos. 403–408 (22, 29 August; 5, 12, 19, 26 September), 718–724, 745–751, 783–788, 807–813, 842–848, 871–876.

______. 1836. "Ueber die Quellen der thierischen Wärme." [Froriep's] *Notizen aus dem Gebiete der Natur- und Heilkunde,* Bd. 49, Nr. 3 (June [= Nr. 1059 der ganzen Reihe]), cols. 39–41.

Williams, L. Pearce. 1962. "The physical sciences in the first half of the nineteenth century: Problems and sources." *History of Science, 1,* 1–15.

Wirth, Johann Ulrich. 1836. *Theorie des Somnambulismus oder des thierischen Magnetismus.* Leipzig and Stuttgart: J. Scheible's Verlags-Expedition.

Wise, M. Norton. 1981. "German concepts of force, energy, and the electromagnetic ether: 1845–1880." In Cantor and Hodge 1981a, 269–307.

Wistinghausen, Constantinus von. 1837. *De calore animali quaedam.* Dorpat: Typis J. C. Schuenmanni.

Wittstein, Georg Christoph. 1846. Review of Mayer 1845. *Repertorium für die Pharmacie,* Bd. 91, Hft. 3, 400–401 (March).

Wöhler, Friedrich. 1840a. *Grundriß der Chemie.* Pt. 1, "Unorganische Chemie." 6th ed. Berlin: Duncker & Humblot.

______. 1840b. *Grundriß der organischen Chemie* (= *Grundriß der Chemie.* Pt. 2, "Organische Chemie"). Berlin: Duncker & Humblot.

Wöhler, Friedrich, and Liebig, Justus. 1837. "Ueber die Bildung des Bittermandelöls." *Annalen der Pharmacie,* Bd. 22, Hft. 1, 1–24 (≈ April).

Woodward, William R. 1975. "The Medical Realism of R. Hermann Lotze." Ph.D. diss., Yale University.

Woodward, William R., and Rainer, Ulrike. 1975. "Berufungs-Korrespondenz Rudolph Hermann Lotzes an Rudolph Wagner (13 Briefe: 1. Dezember 1842–11. April 1844)." *Sudhoffs Archiv. Zeitschrift für Wissenschaftsgeschichte,* 59, 356–386.

Wunderlich, Carl August. 1842. "Erwiderung [auf Duttenhofer's Angriff auf die 'physiologische Medicin' in Nro. 24 und 25 des Correspondenzblatts]." *Medicinisches Correspondenz-Blatt des Württembergischen ärztlichen Vereins,* Bd. 12, Nr. 27 (27 August), 216.

______. 1843. Review of the first two *Hefte* of Schönlein 1842. *Archiv für physiologische Heilkunde,* Bd. 2, 290–306.

______. 1845. *Versuch einer pathologischen Physiologie des Blutes.* Stuttgart: Ebner & Seubert.

______. 1859. *Geschichte der Medicin*. Stuttgart: Ebner & Seubert.

______. 1869. "Wilhelm Griesinger. Nekrolog." *Archiv der Heilkunde*, Jg. 10, 113–150.

Wuttke, Walter. 1972. "Materialien zu Leben und Werk Adolph Karl August von Eschenmayers." *Sudhoffs Archiv; Zeitschrift für Wissenschaftsgeschichte*, 56, 255–296.

Yelin, Julius Conrad von. 1819. *Ueber Magnetismus und Electricität als identische und Urkräfte*. Munich: Ign. Jos. Lentner.

Young, Robert M. 1969. "Malthus and the evolutionists: The common context of biological and social theory." *Past and Present*, no. 43, 109–145.

Zaunick, Rudolph. 1956. "Julius Robert Mayers Essay über medizinische Physiologie." *Sudhoffs Archiv für Geschichte der Medizin und der Naturwissenschaften*, 40, 26–28.

Zeller, Eduard. 1874. *David Friedrich Strauß in seinem Leben und seinen Schriften*. Bonn: Emil Strauß.

______, ed. 1895. *Ausgewählte Briefe von David Friedrich Strauß*. Bonn: Emil Strauß.

______. 1908. *Erinnerungen eines Neunzigjährigen*. Stuttgart: Uhland.

Zöllner, Friedrich. 1878–81. *Wissenschaftliche Abhandlungen*. 4 vols. Leipzig: L. Staackmann.

______. 1879. "Robert Mayer. (†am 20. März 1878.)" *Die Grenzboten. Zeitschrift für Politik, Literatur und Kunst*, Jg. 38, 1. Quartal, [Nr. 2–3], 41–52, 92–106. Published anonymously; the second part reprinted with a few additions in Zöllner 1878–81, *4* (1881), 674–687.

______. 1881. "Robert Mayer aus Heilbronn. (1814–1878.) Eine biographische Skizze aus den 'Grenzboten' 1879. No. 3." In Zöllner 1878–81, *4*, 674–690.

Zott, Regine. 1978. "Justus von Liebig und seine Reflexionen über das Lebenskraft-Problem." *Deutsche Zeitschrift für Philosophie*, 26, 55–66.

# • INDEX •

*Note*: Page and note number in *italics* indicate the location of brief biographical descriptions.